2nd Geoplanning – International Conference on Geomatics and Planning 2017

Geomatic's Application for Disaster Management and Spatial Planning

IOP Conference Series: Earth and Environmental Science
Volume 123

Surakarta, Indonesia
9 – 10 August 2017

ISBN: 978-1-5108-5883-1
ISSN: 1755-1307

Printed from e-media with permission by:

Curran Associates, Inc.
57 Morehouse Lane
Red Hook, NY 12571

Some format issues inherent in the e-media version may also appear in this print version.

Printed by Curran Associates, Inc. (2018)

Institute of Physics
Dirac House, Temple Back
Bristol BS1 6BE UK

Phone: 44 1 17 929 7481
Fax: 44 1 17 920 0979

techtracking@iop.org

Additional copies of this publication are available from:

Curran Associates, Inc.
57 Morehouse Lane
Red Hook, NY 12571 USA
Phone: 845-758-0400
Fax: 845-758-2633
Email: curran@proceedings.com
Web: www.proceedings.com

Table of contents

Volume 123

2nd Geoplanning - International Conference on Geomatics and Planning

9–10 August 2017, Surakarta - Central Java, Indonesia

Accepted papers received: 7 February 2018
Published online: 22 February 2018

Preface

Papers

Disaster Modelling & Management (DMM)

Infrastructure and Settlement (IS)

IOP Conf. Series: Earth and Environmental Science **123** (2018) 011001 doi:10.1088/1755-1315/123/1/011001

PREFACE

The 2nd Geoplanning - International Conference was held on 9-10th September, 2017 in Solo, Central Java, Indonesia. Geoplanning is an international conference covering research and development in the field of applied geomatic's technologies for spatial planning, including GIS, Remote Sensing, and Satellite Image Processing. More than 132 abstract were submitted and after an initial review 80 papers were presented. Through peer review process, 50 papers were accepted for publication in the Geoplanning Conference Proceeding.

The present volume is highlighting major issue of Sustainable Development with focusing topic on "Geomatic's Application for Disaster Management and Spatial Planning." Furthermore, the topic was divided into four themes;

Theme 1: Disaster Modelling & Management

Theme 2: Infrastructure and Settlement

Theme 3: RS-GIS for Spatial Modelling

Theme 4: Urban and Regional Planning

We would like to thank all the reviewers for their time and effort in reviewing the documents. The published papers have passed the process of improvement accommodating the discussion during the conference as well as the reviewers' comments who have guided any necessary improvement. Finally, we would like to thanks to all of the proceeding team who have dedicated their constant support and countless time to bring these scratches into a book.

Chair Committee of Geoplanning

All Participants

Best presenter gift

Coffee Break

Coffee Break 2

Committee Meeting

Conference Gift

Conference gift to Prof Liu

Dr. Agustan

Keynote speakers

Parallel session

Parallel session room 2

Parallel session room 3

Parallel session room 4

Photo Session with All Participants

IOP Conf. Series: Earth and Environmental Science **123** (2018) 011002 doi:10.1088/1755-1315/123/1/011002

Plenary session

Prof Ahmad Nazri

Prof Ahmad Nazri, UTM, Malaysia

Prof Imam and Prof Nazri

Prof Nazri

Prof Nazri Presentation

IOP Conf. Series: Earth and Environmental Science **123** (2018) 011002 doi:10.1088/1755-1315/123/1/011002

Prof Yan Liu Presentation

Prof Yan Liu, Queensland University

Registration of participants

Speakers

Peer review statement

All papers published in this volume of *IOP Conference Series: Earth and Environmental Science* have been peer reviewed through processes administered by the proceedings Editors. Reviews were conducted by expert referees to the professional and scientific standards expected of a proceedings journal published by IOP Publishing.

IOP Conf. Series: Earth and Environmental Science **123** (2018) 012001 doi:10.1088/1755-1315/123/1/012001

Sensitive Land Use Planning, Malinao, Albay, Philippines

A M R Abante[1] and C G R Abante[1]

[1]Bicol University, Philippines

email: anamarie.abante@bicol-u.edu.ph

Abstract. This paper reviews the hazard zone as defined in the zoning ordinance of the Local Government of Malinao. The zonification was completed in accordance with the approved Comprehensive Land Use Plan stipulating the allowed use and regulations of zones to control future land development. This paper brings together an examination of human exposure as well as spatial situations and conditions of their houses within the hazard zone playing with flood risks. *The* purposive selection sample households were based on characteristics of people residing within it, in which the site concurs with the flood forecasted frequent every 5, 25 and 100 years turned to be significant to better understanding 'risks computing' were variables retrieved from the intersecting spaces fused to get the complex interrelationship of the sets of flood hazard, vulnerability and exposure of inhabitants and their place of residence weighted against capability of individual family or household to withstand effects of flooding. The Risk Quotient Object and Field Bases Model were tested in specific location in Malinao. The sample households' individual risk location quotient varies from high to a very high risk distributions ranging from 8 to 125 numerical values. As Malinao stays on to experience flood hazards, changing climate and other natural calamities, the need to understand the six elements of disaster risk computing at household level is becoming crucial in risk reduction meeting the targets and priorities for action as specified in the Sendai Framework.

Keywords: Zoning, Exposure, Risk Computing, Risk Quotient, Household

1. Introduction

The prevailing principle of the Habitat I which was held in Vancouver in 1976 provides importance to water for life and adopt programs for the sanitary disposal of waste water. Similarly the prevailing principles of the Habitat II which was held in Istambul in 1996 arrest the deterioration of global human settlements conditions and ultimately creates the conditions for achieving improvements in the living environment of all the people on sustainable basis. In October 2016, the Habitat III held in Quito, the prevailing principles focus on Housing [1] and Sustainable Urban Development to adopt a New Urban Agenda. This work is seen significant in regulating the use of the land and activities in hazard zones particularly those families within the immediate vicinity of rivers and other waterbodies as well as providing guidance to the residents at risk and decision making in land management, land use planning and monitoring local developments [2,3].

Relatedly, the Sendai Framework outlines seven clear targets and four priorities for action to prevent new and reduce existing disaster risks. However, the management of geospatial information plays a key role in understanding disaster risk, strengthening governance to reduce risk, enhancing preparedness for effective response, and to 'Build Back Better' in recovery, rehabilitation and reconstruction.

Althouh the law in Philippines asserts that land below eighteen percent (18%) in slope generally classified as alinable and disposable land along edge of normal high waterline of rivers and least streams with channel of at least five (5) meters wide are subjected to legal easements or salvages for strips of mangrove or swamp along shoreline facing the ocean, lakes, and other bodies of water, ancient settlements were already built along rivers so as to Filipino houses were also constructed on the river banks or over shallow waters since Spanish era as such form of settlements provided to control of pueblos (town) wherein the principal and the masses reside around the edifices gridiron arrangements or plaza complex. Also, it is a regulation that no person are allowed to stay in environmental critical areas or ecological buffers longer than what is necessary for recreation, navigation, floatage, fishing, and salvage or to build structures of any kind because flood is rise in water level or river overflow of a consistency of water beyond its confines causing inundation of water onto a normally dry adjacent areas. This paper decry the technical and legal interpretation of the three meter margin to the easement of public use in the general interest of navigation, floatage, fishing and salvage (R.A. 386. Art. 368 known as Civil Code of the Philippines), and the conversion from forty or twenty meters easement to three meters easement of public use in the interest of recreation, navigation, floatage, fishing and salvage (P.D. 1067 known as the Water Code of the Philippines.

Also, the empowerment to prepare land use plans and regulations in the Philippines [4] lies with the LGUs but most of these units have not yet developed their own comprehensive land use plans and zoning ordinance. Base on the latest inventory of HLURB, the LGUs with Updated CLUPs as of 1st Quarter 2017 in Bicol Region comprise of the following: Legazpi City, Daraga, Malinao, Libon, Basud, Canaman, Milaor, Minalabac, Pamplona, Sangay, Sipocot, Masbate City, Baleno, Cataingan, Cawayan, Mobo, Placer, Sorsogon City, Castilla, and Juban. Out of 114 LGUs, 21 or 18% have Approved CLUP. However, only 11 LGUs or 10% of total number of LGUs in Bicol mainstreamed DRRM/CCA into CLUP.

However, the lack of geospatial information, although scientific planning is proven effective with the use of technologies such as GIS, Remote Sensing, GNSS and so on, the process or geospatial information modeling to produce a scaled-replica of the real world is tedious and costly oftentimes ends up neglected [5]. Nevertheless, the local officials of the Municipality of Malinao (Malinao) had formulated their CLUP using GIS technology but limited spatial data. It carried out based on the analyses of land use patterns in planning using old topographic maps, tax maps, thematic maps and other available maps all digitized and converted into shapefiles format, in which serve as baseline to plan and address issues on rural and urban settlements, particularly decrease in agricultural crops situated in the alluvial plains near or within the flood and liquefaction prone zones [6]. Through land use space allocation and zonation, spots and places were made available to apportion suitable lands for residential, commercial, farming, timber, infrastructure, and so on. The masses were informed about the potential risks on their households, but they think it was an exaggerated plan and regulations. This uncertainty deferred the approval and ratification of the land use plan and regulations. But when TD Juaning as stated by PAGASA, dumped about 500 millimeters of rain in a span of 15 hours in the downstream of Malinao causing flash flood at the alluvial flats along Quinali 'B' River as high as high as three meters inundation in July 2011, this event prompted the approval and ratification of the CLUP and Zoning Ordinance (ZO) were pushed by the local government then approved in the same year thus it will soon expire by year 2019.

This paper laid emphasis on checking the conformity of social structures such as residential houses, school and so on located within the regulated hazard zone; testing the set-based geometry space risk model proximity with Quinali 'B' River; quantifying the risk and its elements as well as capability of individual sample households; and lastly validating the hazard zone stated in the approved land use plan and zoning ordinance with the newly published flood return periods 100, 25 and 5 years respectively. Now that the needs to maintream disaster risk reduction and climate change in development processes turn into a policy in socioeconomic development and physical plan formulatiton, assessment of trends in the actual land utilization become a challenging work as buffers as defined in Water Code, Civil Code, and the Revised Forestry Code in the end delineate the legal easements suitable for open public use areas or zones adheres to balance of nature and to reduce disaster risks within the 15 years target and priorities of the Sendai Framework.

2. Data and Methods

The first approach, was to identify informants or sample households constantly vulnerable to the physical characteristics and land condition situated in alluvial flat area and basin of the Mayon Volcano, Mount Masaraga and Mount Malinao Watershed streams all flowing to Quinali 'B' River; and contained within an absolute space called Hazard Zone, in which these households were directly affected by 2011 and 2016 flooding resulting to living in makeshift light materials or rebuilding their homes better. The 12 purposely selected sample households were verified in the ground using Mobile GPS and by assigning HPQ IDs as sequential control number encoded in the questionnaires, GIS_ID as collective numbers denoting the provincial (31), municipal (010), barangay (006) numbers combined with household code assigned by the barangay, and lastly latitude and longitude denoting the geographic location of respective samples. The Mobile GIS mapping activity attest the exposure and location of the samples within the HZ. The zoning map as shown in Figure 1 disclose the spot of the study area, in which the Land Use Map in the left reckon dwelling structures denoted by yellow polygons and school campus denoted by blue polygon, while in the Approved Zoning Map the spot was zoned as part of the HZ.

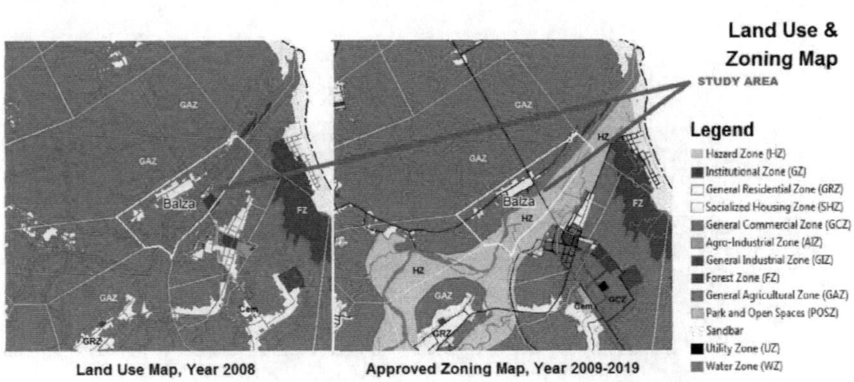

Figure 1. Zoning of Balza, Malinao (2009-2019)

Second, in order to get the desired output, these samples went through GIS overlay analysis to check their specific locations to verify if they are situated in flood hazard zone or not. The technique used in collecting pertinent data are using Mobile GPS to collect the geographic location or latitude and longitude and geo tags of structures being evaluated, assigning identification numbers and codes such as HPQ IDs as sequential control number encoded in the questionnaires, GIS_ID as collective numbers denoting the provincial (31), municipal (010), barangay (006) numbers combined with household code assigned by the barangay, and lastly survey questionnaires to collect the data about families' capability in terms of HH

IOP Conf. Series: Earth and Environmental Science **123** (2018) 012001 doi:10.1088/1755-1315/123/1/012001

members coping capacity, competency and preparedness. With ArcGIS it was easy to create or re-create map layers gathered from Malinao and build attributions necessary to re-create a replica or the Purok 1 located in Balza, Malinao. The 12 'purposely selected sample households' or a set of objects/locations with properties represent the approximately 80 exposed household of Purok 1, Barangay Balza, Malinao. The pre-selected samples as revealed in ArcGIS containing set-based geometry spaces to be constantly vulnerable to the physical characteristics and land condition. Overlays analysis allow the visualization of alluvial flat areas and the basin of the Mayon Volcano, Mount Masaraga and Mount Malinao Watershed streams all flowing to Quinali 'B' River. Although the result of 12 samples seem small proportion of the entire municipality conversely affected by the by 2011 and 2016 flooding, it affected totally washed out residents but choose to stay in makeshift light materials after flooding events. Some households rebuild their homes safer by building new two-storey residential building but remain exposed. In this study, the capability at HH level was treated 'subsets of objects' with spatial properties comprising coping capacity, competency, and preparedness.

Third, to be able to quantify the disaster risks location quotients or estimates, the following approaches were completed:

2.1 Hazard Assessment

This refers to the inspection and ground validation relative to flood occurrences brought by Quinali 'B' River over flows during heavy rains of several hours. Informants were asked to show marks of inundations that appear to be that the flood in year 2011 is twice higher compared to flood occurred in 2016. Informants attest that flood inundation in year 2011 is higher compared to flood occurred in 2016. According to MDRRMO the Quinali River swell up to 3 meters inundation near the bridge thus the same was also spotted by the local residents of Balza. The Figure 2 below denotes the household 008 of Purok 1, Barangay Balza with GIS ID Number 31-010-006-P-008 and located at 576086 Latitude, 1481842 Longitude projected at WGS84. Furthermore, the specific location of household with P-008 code assigned in Balza Spot Map.

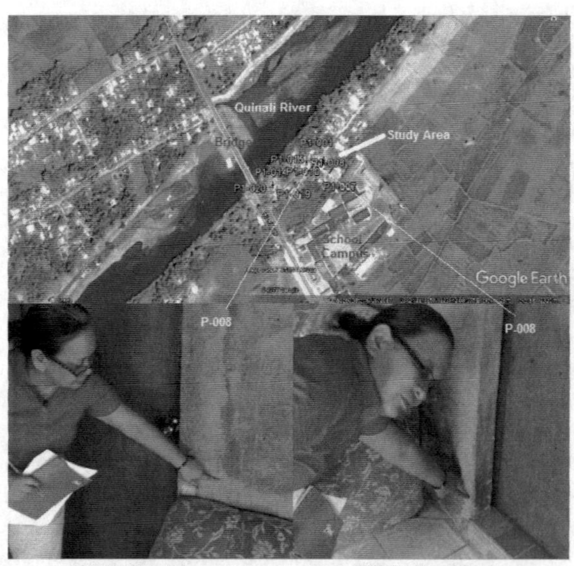

Figure 2. Flood Inundation, Year 2011 and 2016

IOP Conf. Series: Earth and Environmental Science **123** (2018) 012001 doi:10.1088/1755-1315/123/1/012001

2.2 Vulnerability Assessment

The researcher defines the 'vulnerability assessment' [7] as a study of the steady physical environment and the conditions of inhabitants (human ekistics) living in that the same physical space. Vulnerability in this study is a mixture of the 'Permanent Landscape Vulnerability' and the 'Human Vulnerability'.

The 'Permanent Landscape Vulnerability' in this study is defined as 'sets of objects' geographically enclosed inert physical space likely to adverse consequences associated with floods, bank erosion and other hydrological related hazards that may occur when river discharge exceeds its channel's volume causing the river to overflow onto the downstream alluvial flats as well as the coastal area that convey different states of discomfort, security, worries, distress, angst, and more. But the assessment was dependent to the criteria and classification set by the researcher, precision of spatial data and maps, and the data provided by the informants. The 'inert physical factors', such as the drainage network of the 3 mountains enclosing the catchment space dictates how much precipitation the Quinali 'B' River can receive and how rapidly it will come, it appears in the old topographic maps that Quinali 'B' River naturally collects a great volume of water or increasing discharge coming from the mountains, instead farmers converted most of the former river beds to rice lands then build their residence near these critical fields and/or the intrinsic characteristics of alluvial plains where the soil permeability of clay in Malinao prevented water from infiltrating flood water, which make it desirable for planting rice; and more. It likewise applies to the characteristics of 'ekistic household units' in terms of condition of housing units construction materials and/or dendritic housing patterns and its physical layout of the transportation network relative to escape en routes to safe evacuation centers, and/or existing flood mitigation infrastructures intended to lessen vulnerability, the national government constructed a flood control dike along Quinali 'B' River to reduce river overflow so as to protect the lives and properties of the communities at risk.

The 'Human Vulnerability' in this study is defined as 'point objects' referring to 'fields' storing data or so residents of Balza dwelling inside the most susceptible to flooding and bank erosions. It applies to the characteristics of the family members' biological needs, sensation, perception, emotional needs, and moral values.

Figure 3. Vulnerable Informants and Samples

2.3 Exposure Assessment

The researcher defines the 'exposure assessment' as a study of the inhabitants (human ekistics) situated or residing in a hazard or no-built zone. 'Exposure' is symbolized by 'point objects' referring to measured absolute geographical location or projected coordinates of any structure existing in an enclosed polygon subsets denoting the hazard zone. The researcher inventoried only those households situated to be severely exposed households within Purok 1, Balza, Malinao using GIS to validate theirs geographic location relative to flood return periods. Figure 4 display the analytical map presenting the extreme exposure of residents and public school symbolized by yellow and green polygon objects.

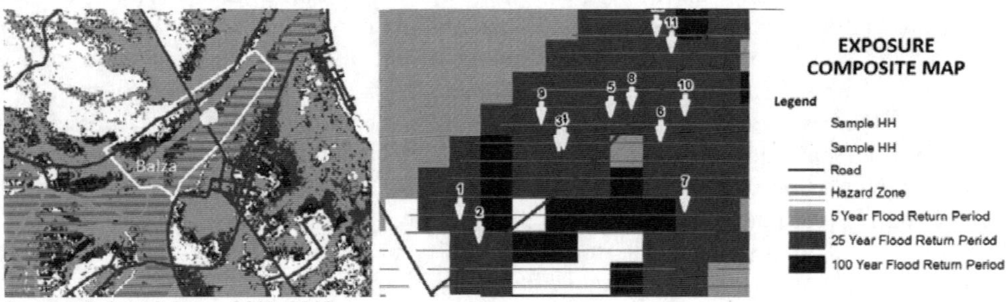

Source: Balza-Malinao, 2009, 2017

Figure 4. Exposure Geo-Objects Model

2.4 Capability Assessment

The researcher treat the 'capability assessment' as a study of coping capacity, competency and preparedness to relocate to safer location or residence or to escape the danger or evacuate/transfer upon announcement of evacuation phase of the Disaster Risk Reduction Management Council of Malinao [8,9]. The 'Capability' is a disintegrated 'subsets of objects' with spatial properties relative to subsets or fields, these are: coping capacity, competency, and preparedness fields, result of analysis of object features (polygon and/or point), attributes and/or field or related databases) and its relationship with information about livelihood, family income, household members age, gender and sensitivity, plans to resettle, evacuation practices, participation in risk reduction related activities, land valuation and ownership, and so on.

2.5 Disaster Risk Assessment (DRA) Modeling

The Risk Quotient Object-Based Model discloses a factor resulting after overlaying analysis of the intersections of hazard, vulnerability, exposure and capability objects and characteristics.

Table 1 is a Risk Sets of Objects or Fields Numeracy Menu that comprise the matrix of object or field properties covering five level scale (5x5 = 25 cells) the 1, 2, 3, 4, and 5 numerical values to convey an ordinal likelihood, these are: 1) classifying risk into five levels denoting: Level 1, Level 2, Level 3, Level 4 and Level 5; 2) grouping risk into five degree denoting: very low, low, moderate, high and very high; 3) sorting the Flood Predictions into five spans denoting: 100 Years or more; >50 to <100 Years; >25 to 50 Years; >5to <25 Years; and <1 to < 5 Years; 4) Symbolizing into five colors for simple visualization denoting: White, Yellow, Orange, Red and Dark Red; and lastly risk sets of objects or fields denoting: hazard; vulnerability; exposure; Capability.

The Risk Quotient Object-Based Model of Balsa discloses a factor resulting after overlaying analysis of the intersections of hazard, vulnerability, exposure and capability objects and characteristics. The first overlay analysis is finding the intersections of the 3 subsets of Risk set, namely: hazard, vulnerability, and exposure. The second overlay analysis is finding the intersections of the 3 subsets of Capability, namely: coping capacity, competency and readiness. The proportion of the 2 resultants, where the 1st result serves as the numerator and the 2nd as the denominator, the end effect is the hazard quotient or risk factor useful for predicting disaster risk to significant for decision making and physical developments limited only to the survey area.

Table 1. Risk Distribution Level Classification Properties of Object Model

Risk Sets Objects or Fields Numeracy Menu	Level 5 Very High <1 to < 5 Years Dark Red	Level 4 High >5to <25 Years Red	Level 3 Moderate >25 to 50 Years Orange	Level 2 Low >50 to <100 Years Yellow	Level 1 Very Low 100 Years or more White
• **h** - Hazard ✓ Flood Susceptibility Geometry Space Object	5.0	4.0	3.0	2.0	1.0
✓ Regular Flooding may inundate to 3 meters or more ✓ Hazard event (h) is an element ($\in$) of Risk denoted as (h $\in$ R)	4.0	4.0	3.0	2.0	1.0
✓ Flood (f) is a subset ($\subseteq$) of Hazard denoted as (f $\subseteq$ h)	3.0	3.0	3.0	2.0	1.0
• **v** – Vulnerability ✓ Permanent Landscape Vulnerability - Physical Inert Land Condition Geometry Space Object ✓ Human Vulnerability - Inherent Human Transferable Condition (Highly Susceptible due to decision to reside inside HZ or homeless, etc. ✓ Vulnerability (v) is an element ($\in$) of Risk denoted as (v $\in$ R)	2.0	2.0	2.0	2.0	1.0
• **e** –Exposure ✓ Building Structure Geometry Space Object ✓ Situated within Hazard Zone/Critical Area ✓ Exposure (e) is an element ($\in$) of Risk denoted as (e $\in$ R) • **C** - Capability denoted as C {Co, Cm,Cp} ✓ Field-based Object ✓ Discrete information about Individual Household member Competency, Coping Capacity, and Preparedness ✓ Identifiable information about Coping capacity (Co) as an element ($\in$) of Capability denoted as (Co $\in$ C) ✓ Identifiable information about Competency (Cm) as an element ($\in$) of Capability denoted as (Cm $\in$ C) ✓ Identifiable information about Preparedness (Cp) as an element ($\in$) of Capability denoted as (Cp $\in$ C)	1.0	1.0	1.0	1.0	1.0

Figure 5 is comprise of risk location quotient after equating the numerator composed of the product of the hazard level of Purok 1, Balza, Malinao , vulnerability and exposure levels of households, divided by the product of coping capacity, vulnerability and exposure levels, where numerical values can be assigned using Figure 5. There are 5 ranges of Risk Location Quotient Distributions Index, these are 1) Very High ranges from greater than 64 to up to 125 maximum numerical values; 2) High ranges from greater than 8 to 64 numerical value; 3) Moderate ranges from greater that 1 to 8 numerical value; 4) Low

IOP Conf. Series: Earth and Environmental Science **123** (2018) 012001 doi:10.1088/1755-1315/123/1/012001

or Balanced with numerical value of 1.0; and lastly 5) Very Low is assumed to be near resilient or resilient.

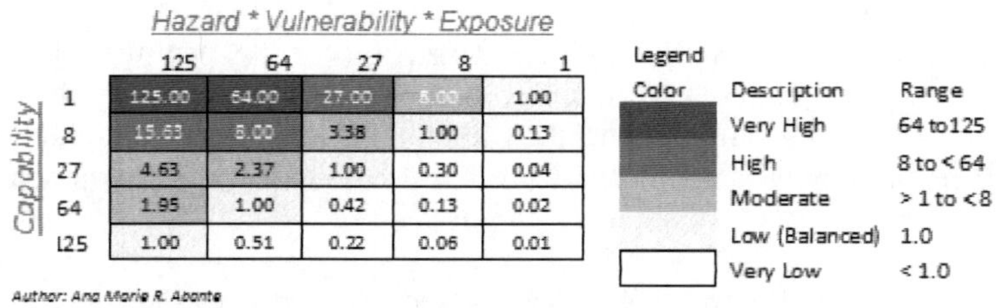

Figure 5. Risk Location Quotient Distribution Matrix

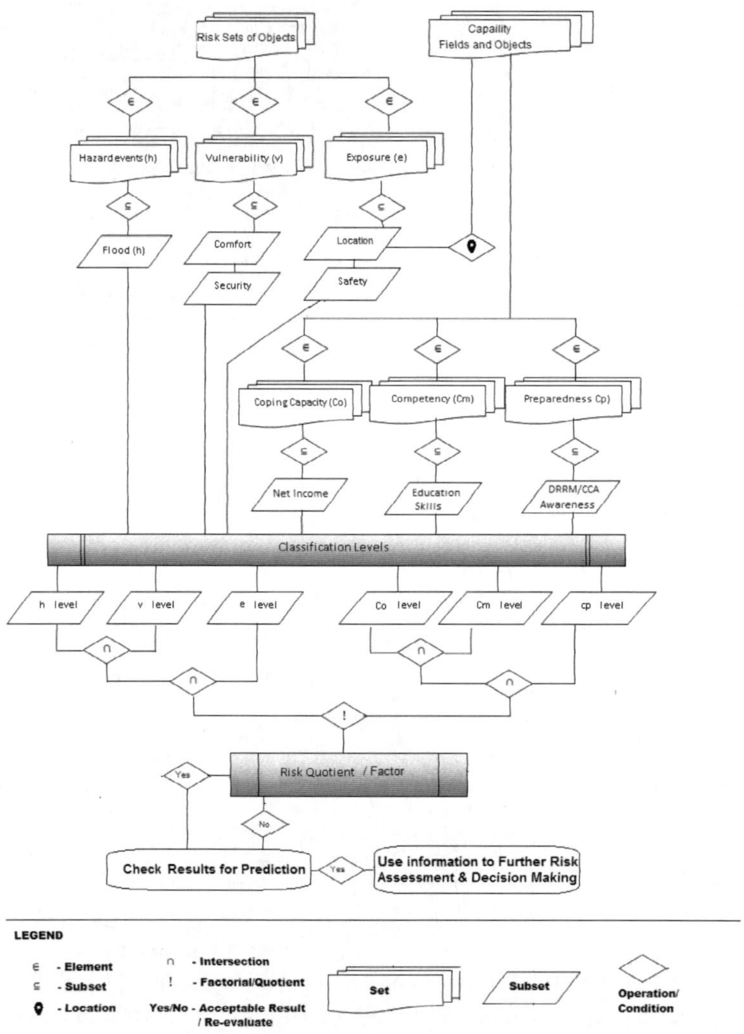

Source: Abante, 2017

Figure 6. Risk Quotient Object and Field Based Model

The universal set theory, the three elements denoted as (∈) of risk, namely: hazard events, denoted as (h), vulnerability denoted as (v) and exposure denoted as (e) comprise the Set of Risk (R) objects. The Flood denoted as (h) is a subset of Hazard denoted as (⊆). The Capability denoted as (C) comprise of a set of {Coping Capacity denoted as (Co), Competency denoted as (Cm), Preparedness denoted as (Cp). The Risk Quotient Object-Based Model discloses the factor of the absolute space existent amongst the intersections of subsets denoted as h level ∩ v level ∩ e level thus resulting to a numerator object. Thence, the intersection of georeferenced objects related to fields intersecting the Coping Capacity level denoted as Co level ∩ Competency denoted Cm level ∩ Preparedness denoted by Cp level, thus resulting to a denominator Capability denoted as C.

Lastly, (4) Validate the HZ with the 100 years Flood, 25 years Flood and 5 Years Bicol Flood forecasted by DOST through the Philipine Lidar Project 1. In July 25, 2011, a flash flood traveled unexpectedly swept away rice land, houses and tree branches by the flood as some cars and personal belongings washed away by floodwater. Malinao was submerged when Typhoon "Juaning" (International Code: Nock-ten) dumped about 500 millimeters of rain in a span of 15 hours resulting to a flash flood as high as three meters. Although the entire Province of Albay truly prepared in which evacuation practices escaping disaster risk is already institutionalized, TD Juaning left 1 casualty in Malinao though cause of dying is the snake bite not by drowning [9]. Fortunately, the flood event occurs at day time if not it can easily kill more person as they may get hit by the debris or submerged. Moreover, in December 28, 2016, Typhoon "Nina" (international name: Nock-Ten) again triggered floods and leveled houses in Malinao, Albay, in which exposed residents fled to safer ground before the storm hit and were forced to spend Christmas in school classrooms that were temporarily converted into evacuation shelters. Based on the report of Albay Public Safety and Emergency Management Office (APSEMO) Report, there are 50 families or 250 persons in Malinao evacuees. In May 2017, the Project NOAH 5, 25 and 100 Year Flood Forecasts published recently by Department of Science and Technology, the result after matching and examining HZ analyzed in 2009 by method of overlaying the three flood return period, implies disaster risk. The study reveal possibility of critically and costly risk reduction on the part on the Local Government of Malinao to evacuate exposed residents every time of calamities and increased risks and hardship on the part of every households. Figure 8 display the return periods 5th, 25th and 100th year flooding overlaid with the Hazard Zone, accessibility and the Barangay (Community) administrative jurisdiction.

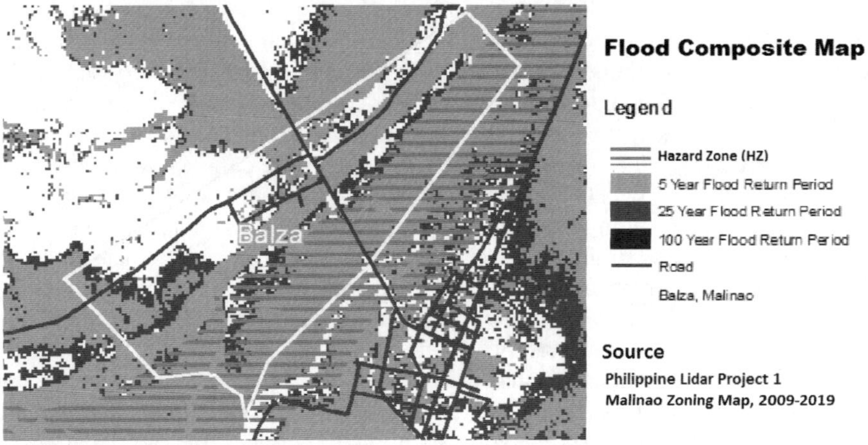

Source: Balza-Malinao

Figure 7. Flood Composite Model

3. Result and Discussion

Disaster Risk valuation (DRE) result reliant on the approach in selecting the informants or exposed-houses that is constantly vulnerable to the physical characteristics and land condition and confined within an absolute space called Hazard Zone. The 2011 and 2016 flooding imply 5 year return period is felt. If so, more flood simulation models is desired to do temporal analysis to truly understand DRE. Locating and storing information on marking of flood inundations as well as exposed-resources using geospatial technologies are significant in valuation and evaluation along with monitoring and evaluating the implementation of zoning ordinance specifically those developments enclosed within the environment critical ecosystems and hazard zones [7]. Most of the researches vulnerability theme generally describe not as a mixture of 'Permanent Landscape Vulnerability' in which resources are fixed or passive in nature, and the 'Human Vulnerability' regarded as active resource capable of lowering vulnerability and escaping danger. In this study, vulnerability and exposure were defined and classified into five categories for the purpose of measuring risk factor. Risk is a factor of numerical values for hazard multiplied by vulnerability and exposure, divided by the capability Assigned value. Capability is measured as the product of the following disintegrated subsets of capability, namely: coping capacity, competency and preparedness of each samples classified based on informants inputs.

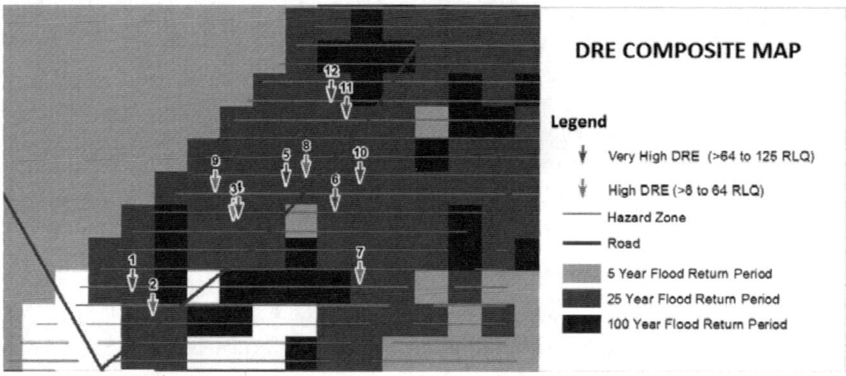

Source: Purok1-Balza-Malinao, 2017

Figure 8. DRE Composite Map

On the sampling, the selection of the study area was purposive and geographically limited within the HZ to collect and examine the likelihood of households at risk. Informants for every sample household as listed in Table 1 and shown in 9 were interviewed and provided data for their respective Household Profile Questionnaires (HPQ), in which it contain facts about their location and household information. The researcher prepared the HPQ, which is a modified questionnaire of the CBMS Questionnaire published by DILG for the purpose of collecting data to measure the capability of every family to escape the danger that may be brought by severe flood resulting from Quinali River overflows. Samples and testing site discloses mean and standard deviation of risk of individual households pointing out risk is measurable at household level. Although the selection of the study area precisely within the hazard zone, it stresses the hazard, permanent landscape vulnerability and exposure to the highest level of classification. Conversely, the mean and standard deviation of: 1) vulnerability of the age dependency group, 2) family income, 3) educational attainment and skills entice competency to get employed or self-employed, and lastly 4) perception on their safety and preparedness to escape flood risk certainly contradict the land use and hazard zone regulations. Calculating the standard deviation individual HH risk location quotient, the confidence intervals based on the estimated standard deviations is illustrated in Table 2.

Table 2. Disaster Risk Estimate

HP Q ID	GIS_ID	Location		Disaster Risk Elements Rating						DRE	Likelihood
		Latitude	Longitude	Hazard	Vulnerability	Exposure	Coping Capacity	Competency	Preparedness		
1	31-010-006-P-020	576033	1481807	5.00	5.00	5.00	1.00	1.00	1.00	**125.00**	**VH**
2	31-010-006-P-019	576039	1481800	5.00	5.00	5.00	3.00	3.00	1.00	**13.88**	**H**
3	31-010-006-P-014	576064	1481829	5.00	5.00	5.00	3.00	3.00	1.00	**13.88**	**H**
4	31-010-006-P-016	576066	1481829	5.00	5.00	5.00	1.00	1.00	1.00	**125.00**	**VH**
5	31-010-006-P-010	576080	1481839	5.00	5.00	5.00	1.00	1.00	1.00	**125.00**	**VH**
6	31-010-006-P-002	576095	1481832	5.00	5.00	5.00	1.00	1.00	1.00	**125.00**	**VH**
7	31-010-006-P-007	576103	1481810	5.00	5.00	5.00	3.00	2.00	2.00	**10.41**	**H**
8	31-010-006-P-008	576086	1481842	5.00	5.00	5.00	3.00	3.00	1.00	**13.88**	**H**
9	31-010-006-P-015	576059	1481837	5.00	5.00	5.00	3.00	3.00	1.00	**13.88**	**H**
10	31-010-006-P-006	576103	1481840	5.00	5.00	5.00	1.00	1.00	1.00	**125.00**	**VH**
11	31-010-006-P-001	576099	1481860	5.00	5.00	5.00	1.00	1.00	1.00	**125.00**	**VH**
12	31-010-006-P-050	576094	1481865	5.00	5.00	5.00	1.00	1.00	1.00	**125.00**	**VH**

Source: Purok1-Balza,July 2017

DRE result is reliant on the approach in selecting the informants or exposed-houses (social structures) that is constantly vulnerable to the physical characteristics and land condition and confined within an absolute space called Hazard Zone. The 2011 and 2016 flooding imply 5 year return period is felt. If this is correct, more flood simulation models is desired to do temporal analysis to truly understand disaster risks and its elements. Locating and storing information on marking of flood inundations as well as exposed social resources using geospatial technologies are significant in valuation and evaluation along with monitoring and evaluating the implementation of zoning ordinance specifically those developments enclosed within the environment critical ecosystems and hazard zones. Most of the researches vulnerability theme generally describe not as a mixture of 'Permanent Landscape Vulnerability' in which resources are fixed or passive in nature, and the 'Human Vulnerability' regarded as active resource capable of lowering vulnerability and escaping danger. In this study, vulnerability and exposure were defined and classified into five categories for the purpose of measuring risk factor in terms of numerical values for hazard multiplied by vulnerability and exposure, divided by the capability Assigned value. The capability is measured as the product of the following disintegrated subsets of capability, namely: coping capacity, competency and preparedness of each samples classified based on informants inputs.

As the validation of HZ extents turned coherent with the flood forecasts and return periods every 5, 25 ad 100 years, the DRE result set in motion the risks that varies from 10.41 (High) to 125 (Very High). The N error of 0.239 is insignificant at this first attempt considering the 12 samples out of 80 potential at risk in their Purok are regarded lacking. For a greater accuracy and reliability the samples for the in all the Purok of Barangay Balza, as well as other barangays of Malinao is desired to calculate the municipal-wide DRE. The result in simulating the Risk Distribution Set-Based Geometry Space Model illustrated the relationships of hazards, vulnerable lives, livelihoods, health, business, communities and so on as stated very similar in the Sendai Framework.

As typical risk assessment models mostly focus on rapid assessment [10,11] of which the six risk elements interrelated in this paper were taken in a broader perspective inevitably 'Risk Location Quotient

Distributions diverges, this paper's contribute the 'Risk Quotient Model and Field Based Model' which implicitly let out the relationships of the 'sets of point objects' denoting spatial or geographic locations of houses or other social related structures overlaid with the 'set-based geometry space' denoting the area of influence of hazard and passive (landscape) vulnerability equated to the 'sets of point objects' denoting the 'fields' or attributes of capability divided into the coping capacity, competency and preparedness at household level.

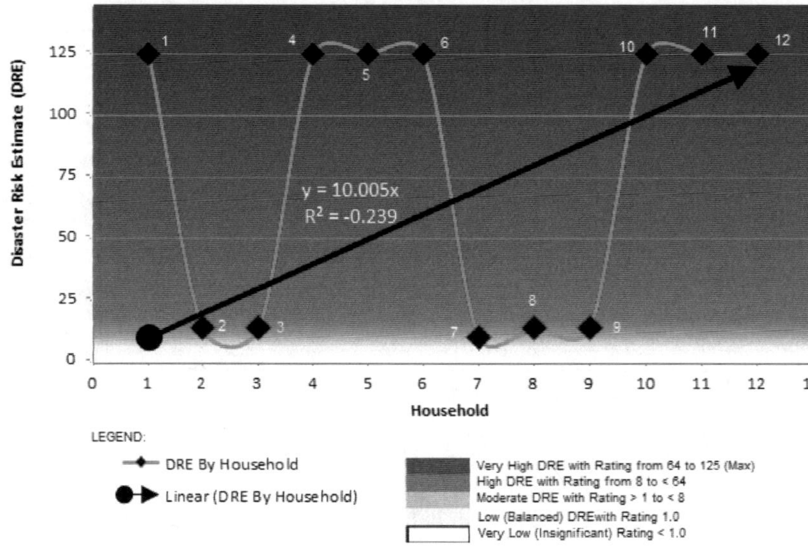

Source: Purok1-Balza-Malinao, 2017

Figure 9. Disaster Risk Estimate Graph

On the DRE, this first attempt in data analysis encompass the risk computing concept of intersections of three objects (polygon features) representing hazard, vulnerability, exposure and capability likelihood. The two selected columns HPQ_ID (or HH Number) and DRE as shown in Table 1 represent the x and y value respectively. In the Ordinary Least Squares (OLS) for DRE per HH, the *x* values (HH#) are fixed, and find the line which minimizes the squared errors in the *y* values of the DRE for each sample households varies from 10.41 (High) to 125 (Very High) as the result of . The OSL for DRE by HH line y = 10.005x fitted the bivariate data sets of regressed in the plot as shown in Figure 9. The mean and standard deviation of DRE reckoned is 78.4 (Very High) and 0.965 respectively. Measurement error of 0.239 is insignificant at this first attempt considering 12 samples are regarded lacking. For a greater reliability of samples, further study and appropriate geospatial technology is wanted.

The outcome of risk computing and GIS overlay exercises demonstrated the likelihood and simply hinting flood risks, reoccurring since ancient times, and will continue to happen. And so wanting geospatial information systems designed for communities turn out to be a platform both in environment protection and development planning. However, further study on informants' perception on the risk they are facing and attitude on preparedness efforts of residents in cooperation with the DRRM managers is seen crucial in DRE as well as in DRRM and CCA as they (exposed residents of Balza) decide where to stay, whichever: (1) designated evacuation center by MDRRMO; (2) chosen temporary space such as two-story residential houses in proximity (located within the same exposed areas where they reside) to stay escaping flood risks; (3) chosen temporary space located outside the exposed areas 'perceived to be safe location' but not necessarily comfortable; or (4) other place safe and comfortable such as residential areas identified as safe and comfortable residential areas.

Of all the risk elements, exposure is the only independent variable where risk transpires only if it holds true. The Disaster Risk Estimates (DRE) is seen significant for decision making in any physical developments and imposing allowed land uses and regulations through a local ordinance. On the capability variable, the study disclosed that the DRE result is dependent on both the researcher's knowledge and skills in GIM as well as the informants' perception on their risk against the coping capacity, competency and preparedness weighted against the scientific findings on risk and its elements. Although lot of governments' efforts on hazard mapping, the variability remains ambiguous on the vulnerability and exposure, but then again significant in fusing all the risk elements in a single integrated information system.

The outcome of risk computing and GIS overlay exercises demonstrated the likelihood and simply hinting flood risks, reoccurring since ancient times, and will continue to happen. And so wanting geospatial information systems designed for communities turn out to be a platform both in environment protection and development planning. However, further study on informants' perception on the risk they are facing and attitude on preparedness efforts of residents in cooperation with the DRRM managers is seen crucial in DRE as well as in DRRM and CCA as they (exposed residents of Balza) decide where to stay, whichever: (1) designated evacuation center by MDRRMO; (2) chosen temporary space such as two-story residential houses in proximity (located within the same exposed areas where they reside) to stay escaping flood risks; (3) chosen temporary space located outside the exposed areas 'perceived to be safe location' but not necessarily comfortable; or (4) other place safe and comfortable such as residential areas identified as safe and comfortable residential areas

4. Conclusion

In pursuit of development, disaster risk and climate change resilient ecosystem, and community-based awareness of environmental protection and wise-use of natural resources, public use areas susceptible to flooding within a confined area delineated as a ROW or floodway adjacent to the legal easements as defined by water code, civil code or the revised forestry code, areas that are ecologically significant water courses must be protected from impacts of adjacent developments and human activities such as construction of permanent structures, classifying and/or converting these floodway or legal easement into urban use or for dwelling or even land titling the ecological buffer for private use.

Changing climate now producing more intense and heavy rains initiating floods or causing inundation onto normally dry adjacent areas. Floodways or right of way (ROW) or legal easement that is located in near their places of work or employment, market shopping malls, place of worship and other basic social services are prone to non-conforming land uses and often times consented and granted with land claims due to absence of approve land use plans and zoning ordinance in the local government units such malpractices in land utilization remains unresolved.

Promoting sustainable growth and resiliency as well as reducing disaster risks through geospatial information systems is fundamental in understanding the natural phenomena affecting settlements and developments. This is also true to speed up land use plan formulation and revisions as well as crafting and amending zoning ordinances opt to lessen the risk of the people and properties although prevention, land tenure and property rights remain boundless challenge to planners, engineers and decision makers and development managers.

5. Acknowledgments

I would like to express my gratitude to my research assistant Mr. Carlo Gabriel R. Abante for helping me carry out the interviews and talk personally to families regularly experiencing floods residing within the hazard zone. My research would not be realized without the technical support and funding from Bicol University and the support from the Bicol University College of Engineering, Research, Development and Management Department and the Office of the Vice President for Research, Development and Extension.

This acknowledge that without the assistance coming from the Local Government Unit of Malinao under the leadership of Former Mayor Billy Ceriola, the present administration of Honorable Mayor Sheryl C. Bilo, the Municipal Planning and Development Officer in the person of Mr. German Gonzaga, Mr. Salvador Carillo, Jr. the staff of the Municipal Disaster Risk Reduction Management Office, Barangay Chairman of Balza in the person of Mr. Nelson Rico together with his Purok leaders, this study cannot be completed.

A special word of thanks to Philippine Lidar Program and DREAM for sharing the copyrighted materials on Flood Maps and its return periods. I appreciate the help of the Geodetic Engineering Department Head and Co-faculties for backing me in pursuing this paper. I am grateful to be my husband EnP. Anthony N. Abante as professional partner in the formulation of the land use plan and zoning ordinance of the Local Government Unit of Malinao, Albay. To my children Paolo, Carlo, Marco and Marietoni and my mother Heidi, thank you for the endless love and understanding. Above all, I humbly offer this piece of work to the Almighty God, to Thee be the Glory!

6. References

[1] Pugh C 2001 The theory and practice of housing sector development for developing countries, 1950-99 *Hous. Stud.* **16** 399–423

[2] Florano E R 2015 Mainstreaming Integrated Climate Change Adaptation and Disaster Risk Reduction in Local Development Plans in the Philippines *Handb. Clim. Chang. Adapt.* 433–456

[3] Mandal R B 1990 *Land utilization: Theory and practice* (Concept Publishing Company)

[4] Strassman P and Blunt A 1993 Land, income, mobility and housing: the case of Metro Manila *J. Philipp. Dev.* **20** 79–115

[5] Aubrecht C, Özceylan D, Steinnocher K and Freire S 2013 Multi-level geospatial modeling of human exposure patterns and vulnerability indicators *Nat. Hazards* **68** 147–263

[6] Doxiadis C A 1959 Ekistics, the science of human settlements *T. Ctry. Plan. Summer Sch. (Outline). Southampt.* **2**

[7] Birkmann J 2008 *Assessing vulnerability before, during and after a natural disaster in fragile regions: Case study of the 2004 Indian Ocean tsunami in Sri Lanka and Indonesia* (Research paper/UNU-WIDER)

[8] UNISDR 2009 2009 UNISDR Terminology on Disaster Risk Reduction *Int. Strat. Disaster Reduct.* 1–30

[9] Cuevas S C, Peterson A, Robinson C and Morrison T H 2016 Institutional capacity for long-term climate change adaptation: evidence from land use planning in Albay, Philippines *Reg. Environ. Chang.* **16** 2045–2058

[10] Ramsey M H 2009 Uncertainty in the assessment of hazard, exposure and risk *Environ. Geochem. Health* **31** 205–217

[11] Montgomery V 2009 *New statistical metliods in risk assessment by probability bounds* (Durham University)

IOP Conf. Series: Earth and Environmental Science **123** (2018) 012002 doi:10.1088/1755-1315/123/1/012002

Modelling Multi Hazard Mapping in Semarang City Using GIS-Fuzzy Method

A L Nugraha[1], M Awaluddin[1], and B Sasmito[1]

[1]Department of Geodetic Engineering, Faculty of Engineering, University of Diponegoro

Email: arief_ln@yahoo.com

Abstract. One important aspect of disaster mitigation planning is hazard mapping. Hazard mapping can provide spatial information on the distribution of locations that are threatened by disaster. Semarang City as the capital of Central Java Province is one of the cities with high natural disaster intensity. Frequent natural disasters Semarang city is tidal flood, floods, landslides, and droughts. Therefore, Semarang City needs spatial information by doing multi hazard mapping to support disaster mitigation planning in Semarang City. Multi Hazards map modelling can be derived from parameters such as slope maps, rainfall, land use, and soil types. This modelling is done by using GIS method with scoring and overlay technique. However, the accuracy of modelling would be better if the GIS method is combined with Fuzzy Logic techniques to provide a good classification in determining disaster threats. The Fuzzy-GIS method will build a multi hazards map of Semarang city can deliver results with good accuracy and with appropriate threat class spread so as to provide disaster information for disaster mitigation planning of Semarang city. from the multi-hazard modelling using GIS-Fuzzy can be known type of membership that has a good accuracy is the type of membership Gauss with RMSE of 0.404 the smallest of the other membership and VAF value of 72.909% of the largest of the other membership.

1. Introduction

Geographically Semarang City is bordered by Java seas in the north, as well as topographical condition of its territory consisting of hilly area, lowland and coastal area. It indicates the existence of various slope and bulge that causes Semarang City area has the potential to be vulnerable to the threat of natural disaster. Semarang city also is one of the major cities in Indonesia that is prone to natural disasters. Frequent natural disasters Semarang city is tidal flood, floods, landslides, and droughts [1]. The disaster has made a loss of economic aspects to fatalities. With this natural condition, the city has been long experiencing with potential multi hazards risks [2].

Geographical information about the disasters is urgently needed in the development of disaster mitigation plans. The presentation of spatial disaster threats is very beneficial because people can directly recognize the conditions of disaster-prone areas. Therefore it is necessary to make a potential map of disaster prone areas. One important aspect of disaster mitigation planning is to conduct hazard mapping. Multi hazard mapping is a map-making activity that represents negative impacts that may result in material and non-material losses in a region in the event of a disaster [3].

Geographical Information System (GIS) is an appropriate method in mapping disaster-prone areas for wide area coverage with a relatively short time. This method is carried out as an effort to analyze risk and hazard mapping through dissemination of hazard information so that the mapping will accelerate the process of delivering information to the community and related institutions and can increase preparedness in taking action to reduce the risk of flood threat in Semarang City. The development of GIS utilization has been quite popular. By integrating GIS technology and precise mathematical methods, it can produce spatial analysis with good precision and accuracy. Spatial analysis can be approached by method of weighting and classification of criteria needed. Various methods to get the weighting and classification, one of them is by doing the method of decision making using Fuzzy Logic method. The uncertainties caused by the insufficient knowledge about the interdependency of various parameters have been considered by means of fuzzy relations [4]. With the use of Fuzzy Logic method can strengthen the hypothesis to be achieved in the hazard mapping in a region.

The focus of this research is how to do a multi hazard mapping in Semarang city with GIS Fuzzy method. The purpose of this research was the formation of Multi Hazard Map in Semarang with the methods and accuracy in accordance with the conditions of Semarang city in order to assist policy makers in deciding policy in anticipation of hazardous occurrences as well as to disaster mitigation planning in city Semarang.

2. Hazard Mapping using GIS

To avoid the impact it causes, people need to first recognize and understand the threats or dangers that exist. One effort to recognize and understand the dangers is to conduct a hazard assessment. Hazard assessment involves analyzing the physical aspects of phenomena or natural events through the collection of historical data, interpretation of topographic, geological, and hydrological data to obtain forecasts of spatial and temporal probability of occurrence and magnitude of hazards. Furthermore, from hazard assessment can be done hazard mapping. Hazard mapping is assessing and presenting hazard information to display hazard characteristics (nature and type of hazards), intensity (time and duration of impacts) and extent of impact areas as hazard zones useful for disaster mitigation activities. Each type of disaster is assessed using a different model, even one type of disaster can be analyzed using different methods [3]. The method used mostly is using probabilistic (quantitative) or deterministic (usually qualitative) analysis. For each type of disaster can be analyzed by several methods. In general it can be said that in the hazard assessment the data relating to the return period and the intensity of the hazard plays an important role of quality assessment results. GIS technology helps a lot of hazard assessment because it is able to combine various data and accommodate the calculation of the occurrence of occurrence and modelling the probability of hazard into the data processing.

3. Assesment Method using Fuzzy Logic

A fuzzy set theory-based risk assessment process realizes the following process [5]. This process uses several rules simultaneously. The attribute of set of rules is that their solution by classical logic can be different or antinomic at the same time. Practically, this inconsistency can be (should be) resolved by the fuzzy logic. This process is a combination of four subprocesses: fuzzification, inference, composition, and defuzzification. The defuzzification subprocess is optional. In the Fuzzification sub process, the membership functions defined on the severities and probabilities of investigated hazards are applied to their actual values, to determine the degree of truth for each rule premise. In the Inference sub process, the truth-value for the premise of each rule is computed, and applied to the conclusion part of each rule. This results in one fuzzy subset to be assigned to each output variable for each rule. In the Composition sub process, all of the fuzzy subsets assigned to each output variable are combined together to form a single fuzzy subset for each output variable. Sometimes it is useful to just examine the fuzzy subsets that are the result of the composition process, but more often, this fuzzy value needs to be converted to a single number — a crisp value. This is what the Defuzzification sub process does. There are more defuzzification methods than you can shake a stick at. Two of the more common techniques are the centroid and maximum methods. In the centroid method, the crisp value of the output variable is computed by finding the variable value of the center of gravity of the membership function for the fuzzy value. In the maximum method, one of the variable values at which the fuzzy subset has its maximum truth value is chosen as the crisp value for the output variable. There are several variations of the maximum method that differ only in what they do when there is more than one variable value at which this maximum truth value occurs. One of these, the average of maxima method, returns the average of the variable values at which the maximum truth value occurs [6].

4. Methods

In this case, the multi-hazard mapping of Semarang city consists of four natural hazards that often occur in the city of Semarang. The four natural hazards are floods, landslides, Tidal floods, and drought. Furthermore, each parameter of the hazard is done weighting and scoring that refers to the regulations in Indonesia, namely PERKA BNPB No. 2 of 2012 on General Guidelines for Disaster Risk Assessment [7]. The method used in mapping each of these hazard threats uses spatial analysis as GIS tools.

The formation of multi hazard map is done by overlaying from each hazard map with weight in accordance with BNPB PERKA no. 2 of 2012 [7] which was modified for four disasters in this study. In this research also conducted a multi-hazard mapping with the classification process using Fuzzy logic type Mamdani method with three types of membership that is Gauss, Trapezium, and Triangle. The next analysis was made a comparison of multi-hazard mapping between classification according to BNPB PERKA no. 2 in 2012 with classification using Fuzzy logic method, so that got conclusion in this research is knowing accuracy of classification of each method is in multi-hazard mapping in Semarang City (Figure 1).

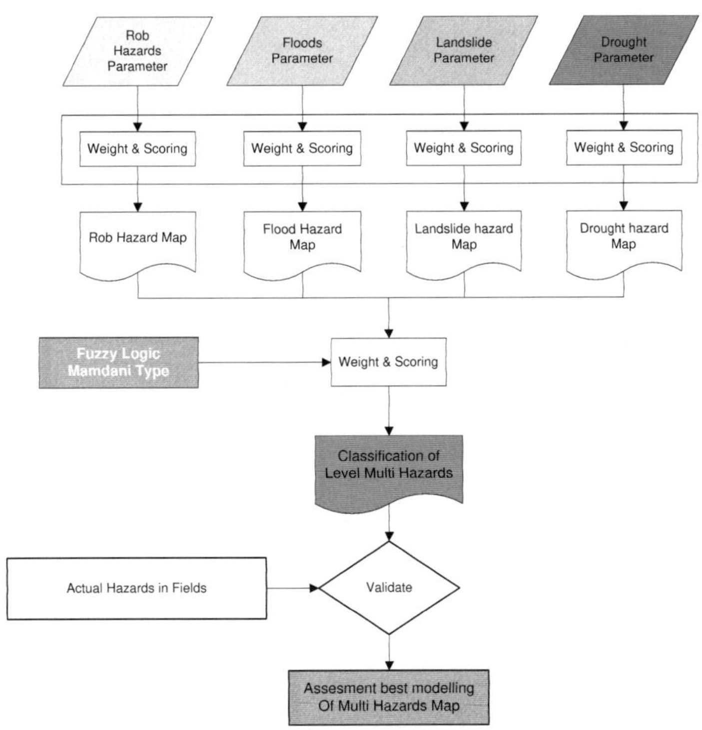

Figure 1. Flowchart of methodology

4.1. Flood Hazard Mapping

This mapping involves the parameters and with the weight of each parameter as in Table 1. The four parameters have the same weight because these parameters have the same effect in the formation of the threat of flood hazard.

Table 1. Parameter Flood Hazard [7]

No	Parameter	Weight
1.	General Flood Zone	0,25
2	Rainfall	0,25
3	Height	0,25
4.	Land Use	0,25

4.2. Tidal Flood Hazard Mapping

Making the tidal flood hazard map is done in several stages of work by different methods, which should prepare the data required for each stage. The data required for mapping the tidal flood is [8]:

a. Data tide that has been processed to obtain the value of MSL and HHWL.
b. Topographic maps which are then processed into DEM topography.
c. Data accelerating decline in soil surface which is then processed into DEM land subsidence.

In general, the tidal flood mapping can be seen in Figure 2.

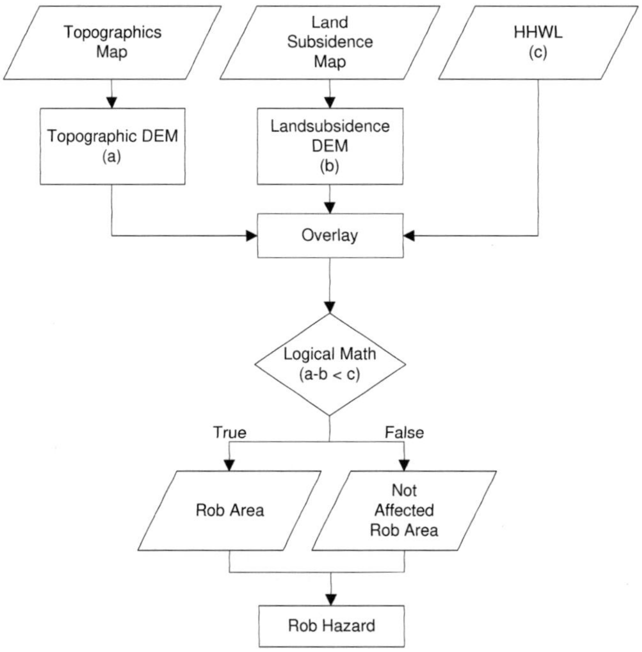

Figure 2. Tidal Flood (Rob) Hazard Mapping Flowchart [8]

4.3. Landslide Hazard Mapping

This mapping involves the parameters and with the weight of each parameter as in Table 2. The four parameters have the different weight because Slope and Geology parameters have more biggers effect than the others in the formation of the threat of Landslide hazard.

Table 2. Parameter Landslide Hazard [9]

No	Parameter	Weight
1.	Slope	0.30
2.	Rainfall	0.20
3.	Geology	0.30
4.	Land Use	0.20

4.4. Drought Hazards Mapping

This mapping involves the parameters and with the weight of each parameter as in table-3. In this case, the four parameters have the different weight because NDVI and Rainfall parameters have bigger effect than the others in the formation of the threat of Drought hazard.

Table 3. Parameter Drought Hazard [9]

No	Parameter	Weight
1.	NDVI	0.35
2.	Rainfall	0.35
3.	Type of soil	0.15
4.	Land Use	0.15

4.5. Multi Hazard Mapping

Once generated map of the threat of floods, tidal flooding, landslides, drought, then weighted and overlay. Referring to the Perka BNPB No. 2 In 2012, the weight of each disaster to disaster multi mapping can be seen in Table 4.

Table 4. Risk Weight Index [8]

No	Hazard Type	Weight (%)
1.	Flood	0.1064
2.	Earthquake	0.1064
3.	Tsunami	0.0638
4.	Fire Building and Housing	0.0638
5.	Drought	0.0638
6.	Extreme weather	0.0638
7.	Landslide	0.1064
8.	Volcano Eruption	0.1064
9.	Extreme waves and Abrasion	0.0638
10.	Forest fires	0.0638
11.	Failure Technology	0.0638
12.	Social conflict	0.0638
13.	Epidemics and disease outbreaks	0.0638

From these provisions, a modification to the disaster that is used is flooding, tidal flood, landslides and droughts using a simple mathematical calculation with a parable a value of one to floods, tidal flood and landslides and the value of half of the drought, so that the weights and scoring for each disaster can be seen in table 5 and table 6.

Table 5. Multi Hazards Weight Index

No	Hazard Type	Weight (%)
1.	Flood	0.286
2.	Tidal Flood	0.286
3.	Landslide	0.286
4.	Drought	0.142

Table 6. Multi Hazards Weight Index

No	Classify	Score
1.	Low	1
2.	Medium	2
3.	High	3

In determining the multi-hazard classification values used for the process of obtaining multi-hazard values and classifications, comparisons of Fuzzy Inference System (FIS) results in three membership functions, namely triangular membership system, trapezium, and gauss, by implementing the FIS stages as mentioned above. Furthermore, validation of the three model performance tests of the three membership functions with reference to the multi-hazard classification assessment process from the scoring system approach and weighting based on the BNPB PERKA no. 2 Year 2012 [7]. In this validation is done by determining the value of RMSE (Root Mean Square Error) and VAF (Variance Accounted For) from the total value of total vulnerability score of the three functions of fuzzy logic membership compared with the value of hazard classification scores through the score and weighting respectively Components of hazard maps.

5. Result and Analysis

5.1. Result of Flood Map

From the results of the weighting parameter overlay and altitude, average monthly rainfall, land use and flood zone generated 20.309% of the area of Semarang get into high threat grade, 43.722% and 35.969% threat was low threat. To map the threat of floods can be seen in Figure 3(a).

Having obtained the mapping of flood threat, validation was done to determine the compatibility between maps threat of flooding to the events on the field. From the results obtained validate the suitability of 52.841%. The discrepancy between the results of the flood threat maps with field validation point is caused by several factors which are already doing to address the problem of floods such as river dredging/times [1]. A further factor is the intensity of rainfall in 2015 that was lower than in previous years.

5.2. Result of Tidal Flood Map

From the data processing DEM topography, DEM soil degradation and tide data that has been dioverlay produced 12.821% of the area of Semarang indicated the threat of tidal flood high class, 3.275% moderate hazard, 2.512% low threat and 81.391% of the area of Semarang no indication hit flood threats rob. To rob flood hazard maps can be seen in Figure 3(b).

Having obtained the threat of tidal flood mapping, validation was done to determine the compatibility between the tidal flood hazard maps with the events on the field. From the results obtained validate the suitability of 85.227%. Some validation point for the tidal flood was not in accordance with the results of the tidal flood hazard map for the ground on some elevated location [8].

5.3. Result of Landslide Map

From the results of the weighting parameter overlay and topography, the average annual rainfall, land use and geology of the region produced 0.252% Semarang enter into the category of high threat, medium threat 19.014% and 80.734% low threat. Map threat of landslides can be seen in Figure 3(c)

Having obtained the mapping of the threat of landslides, validation was done to determine the suitability of landslide hazard map with the events on the field. From the results obtained validate the suitability of 86.932%. Some point the validation does not correspond to the map results threat of landslides due to rainfall intensity in 2015 is not so high [1].

5.4. Result of Drought Map

From the results of the overlay and the weighting parameter NDVI classification, the average monthly rainfall, land use and soil types produced 7.317% of the area of Semarang in the category as high threat, medium threat 43.403% and 49.280% low threat. Map of drought can be seen in Figure 3(d).

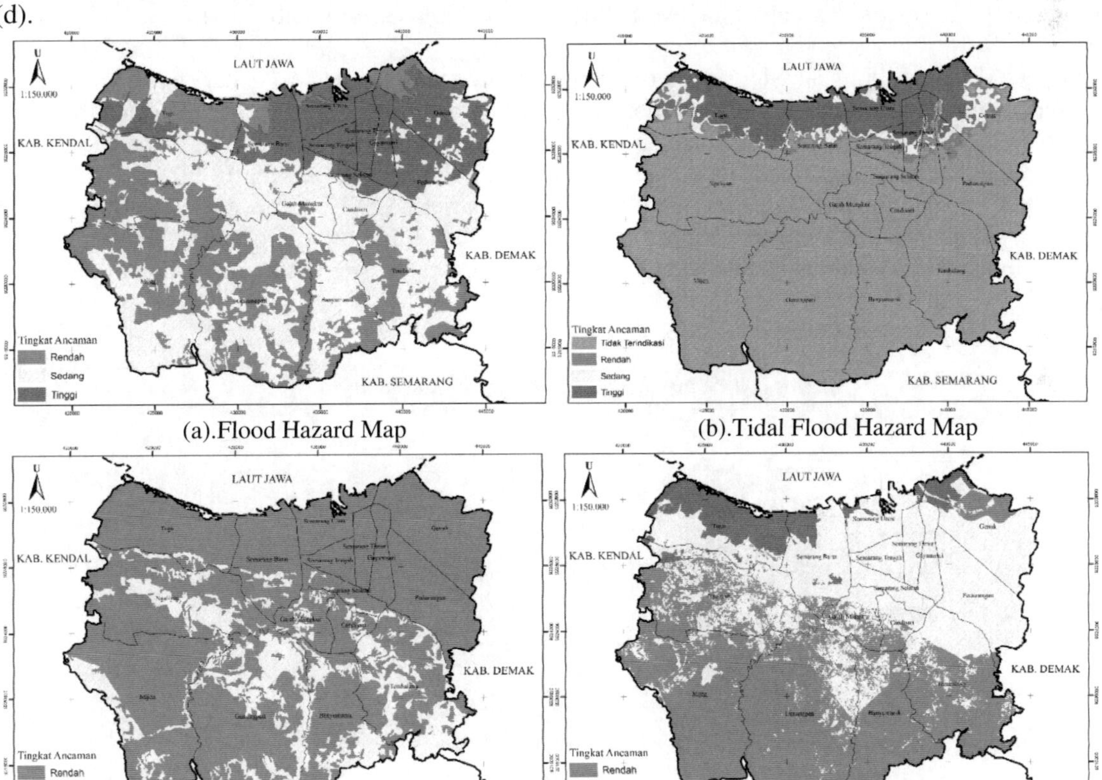

(a).Flood Hazard Map (b).Tidal Flood Hazard Map

(c). Landslide Hazard Map (d).Drought Hazard Map

Figure 3. Hazard Mapping in Semarang City

Having obtained the mapping of flood threat, validation was done to determine the compatibility between maps threat of flooding to the events on the field. From the results obtained validate the suitability of 41.143%. The discrepancy between the map drought with a point validation has been done due to the drought response, which has included the rehabilitation of irrigation networks and also the program PAMSIMAS. So some locations are usually threatened by drought, little by little has been addressed [1].

5.5. Result and Analysis of Multi Hazard Map using GIS-Fuzzy

The results of multi hazard using by PERKA BNPB No. 2 Year of 2012 obtained from areas with a low threat to the disaster area multi 18521.62 ha, accounting for 48.13% of the area of Semarang city and spread over 11 districts in the city of Semarang. 44.73% of the city or area of 17212.58 ha is a region with a medium-class hazard threats that are scattered throughout the districts in the city of Semarang. For the rest, 7.15% of the territory of the city, with an area of 2750.59 hectares is an area that has a high-class hazard threat, spread across eight districts in the city of Semarang.

Then the results of multi hazard using by fuzzy logic obtained from areas with a low-class hazard threat to the disaster area multi 24895.78 ha, accounting for 64.69% of the area of Semarang city and spread over 12 districts in the city of Semarang. 35.31% of the city or area of 13589.01 ha is a region with a medium-class hazard threats that are scattered throughout the districts in the city of Semarang. But there are no area is an area that has a high-class hazard threat in the city of Semarang.In Figure 4 we can see the comparison graph of the two methods in the multi-hazard map result.

For the distribution of disaster points overlapped with each map can be generated distribution of occurrence corresponds to table-7. From the table it can be analyzed that the class presented between the two methods has the same dominant distribution in the medium class, except at the occurrence points on the threat of tidal flood. From the spatial distribution, it can be seen in Figure 5, assuming the smallest spatial unit is sub district, then no one sub district has a hazard event of more than three types of hazards. From these assumptions, it can be concluded that the multi-hazard class generated from the Fuzzy Logic method can provide a realistic picture according to the actual conditions in the field.

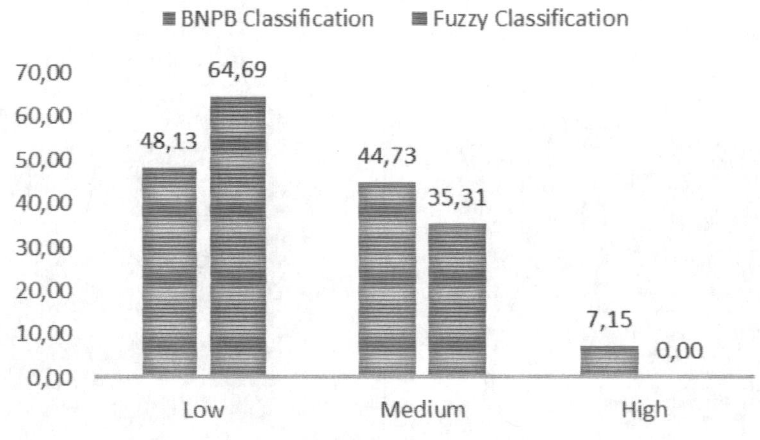

Figure 4. Comparison graph of the two methods in the multi-hazard map result

Table 7. Comparingof Distribution Disaster Point in Multi Hazard Map

Class Hazard	Multi Hazard							
	Flood		Tidal Flood		Landslide		Drought	
Low	9	10*	0	9*	10	9*	9	8*
Medium	19	21*	5	22*	18	22*	11	15*
High	3	0*	15	0*	3	0*	3	0*
Sum	31	31*	20	20*	31	31*	23	23*

* using Fuzzy Logic Method

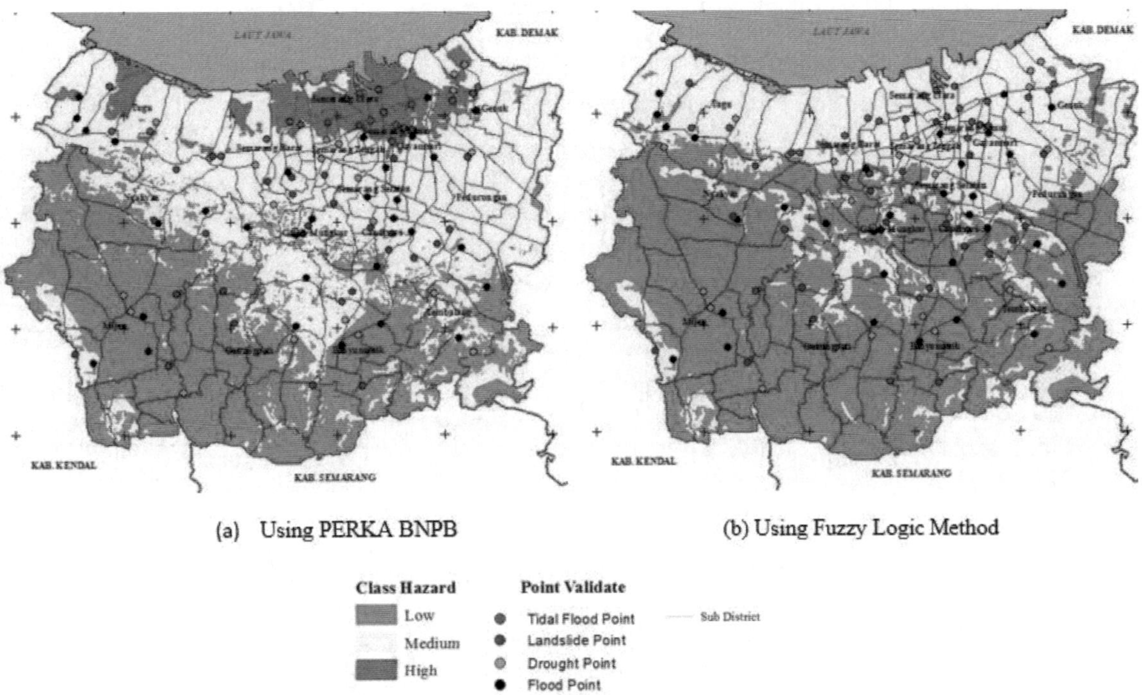

(a) Using PERKA BNPB (b) Using Fuzzy Logic Method

Figure 5. Comparison of Result of Multi Hazard Mapping in Semarang City

Furthermore, to know the type of membership with more accuracy, it can be seen in table-8. From the table it can be seen the best type of membership in determining the classification of multi-hazard map of the city of Semarang is using the Gauss membership system with RMSE of 0.404 the smallest of the other membership and VAF value of 72.909% of the largest of the other membership.

Table 8. Validation Test of Membership Function Fuzzy Logic

Test Variable	Result of Classification of Multi Hazard			
	BNPB Classification	Fuzzy Logic tipeMamdani		
		Gauss	Trapesium	Triangle
Rata-rata	1.481	1.206	1.169	1.189
Variansi	0.183	0.251	0.322	0.279
RMSE		**0.404**	0.465	0.426
VAF (%)		**72.909**	56.813	65.576
Validation Test		In	In	In

6. Conclusion

Based on these results, we can conclude that the multi hazards mapping in Semarang city was done by using GIS and Fuzzy Logic method. With the combination of both methods, it can produce multi hazards modelling in Semarang city that can be concluded in accordance with the conditions in the field. Modelling of multi hazard in Semarang city can be seen there is no significant difference from the result of multi hazard classification using PERKA BNPB and Fuzzy Logic. With the visual analysis of multi-hazard map using fuzzy logic, it can be known that the result gives enough representation of actual situation in the field and with existing disaster event that happened. Then from the multi-hazard modelling using GIS-Fuzzy can be known type of membership that has a good accuracy is the type of membership Gaus with RMSE of 0.404 the smallest of the other membership and VAF value of 72.909% of the largest of the other membership.

7. Acknowledgments

Thanks to BAPPEDA Semarang city, BMKG Semarang, and BPBD Semarang, BIG (Badan Informasi Geospasial) for supporting their data to this research. Thanks to Diponegoro University that has funded this research. And last, thanks to ESRI Indonesia for ArcGIS Desktop license for this research.

8. References

[1] Nugraha A L, Hani'ah and Pratiwi R D 2017 Assessment of multi hazards in Semarang city *AIP Conference Proceedings* vol 1857p 100006

[2] Setyono J S, Yuniartanti R K and others 2016 The Challenges of Disaster Governance in an Indonesian Multi-hazards City: A Case of Semarang, Central Java *Procedia-Social Behav. Sci.* **227** 347–353

[3] Aditya T 2010 Visualisasi Risiko Bencana di Atas Peta (Visualization of Hazard Risk in a Map), Department of Geodetic Engineering *Gadjah Mada Univ. Yogyakarta. Indones.*

[4] Karimi I and Hüllermeier E 2007 Risk assessment system of natural hazards: A new approach based on fuzzy probability *Fuzzy Sets Syst.* **158** 987–999

[5] Pokoradi L 2002 Fuzzy logic-based risk assessment *AARMS, Acad. Appl. Res. Mil. Sci.* **1** 63–73

[6] Pokorádi L 2008 Systems and Processes Modeling *Campus Kiadó, Debrecen* **242**

[7] BNPB RI 2012 *General Guidelines for Disaster Risk Assessment*

[8] Nugraha A L, Santosa P B and Aditya T 2015 Dissemination of tidal flood risk map using online map in semarang *Procedia Environ. Sci.* **23** 64–71

[9] Darmawan M and Theml S 2008 Katalog Methodologi Penyusunan Peta Geo Hazard Dengan GIS *BRR NAD-Nias. Banda Aceh*

Community-Based Disaster Management: A Lesson Learned From Community Emergency Response Management in Banyumas, Indonesia

A Y Pratama[1] **and S Sariffuddin**[1]

[1]Department of urban and regional planning, Faculty of Engineering, Diponegoro University, Indonesia

Email: sariffuddin@live.undip.ac.id

Abstract. This article aimed to review community-based disaster management in terms of its independent coordination and disaster management. Community resilience was tested during disaster emergency. While panic, the community is required to be viable and able to evacuate, manage logistic, collect data on damage and the victim, and coordinate with outsiders independently. The community in Gununglurah Village, Banyumas Regency which was hit by a landslide in 2015 provides a lesson learned about community based disaster management. This research used qualitative descriptive methodology with in-depth interview with 23 informants from the community, donor institution, village officers, and government officers. Through traditional and informal methods, the community implemented disaster management that was categorized into 3 mechanisms that were social, functional, and sequential mechanism. These mechanisms controlled different portion in which social mechanism holds the most important role in disaster management, then functional mechanism and sequential mechanism. Various community activities in the village equipped the community with organizational experience to manage logistic, human resource and other coordination. In 2007, in fact, there was vulnerability risk assessment done by the local government, which recommended efforts to be done by the community to reduce the disaster risk, yet it was not implemented. It was interesting to note that in spite of the independent disaster management there was a scientific assessment neglected. Based on this research, a new discussion on how to synchronize the endogenous knowledge with scientific modern knowledge was opened.

Keywords: CBDM, Emergency Response, Community Resilience

1. Introduction

Indigenous knowledge refers to the methods and experiences developed by a group of community that understands the local environment, which has been established from previous generations [1–3]. This knowledge consists of several different characteristics of other knowledge[4–6]. Such as Komino [7] who conducted endogenous knowledge research in Pakistan, concluded that endogenous knowledge consists of 3 categories namely; social mechanisms, functional mechanisms and sequential mechanisms. Broadly speaking, endogenous knowledge is closely related to the character of the community [8], as well as the character of the region that has different customs, which cannot be generalized [9]. This means that every disaster event must have a local investigation process before disaster management program is carried out.

Indigenous knowledge consists of the community itself, it preserves the shared cultural values established from previous generations, and becomes an embedded adaptation tool as a means of community survival [10,11]. Even before the disaster risk reduction using advanced technology-based early warning system, or using the standard operating procedures countermeasures, many local communities around the world have been equipped with readiness, on how to act and cope with natural disasters using endogenous knowledge passed down from previous generations [1].

The correlation between indigenous knowledge and natural disasters has been growing and increasingly attractive in recent years [8]. Indigenous knowledge is specific culture, and it describes the lifestyle of the society. Thus, the spread and development of indigenous knowledge becomes an interesting issue [1]. Some researches confirm that the principle of indigenous knowledge may be applied to different locations, which require conformity with local culture. Implementation is a process and requires the participation of broader stakeholders and public policy [12,13].

This research investigates the indigenous knowledge in Gununglurah Village, Cilongok District, Banyumas Regency. Indigenous knowledge is important because it contributes not only as a small influence, but has a sustainable role [1]. In completing this research, it is necessary to understand and acknowledge indigenous knowledge as one of the important sources of information and contributors in disaster risk reduction in every part of the world [14,15]. This study aims to raise awareness of indigenous knowledge as an effective method of disaster risk reduction. The results of this study can contribute as a proposal of the Resilience Village program in Gununglurah Village.

2. Research Method

The research method used in this research was qualitative research. The data collection methods in this qualitative research among others were data collection in a natural setting as the primary data sources. Data were mostly collected through in-depth interviews, observation, and documentation. Qualitative data obtained was non-numerical data such as customer satisfaction (very satisfied, satisfied, dissatisfied and so on), so mathematical operations cannot be applied. Qualitative data were data related to categorization. Its characteristics were in the form of questions or in the form of words [16].

The object of this study focused on how communities cope with landslide disaster based on knowledge and experience passed on from previous generations and categories of indigenous knowledge according to Komino [7]. The data were collected by using snowball sampling technique that was by conducting interviews to people/institutions who are competent in their fields and are considered to represent the object of the research (see Figure 1) [16]. This snowball technique is done with the key person of Village Hall of Gununglurah because he has comprehensive data for the chronology of the landslide, the stakeholders involved, and the activity of the community in coping the landslides.

In this qualitative research, the data processing was done by using data reduction, data selection process, data simplification, and rough data transformation emerging from records written on the field [16]. Data reduction was done by creating a summary, searching for themes, and coding and composing verbatim interviews (see table 1). Here are the informants and the target information that was done in the interview. In addition to the interviews, the data collection method was through observation. Observation was conducted to identify the characteristics of landslide disaster and to identify the condition after landslide disaster of Gununglurah Village. On the other hand, the secondary data collection was through the documents study and online data search.

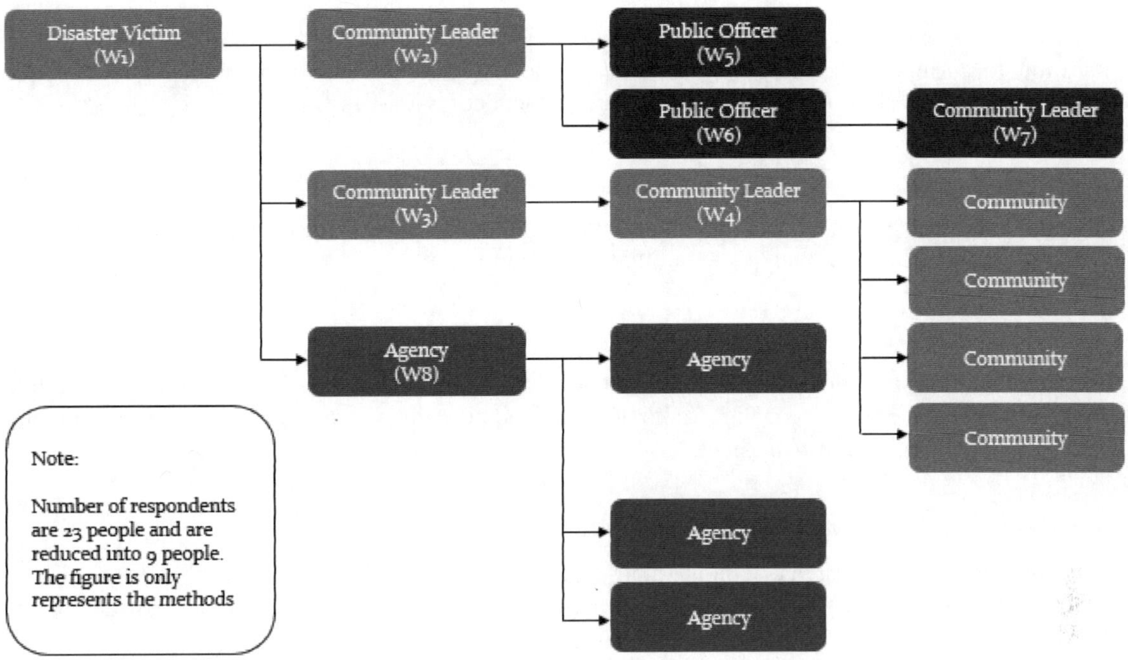

Figure 1. Snowball's Interview Flow

Table 1. Data Collection and Interview Method

No	Informant	Target Information
1.	Landslide Victims	• Characteristics of landslide disaster • Community disaster management procedures • Aid from outsiders of Gununglurah Village • Coordination between communities in coping with landslides disaster • Technical strategy of community in coping with landslide disaster • The traditional method applied by community in coping with landslide disaster
2.	Community Leader I (Small Committee)	• Disaster mitigation mechanism • Division of work of village apparatus and community leaders • Coordination in disaster mitigation
3.	Community Leader II (The Head of RT 05)	
4.	Community Leader III (The Head of RW 01)	
5.	Head of Gununglurah Village	• Work distribution of village officers and community leaders • Coordination with vertical agencies • Disaster financing management • Work distribution of volunteers outside Gununglurah Village community
6.	Village Officers of Gununglurah	
7.	Head of *Dusun* I	
8.	Head of Prevention and Preparedness BPBD Banyumas Regency	
9.	Section Head of Satpol PP Cilongok District	

3. Results and Discussion

Based on the analysis, it is obtained that many stakeholders were involved in disaster mitigation. In addition to stakeholders, there was certainly a mechanism that was done in the prevention of landslide disaster as well as indigenous knowledge of local communities.

3.1. Landslide Risk Level

Figure 2 shows the map of landslides risk zone in Gununglurah Village which was created based on land use map overlay and map of ground movement vulnerability of Gununglurah village [16].

3.1.1. The High-Risk Zones of Landslide Disaster

It was found in the settlement area because there was threat to human's life and to the productive cultivation in the settlement area. High risk zones of landslide disaster are located in the middle and southern part of Gununglurah Village.

3.1.2. The Medium Risk Zones of Landslide Disaster.

Located in residential areas withlow land vulnerability and agricultural lands such as rainfed ricefields, irrigated fields, gardens, and *Palawija* farm located in highland soil susceptibility zones. The landslide risk zone coversmost of Gununglurah Village.

3.1.3. The low risk zones of landslide disaster.

Located on agricultural, plantation and forest lands in low land vulnerability zones. Low landsliderisk zone covers a small part of the north, a small part in the middle, and a small part on the south of Gununglurah Village.

3.1.4. The Very Low Risk Zones of Landslide Disaster.

It is located in settlement areas, agricultural lands, plantations and forests in verylow land vulnerability zones. The very low landslide risk zone covers the central part of Gununglurah Village. The community felt that the landslide which has been occurring has been classified as a small landslide or has a low level of risk because it has never resulted on casualties, as did the landslide disaster that occurred in Mr. Tohari's destroyed house. Please note that the point of landslide disaster that hit Pak Tohari's house was in the medium risk zone of landslide disaster.

The role of parties involved in disaster management. The role of the parties involved in dealing with landslide disaster management activities in Gununglurah Village was very diverse. Many parties, both internal and external, were involved in the management of landslide disaster in Gununglurah Village. The role included three forms, including: (1) Roles related to disaster management, (2) Roles related to evacuation and rehabilitation of buildings, and (3) Roles related to raising public awareness. The role of landslide disaster management was in the form of data collection of victims, data collection of basic needs of the community, coordination with stakeholders related to the decision of the flow of landslide management mechanism in Gununglurah Village. The coordination was a formation of small committees. The small committee was a group of community leaders who had the initiative to manage the incoming aid and provided more assistance for the victims of landslide disaster in Gununglurah Village (see Table 2).

The role of evacuation and rehabilitation of buildings was mostly done by members of the TNI (The Indonesian Military), PMI (the Indonesian Red Cross), police officers, Pemuda Pancasila, members of community protection (linmas) and members of the PKK (the Family Welfare Movement). In the implementation of this evacuation and rehabilitation of building, they served as volunteers and always coordinated with small committees. The activities undertaken was the relocation; evacuation of the people of Gununglurah Village to a safer place (see Table 3).

The role for public awareness improvement was mostly done by BPBD of Banyumas Regency. Activities in the form of socialization and simulation were carried out regularly to ensure the public was ready for landslide disaster. In addition to BPBD of Banyumas Regency, Social Agency and Banser NU (the youth wing of Nahdlatul Ulama) also provided assistance. The efforts included socialization, spiritual refreshment and motivation of spiritual support (see Table 4).

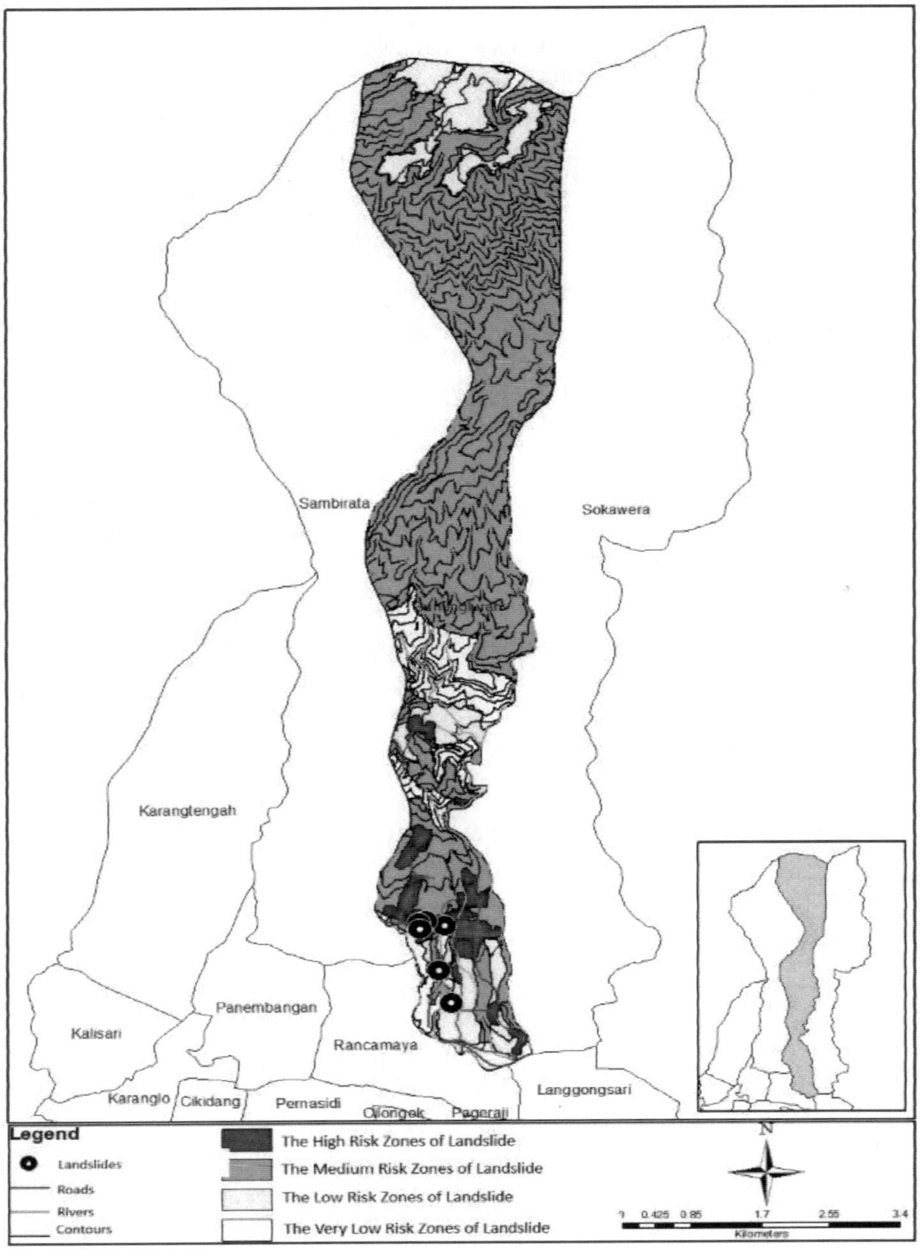

Figure 2. Map of Landslide Risk Zone

Table 2. Duties of the Small Committee

NO.	MAIN DUTY	DETAILS	PERSON IN CHARGE
1.	Data collection	• Listing initial needs • Listing the incoming aid	• Refreshment division • Treasurer
2.	Coordination	• Information landslide disaster • Volunteers • Incoming aid • Fund	Chairman
3.	Proposal submission	Submitting a fund proposal to rural level PNPM (National Program for Community Empowerment)	• Licensing Division • Secretary
4.	Rehabilitation and reconstruction	Recruitment of carpenters or builders	Security division

Table 3. Types of Aids from Various Donors to Gununglurah Village

No.	Types of Aids	Helpers	Beneficiary
1.	Health services	Indonesian Red Cross, BPJS Kesehatan (The Healthcare and Social Security Agency)	Society
2.	Staples	BPBD (Disaster Mitigation Agency) of Banyumas Regency, Indonesian Red Cross	Small committee
3.	Money	Rural level PNPM, Ministry of Religious Affairs, BadanAmil Zakat (alms agency), Banyumas District Head, Gununglurah Village Residents, Schools in Cilongok Sub-district	Small committee
4.	Livestock	Agency Amil Zakat	Society
5.	Building material	BPBD of Banyumas Regency	Small committee
6.	Shelter	Villagers of Gununglurah Village	Society

Table 4. The Role of Institutions in Increasing Public Awareness

No	Types of Aids	Helpers	Beneficiary
1	Socialization	BPBD of Banyumas Regency	Society
2	Spiritual refreshment	Banser NU	Small committee
3	Motivation and psychological support	Social Agency	Society

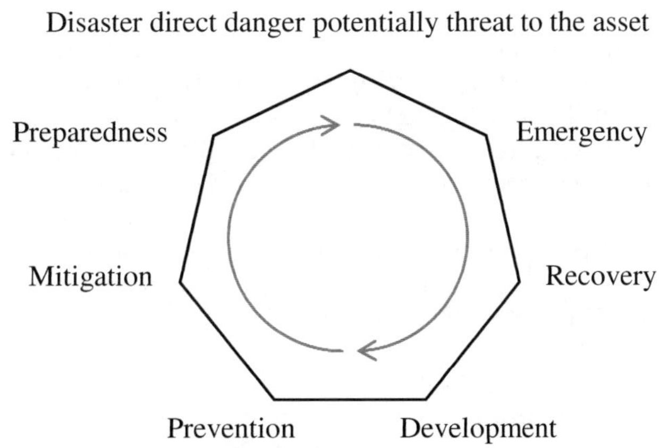

Figure 3. Disaster Management Cycle

Figure 3 shows the flow mechanism of landslide response. The mechanism of landslide disaster management done by people of Gununglurah Village is based on Disaster Management Cycle according to Law Number 24 Year 2007 which is included in the emergency cycle. There is no formal disaster management mechanism by the community, but when there is an emergency status, they use a mechanism of mitigation which is called a logistic mechanism [17]. The logistic mechanism undertaken by the community is categorized into 3 phases, including phase of readiness. It is community readiness to assess rapidly and accurately in the determination of disaster emergency status which consists of several aspects such as location, damage, loss, resources, logistics, and donors. In addition, there is a phase of disaster response that is a community activity related to the rescue and evacuation of disaster victims. The last phase is the phase of reconstruction which is a community activity related to rebuilding or recovering post-disaster victims.

The logistic mechanism carried out by the community in coping with landslide disaster in Gununglurah Village in the first phase needs to be improved. Community readiness only fast assessment of a disaster emergency status in terms of location [17]. There is no assessment of the performance of suppliers, logisticians, and other donors to make the emergency response phase effective. It is necessary to be careful in reviewing the market price during a disaster event in the previous years to prepare the relief supplies as the price of goods changed dynamically.

Second, the phase of landslide disaster management by the community in Gununglurah Village during the phase of emergency response is good. The community has made good efforts during the emergency response and there is a need for accuracy in estimating the inventory during the process of distribution which is stored in the warehouse for the adjustment of needs. Third, the logistic mechanism carried out by the community in coping with landslide disaster in Gununglurah Village in the phase of reconstruction needs to be improved. Aid accountability is needed in order to establish good relationships with donors because the villagers of Gununglurah Village also need assistance from outsiders. The evaluation of emergency response performance is also needed so that the public obtains a lesson from the shortcomings and mistakes that will be useful in the future.

3.2. The Analysis of Community's Indigenous Knowledge

According to Komino [7], indigenous knowledge that applies in Gununglurah Village community is divided into three categories, namely social, functional, and sequential mechanism. Social mechanism can be categorized into two forms, including internal communication and coordination.

IOP Conf. Series: Earth and Environmental Science **123** (2018) 012003 doi:10.1088/1755-1315/123/1/012003

The form of internal communication and coordination is the initiative of the community itself to hold meetings in anticipating disasters and take action when a disaster occurs. There is no special division of roles and it is done internally by Gununglurah Village community aims to reduce the risk of landslide disaster. Next is external communication and coordination. The form of communication and external coordination is the form of communications conducted by villagers of Gununglurah village in whom the village officers report when they really need assistance of the Regional Disaster Management Agency of Banyumas Regency (BPBD of Banyumas Regency) and other formal institutions. In addition, there is coordination of the Small Committee with the National Program for Community Empowerment in Rural Area (Rural PNPM) related to additional funding aids.

The second indigenous knowledge is functional mechanism. Based on the indigenous knowledge, the functional mechanism category that has been implemented by the people of Gununglurah Village consists of three components. Among others is the building structures adjustment. The form of building structure adjustment done by some people is by strengthening the walls to make the house stronger to reducing the risk of damage to the building. The other is repairing the roof of the house so that the water can flow directly to the drainage. In addition, some people no longer make buildings or houses on steep hills. Furthermore, the other form is environmental engineering. Environmental engineering undertaken, among others, are tidying up trees and sloping steep cliffs to reduce the risk of landslide. Besides, the community no longer makes the pond on the hill and also dismantles the pond that has been built. The third component is local technology. Local technology means that the community is doing some activities such as monitoring during rain for its duration and intensity. The other is the construction of permanent drainage channel so that the water does not dribble into the soil that causes the soil to erode. The next is closing the cracks above the mountain cliffs in case of heavy rain for more than two hours and sending a report if there is a crack.

The third indigenous knowledge is sequential mechanism. Indigenous knowledge of sequential mechanism category is indigenous knowledge related to the traditional way of community in disaster management. It includes not to affecting the livelihoods or changing livelihoods although landslide disaster in Gununglurah village is common thing to happen. There is also no migration. There is only change in the consumption and food source after landslide disaster in Gununglurah Village. The type of food given is fast food.

Findings. Indigenous knowledge related to community technical strategies in independent disaster management has some misperceptions such as bamboo tree cutting activities as one of the technical strategies done by the community. It is because of the lack of public knowledge about the environment. In fact, bamboo trees can serve as one of the early warning systems to landslides [18].

The bamboo cutting trees is also one of the findings that should be utilized by the people of Gununglurah Village. Please note that bamboo trees functions to maintain ground water, and its roots are able to withstand landslides, and its leaves are capable of splitting the wind or dampening noise and dust pollution [18]. Indigenous knowledge related to the traditional way of community in disaster management is not always effective as a disaster management strategy and is always subject to change depending on environmental conditions.

4. Conclusion

In accordance to the goals and objectives to be achieved from various analyzes related to the category of community's indigenous knowledge in the landslide disaster management, it can be concluded: Gununglurah Village has 4 levels of landslide risk zone; (1) high risk landslide zone, (2) medium risk landslide zone, (3) low risk landslide zone, and (4) very low risk landslide zone. Most of Gununglurah Village is included in the medium risk landslide zone. The point of landslide that hit Pak Tohari's house was in the zone of medium risk of ground movement and was included in the medium risk landslide zone causing the house to be destroyed. The social mechanism of the community in coping with landslide disaster in Gununglurah Village is divided into 2 forms, namely internal communication and coordination, and external communication and coordination. Social mechanism has the most dominant role compared to other indigenous knowledge mechanisms. This is due to the character of the rural community groups which have a high social spirit. The functional mechanism of the community in coping with landslide disaster in Gununglurah Village is divided into 3 aspects, namely building structure adjustment, environmental engineering, and local technology. There is a need for the community to have a deeper knowledge of the environment and structural mechanisms [19]. There are no prominent community's sequential mechanisms in coping with landslide disaster in Gununglurah Village because the community thinks that the incoming aid is adequate and they feel that the type of landslide that occurred is not risky enough.

Social mechanisms, functional mechanisms, and sequential mechanisms are interconnected and not stand by themselves. In every activity carried out in disaster management area, the community always coordinates, communicates, and works together. There is an informal mechanism that applies in disaster management by communities in Gununglurah Village. The mechanism was initiated by the community by forming a small committee consisting of the head of RW, head of RT, and some ad hoc community leaders. The small committee has an important and dominant role in every disaster management activity in Gununglurah Village and always coordinates with relevant stakeholders.

It is recommended to improve the community's indigenous knowledge and gain social learning [16] in landslide disaster management in Gununglurah village. Then some recommendations are formulated for the public and stakeholders. The recommendations include: (1) Affirmation of strategic policies related to natural disasters in Gununglurah village, for example, by mapping any vulnerable areas and the prohibition of settlement construction in the area [20]. These can minimize the risk of landslide disaster; (2) Preparation of disaster evacuation route in every region which also needs to be done so that it can minimize disaster victim; (3) Enforcement of stricter laws and programs that promote public participation and awareness of the risks of landslides [19]; (4) Improvement of school facilities and infrastructure and socialization of the importance of education to the people of Gununglurah Village to improve the education level; (5) The development of local economy without being dependent on natural resources also needs to be done so that the economic capacity of the region is increasing. Economic capacity affects the ability to adapt because high economic capabilities can support the functional mechanism required in the technical strategy of disaster management; (6) Routine socialization and simulation for the community related to natural landslide disaster in Gununglurah Village, so that people's understanding and awareness of natural disaster is also improved. With the large population of children, the elderly and the low-educated population and the high poverty rate, it needs more understanding towards the community about the prevention capacity so as to improve its adaptability; (7) The community is expected to improve the awareness in maintaining the environment, so as to minimize the occurrence of disasters, especially those caused by climate change.

To examine community activities in landslide disaster management in Gununglurah Village, then some recommendations for researchers are formulated for further research. The recommendations include: (1) Assessing the financial mechanisms in the prevention of landslide disaster in Gununglurah Village, (2) Reviewing the community's indigenous knowledge during pre-landslide disaster in Gununglurah Village, (3) Reviewing the community's indigenous knowledge after the landslide in Gununglurah Village, (4) Assessing gender roles in the management of landslide disaster in Gununglurah Village, and (5) Assessing the level of vulnerability of Gununglurah Village to landslide disaster.

5. References

[1] Shaw R, Uy N and Baumwoll J 2008 Indigenous knowledge for disaster risk reduction: Good practices and lessons learned from experiences in the Asia-Pacific Region *United Nations Int. Strateg. Disaster Reduction, Bangkok*

[2] Mavhura E, Manyena S B, Collins A E and Manatsa D 2013 Indigenous knowledge, coping strategies and resilience to floods in Muzarabani, Zimbabwe *Int. J. Disaster Risk Reduct.* **5** 38–48

[3] Sariffuddin S and Wijaya A P 2014 Pola Adaptasi Masyarakat Pesisir Genuk Kota Semarang (Patterns of Community Adaptation to Environmental Degradation in Genuk Coastal Area, Semarang) *Tataloka* **16** 245–253

[4] Pyles L 2007 Community organizing for post-disaster social development: Locating social work *Int. Soc. Work* **50** 321–333

[5] Joranson K 2008 Indigenous knowledge and the knowledge commons *Int. Inf. Libr. Rev.* **40** 64–72

[6] Smith B M, Basu P C, Chatterjee A, Chatterjee S, Dey U K, Dicks L V, Giri B, Laha S, Majhi R K and Basu P 2017 Collating and validating indigenous and local knowledge to apply multiple knowledge systems to an environmental challenge: A case-study of pollinators in India *Biol. Conserv.* **211** 20–28

[7] Komino T 2008 Indigenous coping mechanisms for disaster management in Mansehra and Battagram districts, North West Frontier Province (NWFP), Pakistan *Indig. Knowl. Disaster Risk Reduct.* 41

[8] Fauzie W Z and Sariffuddin S 2017 The role local initiatives in community based disaster risk management in Kemijen, Semarang City *IOP Conference Series: Earth and Environmental Science* vol 70p 12047

[9] Wright J D 2015 International encyclopedia of the social and behavioral sciences

[10] Mathbor G M 2007 Enhancement of community preparedness for natural disasters: The role of social work in building social capital for sustainable disaster relief and management *Int. Soc. Work* **50** 357–369

[11] Kamarulzaman N H, Vaiappuri S K N, Ismail N A and Mydin M A O 2016 Local Knowledge of Flood Preparedness: Current Phenomena To Future Action *J. Teknol.* **78** 85–89

[12] El-Gack N 2012 Development and indigenous systems: Lessons from north Kordofan, Sudan *Altern. An Int. J. Indig. Peoples* **8** 99–111

[13] Ramnath A 2014 "Indigenous Knowledge"and "Science"in the Age of Globalization *IIM Kozhikode Soc. Manag. Rev.* **3** 101–107

[14] Hutchings S 2014 Introduction: Indigenous knowledges impacting the environment *Altern. an Int. J. Indig. peoples* **10** 445–449

[15] Astuti K D, Farhaeni G, Wahdah L and others 2017 Vulnerability Assessment: The Role of Coastal Informal Settlement Growth to Social Vulnerability in Genuk Sub-District, Semarang City *IOP Conference Series: Earth and Environmental Science* vol 55p 12047

[16] Yunus H S 2010 Metodologi Penelitian Wilayah Kontemporer

[17] Kovács G and Spens K M 2007 Humanitarian logistics in disaster relief operations *Int. J. Phys. Distrib. Logist. Manag.* **37** 99–114

[18] Taufik Wijaya 2016 Jika Menanam Bambu, Mungkin Banjir, Longsor dan Kekeringan Dapat Dicegah *Mongabay*

[19] Castro C P, Ibarra I, Lukas M, Ortiz J and Sarmiento J P 2015 Disaster risk construction in the progressive consolidation of informal settlements: Iquique and Puerto Montt (Chile) case studies *Int. J. Disaster Risk Reduct.* **13** 109–127

[20] Asmat A and Zamzami S Z 2012 Automated house detection and delineation using optical remote sensing technology for informal human settlement *Procedia-Social Behav. Sci.* **36** 650–658

IOP Conf. Series: Earth and Environmental Science **123** (2018) 012004 doi:10.1088/1755-1315/123/1/012004

Local knowledge: Empirical Fact to Develop Community Based Disaster Risk Management Concept for Community Resilience at Mangkang Kulon Village, Semarang City

A B Kapiarsa[1] and S Sariffuddin[1]

[1]Department of urban and regional planning, Faculty of Engineering, Diponegoro University, Indonesia

Email: sariffuddin@live.undip.ac.id

Abstract. Local knowledge in disaster management should not be neglected in developing community resilience. The circular relation between humans and their living habitat and community social relation have developed the local knowledge namely specialized knowledge, shared knowledge, and common knowledge. Its correlation with community-based disaster management has become an important discussion specially to answer can local knowledge underlie community-based disaster risk reduction concept development? To answer this question, this research used mix-method. Interview and crosstab method for 73 respondents with 90% trust rate were used to determine the correlation between local knowledge and community characteristics. This research found out that shared knowledge dominated community local knowledge (77%). While common knowledge and specialized knowledge were sequentially 8% and 15%. The high score of shared value (77%) indicated that local knowledge was occurred in household level and not yet indicated in community level. Shared knowledge was found in 3 phases of the resilient community in dealing with disaster, namely mitigation, emergency response, and recovery phase. This research, therefore, has opened a new scientific discussion on the self-help concept in community-help concept in CBDRM concept development in Indonesia.

Keywords: Local Knowledge, CBDRM, Flood, Community Resilience

1. Introduction

The international attention on local knowledge in overcome decreasing disaster risk still increasing [1]. The first principle that have be noticed are community local knowledge always evolving and often known by local ecological knowledge where that can be used to describing relation between knowledge with environment surround it [2]. Then Colfer and Colchester [3] explaining that local knowledge divided into two type which is scientific knowledge and unscientific knowledge. The scientific knowledge is based on everything thing that can be proved by scientific, while the unscientific knowledge based on everything that cannot be proved by scientific. Unscientific knowledge often considered not to important enough for some people, but sometimes unscientific knowledge have a strong power to be hold and to be applicate by community. The community itself developing knowledge that will build guidelines based on past experiences [4]. That been done by earlier generation through “try and error”. This is happened by none of modern technology at that time. Every guideline and that knowledge are given by generation through generation [5] and sometimes some knowledge had a religion influence to protect sustainability of knowledge itself.

Therefore, the community are strongly implemented unscientific local knowledge. The knowledge is not necessarily accurate on science but can be created as input for guideline.

The unscientific local knowledge containing a wide number of live guidelines, manners etc [6]. Some of guideline containing disaster management that often happen in the local settlement. Natural disaster that happen in each settlement have a different kind and intensity. The knowledge about disaster management has been considered very important to understand by local community and valued so priceless to share among other community [7,8]. The lesson through experience in overcome disaster can be made as references to other community that facing the same condition. Therefore, they can build a resilience community in encounter local disaster. Local knowledge in general already discussed in the community based disaster risk management guideline since 1996 until 1998 [9]. The term of local knowledge and community based disaster risk management has been developed by international activist. If in Indonesia the terms are known since one of figure in Komunitas Pecinta Alam dan Pemerhati Lingkungan (Kappala) Indonesia which is Dr. Eko Teguh Paripurno also researcher from UPN Veteran Yogyakarta publishing the Participatory Rural Appraisal (PRA) book. Since then arise much of movement that actively have a role in overcome disaster, for example when Nusa Tenggara Timur province in 1998 hit by drought when El-Nino comes. In the beginning, this activity led by Pusat Informasi Rawan Pangan (PIRP) until in the 1999 changing name into Forum Kesiapan dan Penanggulangan Bencana (FKPB) which can do discourse about community based disaster risk management. Since 1998 until 2000 FKPB followed training about community based disaster risk management in Bangkok, Thailand to infiltrate any guide and new knowledge that can be applied in Indonesia. Until 2000 the training is held in Nusa Tenggara Timur by Oxfam GB, since then community based disaster risk management in all around Indonesia are really developed. Then appear regulation about disaster management enforcement within 2007 Indonesian law.

Every disaster management practice based on local knowledge keep on developing and becoming input to construct resilience community [10] in all around Indonesia. One of community location that shown uniqueness in facing flood is Mangkang Kulon Village. Based on previous survey has been found nets that tied on bamboo piece and installed around embankment. This is used for avoiding fishes and shrimps go out from embankment when flood came. Beside every villager also built fence wall and embankment (short stairs from cement) in their yard for preventing floods overflow crashing their house directly that came from river in front their settlement. For overcome disaster villager built barrier (wall in the riverbank) by their own also pile up sand sac for unprotected riverbank. These disaster management practices are a real applied community local knowledge that in enhanced by technology to adapt from flood [11].

The Mangkang Kulon Village is in coastal area of Semarang City and be passed by Plumbon river. The Plumbon river sometimes unable to accommodate overflow from upstream that can cause flood in Mangkang Kulon Village. Highest intensity of rain, wrong river management, also behavior of throw garbage into river became one of reason why flood occur in that area. Mangkang Kulon Village is one of region that have been hit by very bad flood in the Semarang city history [12]. The big flood happened in 2009. After that flood happened again in 2010 and make the wall of the bank in that area broken down. Flood happened again in 2014 and resulting one casualty. Then flood came also in 2015 and drowning the rice field villager about 80 cm high. In the next year in June 2016 flood happened really bad because of the big force that can break down the embankment about 15 meters in length, also repeal the plants along the river. With these events and a lot of loss that have been caused by flood, we need to research the knowledge of Mangkang Kulon villager in facing disaster also disaster management practices starting with mitigation, preparedness, response, until recovering. From every knowledge and practice, we need to know how wide it was being spread in this community for identifying the type of local knowledge. Identifying the type of local knowledge is very important role to know how resilience the community in dealing with flood [4] and how to reduce social vulnerability [13]. If the knowledge and practice in 4 phase of resilience community just been known by small of people in the community means that those people in the community still far from resilience. While the knowledge or practice in 4 phase of resilience community known by almost event all the people in community then that is mean they can be resilience community. That is explanation that the conclusion of this research is important as input for community or stakeholder for more strongly managing the flood in the future.

2. Research Method

The approach that has been used in this research is mix method because using two analysis which are qualitative and quantitative. Mix method in this research include into sequential mix method with using qualitative interview and then followed with quantitative to make a result [14]. The qualitative method in this research include as narrative qualitative research, which is research strategy that investigating life each respondent and asking someone or group of people to tell their story live, then the researcher can be retelling information in narration form. While quantitative method is included as survey to resuming opinion, manner, or trend quantitatively some population using questionnaire [15].

3. Result and Discussion

3.1. The Relation Between 4 Phase of Resilience Community and Respondent Background

Based on accumulation of answers that has been given by respondents makes researcher want to know relation between respondent answer (25 resilience community variables) and background each respondent (respondent background variables). The respondent background is determined by 6 variables, which are gender, role in community, age, location of stay, length of stay, and last education. To looking for relation between two variable we are using IBM SPSS Statistics application through Crosstabs analysis. In this analysis, we need hypothesis and decision-making base for finding 4 phase of resilience community variable that have relation with respondent background variable. The hypothesis in this research like down below.

H_o : If there is no relation between 4 phase of resilience community variables with 6 respondent background variables
H_i : If there is relation between 4 phase of resilience community variables with 6 respondent background variables

While for decision making we are using probability where can be seen in third output table (Chi-Square test) and Asymp. Sig column which is mean:

If probability > 0,05, then H_o will accepted
If probability < 0,05, then H_i will unaccepted

The score 0,05 as base of decision making in probability obtained from significance level (α) 5% with degree of freedom (Df) 2 and there is Asymp.Sig table that containing result of crosstabs analysis between two variables in down below.

Based on table 1, we can know that yellow column has Asymp. Sig value in the below of lambda value (α) about 0,05. Which H_o is unaccepted that have mean there is a relation between two variables. From the result of analysis there is found 16 variables from 25 overall resilience community variable that have relation with 6 respondent background variables. From 16 variables, there are 11 variable that have relation with role in community variable, which the most dominant relation between these two groups variables and giving mean that role in the community really determine people acknowledge about resilience community to overcome the flood. Followed by 8 resilience community variable which have a relation with length of stay variable, then 5 resilience community variable which have a relation with age variable, then 4 resilience community that have a relation with gender variable, also 2 different resilience community variable that have a relation with location of stay and last education variable in each.

Table 1. Relation of 4 phase of resilience community variables with 6 respondent background variables

Resilience Community Variable	Asymp. Sig Value / Respondent Background Variables					
	Gender	Role in Community	Age	Location of stay	Lenght of stay	Last education
Mitigation Phase						
Flood definition by community	0.022	0.515	0.083	0.249	0.419	0.004
Flood season	0.414	0.204	0.245	0.627	0.022	0.747
Save area	0.026	0.863	0	0.697	0.046	0.648
Flood cause	0.148	0	0.524	0.765	0.315	0.771
Effect of flood	0.63	0.107	0.006	0.271	0.261	0.878
Flood history	0.708	0.001	0.001	0.564	0	0.552
Natural sign before flood came	0.875	0.014	0	0.387	0.752	0.65
Transmission of flood information system	0.006	0	0.217	0	0.74	0.851
Act after knowing flood came	0.703	0	0.511	0.835	0.794	0.403
Element who concern in avoid the flood	0.243	0.007	0.961	0.594	0.714	0.489
Element who responsible in managing river environment	0.606	0	0.235	0.602	0.148	0.69
Element who concern in overcome the flood	0.601	0	0.317	0.336	0.281	0.119
Preparedness Phase						
Preparedness action in facing the flood	0.193	0	0.406	0.507	0.227	0.557
Respond Phase						
Human requirement while flood hit	0.702	0.003	0.084	0.72	0.03	0.92
Recovery Phase						
Recovery	0.024	0	0.462	0.57	0.049	0.707
Rebuild	0.41	0.198	0.021	0.491	0.002	0.27

3.2. The 3 Type of Community Local Knowledge

Based on accumulation answers from respondents from four community association, we found 3 kind of local knowledge that evolve surround community which are common knowledge, shared knowledge and specialized knowledge (see figure 1). The common knowledge collected from 139 respondent answer which have percentage about 8% from all answer. Then shared knowledge built by 1.381 respondent answer and have percentage about 77%. The third is specialized knowledge collected by 264 respondent answer and have percentage about 15%. So shared knowledge become a dominant knowledge that have been spread around Mangkang Kulon community.

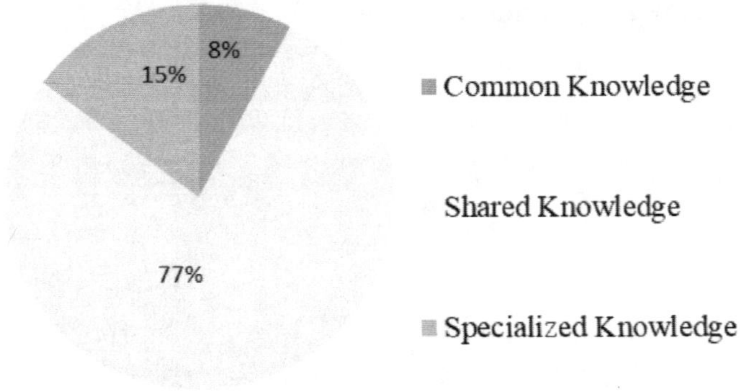

Figure 1. Diagram of Percentage 3 types of community local knowledge

3.3. Common Knowledge

The common knowledge is held by most people in the community [16]. Based from the definition researcher itself divided group of those 3 types of knowledge from accumulation respondent and which indicator that had 66 until 73 respondent answer (91%-100%) makes that variable where the indicator belongs include as common knowledge. Therefore, we found 2 variable that include in the common knowledge which are process when flood comes and element who responsible in monitoring flood when it is comes. Those two variables are included in mitigation phase.

3.4. Shared Knowledge

The shared knowledge is held by many, but not all, community members [16]. Therefore, researcher decided the accumulation of respondents answer in one indicator that have length about 19 until 65 respondents answer (26%-90%) with no other indicator has more than 65 respondents answer in the same variable, makes the variable include as shared knowledge. Then we found 17 variable that include as shared knowledge. Those 17 variables in the 3 from 4 phase of resilience community which is mitigation, respond and recovery. The Shared knowledge become the most dominant and make statement that Mangkang Kulon villager have a middle stage in knowing the disaster management, also make a good step to build community that can be resilience in deal with disaster especially flood.

3.5. Specialized Knowledge

The Specialized knowledge is held by a few people in the community [16]. So, the researcher divided the accumulation of respondents answer in one indicator that have length about 1 until 18 respondents answer (1%-25%) with no other indicator has more than 18 respondents answer in the same variable, makes the variable include as shared knowledge. Therefore, we found 2 variable that include as specialized knowledge which are natural sign before flood came and preparedness in facing the flood. Those variables are included in 2 phase of resilience community that which is mitigation and preparedness phase. After knowing those result data, we made following a vertical bar diagram from the least answer until the lots answer each variable in down below:

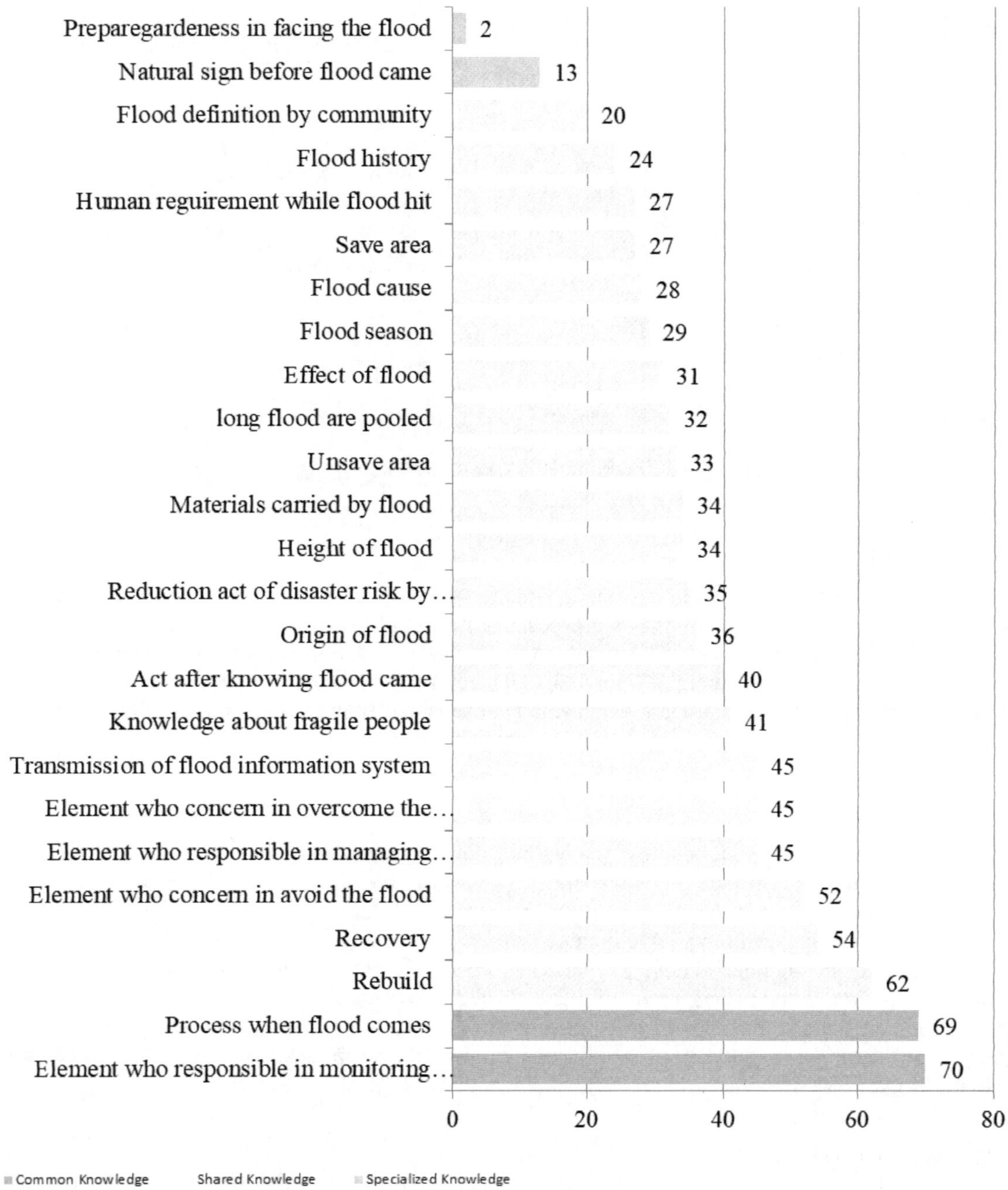

Figure 2. Variables Stage of 4 Resilience Community Phase In 3 Local Knowledge Type of Mangkang Kulon Community Diagram

In the figure 1, the finding was distinguishing the variables into 3 based on 3 types of local knowledge, which the green is specialized knowledge, the yellow is shared knowledge and the orange is common knowledge. From that diagram, we knew that the most dominant local knowledge that had by Mangkang Kulon community is shared knowledge. Followed by common knowledge and specialized knowledge.

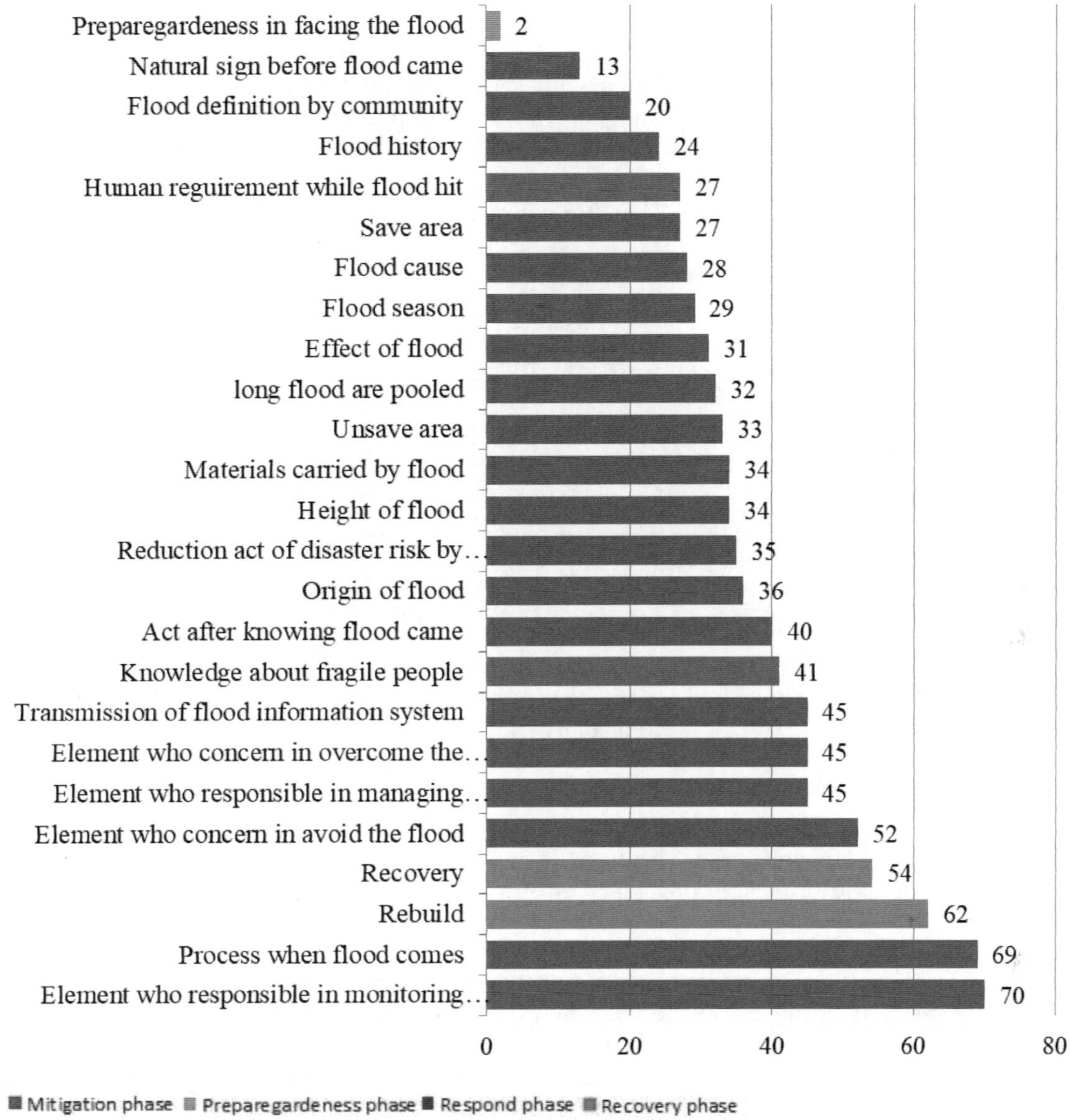

Figure 3. Variables Stage of 4 Resilience Community Phase of Mangkang Kulon Community Diagram

In figure 2, we are distinguishing the variables into 4 phase of resilience community, which the dark blue is mitigation phase, the light blue is preparedness phase, the dark purple is respond phase and the light purple is recovery phase. If we correlate the second diagram with first diagram we know that all of variable that in respond and recovery phase belongs in shared knowledge, also with almost all of mitigation phase variables (see Figure 4). While one and only preparedness phase variable with one mitigation phase variable include as specialized knowledge. Then the last knowledge is common knowledge filled with two mitigation phase variables.

Those diagrams making a half pyramid form if we see them from big picture and that is giving an idea to researcher to make another diagram as conclusion. That diagram having a pyramid form which specialized knowledge in the top of pyramid and giving mean that just a small number of people of community who know certain knowledge. Therefore, specialized knowledge belongs in individual stage based on 7 stage of resilience community by IFRC [17]. Then followed by shared knowledge in the middle of pyramid which giving meaning that a lot of people in the community knowing certain knowledge, therefore this type of knowledge spread among household and community and clear that this knowledge belongs to household and community stage based on 7 stage of resilience community. The last is common knowledge in the bottom of pyramid which giving meaning that knowledge known by all people in the community. Therefore, this knowledge belongs in the community and local association stage based on 7 stage of resilience community. Then this is the pyramid diagram that shown 3 types of local knowledge and it relation with 7 stage of resilience community in down below.

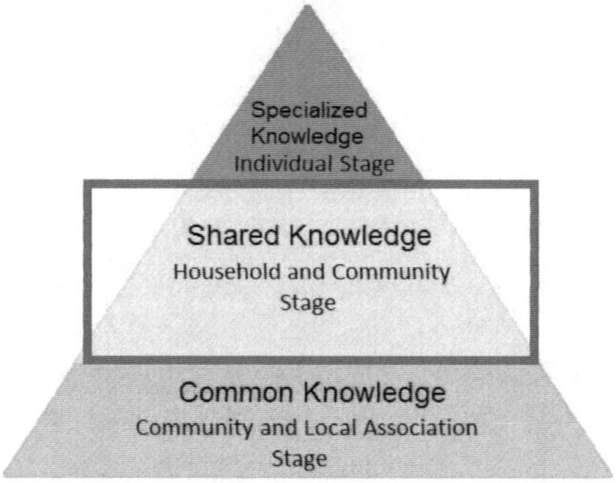

Figure 4. Type of Local Knowledge and the stage of Resilience Community of Mangkang Kulon Community Diagram

4. Conclusion

The dominant local knowledge that had by Mangkang Kulon community belongs to Shared knowledge which in the middle stage. Therefore, Mangkang Kulon community have a good asset to make a forward in build resilience community in overcome disaster, especially in flood. Although Mangkang Kulon community still have a weakness in preparedness phase. The community already have an awareness in overcome the flood, it is shown in the way they try mitigating and responding flood by themselves, although they had limited knowledge. This is one of sign that villager want to resilience but not have a full capability yet. The role in community have a significant power to determine that person is having a high knowledge or not about flood and how to overcome it, especially everything that needs to build a resilience community, beside length of stay and age every person in Mangkang Kulon community. While the least power is gender, location of stay and last education that have been taken by every person, but still have a relation in between.

5. References

[1] Steele M Z and Shackleton C M 2010 Using local experts as benchmarks for household local ecological knowledge: Scoring in South African savannas *J. Environ. Manage.* **91** 1641–6
[2] Cumming G S 2011 A Theoretical Framework for the Analysis of Spatial Resilience *Spatial Resilience in Social-Ecological Systems* (Springer) pp 35–66
[3] Colfer C J P, Colchester M, Joshi L, Puri R K, Nygren A and Lopez C 2005 *Traditional Knowledge and Human Well-Being in the 21st Century.* vol 17(IUFRO-World Forests Society and Environment Project)
[4] *Fauzie* W Z and Sariffuddin S 2017 The role local initiatives in community based disaster risk management in Kemijen, Semarang City *IOP Conference Series: Earth and Environmental Science* vol 70p 12047
[5] Nakmofa Y and Lassa J 2009 Community based approach to disaster risk reduction and climate change adaptation towards sustainable livelihood: ten years experiences from PMPB Kupang

J. Ntt Stud. **1** 136–45
[6] Ikeda S 2004 Urban interventionism and local knowledge *Rev. Austrian Econ.* **17** 247–64
[7] Wright J D 2015 International encyclopedia of the social and behavioral sciences
[8] Sariffuddin S and Wijaya A P 2014 Pola Adaptasi Masyarakat Pesisir Genuk Kota Semarang (Patterns of Community Adaptation to Environmental Degradation in Genuk Coastal Area, Semarang) *Tataloka* **16** 245–53
[9] UNESCAP 2008 *Builidng Community Resilience to Natural Disasters Through Partnership: Shaing Experience and Expertise in the Region* (New York)
[10] Cutter S L, Barnes L, Berry M, Burton C, Evans E, Tate E and Webb J 2008 A place-based model for understanding community resilience to natural disasters *Glob. Environ. Chang.* **18** 598–606
[11] Marfai M A, King L, Sartohadi J, Sudrajat S, Budiani S R and Yulianto F 2008 The impact of tidal flooding on a coastal community in Semarang, Indonesia *Environmentalist* **28** 237–48
[12] Marfai M A and King L 2008 Coastal flood management in Semarang, Indonesia *Environ. Geol.* **55** 1507–18
[13] Astuti K D, Farhaeni G, Wahdah L and others 2017 Vulnerability Assessment: The Role of Coastal Informal Settlement Growth to Social Vulnerability in Genuk Sub-District, Semarang City *IOP Conference Series: Earth and Environmental Science* vol 55p 12047
[14] Creswell J W and Poth C N 2017 *Qualitative inquiry and research design: Choosing among five approaches* (Sage publications)
[15] Babbie E R 2015 The practice of social research: Nelson Education
[16] FAO 2005 Building on Gender, Agrobiodiversity and Local Knowledge
[17] IFRC 2014 Framework *for community resilience Geneva: The International Federation of Red Cross and Red Crescent Societies*

Investigation of Potential Landsubsidence using GNSS CORS UDIP and DinSAR, Sayung, Demak, Indonesia

B D Yuwono[1], Y Prasetyo[1], and L J F Islama[1]

[1]Diponegoro University, Indonesia

Email: bdyuwono92@gmail.com

Abstract: The coastal flooding induced by land subsidence is one of major social problems in the coastal area of Central Java, especially North Demak. Recent advance technology Global Navigation Satellite System Continuously Operating System (GNSS) and Differential Synthetic Aperture Radar Interferometry (DInSAR) is already increased our capability to identify of land subsidence processes. DInSAR required not only availability of good quality input data but also rigorous approaches. In this research we used DInSAR analysis with focusing on landsubsidence phenomena. Tests were done with geodetic GPS survey with GNSS CORS UDIP as base station. Performance assessment of development method was conducted on study area affected by land subsidence. The results of this study indicate land subsidence spreads in study area with varying degrees of subsidence.

Keywords: GNSS CORS, DInSAR, Land Subsidence

1. Introduction

Land subsidence is definite as downward displacement of the land surface relative to certain reference surfacel, such as mean sea level (MSL) or reference ellipsoid. Land Subsidence mainly caused by the combination of natural consolidation of alluvium soil, excessive groundwater extraction and loading constructions [1]. Natural-anthropogenic hazard occurs quite many in large urban and the coastal area including coastal Semarang Demak [2].

There are several methods for identifying subsidence e.g. GPS [1], Gravity survey [3], Levelling and InSAR [4]. This research mainly discuss about identify subsided area in Sayung using DInSAR and GPS method. The use of SAR interferometry technique can explore and map deformation area especially land subsidence with good precision. SAR techniques have been tools successfully applied for the study deformation as for example in Jakarta [5], Semarang [6].

2. DInSAR Monitoring of Height Change

2.1. Studi Area

Demak is one of the regencies located in central java province. Demak Regency has an area of 897.43 km². Geographically Demak Regency is located at 6º43'26 "- 7º09'43" LS and 110º48'47 "BT (Wikipedia, 2015) Demak Regency is bordered by Java Sea in the west, Regency of Jepara in the north, Regency of Kudus in the east, and Regency Grobogan in the southeast The coastal area of Sayung District Demak is often affected by tidal floods or better known as rob floods. The rob flood is an ever-present phenomenon in the northern Demak district since 1997. The flood of rob in the village Bedono has drowned 2 villages namely Tambaksari and Senik [7].

2.2. DInSAR

Deformation especially landsubsidece phenomenon can be studied by using geodetic method, e.g Global Positioning System (GPS), Differential Interferometri Aperture Syntetic Radar (DinSAR). Both methods have been utilized to study landsubsidene in Jakarta, Semarang, Bandung and Surabaya [8]. This research had been done by using both method In order to compare and to complete earch other, because each of method has disadvantages.

Sentinel-1 is the first satellite series of seven satellite missions launched as part of the Copernicus program. Copernicus is the new name of GMES (Global Monitoring for Environment and Security), an earth observation program initiated by the European Commission (EC) and the European Space Agency (ESA). This observation program aims to provide accurate, timely and accessible information to improve environmental management, understand and mitigate the impacts of climate change and ensure civilian security [9].

Sentinel-l has four observation modes. The main modes are Interferometric Wide-swath (IW) mode, and Wave mode (WV). It also features Stripmap (SM), Extra Wide-swath (EW) modes for data continuity reasons and usage requirements for certain situations.

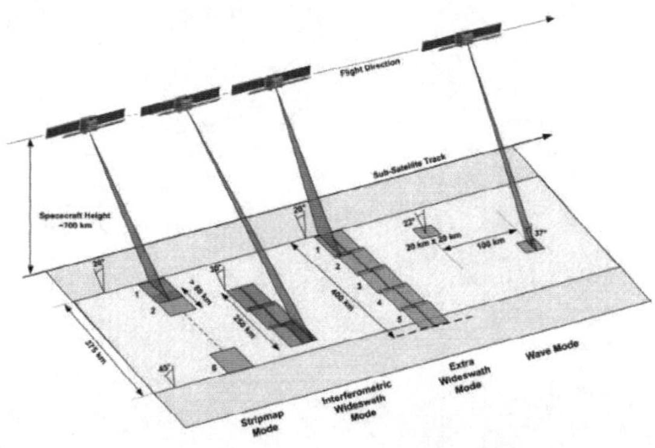

Figure 1. Mode Operation of Sentinel-1 (ESA, 2012)

The basic concept of the InSAR and DInSAR methods is to utilize coherence (the measure of correlation or suitability for the phase information of the corresponding signals) in phase measurement to obtain distance difference and distance change from two or more SAR images having complex values from the same area. If there are two images in a single complex observation format of the same area then the phase information of the two images in the form of such complex numbers can be combined. The result of this phase difference produces a phase difference image called interferogram, where the fringes indicate all relative geometry information [10].

In this research there are two pairs of images were processed. The first was Master 29 October 2015 and the second was Slave 29 October 2016. In this case, the subsidence is derived using differential method, and data SRTM was used for generating the DEM for the area. In the interferogram, the phase difference is described as the distance difference measured by the line of sight (LOS) radar and includes topography, orbit shifts, surface deformation and atmospheric effects [11]. The phase difference in the interferogram can be written by the following equation I:

. . . .(1)

φ = Difference Phase
φ_{topo} = Topographic Phase
φ_{defo} = Deformation Phase
φ_{atm} = Atmosphere Phase
φ_{orb} = Orbit Phase

DInSAR is the process whereby the phase component caused by ground surface displacement is extracted by somehow eliminating or minimizing the phase difference magnitudes attributable to other sources. [12].

This research is done in accordance with the figure 2. With using DInSAR and GPS method including data collecting, data processing and data analyzing. It would be compared result deformation in both methods.

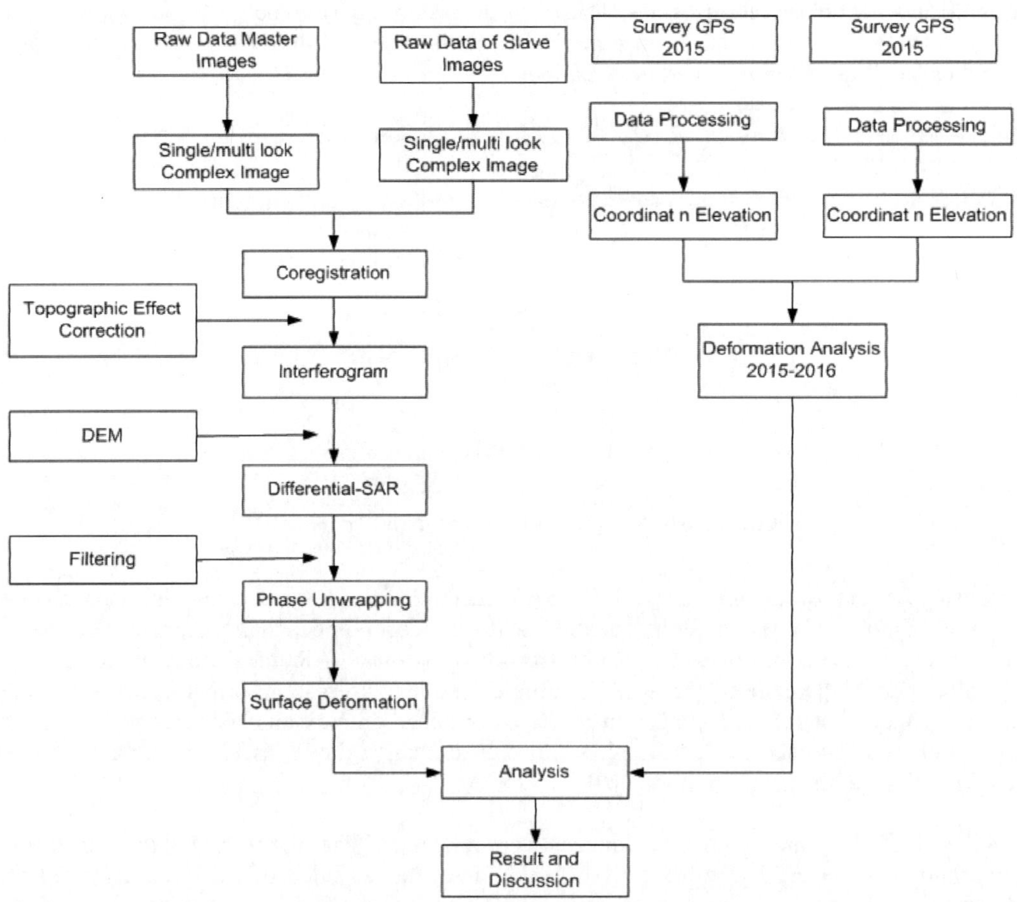

Figure 2. Research methodology

2.3. Baseline Estimation

At this stage, the image pair is evaluated to select the best image pair. As can be seen in Figure III.3, information obtained is acquisition date, sensor mode, perpendicular and temporal baseline, estimation of coherence model and Doppler centroid frequency difference between the two images. However, the coherence information (γ) obtained is not the result of the calculation of complex data but the calculation of the metadata of both images. So the coherence value will be different after further processing.

2.4. Coregistration

At this stage, the image pair is evaluated to select the best image pair. As can be seen in Figure III.3, information obtained is acquisition date, sensor mode, perpendicular and temporal baseline, estimation of coherence model and Doppler centroid frequency difference between the two images. However, the coherence information (γ) obtained is not the result of the calculation of complex data but the calculation of the metadata of both images. So the coherence value will be different after further.

2.5. Interferograms

If two images exist in a single complex observation format from the same area, then the phase information of the two ideals in the form of such complex numbers can be combined. The formation of the inteferogram is done by computing the complex form of the conjugate complex product (*) between the main image of C1 (master) with the second image of C2 (slave), through the following relationship [13].

$$I = C_1 \cdot C_2 \quad (2)$$

$$\varphi\text{^} = \arctan\frac{I_m(C_1C_2^*)}{R_e(C_1C_2^*)} \quad (3)$$

$$\gamma = \left|\frac{\Sigma C_1.C_1^*}{\sqrt{\Sigma C_1\, C_1^*\, \Sigma C_2 C_2^*}}\right| \quad (4)$$

I = Interferogram I_m = Komponen imaginer γ = coherence.
C1 = Master R_e = real component
C2 = Slave φ^ = different phase

In addition to forming interferogram, equation 2) in this stage also carried out the process of coherence formation. This coherence value (γ) is closely related to the match that will occur between the master image and the slave image. The higher the value of the coherence the greater the compatibility of the two images and the smaller the coherence value the lower the match rate of the two images. This coherency value estimate is expressed in equation 4. Coherence between C1 and C2 is defined as [13].

The coherence value (γ) of an image pair has a range of values from 0-1. If the image pair is really identical then the value γ = 1 and vice versa if the image pair is different or the image is only a noise then γ = 0. In this study the determined coherence limit is 0.2 which refers to the minimum value of the formation of the surface of the digital model in Set by ESA. The author does not set a smaller value because generally PMT in Semarang city is not too large to cause the coherence value is below the value of 0.2. In addition fringes also cannot be formed if the coherence is too small where the fringes appearance is required for the next stage. Interferogram can be distributed by removing the topographic phase. Topographic phase removal operations in SNAP software will create an interferogram simulation based on the DEM reference and substrate the topographic phase of the processed interferogram. This stage will produce an interferogram with deleted topography and an additional topographic phase channel such as Figure 3 and Figure 4.

Figure 3. Before topographic removal

Figure 4. After topographic removal

2.6. Phase Unwrapping

Interferogram is a different phase information. This information is directly related to the form of topography. This information is limited between 0 and 2π, thus raising the ambiguity problem in calculating the phase cycle required to obtain the correct slash [13]. To solve this ambiguity is the processing of phase unwrapping, ie to obtain the absolute phase, through the relationship of formula (5) .

$_{abs}$ = ^ + k. 2..(5)

$_{abs}$ = fase absolut and ^ = relative phase

2.7. Surface Deformation

After importing the results of phase unwrapping data, the next step is to change the phase to shift to determine the Los quantity. The result of the phase change process to the shift. Results from DInSAR processing shown in Figure 2 show Sayung district tend to subsided with a max value of 4.7. cm in the village Pilangsari with the lowest magnitude value in the village Sayung. The magnitude of the decline varies in every region of Sayung district.

Figure 5. Ground Deformation observed by DInSAR

IOP Conf. Series: Earth and Environmental Science **123** (2018) 012005 doi:10.1088/1755-1315/123/1/012005

2.8. Global Potioning System Survey

Global Positioning System (GPS) surveying methods were used to observe the location, extent, and magnitude of the vertical land-surface changes in the Northern Sayung Demak [2]. GPS measurements made at 7 geodetic monuments in 2015 and in 2016. Monuments. were located in Sidodadi, Gemah, Tugu, Purwosari, Bedono and Timbulsloko. Distribution of GPS Survey is shown at figure 6. GPS surveys were conducted in May – June 15, and the second observation in April - May 2016. GPS survey done using dual frequency geodetic-type GPS receivers Hiper II and Hiper Gb.

Figure 6. GPS Distribution Survey (Yuwono, 2017)

2.9. GPS Data Processing

Based on ellipsoidal height differences defined from GPS processing between two consecutive observations, epoch can be calculated. To obtain subsidence rate it should be defined in different epoch. This research was based on the GPS Surveys conducted in 2015 and 2016 Periods.

For statistically [2] check the significance of land subsidence values, It used the general linear hypothesis test {11) Reference Height change defined as dhij and its rate dhij at each station are derived using the following relation.

dhij = dh(tj) – dh(ti) .... (6)

dhij = dhij / (tj – ti) .... (7)

Null hypothesis Ho : dhij = 0 .... (8)
Alternative hypothesis dhij 0 .... (9)
Test statistic is

$$t = \frac{\Delta dh_{ij}}{\sigma\left(dh_{ij}\right)} \qquad \text{.. (10)}$$

if Ho is true. The null hypothesis is rejected if.

$$|t| > t_{df,\alpha/2} \quad \text{...(11)}$$

At a confidence level of 99% (i.e. =1%), the critical value is $t_{\infty,005=2.576}$. The Estimated land subsidence at several locations in Sayung Demak is given in Table 1.

Table 1. Landsubsidence Observation GPS Survey (Yuwono, 2017)

Station	Years		Landsubsidence			
	h 2015	h 2016	Dh12 cm	s(Ddh12) (mm)	12 cm	t-test
WNSR	27.8	27.72	-8.1	5	-8.4	-16.3
SIDO	27.1	26.99	-11.9	4.9	-12.2	-24.1
GEMA	27.45	27.33	-12.3	4.9	-12.6	-24.9
TUGU	27.62	27.56	-6.5	5	-6.8	-13.1
TMBL	27.28	27.16	-11.7	5	-12	-23.6
SURO	27.04	27.04	0.3	4.9	0.3	0.7

Result from the GPS observation show that subsidence area has distributed in spatial variation, ranging between 0.3 cm to 12.6 cm (see table 1). Base on the result in table 1, it could be statistically concluded that with 99% confidence lever there were significant ellipsoid changes observed by GPS surveys at all the stations during the period between 2015 and 2016 except suro. Figure 7 shows that the Middle region on Sayung district indicates higher rates of subsidence compare to it north region.

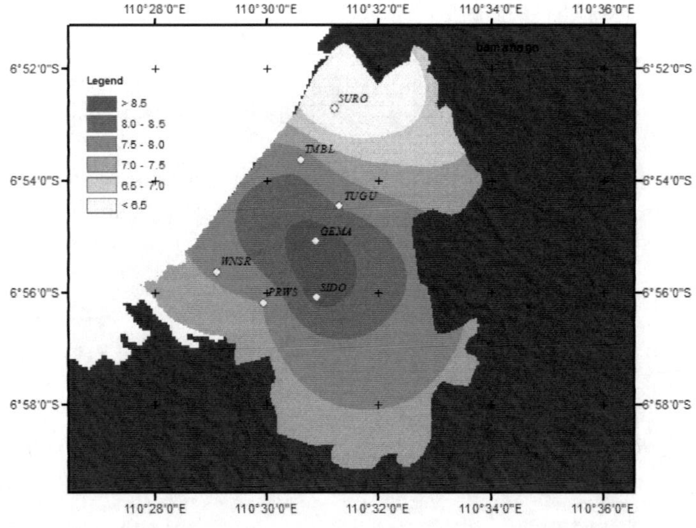

Figure 7. Ground Deformation in Sayung Demak Observed by GPS (Analysis 2017)

3. Result and Discussion

Figure 8 and Figure 9 shows the section profile of the DInSAR in the north area of Sayung. They indocate taht subsidence in Sayung is generally occurring with the maximum rates of about 4.7 cm. The results are more or less similar with the aforementioned results derived by other methods. There are should be noted that the uplift presece in figure 8 is mostly caused by the very low correlations of images, and the uplift presence should not be considered as a 'real' uplift.

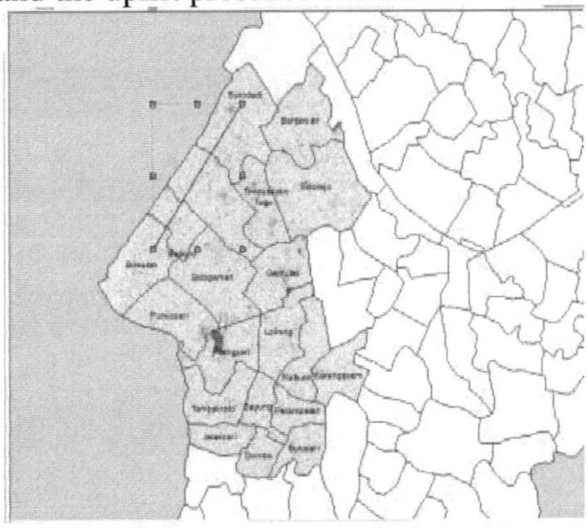

Figure 8. Section of Landsubsidece

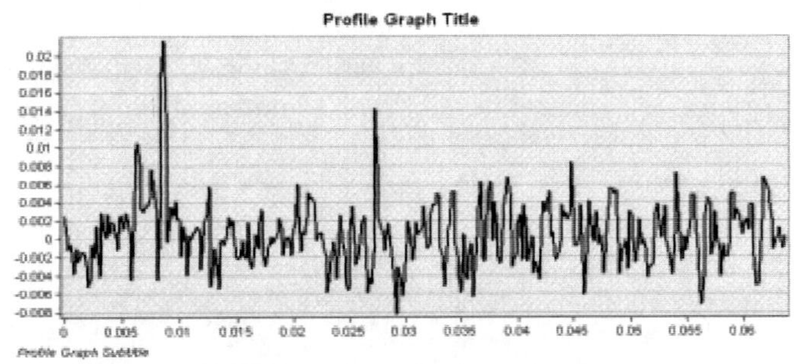

Figure 9. Profile subsidence

Analysis for DInSAR processing show Sayung district tend to subsided with the average value Sabsidence value in each villages shown in table2 , which the highest subsidence level was located in tugu village with the subsidence level 3 cm.

Table 2. Characteristics of landsubsidence at Sayung Demak (Analysis 2017)

No.	Village	Average (cm)	Minimal (cm)	Maximal (cm)
1	Jetaksari	0.01	-2.17	2.34
2	Loireng	0.04	-3.37	3.23
3	Gemulak	0.02	-3.18	2.39
4	Sidogemah	0.08	-3.13	2.64
5	Purwosari	-0.01	-3.65	2.75
6	Sriwulan	0.01	-2.86	3.05
7	Bedono	0.00	-2.96	2.77
8	Timbulsloko	0.00	-2.77	2.80
9	Tugu	-0.03	-2.71	3.07
10	Sidorejo	0.00	-2.26	3.20
11	Banjarsari	0.01	-2.73	2.12
12	Dombo	0.04	-2.03	1.75
13	Surodadi	0.01	-2.23	1.73
14	Bulusari	0.01	-2.46	2.12
15	Perampelan	0.02	-2.09	1.73
16	Karangasem	-0.03	-2.43	2.49
17	Kalisari	0.00	-2.05	2.81
18	Sayung	0.03	-1.53	2.09
19	Tambakroto	-0.02	-2.78	2.78
20	Pilangsari	-0.15	-4.57	2.71

IOP Conf. Series: Earth and Environmental Science **123** (2018) 012005 doi:10.1088/1755-1315/123/1/012005

The comparison of DInSAR and GPS observations indicates that the study area has spatial variation of landsubsidence. The subsidence value of DInSAR and GPS has not shown correlation, due to different observation time intervals, data accuracy, and accuracy typical of limitation seen in Table 3.

Table 3. Subsidence level observed by GPS and DInSAR (Analysis 2017)

No	Village	Latitude	Longitude	GPS (cm)	DInSAR (cm)
1	WNSR	-6.936	110.499	-5.29	-0.50
2	TMBL	-6.894	110.510	-4.74	-2.00
3	GEMA	-6.918	110.515	-5.87	-1.50
4	SIDO	-6.935	110.515	-8.37	-2.80
5	SURO	-6.879	110.521	-0.66	-3.00
6	TUGU	-6.908	110.522	-5.07	-1.00
7	DADI	-6.869	110.527	2.078	-1.00
			Standard Deviation		0.95
			Average Difference		1.68

4. Conclusion

Land subsidence characteristic in Sayung can be estimated by using geodetic measurement techniques such as DInSAR and GPS. And each method has strengths and limitation. It cannot be stated that the existence of land subsidence in Demak is due to a fairly short range of observations. It is necessary to study land subsidence through periodic measurement in the long term so that will give better conclusion about deformation was occurred.

5. Acknowledgments

In this research GPS surveys has been conducted using research RPP grants form from Diponegoro University and were conducted by staffs of Survey and Mapping, Remote Sensing Laboratory of Diponegoro University and student of Geodetic Diponegoro University. We thank to PT Adhimulia Interniagatama has supported GNSS CORS.

6. References

[1] Abidin H Z, Andreas H, Gumilar I, Sidiq T P and Gamal M 2015 Environmental impacts of land subsidence in urban areas of Indonesia *FIG Working Week*

[2] Yuwono B D, Abidin H Z, Gumilar I, Andreas H, Awaluddin M, Haqqi K F and Khoirunisa R 2016 Preliminary survey and performance of land subsidence in North Semarang Demak *AIP Conference Proceedings* vol 1730p 60004

[3] Sarkowi M, Kadir W G A and Santoso D 2005 Strategy of 4D microgravity survey for the monitoring of fluid dynamics in the subsurface *Proceedings World Geothermal Congress* pp 1–5

[4] Murdohardono D, Sudradjat G M, Wirakusumah A D, Kühn F and Mulyasari F 2009 Land subsidence analysis through remote sensing and implementation on municipality level; case study: semarang municipality, Central Java Province, Indonesia *BGRGAI-CCOP Workshop on Management of Georisks* pp 23–5

[5] Chatterjee R S, Thapa S, Singh K B, Varunakumar G and Raju E V R 2015 Detecting, mapping and monitoring of land subsidence in Jharia Coalfield, Jharkhand, India by spaceborne differential interferometric SAR, GPS and precision levelling techniques *J. Earth Syst. Sci.* **124** 1359–1376

[6] Abidin H Z, Andreas H, Gumilar I, Sidiq T P, Gamal M, Murdohardono D and Supriyadi Y F 2010

Studying land subsidence in Semarang (Indonesia) using geodetic methods *Proceedings of the FIG Congress 2010*

[7] Damaywanti K 2013 Dampak Abrasi Pantai terhadap Lingkungan Sosial (Studi Kasus di Desa Bedono, Sayung Demak

[8] Abidin H Z, Djaja R, Andreas H, Gamal M, Hirose K and Maruyama Y 2004 Capabilities and constraints of geodetic techniques for monitoring land subsidence in the urban areas of Indonesia *Geomatics Res. Aust.* **81** 45–58

[9] Spatiale européenne A 2012 *Sentinel-1: ESA's Radar Observatory Mission for GMES Operational Services* (ESA communications production)

[10] Prasetyo Y 2014 Analisis Optimisasi dan Estimasi Penentuan Penurunan Muka Tanah Menggunakan Teknik Light Permanent Scatterer Interferometric Synthetic Aperture Radar (LPS-InSAR)(Studi Kasus: Cekungan Bandung) *Disertasi Dr. Geod. dan Geomatika Inst. Teknol. Bandung*

[11] Castaneda C, Pourthie N and Souyris J-C 2011 Dedicated SAR interferometric analysis to detect subtle deformation in evaporite areas around Zaragoza, NE Spain *Int. J. Remote Sens.* **32** 1861–1884

[12] Chang H-C, Ge L and Rizos C 2004 Environmental impact assessment of mining subsidence by using spaceborne radar interferometry *3rd FIG Regional Conference Jakarta, Indonesia* pp 1–13

[13] Ismullah I H 2002 Model Tinggi Permukaan Dijital Hasil Pengolahan Radar Interferometri Satelit Untuk Wilayah Berawan (Studi Kasus Gunung Cikurai-Jawa Barat) *Disertasi Progr. Doktor, Inst. Teknol. Bandung, Bandung* 8–97

Landslide Hazard Analysis and Damage Assessment for Tourism Destination at Candikuning Village, Tabanan Regency, Bali, Indonesia

I N Sunarta[1], K D Susila[2], and I N Kariasa[2]

[1] Faculty of Tourism, Universitas Udayana, Bali, Indonesia
[2]Center for Spatial Data Infrastructure Development (PPIDS) Universitas Udayana, Bali, Indonesia

Email: cairns54@yahoo.com

Abstract. Landslide is a movement down the slope by the soil mass or slope constituent rock, a result of disturbance of the stability of the soil or rocks that make up the slope.Bali as one of the best tourism destinations in the world, also has landslide prone areas. Tourism attraction in Bali that is prone to landslides are Lake Beratan and Pura Ulun Danu Beratan in Candikuning Village, Tabanan Regency, Bali Province, Indonesia. Candikunig village area has tourismdestination, settlements and agricultural land. This study aims to analyze landslide-prone areas and the losses caused by landslides include damage analysis for the attractions of Beratan Lake and Ulun Danu Beratan Temple and settlements. The method used is matching and scoring with parameters of rainfall, soil type, slope and land use.The result is, Beratan Lake area has moderate to high landslide prone areas in the eastern and southern parts where most of the settlements in Candikuning Village are located in areas prone to moderate and high landslides hazard.

Keywords: Landslide Hazard, Damage Assessment, Tourism Destination

1. Introduction

Landslide is a movement of soil down the slope, resulting from the disturbance of the stability of the soil or the rock that makes up the slope. If this moving mass is dominated by soil mass and its movement down a slope, then the process of movement is called a landslide. Landslide has a very complex impact on landuse or on the environmental conditions. Landslide will occur if there are three conditions, (1) the slope is quite steep so that the soil mass can move or slide downward quickly (2) has a semi permeable glide layer below the soil surface (3) the presence of a mass water content that gives weight to the soil, so that the ground mass will be able to move or glide downwards [1]. Landslides occur in mountainous area with steep slope with a slope of 15° - 45° and high rainfall intensity. Other factors causing landslides are the morphology of the Earth's surface ie land use, lithology, geological structures, and earthquakes. In addition to natural factors, there's also factors of human activity that affect a landscape, such as the conversion of forest to agricultural land, settlement, mining and slope cutting [2].

One example of the phenomenon of landslide problems occurred in the area of Bali, especially in Candikuning Village, Tabanan Regency. Tabanan Regency physiographically has two regional models, lowland or coastal areas in the southern part and mountainous area in the northern part. One of the mountainous area in Tabanan Regency is Candikuning village in Baturiti Sub-district.

Candikuning Village is one of the villages in Baturiti Sub-district that lately had been struck by landslide in December 21, 2016. The damage from landslides in this area includes damage of settlements, agricultural land, the Denpasar – Singaraja main road also tourism destination such as Lake Beratan and Pura Ulun Danu Beratan. The tourism destinations in Candikuning Village has already well-known worldwide due to its uniqueness and the beauty of the traditional architectures. These tourism destinations, Lake Beratan and Pura Ulun Danu Beratan, are exposed to landslides hazard due to the topographic conditions and rainfall conditions of Candikuning Village.

The spatial distribution of landslide potential areas can be identified more easily through mapping process. Visual representation with maps is necessary to provide a spatial information of the lands affected by landslides. Responding from the absence of information about the map of landslides hazard prone areas, this research aims to spatially map potential landslide-prone areas especially to the tourism destination in Candikuning Village and to estimate damage for the settlements in Candikuning Village. The map of disaster prone areas in Baturiti Sub-district is expected to be used as a reference for the government, especially in tourism planning in Candikuning Village and make future disaster mitigation efforts so that the risk of landslide disaster can be minimized.

2. Literature Review

2.1. Landslide

Landslides also referred as land movements. It is defined as the soil mass or clay, gravel, sand, and crusty material and lumps and mud, moving along the slopes or out of the slopes due to the earth's gravitational factors [3].Movement of the soil is a product of the process of disturbance of the slope balance which causes the movement of the soil and rock mass to a lower area. The force that trigger the soil mass movements along the slope is the physical properties of the soil and the angle in the soil shear resistance acting along the slope. The change of forces is caused by the influence of natural change and human action. Changes in natural conditions can be caused by earthquakes, erosion, change in moisture slopes due to rainwater absorption, and changes in surface flow. Human influence on the changes of forces, among others, is the addition of loads on the slopes and the edge of the slope, excavation of soil at the edge of the slope, and sharpening of the slope angle. The pressures of the population that many land use change of the slopes into settlements or cultivated land greatly affect the increased landslide hazard [4].

Type of landslide based on the speed of movement can be divided into 5 (five) types, namely [5]: (1) Stream; Landslides simultaneously / suddenly at high speed; (2) Avalanches; The avalanche material moves slowly with the former horseshoe-shaped slide; (3) Collapse; Generally landslide material in the form of stone and soil move quickly to very fast on a cliff; (4) Compound; An avalanche that develops from collapse or landslides and progresses further into streams; (5) Land Subsidence; Occurs in underground mining, excessive ground water suctioning, soil erosion processes and in areas where soil compaction processes occur.

Land subsidence may occur due to consolidation process, i.e. the subsidence of the soil surface due to the compaction process or the change in the volume of a soil layer. This process can take place faster if there is a load that exceeds the carrying capacity of the soil or excessive groundwater extraction. Excessive groundwater retrieval may result in lower groundwater levels and lower hydraulic pressures, while the pressure between rocks increases. Land subsidence generally occurs in the plains areas of soft rocks / soils [5].

Soil types are critical to the potential for landslides. Loose soil will be more potential to landslides than the massive soil like clay. This can also be seen from the sensitivity of soil erosion. The soil value of soil erodibility (K) indicates whether or not soil is eroded, determined by soil physical and chemical properties. The smaller K value becomes less sensitive to the soil against erosion [5].

The depth or soil solum, texture, and soil structure determine the size of the surface runoff water and the saturation rate of the soil. In deep soil (> 90 cm), loose structures, and dense land cover, most of the rainwater is infiltrated into the soil and only a small part of it becomes surface runoff water. Conversely, in shallow soil, solid structures, and less dense land cover, only a small percentage of infiltrated rainwater and much of it becomes surface flows which triggers landslide [7].
Rain with high intensity, for example 50 mm which lasts longer than 6 hours has the potential to cause landslide, because in that condition there still saturation process of soil by which increase soil mass. There are two types of rain, the heavy rainfall that can reach 70 mm / h or more than 100 mm / day. This type of heavy rain is very effective in triggering landslide, especially in clay and sandy soil. While the normal rainfall which less than 20 mm / day can cause landslides on slopes if rain lasts for several weeks to over a month [6].

2.2. Tourism

Tourism is a travel activity undertaken by a person or group of people by visiting a particular place for recreational purposes, personal development, or studying the unique attractions of the visited tourist in the interim period. Tourism industry is a collection of tourism businesses are inter-related in order to produce goods and services for the needs of tourists in the implementation of tourism. Tourism Business is a business that provides goods and / or services for meeting the needs of tourists and the implementation of tourism. Tourism facilities are all types of facilities specifically intended to support the creation of convenience, comfort, safety of tourists in making visits to tourism destinations. Tourism facilities are products and services needed to support the fulfillment of the needs of tourist travel [7].

The relationship between tourism and the environment might initially be characterized as one of '*coexistence*'[8].The environment, be it predominantly natural or largely human-made, is one of the most basic resources for tourism and a core element of tourism products [9].Tourism is an entire activity related to tourism and is multidimensional and multidisciplinary that emerges as a manifestation of the needs of every person and country as well as the interaction between tourists and the local community, fellow travellers, government, local government, and entrepreneurs. Tourists are people who do tours [10].An area can be developed into tourism area if there's natural resources in the form of minerals, water sources, fertile land, animal and vegetation sources or adequate human resources, high level of culture, technological level and sufficient capital to be able to develop the natural resources available for the prosperity for mankind. Tourism potential can be grouped into 3 namely the potential of nature, cultural potential, and human potential. The potentials are described below [11]: Natural Potential, consisting of physical potential, flora and fauna. These three natural potentials can be a tourist attraction that plays a similar role, but one of the attractions can be more prominent. In general, tourists are more interested in the open like the mountains, forests and beaches; Culture Potential, i.e. culture in the broad sense, not only includes high culture but includes customs and all activities that live in the midst of society.; Human Potential, namely the ability that exists in human beings that can be utilized for the benefit of tourism.

Tourism planning is closely related to the physical geography and environmental components. The initial effects of trampling of vegetation by tourists become compounded through related processes of environmental change that may, in extreme circumstances, culminate in the collapse of local ecosystems.

In physical planning, the issues of concern may be control over land development by both public and private sectors, zoning of land use. In the environment, issues of concern such as control of pollution and assessment of hazards [12].

IOP Conf. Series: Earth and Environmental Science **123** (2018) 012006 doi:10.1088/1755-1315/123/1/012006

3. Methodology

3.1. Area of Study

Area of study is located in Candikuning Village, Baturiti Sub-district, Tabanan Regency, Bali Province, Indonesia. Geographically, Candikuning Village located in UTM Zone 5o S, with easting 297258.12 m E and northing 9084770.92 m S. Candikuning Village bordering with Sukasada Village, Sawan Village and Kubutambahan Village of Buleleng Regency in the north, Belok Village of Badung Regency in the east, Batunya Village of Baturiti Sub-District in the south and Pancasari Village of Buleleng Regency in the west.

3.2. Tools and Material

The material used in this research are spatial datas as follows: (1) Slopes Map Scale 1: 25,000; (2) Soil Map Scale 1: 25,000; (3) Landuse map Scale 1: 25,000; (4) Rainfall Map (isohyet) Scale 1: 25,000. While the tool used are: (1) Computer or laptop used to process the spatial data; (2) GPS (Global Positioning System) to determine the coordinate point of observation.; (3) Camera; (4) Stationary.

3.3. Data collection

Data collection method used in this research is primary data and secondary data method: (1) Secondary data collection in the form of slope map, rainfall data, soil type and land use. To create a landslide hazard map in the Candikuning Village. Then documentation techniques used in the research are conducted to obtain almost all the required data, ie slope map, rainfall map, soil type map, geological map and land use map; (2) Primary data collection through is by conducting a field survey. Field surveys were made usingland unit samples made from intersections of slope map, soil map, land use map and isohyet map.

3.4. Data Analysis

Data analysis of the data in this study used the matching and scoring system from soil data, slope data, landuse data and rainfall data. Then the analysis technique using overlay method using geographic information system to generate new landslide hazard area from slope map, rainfall map, soil type map, and land use mapwhich has been scored.

Scoring was used to determine or assess the landslide hazard in the research area. The scoring value was based on the size of the influence of the supporting variable of landslide in the research area. The landslide hazard is indicated by the total number of scores from each of the supporting parameters of landslide, ie slope map, rainfall map, soil type map, geological map and landuse map [13].

3.4.1. Rainfall scoring

The rainfall score, divided into five classes as listed in Table 1 [13].

Table 1. Rainfall Score

No	Rainfall (mm/yr)	Score
1	< 1.500	1
2	1.500 - 1.800	2
3	1.800 - 2.100	3
4	2.100 - 2.400	4
5	> 2.400	5

IOP Conf. Series: Earth and Environmental Science **123** (2018) 012006 doi:10.1088/1755-1315/123/1/012006

3.4.2. Slope Scoring

In the landslide hazard, slope factor is very influential, the steeper slope then the possibility of the occurrence of landslide is higher. The slope score, listed in Table 2[13].

Table 2. Slope Score

No	Slope (%)	Score
1	0 – 8	1
2	8 – 15	2
3	15 – 30	3
4	30 – 45	4
5	> 45	5

3.4.3. Landuse Score

Settlements have a smaller share because runoff is more prevalent and infiltration because of the impermeable surface properties. The landuse score, listed in Table 3 [13].

Table 3. Landuse Score

No	Landuse	Score
1	Forest, Mangrove, swamp, irrigated rice field, ponds, salting, sand	1
2	Rice Fields non irrigated	2
3	Building, settlement	3
4	Shrubs, gardens / plantation	4
5	Grass, bare land, moor	5

3.4.4. Soil Score

Land is more sensitive to landslide then the higher the score given. The degree of sensitivity to landslide is related to the level of soil's ability to release incoming water, ie, soils with very slow permeability. The soil score listed in Table 4 [13].

Table 4. Soil Score

No	Soil type	Score
1	Brown mediteran, red brownis mediteran	1
2	Brown yellowish latosol, red brownis Latosol, brownish latosol andlitosol,	2
3	Aluvial brownish grey, alluvial hidromorf	3
4	Brownish regosol, brown yellowish regoso, grey brownish regosol, Regosol Humus, greyish regosol	4
5	Greyish brown andosol	5

3.4.5. Landslide score

The score for landslide hazard is the total of slope score, rainfall score, soil type score, and landuse score. The total landslide score based on the following equation [13]:

Landslide Score = (5 Sl) + (5 R) + (4 LU) + (2 St)

Sl= slope score; R = rainfall score; LU = Landuse score; St = Soil score

The landslide classes listed in Table 5 [15].

Table 5. Landslide Hazard Class

No	Landslide Hazard Score	Landslide Hazard Class
1	<120	Very low
2	120 – 150	Low
3	150 – 195	Medium
4	> 195	High

Analysis of damage assessment of settlements and analysis the tourism object that affected by landslide is conducted by the result of scoring and overlay of landside hazard parameter. The summary of methodology in this research featured in Figure 1.

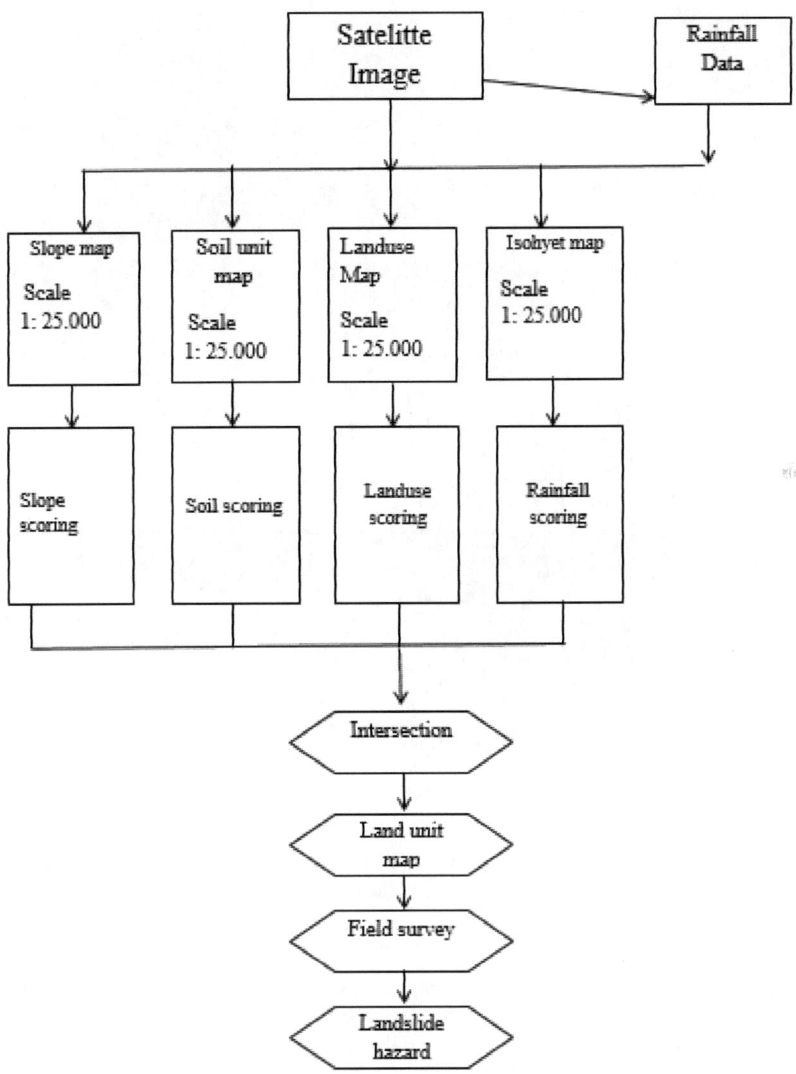

Figure 1. Methodology Diagram

4. Results and Discussion

4.1. Landslide Hazard Analysis

The Landslide hazard in Candikuning Village is divided into three classes which are high, medium and low. Landslide hazard map is created by using overlay method and scoring methods to degree of landslides hazard Physical condition in Candikuning Village which triggers landuse are slope, rainfall, soil and landuse. Slope in Candikuning village varies between 0%-8%, 15%-45% and more than 45%. Rainfall intensity is very homogen with intensity between 2500-3000 mm/ year. Meanwhile the soil in the research area is andosol which has the characteristic of sensitivity in erosion and landslides. Majority of landuse in Candikunging village is forest (54.69%) and ricefields (23.40%) followed by plantation (14.94%), Settlements (6.67%) and moor (0.30%). Steep to very steep topography with a slope of more than 30%, heavy rainfall which occur more than 2500 mm/ year is the main factors that triggers landslide in Candikuning Village. Medium hazard until high hazard are located in the steep sleep east and west of Beratan Lake. Meanwhile low hazard located along the Denpasar – Singaraja highway. Figure 2 shows the landslide hazard area in Candikuning Village.

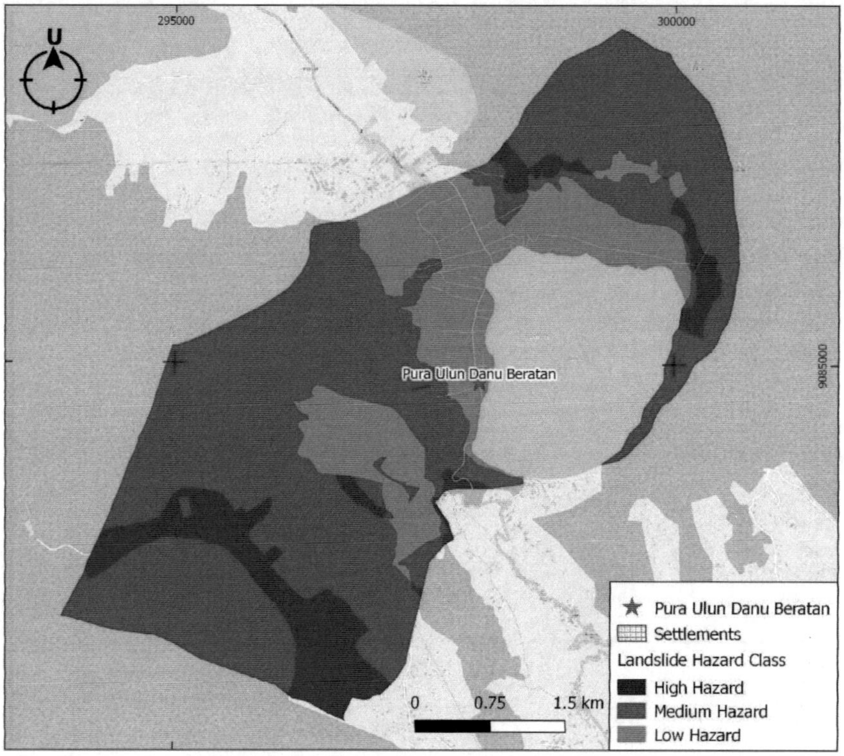

Figure 2. Landslide Hazard Class

Maps of the landslide hazard were created by using overlay method and scoring methods to degree of landslides hazard. Low landslide hazard area in total is 394.14 Ha or occupied 20.92% of total area of Candikuning Village. Low hazard consist of rice field, forest and settlement which located on 0%-8% slope and andosol soil. The rainfall intensity is 2500 – 3000 mm/ year. Medium landslide hazard area in total is 1258.88 Ha or occupied 66.82% area of Candikuning Village. Medium landslide hazard located at various landuse namely rice field, forest, settlement, plantation and moor. The slope is varies between 0% until 45% with 2500 mm/year until 3000 mm/ year of rainfall and andosol soil. Meanwhile the high hazard is occupied 230.89 ha area of Candikuning Village or 12.25% of total area. High hazard consist of more than 15% slope until more than 455 slope. The slope steepness is mainly the triggering factor of landslide. The landuse in the high hazard are settlements, plantation and moor.

The rainfall intensity is 2500 mm/ year until 3000 mm/ year and the soil type is andosol. Landslide hazard area is listed in Table 6.

Table 6. Landslide Hazard Area (a)

No	Hazard Class	Slope	Landuse	Soil Type	Rainfall Intensity	Area (Ha)	Area (%)
1	Low Hazard	2 - 5%	Ricefield	Andosol	2500 - 3000 mm/yr	212.03	11.25
2	Low Hazard	2 - 5%	Forest	Andosol	2500 - 3000 mm/yr	105.90	5.62
3	Low Hazard	2 - 5%	Settlements	Andosol	2500 - 3000 mm/yr	76.21	4.05
4	Medium Hazard	15 - 40%	Settlements	Andosol	2500 - 3000 mm/yr	42.18	2.24
5	Medium Hazard	2 - 5%	Moor	Andosol	2500 - 3000 mm/yr	3.11	0.17
6	Medium Hazard	2 - 5%	Plantation	Andosol	2500 - 3000 mm/yr	53.39	2.83
7	Medium Hazard	> 40%	Ricefield	Andosol	2500 - 3000 mm/yr	209.64	11.13
			Table 6 (b)				
8	Medium Hazard	> 40%	Forest	Andosol	2500 - 3000 mm/yr	950.56	50.46
9	High Hazard	15 - 40%	Moor	Andosol	2500 - 3000 mm/yr	2.54	0.13
10	High Hazard	> 40%	Plantation	Andosol	2500 - 3000 mm/yr	228.35	12.12
			Total			1883.91	100

4.2. Damage Assessment

Damage for settlements were calculated by using the scoring result of the landslide hazard calculation. The damage are basically caused by the soil mass that struck the settlement's building. The damage divide into three class, which are minor damage, moderate damage and severe damage. Minor damage apply to settlements with low hazard along the Denpasar – Singaraja highway which has flat to undulating topography with 0% to 8% slope. The moderate damage will occur to settlements at medium hazard at east and west side of Beratan Lake with hilly until mountainous topography with 15% to 45% slope. Figure 3 shows the damage class to settlements which exposed to landslide hazard in Candikuning Village.

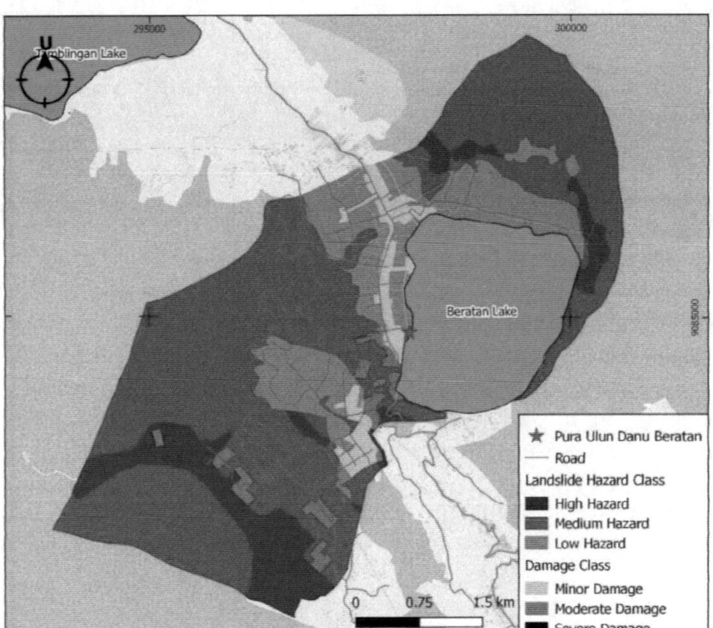

Figure 3. Damage Class

IOP Conf. Series: Earth and Environmental Science **123** (2018) 012006 doi:10.1088/1755-1315/123/1/012006

The total area of settlements that has minor damage if landslide occur is at 83.59 Ha or 33.53% of total area. Meanwhile total area of moderate damage is at 42.18 Ha or 66.46% of total area of settlements in Candikuning Village.

4.3. Landslide Impact to Tourism Destination at Candikuning Village

Tourism destination in Candikuning Village is Ulun Danu Beratan Temple and Beratan Lake. These two tourism destinations are a very popular destination for local tourist and foreigners. Ulun Danu Beratan Temple is one of nine "Kahyangan Jagat Temples" which is surrounding Bali Island that makes it becomes one of the most important Temples for the Balinese, especially for Hindu. It consists of five compounds of temples and one Buddhist Stupa. Ulun Danu Beratan Temple is both a famous landmark and a significant temple complex located on the western side of the Beratan Lake in Candikuning Village, or well-known as Bedugul area. The whole Bedugul area is actually a hilly and mountainous area with cold weather and heavy rainfall which became favourite holiday destination for locals and foreigner. The reflective surface of the lake surrounding most of the temple creates a unique panorama, while the mountain range of the Bedugul area encircling the lake provides the temple with a very beautiful panorama. Figure 4 shows the image of Ulun Danu Beratan Temple and Beratan Lake.

Figure 4. Ulun Danu Beratan Temple and Beratan Lake

Located in hilly and mountainous area makes Ulun Danu Beratan Temple and Beratan Lake becomes vulnerable to landslide hazard. The location of Ulun Danu Beratan is located road side of Denpasar - Singaraja highway. Ulun Danu Beratan Temple is included in low hazard class. The challenge for tourism activities in Ulun Danu Beratan Temple and Beratan Lake related to landslides that the potential of landslide to cover Denpasar – Singaraja Highway. Denpasar – Singaraja Highway is the only major road to accesUlun Danu Beratan Temple and Beratan Lake. The south part of Denpasar – Singaraja Highway in Candikuning Village is located in medium hazard and high hazard to landslides. This condition makes the Denpasar – Singaraja Highway become prone area to landslides.

Landslide that occurred on the eastern slope of Lake Beratan impacts on the preservation of the lake as a potential for tourism. Landslide material can conflate the lake water. In addition, landslide material can also reduce lake water volume. This will impact on the preservation of Pura Ulun Danu

Beratan as a tourism icon in Bedugul area. The existence of spillway buildings in Lake Beratan will reinforces the fear of iconic loss in this area. The view of Pura Ulun Lake that flooded the lake water will be damaged and will be difficult toestored as before. Therefore, landslide that occurred in Candikuning Village has negatively impact on Bedugul tourism destination.

5. Conclusions

Landslide hazard areas in Candikuning village are divided into three classes.Low hazard located along the Denpasar – Singaraja highway, medium hazard until high hazard are located in the steep sleep east and west of Beratan Lake. Low hazard area in total is 394.14 Ha (20.92%), Medium landslide hazard is 1258.88 Ha (66.82%) and the high hazard 230.89 ha (12.25%) of Candikuning Village.

Two damage class for settlements are found due to landslide hazard in Candikuning Village. Minor damage area is 83.59 Ha (33.53%) and moderate damage is at 42.18 Ha (66.46%). Meanwhile the tourism destinations in Candikuning Village, namely Ulun Danu Beratan Temple and Beratan Lake, is located at low hazard class, but the access to reach the destinations, the Denpasar – Singaraja Highway, is located in medium hazard and high hazard to landslides. Which makes the accessibility to reach the tourism destination in Candikuning Village is a prone area to landslides.

Landslide that occurred in Candikuning Village has a negative impact on tourism potential in this village. Candikuning village is the location of Pura Ulun Danu and Lake Beratan which is the icon of tourism area Bedugul. So that the destruction of tourism potential in Candikuning Village will damage the tourism potential in this area.

6. Acknowledgments

Writers would like to thank the Rector of Universitas Udayana, Dean of Agriculture Faculty of Universitas Udayana, the Dean of Tourism Faculty of Universitas Udayana, Head of Center for Spatial Data Infrastructure Development (PPIDS) Universitas Udayana, fellow staf, researchers and professors at Universitas Udayana for all the support, comments and suggestions that makes this paper finished.

7. References

[1] Arsyad S 1989 Konservasi tanah dan air
[2] Alhasanah F 2006 Pemetaan dan Analisis Daerah Rawan Tanah Longsor Serta Upaya Mitigasinya Menggunakan Sistem Informasi Geografis (Studi Kasus Kecamatan Sumedang Utara dan Sumedang Selatan, Kabupaten Sumedang, Provinsi Jawa Barat) Tesis *Sekol. Pasca Sarjana. Inst. Pertan. Bogor. Bogor*
[3] Vulkanologi D and Geologi M B 2013 Manajemen Bencana Tanah Longsor
[4] Geoteknologi-LIPI P P 2001 Aplikasi Citra Satelit Dalam Penentuan Daerah Rawan Bencana Longsor *Pus. Penelit. Geoteknologi-LIPI, Rep.*
[5] Sitorus S R P 2006 Pengembangan Lahan Berpenutupan Tetap Sebagai Kontrol Terhadap Faktor Resiko Erosi dan Bencana Longsor *Direktorat Jenderal Penataan Ruang Dep. Pekerj. Umum. Jakarta*
[6] Pertanian L D 2006 *PedomanUmumBudidayaPertanian di LahanPegunungan* (Jakarta)
[7] Anon 2016 *Peraturan Daerah Kabupaten Badung tentang Rencana Induk Pembangunan* Kepariwisataan *Kabupaten Badung Tahun 2017- 2025* (Indonesia)
[8] Wall G and Mathieson A 2006 *Tourism: change, impacts, and opportunities* (Pearson Education)
[9] Page S J, Dowling R K and others 2001 *Ecotourism.* (Pearson Education Limited)
[10] Pendit N S 2002 Ilmu Pariwisata *Jakarta: Pradnya Paramita*
[11] Suwantoro G 2002 1997 *Dasar-dasar Pariwisata*
[12] Stephen W 2009 *Tourism Geography: A New Synthesis* (Taylor & Francis)
[13] UGM P S B A 2001 Penyusunan Sistem Informasi Penanggulangan Bencana Alam Tanah Longsor di Kabupaten Kulon Progo *Yogyakarta PSBA UGM*

Beyond Vulnerability Assessment: Impact of Developments toward Local Adaptive Capacity in Kemijen City Village, Semarang City

M E Jayanimitta[1], D A Puspasari[1], R Widyahantari[1], D Kristina[1], T Ratnaningtyas[1], A Setionurjaya[1], and Y A Anindita[2]

[1]Master of Urban and Regional Development, University of Diponegoro
[2]Development of Environmental Sanitation System of Settlement, Ministry of Public Work Indonesia

Email: maria.ekacarini@gmail.com

Abstract. Vulnerability Assessment is usually used for assessing the ability of an area on facing disaster. In previous studies, the study of Vulnerability Assessment applied only quantitative method to show the vulnerability level. Therefore, this study attempts to add information reviews using qualitative method. Kemijen City Village is one of the administrative areas in the northern part of Semarang City affected by climate change. The residents have to adapt it by renovating and elevating their houses and other infrastructures to avoid floods. There are some development programs held by government, NGOs, and corporations such as Banger Polder Development, PLPBK, etc. It is interesting to know how big the vulnerability level of Kemijen on facing flood disasters, then how the projects can affect local adaptive capacity. To answer it, this research uses mixed-method approach. Vulnerability Assessment uses quantitative method by scoring indicators of Exposure, Sensitivity, and Adaptive Capacity, while the development impact uses qualitative method. The data were collected through interviews and FGD conducted in Joint Studio Course between Diponegoro University and University of Hawaii in October 2016. Non-physical programs such as community empowerment have more positive impacts on local adaptive capacity in Kemijen. Community participation is important for environmental sustainability that can not be done in a short time to educate the people.

Keywords: Vulnerability assessment, local adaptive capacity, Semarang City

1. Introduction

Kemijen City Village is located in the northeast part of East Semarang Sub District which is one of the slum areas in Semarang City. This is accordance with the Decree of Mayor of Semarang No. 050/801/2014 about The Location Determination of Housing and Slum Area in Semarang City. The area has many environmental and social degradations, including floods, land subsidence, waste, etc.

The floods in Kemijen were caused by the surge of sea water/ tidal floods [1] due to climate change. As Yousef & Ghonem [2] mentioned that climate change is a situation when the climate patterns of the world are permanently shifting either upwards or downwards of the average. This phenomenon is causing the changes in season length. Some climate change impacts are the rise in sea levels and erratic rainfall. There are two factors causing floods in Kemijen i.e lack of water catchment areas that made high density of the water in a watershed and local floods caused by sedimentation from the accumulation of garbage and poor drainage system [3]. It happened every day and damaged infrastructures especially roads [4]. In terms of health, it can cause diseases, and in terms of economy, it can disrupt community activities that can lead to financial losses for both the government and the community side [5]. Beside floods, land subsidence also be one of the problems in Kemijen. The land subsidence in Kemijen is approximately about 6-8 cm/year [6] which generally caused by natural soil consolidation and the excessive ground water extraction. This condition worsens the floods. The waste problems in Kemijen also decrease the environment and need special handlings.

This location is unique because most of the population do not want to move to another place and keep on staying in that location with all of the risks [7]. They have their own adaptive capacity to ensure their life sustainability, social activity and quality of life[8,9], and minimize climate change vulnerability [10, 11]. Adaptive capacity (AC) is an ability to adjust, take advantage of opportunities, or cope with consequence [10,12]. The adaptive capacity in Kemijen is formed by institutional and social factors, including strength, culture and the ability of people to engage in collective strategies and actions to address environmental problems [13]. Successful adaptation is closely connected with social groups' ability to act collectively, but many social-ecological challenges exceed local adaptive capacity which needs help from the government [13]. As Wolfson [14] mentioned, the main contribution in developing an area is providing a mechanism for developing, sharing, and consider information. In this case, the local government has been running the programs to help people in preparing and recovering from floods, especially in the East Semarang Sub-District. The programs are the Community Disaster Alert, Banger Polder System, Neighborhood Development (PLPBK), and other programs. The government program that has an enormous impact is the Banger Polder System.

Polder as a system which serves to prevent floods should consist of retention and drainage system [15]. On the polder system, its condition doesn't let the water flows from the outside area [16]. According to the results of interviews, before Banger Polder System was operated (approximately on August 2016), the floods happened almost everyday as high as 10 to 40 centimeters. These floods could be defined as a small scale flood which was related with low water depths (i.e. < 0,5 m). These floods could occur when there was insufficient retention for water in the polder [15]. But after it was operated, the floods never happened in the area that protected by the polder system. The Banger Polder should protect the East Semarang Sub District, including Kemijen City Village, but now still only limited areas have the benefit of the polder. Those are RW 6, RW 7, RW 8, RW 9, and RW 10 in Kemijen City Village.

The Banger Polder System has been started since 2001 through socialization to the community and has been built on 2010. Although it was operated on August 2016, but not all of the components have built such as retention basin (planned to build in RW 4 and RW 5); the permanent closing dam of the Banger River; and the secondary canal improvements along the Banger Polder System project. The main obstacles of this development are the land ownership and community empowerment.

A development has to be measurable, which will be very significant and will be the presence of uncertainty. The uncertainty will play an important role in the predevelopment and the development stage of a project [17]. In some previous studies, the study of Vulnerability Assessment applied only quantitative method to show the vulnerability level. Therefore, this research tries to add information reviews using qualitative method. It is expected to find how the development program that has been running affected the local community adaptive capacity. This research also will compare the before and after conditions to know the impact of development program. In addition, will be known the factors that affect the development and the recommendations for the sustainability of these conditions.

IOP Conf. Series: Earth and Environmental Science **123** (2018) 012007 doi:10.1088/1755-1315/123/1/012007

2. Methodology

This article applies the scoring analysis and descriptive qualitative. The scoring analysis was used to assess the vulnerability of Kemijen city village area, while the descriptive qualitative was used to determine in-depth information related to the impact of development program toward local adaptive capacity. The research variables refer to the statement mentioned by USAID [18], consisting of exposure, sensitivity and adaptive capacity. Table 1 below shows the variables and indicators to assess the vulnerability. Therefore, the variable of the development program impact is referring to Adger [19] statement that the social adaptive capacity of the government's role in adaptive capacity building have an important role and supported by local potentials that can help accelerating adaptive capacity of the communities.

Table 1. Variables, Indicator and Methodology

Variables	Indicators	Methodology
Exposure	Flood (F)	Quantitative Method
	Land Subsidence (LS)	
	Waste (W)	
Sensitivity	Near River	
	Near dumping side	
	Flood area	
	Slum area	
	High rate of Land Subsidence	
Adaptive Capacity	Flood	
	Land Subsidence	
	Waste	
	Experience	Qualitative Method
	Knowledge	

Vulnerability assessment is based on the formula presented by Turner and Kasperson [20]

$$\textbf{Vulnerability} = \frac{\textbf{Exposure x Sensitivity}}{\textbf{Adaptive Capacity}}$$

The assessment of exposure was assessed based on three main problems in the region which are floods, land subsidence, and waste. Each region was assessed with a score range from 1 to 3 with score 1 for low exposure, score 2 for medium exposure and score 3 for high exposure. Elements of sensitivity based on how many assets, regions, or groups, influenced by the exposure to danger. In this study, each region was assessed according to the following location categories which are near the river, near the dumping site, flooded area, slum area and the area that has high rate of land subsidence. The sensitivity is high when locals tend to have a dependence on natural resources or ecosystems. The Adaptive Capacity Assessment was classified into individual, community and government capabilities on 3 main issues. This stage also equipped by qualitative method with in-depth interview to explore experience and knowledge from community leaders. This method is used to find out how much vulnerability projects implemented by the government can affect the adaptive capacity of the community.

The quantitative data were collected based on the questionnaires. The sampling provides a numeric description of trends and opinion of Kemijen populations. This research used **random**

sampling, where each individual in the population has an equal probability of being selected. The number of population in Kemijen area are 11.936 inhabitants which consist of 3.603 households and spreads into 11 RWs. The sampling was using random sampling on RW's household. According to Frankel and Wallen [21], the minimum number of samples taken in a study is 100 samples and the detail of the number of household's sampling can be shown in the table below.

Table 2. Sample in each RW

No.	RW	Number of Households	Number Of Sampling
1	I	451	13
2	II	386	11
3	III	370	10
4	IV	75	2
5	V	498	14
6	VI	328	9
7	VII	435	12
8	VIII	310	9
9	IX	229	6
10	X	286	8
11	XI	235	6
Total		3.603	100

Source: Analysis Result, 2017

Qualitative data were obtained through FGD results conducted on communities in Kemijen. In FGD activities, the community is directed based on the aspects of their experience and knowledge about the flood disaster that occurred in Kemijen. In addition, an in-depth interview was conducted at BPP SIMA (the initiator of the Polder Banger development) to define adaptive capacity conditions of communities before and after the construction of the Banger Polder. Basically, there are two types of data used in this research. The data are the primary data obtained from the observation and structured interviews. In this method, the sampling technique used is the *snowballing* technique.

3. Data and Analysis

The terminology of vulnerability in this article refers to the notion by Turner [20], focusing on the vulnerable area that affected by vulnerabilities as the result of the disaster risk, and for the impact of vulnerability in social aspect such as experience of the community and knowledge is referring to Folke [22]. The discussion in this article will be divided into 2 parts. The first is the explanation about the vulnerability assessment based on the components of exposure, sensitivity and adaptive capacity. The second is the Kemijen development, describing about the impact of the development in Kemijen city village after the programs were finished toward the community adaptive capacity.

3.1 Vulnerability Assessment

3.1.1. Exposure

Exposure refers to the inventory of elements in an area in which hazard events may occur. Hence, if population and economic resources were not located in (exposed to) potentially dangerous settings, no problem of disaster risk would exist [23]. Furthermore, Cardona [24] explained the difference between exposure and vulnerability. Semarang City has historically faced hazards such as drought, land subsidence, landslides, and floods, many of which are likely to become more severe and frequent as a result of climate change [25]. This variable classified into three types of exposure :

a. *Flooding*. Based on the history, floods in Kemijen were caused by the disturbance in the upstream area, sedimentation by the accumulation of garbage and poor drainage system, sea level rise and land subsidence. Tidal period in Kemijen occurs every day where the inundation can be as high as an adult's knee, indubitably affecting their socio-economic set-ups and daily activities. A study from Anggraini [26] found that based on personal interview with the locals, most of the respondents recalled the large floods which possibly caused a total shut down in the area were happened in 2006, 2005 and also in 1990 even though the tidal flood happens every year. With the height of 50 cm until 2 meters, depending on the location and the condition of drainage. With such disruption happens every day and every year, some locals could not continue their daily activity effectively since the access to workplace is frequently flooded. This condition was worsened by the bad sanitation and water supply in the area.

The social condition in Kemijen have low level of education, which proven by 34% of the people was not graduated from primary school [27]. Whereas the low education leads to low knowledge about how should they handle the risk of tidal flood thus provision of knowledge from government and/or NGO become important. For example, before the announcement and socialization of the Banger Polder project, people in Kemijen did not seem to be disturbed as long as it (the flood) did not affect directly, in this case was the intrusion to their houses [28].

While the economic condition in Kemijen are dominated by the low-income communities (MBR) with the income is below Rp 2.100.000,- per month with a percentage of 84%, tidal floods threats perceived as nuisances which the negative impacts seen to have damaged their properties rather than directly threaten their lives [26]. Because of the increasing of the tidal floods in Kemijen, the burden of public expenditures were higher. This is due to the additional needs on improving their homes and vehicles. Based on the conditions, the scoring results show that some part of Kemijen such as in RW 3,4, and 5 are having higher exposure which experiencing floods with height more than 25 cm.

b. *Land Subsidence.* In Kemijen City Village, land subsidence is generally caused by natural soil consolidation and the excessive ground water extraction. More towards the North, the level of land subsidence is getting higher, so the worst area that affected by land subsidence is the northernmost of Kemijen, which is RW 4, 5, and 6. Land subsidence in Kemijen affected the people who live in that area and the infrastructure such as streets and houses.

c. *Waste and Sewerage.* There are two types of waste, i.e. solid waste and water waste. Solid waste in Kemijen City Village comes from several places: the textile factory outside Kemijen city village, upstream communities, the market on the southern border of Kemijen City Village, and the communities inside Kemijen City Village itself. Solid waste management in Kemijen city village is very good at the household level. Most people reported that waste was collected from their home every 1-2 days. This is done informally by community members for an IDR 7000-10000 a month. But people also reported that the waste will not be collected by the government and moved out of Kemijen city village. People in Kemijen City Village thought that the drainage should be cleaned more frequently because it makes the flood worse. Also, this water is not treated before it is pumped into the neighboring East Flood Canal and so far there has been no environmental impact assessment of the polder system. It is still dangerous as untreated water waste has a negative impact on environment and ecosystem. The High exposure of waste area in Kemijen, which doesn't have retrieval facility so still put the wastes in the neighborhood area, are located in RW 5 and 7.

Based on the three elements of exposure (i.e. Flood, Land Subsidence, and Waste and Sewerage), were found several places which are exposed in term of vulnerability assessment in Kemijen. These places are related to the level of the exposure. Table 3 shows the scoring analysis for the Exposure assessment in Kemijen.

Table 3. Scoring Analysis for Exposure

RW	Floods Exposure Level	Land Subsidence Exposure Level	Wastes Exposure Level	Total Score of Exposure
1	2	1	1	4
2	1	1	1	3
3	3	2	2	7
4	3	3	1	7
5	3	3	3	9
6	1	3	1	5
7	1	2	3	6
8	1	1	1	3
9	1	1	1	3
10	2	1	1	4
11	2	1	1	4

Source: Analysis Result, 2017

3.1.2. Sensitivity

Sensitivity is always linked with exposure in the respect of vulnerability assessment. Measuring sensitivity is rather difficult than measuring exposure. Sensitivity measures the effects of perturbations or external stresses to a system directly and/or indirectly, as referred by Adger [29]. Other researchers describe sensitivity as susceptibility/fragility that influences the tendency of environmental physical, human, and infrastructure in the region [30]. It means when a dangerous phenomenon (e.g. floods) struck one region then the system within this region (i.e. environmental physical, human, and infrastructure) will collapse or experience major harm and damage since its lack of resistance and predisposition of society and ecosystems. Thus, sensitivity assessment in Kemijen based on three elements: 1) location; 2) people, and 3) access to infrastructure.

a. *Location.* In term of location, the impact of climate change and all the problems that are happened in Kemijen possibly caused by land use change in Kemijen. Land use change is one of the aspects that is sensitive and could indirectly affect due to the frequent floods occurred in Kemijen. For example, areas that are very prone to flood-affected land use may be changed from settlement become other land use. Related to Banger Polder plan, some parts of the nearby swamps, as on the RW 3,4, and 5, are planned to be used as retention basin. Due to this plan, it will change the existing land use from settlements become retention basin by relocating houses.

b. *People.* Sensitivity assessment may focus on a few specific segments of the population. Sensitivity that affects people is related to their socio-economic aspect because this aspect also come into affected indirectly. This socio-economic aspect will talk about people who are sensitive in Kemijen based on age groups and levels of their economy. In term of social aspect, it can be known that the residents who are sensitive to problems in Kemijen are the residents in a non-productive age group, i.e., the age groups of children (0-14 years) and the age group of the older population (65 years and above) rather than productive age (15-64 years). Specifically, more sensitivity felt by old women.

   Meanwhile, sensitivity to People element in term of economic aspect related to livelihoods. The population with the poor economic condition are judged more sensitive to the problems occurred in Kemijen. As the chart shown below, the inhabitants of Kemijen are rated quite sensitive in terms of the economy because the majority of the population worked as laborers and merchants who have income under the UMR (Regional Minimum Salary). Thus, based on the combination of social and economic aspects, is known that old women with low economic condition are relatively the most sensitive people in Kemijen.

c. *Access to Infrastructure.* Physically, the infrastructure assets utilized by the inhabitants of Kemijen are having a direct exposure as already described in the previous discussion, while the sensitivity is the access to the infrastructure because the residents will feel difficult indirectly in accessing infrastructures such as roads and water supply networks.

Based on the three elements of sensitivity (i.e. Location, People, and Access to Infrastructure), were found several places which are sensitive in term of vulnerability assessment in Kemijen. These places located near the river, near dumping site (both formal and informal), flooded area, slum area, and the area with high rate of land subsidence. Table 4 shows the scoring analysis for Sensitivity assessment in Kemijen.

Table 4. Scoring Analysis for Sensitivity

RW	Near River	Near Dumping Site	Flooded Area	Slum Area	High LS	Total Score of Sensitivity
1	V		V			2
2		V	V			2
3		V	V	V	V	4
4		V	V	V	V	4
5	V	V	V	V	V	5
6	V		V		V	3
7	V		V			2
8	V	V	V			3
9	V	V	V			3
10	V		V			2
11	V		V			2

Source: Analysis Result, 2017

3.1.3. Adaptive Capacity

Adaptive capacity will provide an analysis to show the level of adaptability in Kemijen City Village to overcome the problems. There are several adaptive capacity levels according to Adger and Vincent [19], ranging from individual, community, and nation level. In its implementation, adaptive capacity is more easily done at the individual level than the national level. Adger and Vincent [19] mentioned that the ability of adaptation of society is determined by the economic ability, knowledge, experience and community response to environmental conditions. According to the exposure mentioned before, there are 3 problems in Kemijen Area, which are floods, land subsidence, and waste that classified into three types of adaptive capacity:

a. *Individual Adaptive Capacity (IAC)*
Individual adaptive capacity is the act of an individual or family in an effort to protect the livelihoods and assets from the possible problem of climate change.
b. *Community Adaptive Capacity (CAC)*
Community Adaptive Capacity is an ability or action by certain groups. The emphasis is on the collective rather than individual or family.
c. *Government Adaptive Capacity (GAC)*
Government Adaptive Capacity is the ability of adaptation refers to the organizational ability. The realization of this adaptation can be a program, policy, regulatory, human resources and technical expertise from both local and central government, and civil society groups. The scale of the adaptation is quite spacious, and very systemic aims, as well as the long term.

The big problems found in Kemijen were floods, land subsidence, and waste. Adaptive attitude is needed to solve the problems. Each adaptive capacity to overcome the floods, land subsidence, and waste problems is shown in the table below.

Table 5. Adaptive Capacity Activities in Kemijen

<table>
<tr><th></th><th>FLOODS</th><th>LAND SUBSIDENCE</th><th>WASTE</th></tr>
<tr><td>IAC</td><td><ul><li>The ability of household to prevent water from entering into the house by giving sand bags on their doorsteps.</li><li>The ability of people to buy the individual pumps for their house</li><li>The ability of household to elevate their houses.</li><li>The ability of household to raise their house into 2 or more floors.</li></ul>All the people have own adaptive capacity to cope the floods.</td><td>There is no specific individual adaptive capacity for land subsidence.</td><td>The waste management is very good on household level. Every household keeps clean their surroundings such as house and street, but they dump the waste in open area which are close to the river or into the fish ponds.

In Kemijen City Village, all the people have own adaptive capacity to cope waste problems.</td></tr>
<tr><td>CAC</td><td><ul><li>The community set up their pumping houses in each RT. Some households in Kemijen City Village buy a pump that aiming to pump water from their RT.</li><li>RT has a contribution from the community used for maintenance and operation of that pumps. The contributions are varied, among IDR 2.500 – IDR 15.000. There are also neighborhood who do not contribute regularly, so they will contribute only if the pump was broken and in need of repair</li><li>The community initiates to help the others, because of their sense of mutual help for unable people. The community collect money to buy individual pumps for unable people.</li></ul>From 11 RWs in Kemijen, there are 2 RW that did not have that CAC, there are RW 3 and RW 4</td><td>There is no specific community adaptive capacity for land subsidence:</td><td>In RW 1, RW 2, RW 3, RW 4, RW 6, RW 8, RW 9, RW 10, and RW 11, most of the waste collection system on the RT level is actually very good. Most people reported that waste was collected from their home every 1-2 days. This is done informally by community members for an IDR 7.000-10.000 a month.

But in RW 5 and RW 7, they have no retrieval facility. So they still put the waste in the neighborhood area.</td></tr>
<tr><td>GAC</td><td><ul><li>Community disaster alert
The government formed a community disaster alerts.</li><li>Banger Polder System
It will protect more than 84.000 inhabitants in the area of 527 Ha. It will also protect the assets of some prominent stakeholders such as Indonesia Railway Company, state owned oil company Pertamina, and other major companies inhabiting Banger area. The companies as well as the inhabitants of Banger area will be benefited from the project, particularly to be free from the daily floods which has been enormously disturb their assets.</li></ul>Currently the implementation of Banger Polder System is about 25%, in which only affected for RW 6, RW 7 RW 8, RW 9, and RW 10.</td><td>Government adaptive capacity for land subsidence is by making policy or regulation to limit the ground water extraction</td><td>The GAC for waste is just in RW 9, which The Government take the trash from the temporary disposal to the final disposal</td></tr>
</table>

Source: Analysis Result, 2017

The scoring calculation is based on the indicators of these variables. All RWs in Kemijen City Village has adaptive capacity, although there are some RWs that have low abilities of adaptive capacity. Here is the summary of adaptive capacity from 3 problems in each RW:

Table 6. Scoring Analysis for Adaptive Capacity

RW	**Adaptive Capacity**									**Total Score Of Adaptive Capacity**
	Individual AC			Community AC			Government AC			
	Flood	Land subsidence	Waste	Flood	Land subsidence	Waste	Flood	Land subsidence	Waste	
1	1	0	1	1	0	1	0	1	0	5
2	1	0	1	1	0	1	0	1	0	5
3	1	0	1	0	0	1	0	1	0	4
4	1	0	1	0	0	1	0	1	0	4
5	1	0	1	1	0	0	0	1	0	4
6	1	0	1	1	0	1	1	1	0	6
7	1	0	1	1	0	0	1	1	0	5
8	1	0	1	1	0	1	1	1	0	6
9	1	0	1	1	0	1	1	1	1	7
10	1	0	1	1	0	1	1	1	0	6
11	1	0	1	1	0	1	1	1	0	6

Source: Analysis Result, 2017

Human adaptive behavior toward things can be formed after going through some of the events they have experienced. The experience of the flood disaster has been felt by the community in Kemijen since years ago. The elevation of the houses and roads is done every five years as a form of adaptation of local communities. Affected by floods especially during the rainy season also makes people have pumps to pump water out of the settlement environment. Up until now, there is at least one pump in each RT to pump water from houses to rivers. By looking at these development needs, communities are also increasingly experienced in finding development funds and submitting aid proposals to government or CSR firms around Kemijen.

Initially, the people of Kemijen did not have sufficient knowledge about disaster response and how to adapt to the flood. The level of education is also still relatively low, by seeing that the majority of the population of Kemijen is an elementary school graduate. After some development in Kemijen City Village involving the community, adaptive capacity of the community has increased. The development is not only from the government, but also from NGOs and government CSR programs. The development involving the community includes the construction of Banger Polder, PLPBK (Neighborhood Development), village improvement program by Hysteria, etc.

3.1.4. Scoring Result

After being discussed and assessed at every stage of Exposure, Sensitivity, and Adaptive Capacity, then they are summed up to obtain an assessment of identified vulnerability. Based on the all scoring calculations in each component of vulnerability assessment, then being calculated the vulnerability assessment using the previous formula. The result can be seen in the table below:

Table 7. Scoring Result to Identified Vulnerability in Kemijen

Area	Exposure (E)	Sensitivity (S)	Adaptive Capacity (AC)	Vulnerability = (ExS)/AC
RW 1	4	2	5	1.6
RW 2	3	2	5	1.2
RW 3	7	4	4	7
RW 4	7	4	4	7
RW 5	9	5	4	11.25
RW 6	5	3	6	2.5
RW 7	6	2	5	2.4
RW 8	3	3	6	1.5
RW 9	3	3	7	1.3
RW 10	4	2	6	1.3
RW 11	4	2	6	1.3

Source: Analysis Result, 2017

The scoring result stated that RW 5 is the most vulnerable region compared to other neighborhoods. Additionally, RW 3 and RW 4 also have a fairly high susceptibility values. Because of Exposure and Sensitivity assessment, this area is quite exposed and sensitive to the issues, while it is still lack of Adaptive Capacity.

Figure 1. Map of Vulnerability Level in Kemijen

IOP Conf. Series: Earth and Environmental Science **123** (2018) 012007 doi:10.1088/1755-1315/123/1/012007

3.2 Kemijen's Development

Some of the developments are carried out by the government, NGOs, and CSRs, that bring impact to the local people of Kemijen. The impact is felt by the community starting from the time of development of Polder Banger System which mostly exists in Kemijen. Afterward, Kemijen became one of the pilot projects of PLPBK (Neighborhood Development) in Semarang City on 2014. There are pros and cons on such physical development, such as road elevation activities in PLPBK. In terms of aesthetics, they look good, but less supportive on environmental. The elevation of roads leads to drowning homes and not providing solutions for floods. In order to flood prevention, it is important to pay more attention to the drainage network of settlements rather than merely elevating the road periodically. On the other hand, non-physical programs such as the establishment of Disaster Preparedness Group (KSB), trainings and socializations are considered to be more supportive and encourage community participation to engage in environmental conservation in Kemijen. The establishment of the Banger Polder Management Board (BPP Banger SIMA) also has an impact on the formation of local adaptive capacity. Residents are involved in the Banger Polder Board and in the management of Banger Polder System. This is done to increase knowledge and empower the community so that people aware to protect their environment.

The Banger Polder Management Board created a mission and vision of "dry feet for all" with the auspicious goal of ridding the residents of Kemijen from coastal inundation and rainwater floods. In order to achieve this goal, the board did socialization to the residents of Kemijen since 2001. Due to the board is a non-profit organization, they sought a way how to held socialization to the residents in the effective and smart way. In this case, they approached the local women's organization in Kemijen, called Empowerment of Family Welfare or Pembinaan Kesejahteraan Keluarga (PKK). Every month, PKK held a regular meeting in order to increase knowledge and educate women in the neighborhood.

Based on the interview, the board explained the reason they choose the local women's organization as following: 1) generally, women have responsibility to keep their family and children; 2) women also have responsibility to manage their family's financial; 3) women hold responsibility to keep the cleanness of their neighbourhood (e.g. person usually throw garbage to the river is woman); 4) women usually work to gather neighbourhood's data voluntarily; and 5) empower women to participate maintaining the environment in their neighbourhood.

Since then, the municipality and NGOs also held a program in order to increase the involvement of local community in the planning process. Each city village has one community organization called Badan Keswadayaan Masyarakat (BKM). This organization acts as leading sector and coordinator of the other organizations in the city village (i.e. PKK, youth community group, disaster preparedness group). In Kemijen, BKM already well functioned and have regular activities in order to improve their neighborhood. For example, when the municipality gave fund for improving the infrastructure in Kemijen, all the community groups held a discussion about the locations that need to improve and how to do it. Thus, all of the neighborhood will get the beneficiaries.

4. Conclusion

The assessment of disaster-affected areas is to provide an understanding of disaster risk and management in decision-making for regulation and disaster management. Based on the results of the research, Kemijen is a vulnerable area due to the flood disaster and rain. After the developments were done in Kemijen area, the impact is felt by the community. But it is not affect much in some RWs such as RW 3,4,5 in Kemijen city village which still has a high level of vulnerability. This applies to environmental conditions that are still susceptible to flooding and low adaptive capacity. One of the physical development programs that have been done so far is the elevation of roads, but the program will be less effective from the environmental aspect because it is less support to reduce the impact of the flood.

The more influential impact in Kemijen city villages is non-physical aspects through the socialization and involvement of community participation in disaster management. The existence of such non-physical development has helped to increase public participation not within a short time. Community participation in Kemijen City Village involves more on women, that can be seen from the role of woman local organization (PKK). It conducts regular meetings in order to increase knowledge and educate women in the neighborhood. To maintain sustainable environmental conditions, it needs community participation and cooperation from government, corporations, NGOs in the decision-making of development activities.

5. Acknowledgement

The authors would like to thank the Master of Urban and Regional Development Diponegoro University and University of Hawaii at Manoa who have provided experience and sharing knowledge in joint studio program which was held in October 2016.

6. Reference

[1] Marfai M A and King L 2008 Tidal inundation mapping under enhanced land subsidence in Semarang, Central Java Indonesia *Nat. Hazards* **44** 93–109
[2] Ghoneem M Y M 2016 Planning for Climate Change, Why does it Matter? (From Phenomenon to Integrative Action Plan) *Procedia - Soc. Behav. Sci.* **216** 675–688
[3] Marfai M A, King L, Sartohadi J, Sudrajat S, Budiani S R and Yulianto F 2008 The impact of tidal flooding on a coastal community in Semarang, Indonesia *Environmentalist* **28** 237–248
[4] Espinet X, Schweikert A, Heever N Van Den and Chinowsky P 2016 Planning resilient roads for the future environment and climate change : Quantifying the vulnerability of the primary transport infrastructure system in Mexico *Transp. Policy* **50** 78–86
[5] Berdica K 2002 An introduction to road vulnerability: What has been done, is done and should be done *Transp. Policy* **9** 117–127
[6] Caaij T Van Der, Geografie B, Managementwetenschappen F Der and Nijmegen R U 2016 Semarang ' s Sunken Neighborhood East Kemijen
[7] Ramaker S, Geografie B and Managementwetenschappen F Der 2013 Access to information and social vulnerability in flood prone neighbourhood Kemijen
[8] Gallopín G C 2006 Linkages between vulnerability, resilience, and adaptive capacity *Glob. Environ. Chang.* **16** 293–303
[9] Ford J D, Smit B, Wandel J and MacDonald J 2006 Vulnerability to climate change in Igloolik, Nunavut: what we can learn from the past and present; Vulnerability to Climate Change In Igloolik, Nunavut; Ford, Smit, Wandel, and Macdonald *Polar Rec. (Gr. Brit).* **42** 127–138
[10] Araya-Muñoz D, Metzger M J, Stuart N, Wilson A M W and Alvarez L 2016 Assessing urban adaptive capacity to climate change *J. Environ. Manage.* **183** 314–324
[11] IPCC 2014 Summary for Policymakers *Climate Change 2014: Impacts, Adaptation, and Vulnerability. Part A: Global and Sectoral Aspects. Contribution of Working Group II to the Fifth Assessment Report of the Intergovernmental Panel on Climate Change* pp 1–32
[12] Williges K, Mechler R, Bowyer P and Balkovic J 2016 Towards an assessment of adaptive capacity of the European agricultural sector to droughts *Clim. Serv.*
[13] Karlsson M and Hovelsrud G K 2015 Local collective action : Adaptation to coastal erosion in the Monkey River Village , Belize *Glob. Environ. Chang.* **32** 96–107
[14] Wolfson D J 2014 Who gets what in environmental policy ? *Ecol. Econ.* **102** 8–14
[15] Lendering K T, Jonkman S N, Gelder P H A J M Van and Peters D J 2015 Risk-based optimization of land reclamation *Reliab. Eng. Syst. Saf.* **144** 193–203
[16] Makarov A, Mihailova A, Arefiev N, Pavlov S, Chashchina T, Terleev V and Badenko V 2015 Country Area Territory Protection from Flooding ; Construction Conditions , Problem Definition and Solution *Procedia Eng.* **117** 225–231

[17] Marx A 2002 Uncertainty and social impacts A case study of a Belgian village **22** 79–96
[18] USAID 2014 Kajian Kerentanan terhadap Perubahan Iklim Kota Manado *United States Am. Int. Dev. Agency*
[19] Adger W N and Vincent K 2005 Uncertainty in adaptive capacity *Comptes Rendus - Geosci.* **337** 399–410
[20] Turner B L, Kasperson R E, Matson P A, McCarthy J J, Corell R W, Christensen L, Eckley N, Kasperson J X, Luers A, Martello M L, Polsky C, Pulsipher A, Schiller A, Turner II B L, Kasperson R E, Matson P A, McCarthy J J, Corell R W, Christensen L, Eckley N, Kasperson J X, Luers A, Martello M L, Polsky C, Pulsipher A, Schiller A, Turner B L, Kasperson R E, Matson P A, McCarthy J J, Corell R W, Christensen L, Eckley N, Kasperson J X, Luers A, Martello M L, Polsky C, Pulsipher A and Schiller A 2003 A framework for vulnerability analysis in sustainability science *Proc. Natl. Acad. Sci.* **100** 8074–8079
[21] Fraenkel J R, Wallen N E and Huyn H H 1993 How to Design and Evaluate Research in Education *PsycCRITIQUES* **38** 710
[22] Folke C 2006 Resilience: The emergence of a perspective for social-ecological systems analyses *Glob. Environ. Chang.* **16** 253–267
[23] UNISDR 2009 2009 UNISDR Terminology on Disaster Risk Reduction *Int. Strat. Disaster Reduct.* 1–30
[24] Cardona O D, van Aalst M K, Birkmann J, Fordham M, McGregor G, Perez R, Pulwarty R S, Schipper E L F and Sinh B T 2012 Determinants of Risk : Exposure and Vulnerability Coordinating *Manag. Risks Extrem. Events Disasters to Adv. Clim. Chang. Adapt.* 65–108
[25] Mulyana, Wahyu; Daniel D D & S Z S 2013 Climate vulnerability and adaptation in the Semarang Metropolitan Area: a spatial and demographic analysis *Tech. Brief.* **October 20** 4–7
[26] Dewi A 2007 Community-Based Analysis of Coping with Urban Flooding: a Case Study in Semarang, Indonesia *Int. Inst. Geo-Information Sci. Earth Obs. Enschede, Netherlands* 79
[27] BPS 2005 *Semarang Dalam Angka 2005*
[28] Puslitbang SDA B L K P U 2009 Pedoman Polder Perkotaan Volume 4: Studi Kasus Polder Banger, Semarang Semarang, Maret 2009 **4**
[29] Adger W N 2006 Vulnerability *Glob. Environ. Chang.* **16** 268–281
[30] Tan B, Uk V M, Uk M P and Germany J P 2012 Determinants of Risk : Exposure and Vulnerability Coordinating Lead Authors : Lead Authors : Review Editors : Contributing Authors : 65–108

Assessing Evacuation Route Against Mount Merapi Hazard By Using Least Cost Path Method in Mriyan-Boyolali, Indonesia

L K R Putri[1] and M Maryono[1]

[1]Department of Urban and Regional Planning, Diponegoro University, Semarang 50275, Indonesia

Email: laraskrputri@gmail.com

Abstract. This research aims to analyze the Least-Cost Path, an ArcGIS raster analysis, as a tool to find the evacuation route. The evacuation route of mount Merapi Eruption in Mriyan-Musuk of Boyolali regency in the scenario of sister village with Kiringan Village was selected as case study. There are 4 steps of analysis in this method such as; cost surface, cost distance cost backlink and cost path. This study analysis found 3 alternative routes which estimates 25-30 minute. These alternatives evacuation route are useful in distributing the flow of people when the eruption occurs.

Keywords: Evacuation route, GIS, Least cost path analysis

1. Introduction

Learning from the evacuation during Merapi volcano eruption 2010, fostering preparedness in for evacuation process in disaster management very needed [1]. Moreover, preparedness research to foster disaster resilience has been conducted and more interesting in Disaster management. It is not only for academic view side but for practical. However, in Indonesia, the study tends to focus on tsunami and earthquake. The topic mostly in general preparedness [2–4] and community preparedness [5–8]. The study concerning to the volcano eruption preparedness is still very limited [9,10]. With the high potential in volcano hazard, finding the optimum evacuation route for volcano victim not yet discussed [1].This study aims to fill the gap by exploring the alternative routes to foster volcano evacuation preparedness. The route will be explored to distribute the flow of people, vehicle and farm animals during evacuation. In addition, community surrounding needs a specific route and shelters for their farm animal.

The Least-Cost Path analysis is a raster analysis to find route from one point to another. Several research have been utilized this method for example utilize this method for assessing the corridor redundancy [11], routeing of power line [12], and greenway planning [13]. Different with the previous study, in this analysis variable of 'road width' will be proposed and examined due to it is not only for examined about fasting time, but also account of the number of people. Moreover, the variable of 'road width' also reflects the road capacity. As of the previous study, the time needs to evacuate is reflected in variable of 'edge length', 'road damage', and 'surface type'.

The goal of this study is to examine the accuracy of the Least Cost Path Analysis method to find alternative evacuation route. The specific parameter that will be examined is road capacity. Mriyan-Musuk of Boyolali regency in the scenario of sister village with Kiringan Village is selected as case study.

2. Method

This study utilized GIS approach that have been used on previous study in flood evacuation route in Bengawan Solo flood [14] and volcano mudflow evacuation [15]. The concept of this analysis is choosing the least-accumulated value of cells to go from one origin point to the destination point. To run the analysis, there are 4 steps to do [16,17]: (1) analyze the cost surface, (2) analyze the distance cost, (3) analyze the cost backlink, and (4) analyze the cost path. The first step was done by running the 'weighted-overlay' tool under the 'Overlay' toolset under the 'Spatial Analyst' toolbox. The cost surface was used as the base calculation. In this study, the variables were converted from polygon to raster with conversion tool. After that, the variable rasters were summed with weighted-overlay. The weight and score for the variables are described in the following table:

Table 1. Weight and Score of Variable

No.	Variables	Weight	Class	Score
1.	Road width	0.35	0.5 – 2.5 m (1 motorcycle)	80
			3 – 4 m (2 motorcycles or 1 car)	60
			4.5 – 5 m (1 car and more than 1 motor cycle; 2 cars)	40
			>5 - <= 6 meter (2 cars and more)	30
			>6 meter (more than 2 cars)	10
			not road	1000
2.	Road	0.3	0	10
	damage		1 – 3	20
			4 – 6	30
			7 – 9	50
			10 – 12	70
			13 – 15	80
			not road	1000
3.	Edge length	0.25	<=100	10
			>=100 - <200	20
			>=200 - <300	30
			>=300 - <400	40
			>=400 - <500	50
			>=500 - <1000	60
			>=1000 - < 1500	70
			>1500	80
			not road	1000
4.	Surface type	0.1	Asphalt	20
			Concrete	30
			paving block	40
			asphalt/concrete with grass/soil	60
			unpaved (soil, grass, and gravel)	80
			not road	1000

Sister village policy is a community-based disaster risk management where people are encouraged to carry out their evacuation independently. Thus, most people prefer to use motorcycle and personal car to carry themselves and neighbours. Hence, the road capacity should be considered which is reflected in the use of width road variable. Manual Capacity of Indonesia Road [18] method is carried out to give the score of the road damage variable every 100 meter segment. The damage for roads having a less than 100 meter is calculated with the same method.

The second step of the analysis is calculating the distance cost. This was done by running the 'Cost Distance' tool under the 'Distance' toolset under the 'Spatial Analyst' toolbox. This tool calculated the total value (total cost) of each cell from the origin point. From here, the sum of value of cells is gained to determine the chosen cells to reach the destination from the origin point. The third step was calculating the cost backlink. This was done by running the 'Cost Backlink' tool under the same toolset and toolbox. The result of this step was a raster output showing the cells direction in 8 colours. This tool showed the direction from one cell to the next cell from an origin point until reach the destination point. The last step is calculating the 'Cost Path'. This was done by running the 'Cost Path' tool under the same toolset and toolbox. The input of this tool is the output of 'Cost Distance' and 'Cost Backlink'. The result of this step is a raster output showing the cells choose forming a route from the origin point to the destination.

3. Study Area

To explore the model within least cost path, Mriyan Village, Musuk District, having Kiringan village was as the pair selected as case study. Mriyan people understand and have learned the socialized evacuation process and capable to evacuate such as in 2010. In the previous event, people took turns in moving downward during five to seven days. When the activity of Merapi increases, kentongan (drum made from bamboo or wood) in mosques are sounded as early warning system. People then gather in the mosque to allocate the vehicle, then move to the assembly point. In here, people allocate the vehicle to carry the family and neighbors.

Mriyan village is separated by the cliff resulting two areas, the southern and the northern part. Each part has an assembly point and an evacuation route. The assembly point in the southern part is the Mriyan village Hall while the northern part is the Mriyan 2 Elementary School. The both evacuation routes come together in the node named Drajidan. From Drajidan, the both start to pass the same route to the destination. This study only takes the northern part for running the analysis. Figure 1 below shows the map of the study area.

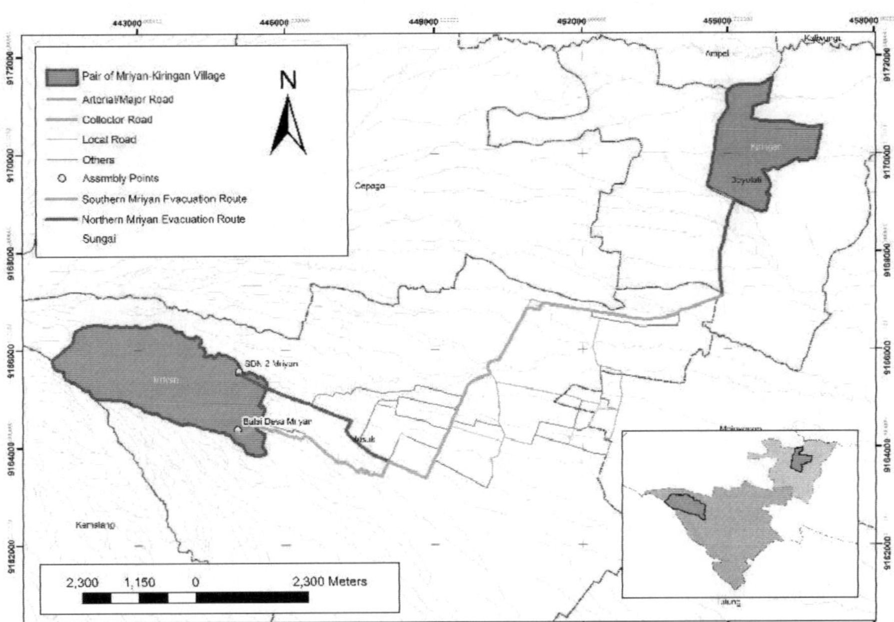

Figure 1. Study Area, Mriyan Assembly Point, and Mriyan Existing Evacuation Route (Interview with Village Chief, 2017).

4. Data

The road data were obtained through field observation. Generally, roads from Mriyan village to the bottom village tend to be straight. There is not so much bent/turning. The most road pavements are asphalt and concrete. The roads in the upper part are mostly in good condition. The damages are cracking and some ravelling. The road with the worst condition is in Sruni. It is the only access to the Southern Mriyan. The local road condition from Drajidan to Kiringan to Musuk is mostly good enough, but there's a heavy damage in the form of potholes, collapsed, and cracking at once. Mostly, the road width is 3.5 – 4 meter, which is enough for a car and two motorcycles to pass the road side by side safely.

5. Analysis

The analysis started by filling the attribute table in the road polyline dataset. The polyline was then buffered, becoming polygon. To gain the 'not road' polygon, the administration polygon is cut in a hole with the dissolved buffer polygon. And then the two was merged. The 'not road' features were given score. After the attributes are complete, the polygon was then converted to raster with a 'Polygon to Raster' tool under the under the 'To Raster' toolset under the 'Conversion' toolbox. The output raster of the cost surface, cost distance, cost backlink, and cost path are shown in table 2.

Table 2. Route Evacuation Analysis by Using Least Cost Path Method

	Alternative 1	**Alternative 2**	**Alternative 3**
Cost surface	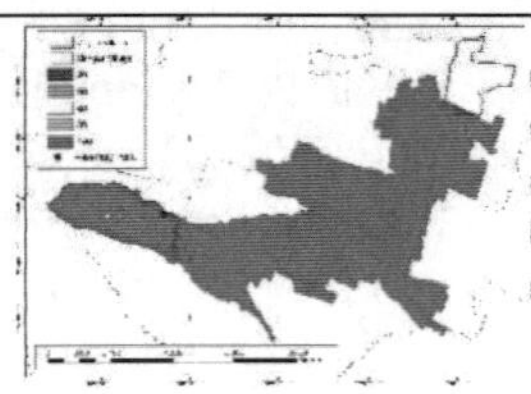	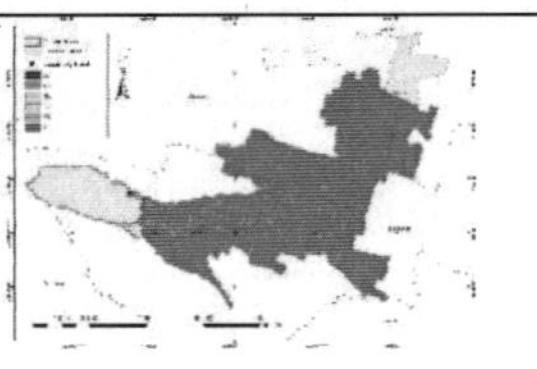	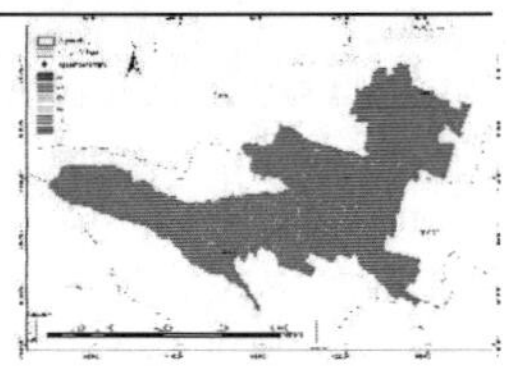
Cost distance	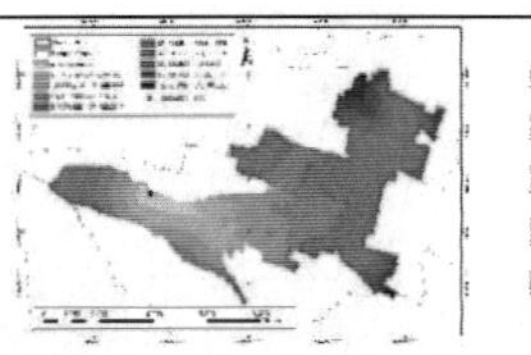	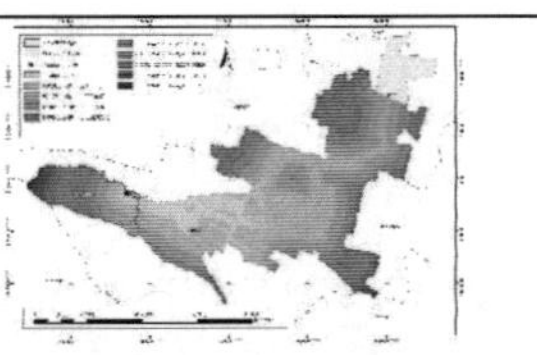	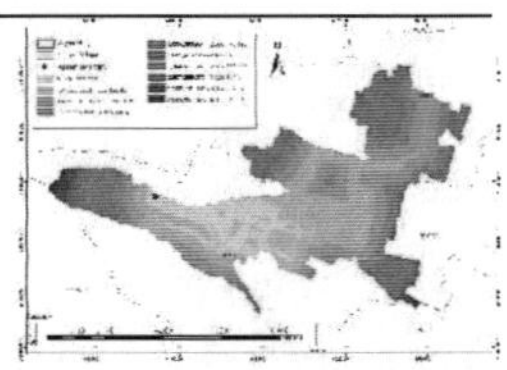
Cost backlink	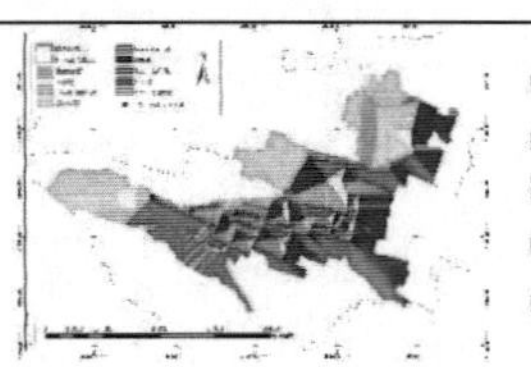	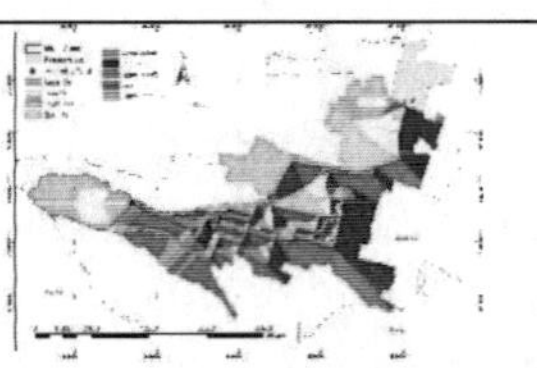	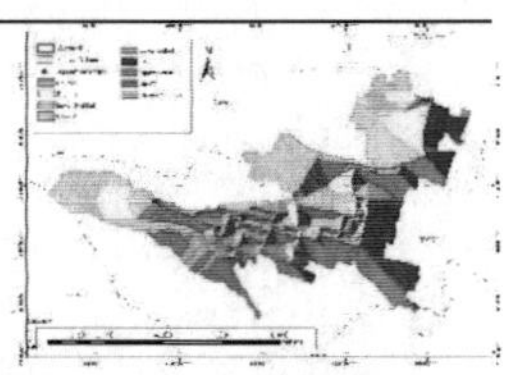
Cost path	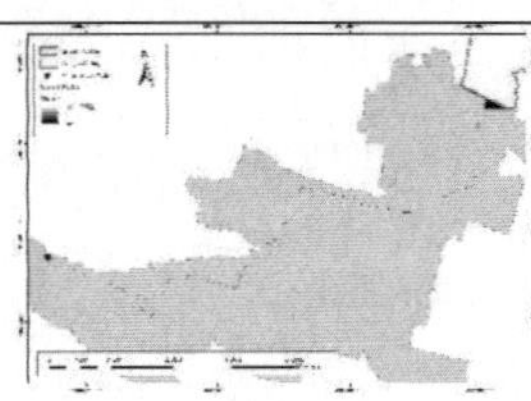	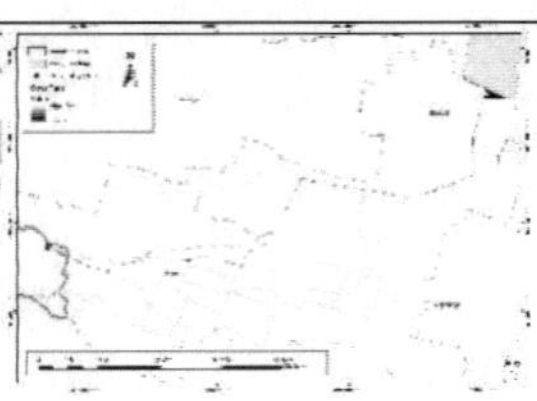	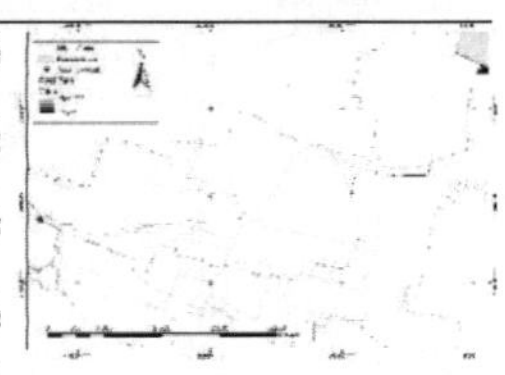

IOP Conf. Series: Earth and Environmental Science **123** (2018) 012008 doi:10.1088/1755-1315/123/1/012008

6. Result and Discussion

The first result (alternative 1) of the tool shows that the recommended evacuation route is similar with the existing. This shows that this analysis is effective to find evacuation route. The second result (alternative 2) is an unpaved moor road, the concrete-grass road, and concrete road. This result provides recommendation to construct the road. In normal situation, the road would help the villager when they're working. In disaster situation, the road would help the people to evacuate. The third result (alternative 3) is moor road. The moor roads are in good condition and already paved with concrete. The result could be seen in the Table 3 as follows :

Table 3. Evacuation Route Iteration and Result

Alternative	Cost Path
Alternative 1	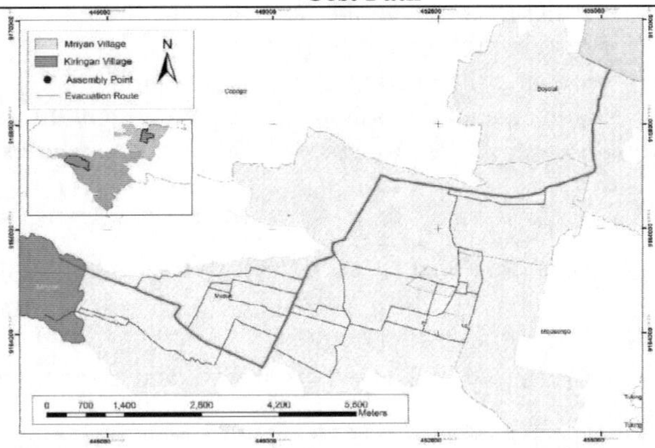
Alternative 2	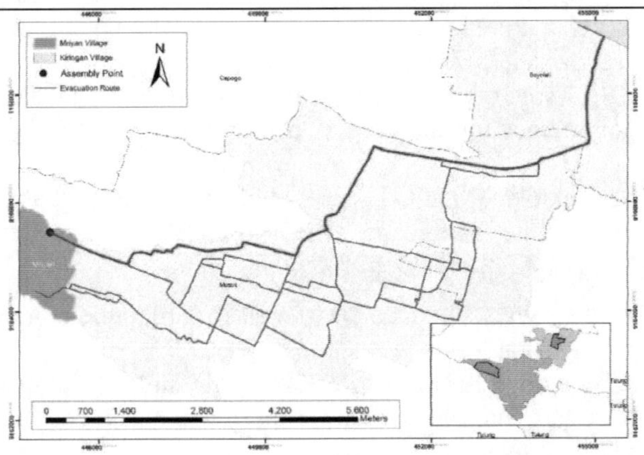
Alternative 3	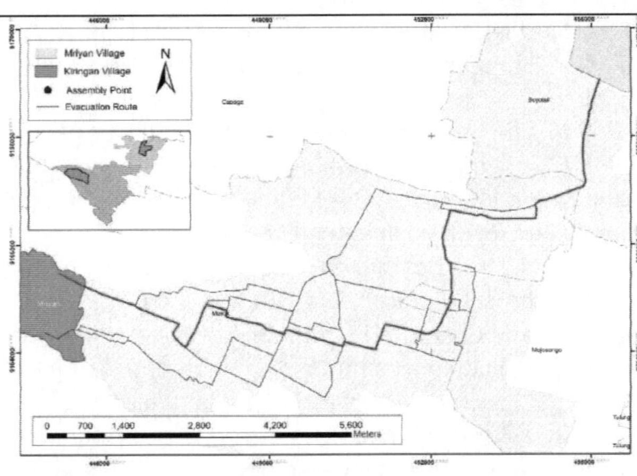

7. Conclusion

The Least-Cost Path is an analysis used to find evacuation routes. By using this method, 3 alternative route from Mriyan to Kiringan is founded as the best alternative route in sister village scenario. This study found analysis 3 alternative routes which estimates 25-30 minute.

8. Acknowledgement

This research is funded Annual Year of Research Budget of Faculty Engineering Diponegoro University by contract No. 170/SK/UN7.3.3/V/2017. Author would like to thank to the Govermentl of Boyolali Regency for giving opportunity to conduct this research, Mr Suwadi Mriyan Village in chief for the data and information during volcano eruption, Mr Sukardi for showing the road network and Suparmi for help of housing andfor temporary studio during data assessment dan motorcycle during collection data, Intan Hapsari Hasmantika and Rakan Priamoe Izdihar, during discussion and collection of data. We also express our sincere thanks to Mr. Isnaini for providing a ride to do the survey, Mr. Darno in providing the lodging; Adito Maulana in helping to understand the failed tool in ArcGIS; Muharar Ramadhan and Endo Foury Lovenda in giving understanding about the raster; Novita SL and Ad'dini Syarafina Hanifa in providing me extra batteries.

9. References

[1] Mei E T W, Lavigne F, Picquout A, De Bélizal E, Brunstein D, Grancher D, Sartohadi J, Cholik N and Vidal C 2013 Lessons learned from the 2010 evacuations at Merapi volcano *J. Volcanol. Geotherm. Res.* **261** 348–365

[2] James E 2008 Getting ahead of the next disaster: recent preparedness efforts in Indonesia *Dev. Pract.* **18** 424–429

[3] Esteban M, Tsimopoulou V, Mikami T, Yun N Y, Suppasri A and Shibayama T 2013 Recent tsunamis events and preparedness: Development of tsunami awareness in Indonesia, Chile and Japan *Int. J. Disaster Risk Reduct.* **5** 84–97

[4] Seng D S C 2013 Tsunami resilience: Multi-level institutional arrangements, architectures and system of governance for disaster risk preparedness in Indonesia *Environ. Sci. Policy* **29** 57–70

[5] Nakayama H, Shimaoka T and Others 2015 Identification of Factors Affecting Stakeholders' Intentions to Promote Preparedness in Disaster Waste Management: A Structural Equation Modeling Approach 九州大学工学紀要 **74** 79–98

[6] Hidayati D 2012 Striving to reduce disaster risk: vulnerable communities with low levels of preparedness in Indonesia *J. Disaster Res. Vol* **7** 75

[7] Marhaento H and Kurnia A N 2015 Refleksi 5 Tahun Paska Erupsi Gunung Merapi 2010: Menaksir Kerugian Ekologis Di Kawasan Taman Nasional Gunung Merapi *Geoplanning J. Geomatics Plan.* **2** 69–81

[8] Marhaento H 2016 GIS-Based Analysis for Assessing Landslide and Drought Hazard in The Corridor of Mt. Merapi and Mt. Merbabu National Park, Indonesia *Geoplanning J. Geomatics Plan.* **3** 15–22

[9] Peterson D W 1996 Mitigation measures and preparedness plans for volcanic emergencies *Monitoring and Mitigation of Volcano Hazards* (Springer) pp 701–718

[10] Sagala S, Okada N and Paton D 2009 Predictors of intention to prepare for volcanic risks in Mt Merapi, Indonesia *J. Pacific Rim Psychol.* **3** 47–54

[11] Pinto N and Keitt T H 2009 Beyond the least-cost path: evaluating corridor redundancy using a graph-theoretic approach *Landsc. Ecol.* **24** 253–266

[12] Bagli S, Geneletti D and Orsi F 2011 Routeing of power lines through least-cost path analysis and multicriteria evaluation to minimise environmental impacts *Environ. Impact Assess. Rev.* **31** 234–239

[13] Teng M, Wu C, Zhou Z, Lord E and Zheng Z 2011 Multipurpose greenway planning for changing cities: A framework integrating priorities and a least-cost path model *Landsc. Urban*

Plan. **103** 1–14
[14] Harsini S R I 2014 Luapan Sungai Bengawan Solo Di Kota
[15] Ardana D M S and Purwanto T H 2013 Penentuan Jalur Evakuasi dan Dampak Banjir Lahar Dingin Gunung Merapi Magelang, Jawa Tengah *J. Bumi Indones.* **2**
[16] Chang K-T 2006 *Geographic information system* (Wiley Online Library)
[17] Etherington T R and Holland E P 2013 Least-cost path length versus accumulated-cost as connectivity measures *Landsc. Ecol.* **28** 1223–1229
[18] Ministry of Public Works 1997 *Manual Indonesia Road Capacity*

Spatial Analysis of Land Subsidence and Flood Pattern Based on DInSAR Method in Sentinel Sar Imagery and Weighting Method in Geo-Hazard Parameters Combination in North Jakarta Region

Y Prasetyo[1], B D Yuwono[2], and Z Ramadhanis[3]

[1]Remote Sensing and Photogrammetry Lab, Geodesy Engineering Dept, Diponegoro University, Indonesia
[2]Geodesy Lab, Geodesy Engineering Dept, Diponegoro University, Indonesia
[3]Remote Sensing and Photogrammetry Lab, Geodesy Engineering Dept, Diponegoro University, Indonesia

Email: yudo.prasetyo@ft.undip.ac.id

Abstract: The reclamation program carried out in most cities in North Jakarta is directly adjacent to the Jakarta Bay. Beside this program, the density of population and development center in North Jakarta office has increased the need for underground water excessively. As a result of these things, land subsidence in North Jakarta area is relatively high and so intense. The research methodology was developed based on the method of remote sensing and geographic information systems, expected to describe the spatial correlation between the land subsidence and flood phenomenon in North Jakarta. The DInSAR (Differential Interferometric Synthetic Aperture Radar) method with satellite image data Radar (SAR Sentinel 1A) for the years 2015 to 2016 acquisitions was used in this research. It is intended to obtain a pattern of land subsidence in North Jakarta and then combined with flood patterns. For the preparation of flood threat zoning pattern, this research has been modeling in spatial technique based on a weighted parameter of rainfall, elevation, flood zones and land use. In the final result, we have obtained a flood hazard zonation models then do the overlap against DInSAR processing results. As a result of the research, Geo-hazard modelling has a variety results as: 81% of flood threat zones consist of rural area, 12% consists of un-built areas and 7% consists of water areas. Furthermore, the correlation of land subsidence to flood risk zone is divided into three levels of suitability with 74% in high class, 22% in medium class and 4% in low class. For the result of spatial correlation area between land subsidence and flood risk zone are 77% detected in rural area, 17% detected in un-built area and 6% detected in a water area. Whereas the research product is the geo-hazard maps in North Jakarta as the basis of the spatial correlation analysis between the land subsidence and flooding phenomena.double point.

Keywords: DInSAR, Flood, Geo-Hazard, Land Subsidence, North Jakarta

1. Introduction

The reclamation program carried out in most cities in North Jakarta is directly adjacent to the Jakarta Bay. Beside this program, the density of population and development center in North Jakarta office has increased the need for underground water excessively.

IOP Conf. Series: Earth and Environmental Science **123** (2018) 012009 doi:10.1088/1755-1315/123/1/012009

Office infrastructure development which is dominated by high buildings has made enormous pressure on land consolidation. As a result of these things, land subsidence in North Jakarta area is relatively high and so intense. In addition to this situation, the emergence of flood disasters that periodically and tends to expand in all around of the area of North Jakarta to be one of the main issues to be solved in this study. The research methodology was developed based on the method of remote sensing and geographic information systems, expected to describe the spatial correlation between the land subsidence and flood phenomenon in North Jakarta. The previous research that examined the land subsidence in Jakarta used the method of measurement Levelling, GPS and InSAR [1]. The research was conducted in 1982 until the 1991 years which was obtained the land subsidence results in Jakarta reaches 3-10 cm per year.

The DInSAR (Differential Interferometric Synthetic Aperture Radar) method with satellite image data Radar (SAR Sentinel 1A) for the years 2015 to 2016 acquisitions was used in this research. It is intended to obtain a pattern of land subsidence in North Jakarta and then combined with flood patterns. For the preparation of flood threat zoning pattern, this research has been modeling in spatial technique based on a weighted parameter of rainfall, elevation, flood zones and land use (Geo-hazard modelling) [2–4]. In the final result, we have obtained a flood hazard zonation model then do the overlay against DInSAR processing results.

2. Data and Methods

2.1. Research Area

The study area of this study is North Jakarta City at latitude 6° 8' 18.29° S and longitude 106° 51' 50.24"E with a width research area around 146.7 km2 and covers 6 districts. The research area can be seen in the Figure 1.

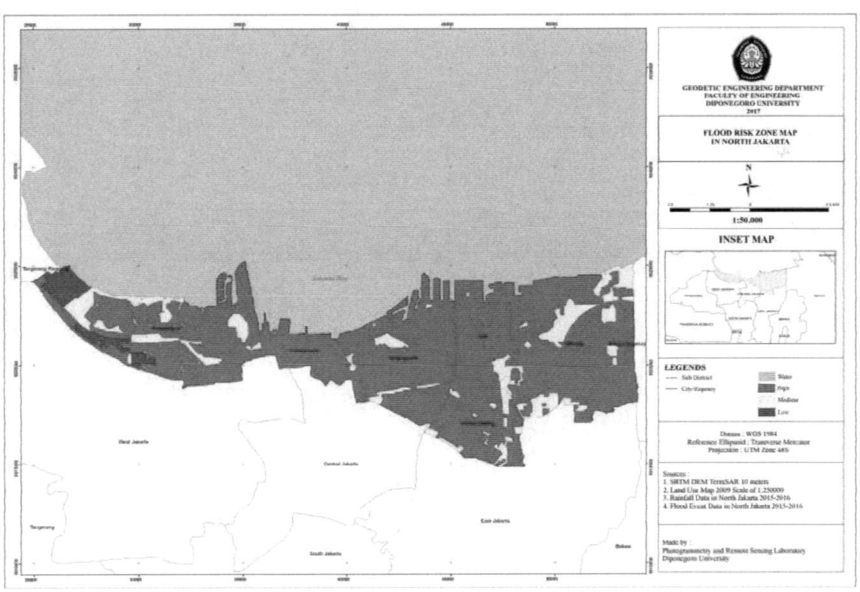

Figure 1. The Flood Risk Zone Map

2.2. Research Data

The data needs related to the research using a combination of satellite image data including Sentinel and SRTM images combined with attribute data related to the flood phenomenon in Jakarta can be seen in the Table 1.

IOP Conf. Series: Earth and Environmental Science **123** (2018) 012009 doi:10.1088/1755-1315/123/1/012009

Table 1. Research Data

No	Data	Acquisition Time
1.	Sentinel Imagery-1A IW level 1 SLC	2015 – 2016
2.	DEM SRTM TerraSAR	2011
3.	Land Use Map of DKI Jakarta Area	2009
4.	Rainfall Intensity Data	2015 – 2016
5.	Flood Intensity Record	2015 – 2016
6.	GPS Survey Data in Jakarta Area	1997-1999

2.3. DInSAR Method

DInSAR is an acquisition technique of two paired SAR images in combination of complex image data at the same spatial position or slightly different position in the same area by multiplication of multiple conjugations resulting in digital elevation model (DEM) or shifting of the earth surface [5].

The DInSAR method utilizes coherent in phase measurements to obtain distance differences and distance changes from two or more SAR images that have complex values from the same surface. The result of the phase difference produces a new type of image called an interferogram. The interferogram will show whether the area under study is experiencing a decrease in the face of the soil or soil rise. The flowchart of DInSAR data processing can be seen in Figure 2.

2.4. Geohazard Modelling Method

Table 2. Geohazard Parameter and Score

Parameter	Criteria	Score	Weight	Final Weight
Rainfall Intensity (mm/year)	>3000	5	0,25	1,25
	2500-3000	4	0,25	1,00
	2000-2500	3	0,25	0,75
	1500-2000	2	0,25	0,50
	<1500	1	0,25	0,25
Topography Height (m)	< 10	5	0,25	1,25
	10-50	4	0,25	1,00
	50-100	3	0,25	0,75
	100-200	2	0,25	0,50
	>200	1	0,25	0,25
Flooding Deep Zone (m)	< 0,76	1	0,25	0,25
	0,76-1,5	2	0,25	0,50
	> 1,5	3	0,25	0,75
Land Use	Rural Area	5	0,25	1,25
	Building	5	0,25	1,25
	Paddy Field	4	0,25	1
	Paddy Wet Field	4	0,25	1
	Hujan			
	Farm	3	0,25	0,75
	Field	2	0,25	0,50
	Rock and Hill Area			
	Forest	1	0,25	0,25
	Grass	1	0,25	0,25
	Savannah	1	0,25	0,25
	Land Water Area	1	0,25	0,25
		0	0,25	0

For this Geo-hazard modeling was used 4 parameters (Rainfall Intensity, Topography Height, Flooding Deep Zone and Land Use) based on previous research [6] and can be explained in detail related to the assessment and weighting in the Table 2.

2.5. Research Methodology

The research stages can be explained based on the sequence of data processing in Figure 2.

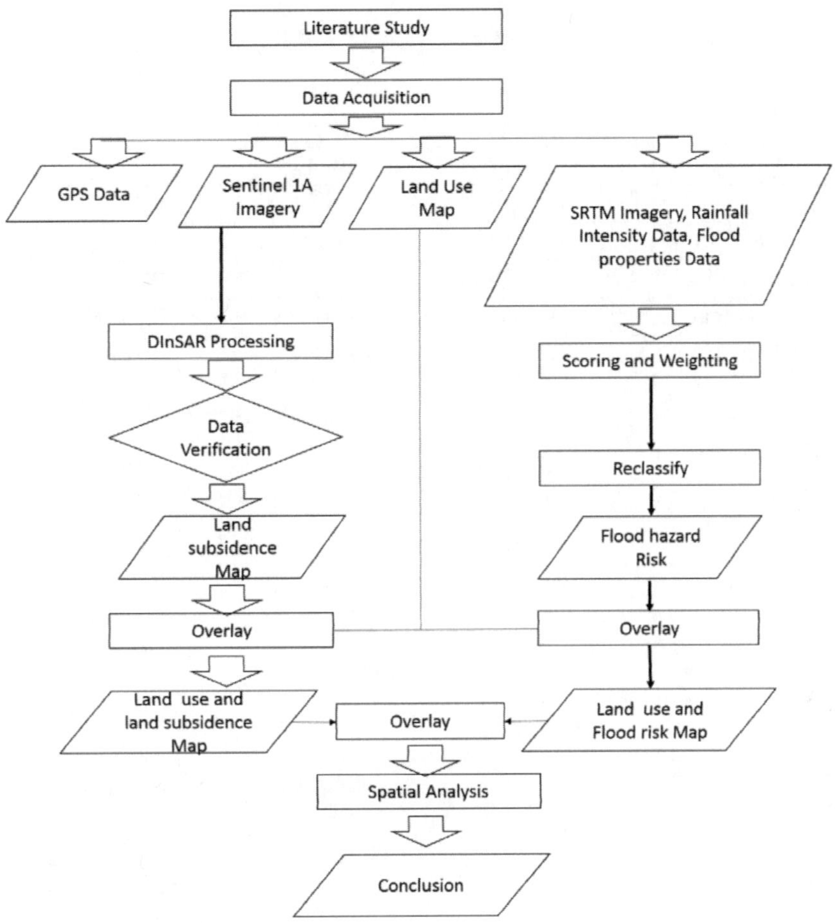

Figure 2. Research Methodology Flowchart

Stages of this research are generally divided into four parts that include the stages of data collection and literature studies, stages of data processing, stages of data validation and stages of data analysis. At the stage of data collection is done by collecting the required data (refer to Table 1) which includes four main parameters namely rainfall intensity data, topography data, flooding deep zone and land use data. Furthermore, at the stage of data processing apply three methods of data processing include DInSAR method, DEM data extraction method, weighting method and interpolation methods of rainfall intensity. The processing result is then validated using GPS data to DInSAR processing data. For data on flood inundation and land subsidence are also verified using field survey. The result of data processing that has been validated and verified then overlaid with land cover data to get spatial pattern of flood inundation and land subsidence. The next stage of data analysis to be drawn general conclusions in related to the correlation between flood inundation and land subsidence in the study area.

3. Result and Discussion

There are several results and analysis based on the stages of data processing which can be explained as follows:

3.1. Land Subsidence Analysis Based On DInSAR Method Processing

Based on the result of processing, the result of the highest land subsidence occurred in Penjaringan sub district in North Jakarta, with the most highest land subsidence around -0.0437 meter/year where the details of the results can be seen in the Table 3 and Figure 3.

Table 3. Land Subsidence Velocity in Sub District on North Jakarta Based on DInSAR Method

Sub District	Average (m/yr)	Minimum (m/yr)	Maximum (m/yr)
Kelapa Gading	0,0821	0,0247	0,1294
Koja	0,0823	0,0331	0,1270
Tanjungpriok	0,0573	-0,0245	0,1082
Cilincing	0,0829	-0,0018	0,1566
Penjaringan	0,0469	-0,0437	0,1118
Pademangan	0,0522	-0,0042	0,1186

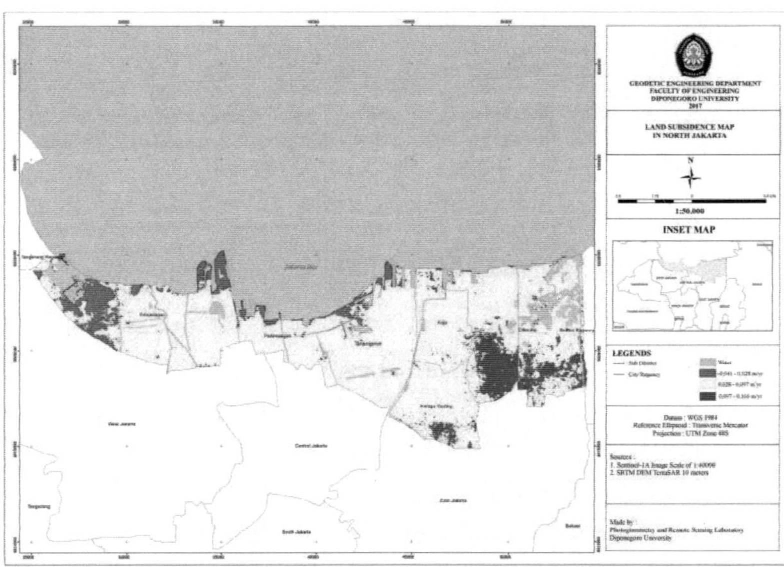

Figure 3. Land Subsidence Map Based on DInSAR Method

3.2. Flooding Hazard Risk Analysis Based On Geo-Hazard Modelling

Based on Geo-hazard modelling [2,3,7], the result of flooding hazard risk can be resulted from 4 parameters (rainfall intensity, topography height, land use and flood deep zone) can be seen in Figure 1 . From that map it can be seen that is the red color representation is a high flood threat zone reaching 82% or 115,706,032.444 m2, the yellow is a zone of medium flood threat that reaches 15% or 21,229,545.740 m2 and the color green Is a low flood threat zone that reaches 3% or an area of 4,350,438.643 m2.

IOP Conf. Series: Earth and Environmental Science **123** (2018) 012009 doi:10.1088/1755-1315/123/1/012009

From the results obtained, that almost all areas of North Jakarta City has high potential for flood threats. This is because almost the whole area of North Jakarta is used as a settlement, while the largest score on land use parameter is settlement with score value of 5. Thus, modeling flood disaster threat generated by weighting method tend to have high flood threat class. This is also evidenced by the percentage of high flood threats that reached 82%.

3.3. Spatial Analysis Between Land Subsidence and Land Use Combination Data

The results of the correlation map between Land Subsidence and Land Use of North Jakarta can be seen in Figure 4.

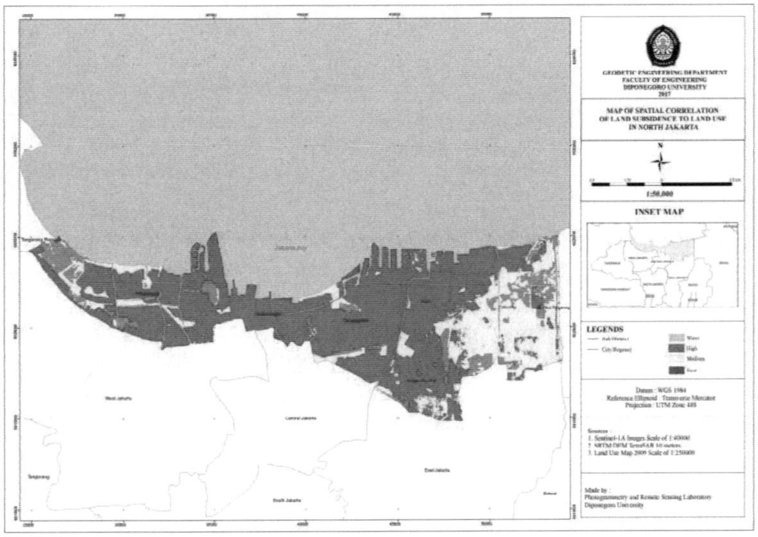

Figure 4. Land Subsidence and Land Use Correlation Map

Based on the results of data processing obtained that the percentage of spatial correlation of the highest land subsidence with residential, industrial, office building and rural area is the highest suitability class (red color indicator), reaching 61% or 86,762,837.094 m2 spread evenly in all districts in North Jakarta City. Meanwhile, for percentage of spatial correlation area of land subsidence with farm, garden, forest and unbuilt area, the second highest land subsidence is medium grade suitability, which is 25% or 34,703,273.64 m2. Meanwhile, for the percentage of the last is the low grade suitability between land subsidence with open water area, reaching 14% or 19,819,906.090 m2.

3.4. Spatial Analysis Between Flooding Hazard Risk and Land Use Combination Data

The purpose of making this map is to analyze the spatial impacts arising from the phenomenon of flood hazard threat againts land use in North Jakarta [4]. The results can be seen in Figure 5. The red color is a representation of the high suitability of spatial phenomena in the form of built areas, such as settlements, industrial estates, transport areas, government areas, and so on. Then, the yellow color is a representation of the moderate suitability of the spatial flood phenomenon in the form of non-built areas, such as: agricultural land and livestock, empty land, mangroves, and so on. Furthermore, the green color is a representation of the low suitability in the form of open water areas, such as: rivers, reservoirs, and so forth.

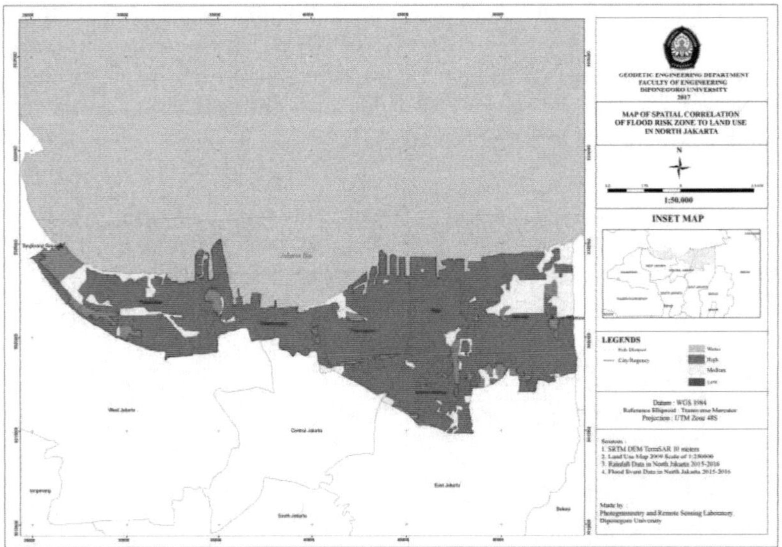

Figure 5. Flood Hazard Threat Zone and Land Use Correlation Map

In Figure 5, it can be seen that the percentage of spatial correlation area of the highest flood hazard threat is the high suitability class, reaching 81% or 114,244,882.530 m2, medium grade suitability class, reaching 12% or 17,365,058.16 m2 and the lowest class, reaching 7% or 9.636.076,131 m2.

3.5. Spatial Analysis Between Land Subsidence and Flooding Hazard Risk Combination Data

The purpose of making this map is to analyze the spatial impacts arising from the phenomenon of land subsidence against flooding hazard threat [8]. The results of this spatial analysis can be seen in Figure 6.

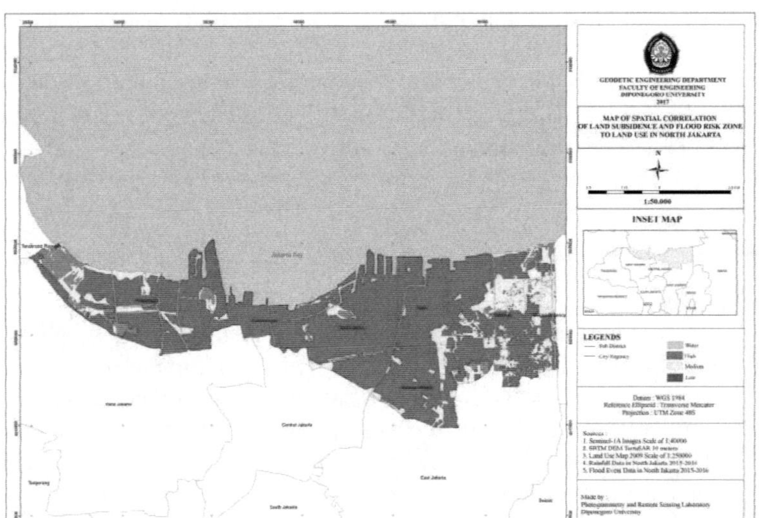

Figure 6. Flood Hazard Threat Zone and Land Subsidence Correlation Map

The red color is a representation of the high spatial suitability of the phenomenon of land subsidence and flooding hazard threat of spatial form of rural areas, such as settlements, industrial estates, transport areas, government areas, and so on. Then, the yellow color is a representation of the moderate conformity of the phenomenon of land subsidence and flooding to spatial form of non-built areas, such as: agricultural land and livestock, empty land, mangroves, and so on. Furthermore, the green color is a representation of the low suitability of open water areas, such as: rivers, reservoirs, and so forth. In Figure 6, it can be seen that the highest percentage of spatial correlation of the highest decrease in soil surface is the high grade suitability, reaching 77% or 108,182,957.001 m2, medium grade suitablity, reaching 17% or 24,392,097.28 m2 and low grade suitability, reaching 6% i or 8,710,962.545 m2.

4. Conclusion

The conclusion in this research based on data results and analysis, it can be explain as follows:

1. The phenomenon of land subsidence can be examined based on the DInSAR processing that is occurred in four sub-districts in North Jakarta. In Specifically the results of the land subsidence measurements can be described as follow: Tanjungpriok at -2.45 cm/year, Cilincing at -0.18 cm/year, Penjaringan at 4.37 cm/year and Pademangan District at -0.42 cm/year. This phenomenon of land subsidence, because most of North Jakarta area is built or rural area. It can be seen from North Jakarta Land Use Map which consists of 71.6% of rural area. The more dense *settlements*, the more densely populated in North Jakarta, so that the occurrence of continuous water taking and causing the land subsidence. In addition, the number of skyscraper building in the form of apartments and industrial sectors will be one big problem in a compaction soil in North Jakarta. It is always accelarate a speed of land subsidence in North Jakarta.
2. The result of the spatial correlation of land subsidence to flooding is a largely correlated. This is because almost all zones of high threat flood disaster detected as land subsidence area. The percentage of high risk flood disaster zone in North Jakarta is 82% with an area of 115,706,032.444 m2 and land subsidence is found in four of the six sub-districts of North Jakarta. In addition, areas that have a corresponding between land subsidence and flood is around 77% with an area of 108,182,952.001 m2.
3. The impact area of the land subsidence and the flood zone is 77% in higher correlation class with an area width around 108,182,952.001 m2, 17% in normal correlation with an area width around 24,392,097.28 m2 and 6% in lower correlation with an area width around 8,710,962.545 m2. That is, because most of the North Jakarta area consists of 71.6% rural area which is compose a solid majority of the residential area, industrial areas, transportation area, district administration area, district office area and social facilities.

5. Acknowledgments

The authors would like to thank the reviewers for their valuable comments. The authors would also like to thank Prof. Ketut Wikantika and Prof Hasanudin Zainal Abidin at Geodesy Engineering Dept-Institute of Technology Bandung for a valuable comment and discussion. And also all staff and researcher in Remote Sensing and Photogrammetry Lab in Geodesy Engineering Dept-Diponegoro University for a field survey and data processing support in this research.

6. References

[1] Abidin H Z, Djaja R, Darmawan D, Hadi S, Akbar A, Rajiyowiryono H, Sudibyo Y, Meilano I, Kasuma M A, Kahar J and others 2001 Land subsidence of Jakarta (Indonesia) and its geodetic monitoring system *Nat. Hazards* **23** 365–87
[2] Abdhika I R 2016 Pemanfaatan Sistem Informasi Geografis Dengan Metode Analytical Hierarchy Process (AHP) Untuk Prediksi Daerah Rawan Banjir Di Kota Semarang *Skripsi, Fak. Ilmu Komput.*
[3] Marfai M A and King L 2006 Impact of the escalated tidal inundation due to land subsidence in a coastal environment *Geophys Res Abstract* vol 8

[4] Pratiwi R D, Nugraha A L and others 2016 Pemetaan Multi Bencana Kota Semarang *J. Geod. Undip* **5** 122–131
[5] Prasetyo Y, Wikantika K and Sumantyo J T S 2012 Data Optimalization In Permanent Scatterer Interferometric Synthetic Aperture Radar (PS-INSAR) Technique For Land Subsidence Estimation *In Proceedings of the AGSE. 221-229. Johor, Malaysia: Applied Geoinformatics for Society and Environment*
[6] Darmawan M and Theml S 2008 Katalog Methodologi Penyusunan Peta Geo Hazard Dengan GIS
[7] Sekretariat Negara Republik Indonesia 2007 *UU No 24 tahun 2007 tentang Penanggulangan Bencana*
[8] Muhamad D S S 2015 SIG untuk memetakan daerah banjir dengan metode skoring dan pembobotan (studi kasus kabupaten Jepara) *Skripsi, Fak. Ilmu Komput.*

Living in Prone Flooding Area: in Coastal Areas of Semarang

W P Tyas[1]

[1]Department of Urban and Regional Planning, Diponegoro University, Indonesia

Email: w.p.tyas@pwk.undip.ac.id

Abstract. When settlements are not able to provide a comfort area to live in, in this case because of a periodic threat of tidal flood coming to certain settlement areas, it is likely that the people still cannot leave the area. This paper explores the leading factors of the attachment of people to the areas, from economic, physical, social and psychological factors, including a place attachment. Therefore, the approach of the problem solution to tackle the tidal flooding in the areas should be also concern and have considerations relate to the factors.

Keywords: Tidal Flooding, Adaptation, Place Attachment

1. Introduction

Semarang is a city that often experienced the flood of tidal flood. The causes are as follows: (1) decreased ground level due to uncontrolled drilling wells; (2) at the same time climate change occurs; (3) In the eastern part of Semarang City's area experiencing faster surface decline due to the industrial area to the east of Semarang has the type industries that requires a lot of water, such as textile; (4) As with routine frequency, the tidal flood are pooled for several hours and continuously. If raining, the flooded will be worsen [1].

The consequence of tidal flood is greatly reduced by decreasing of accessibility level due to congestion, also low level of productivity due to flooding in ports, bus terminals and industrial estates and also activity disruption in settlements in the nearby coastline. The tidal flood has triggers responses both by government and society. For example in Tanah Mas Real Estate, Semarang City which designed for higher income households, the problems are handling tidal flood with the construction of polders, integrated systems of pumping systems in the area affected by the tidal flood.

However, the efforts are still insufficient, the tidal flood is still occured. Therefore, this study explores the communities how they survive in flood prone areas. Many factors seem affect to the adaptability. One of them is place attachment as an emotional binding factor to the place of birth can be a factor that binds the community still want to live in a certain places [2].

This paper explores the determinants of the local people who want to stay in the tidal flood prone area in Tambak Lorok Kampong, Tanjung Mas and Tambakrejo, Semarang will be explored, both economically, physically, socially and psychology. The organization of writing in this paper is the methodology, case studies, discussion, and conclusion.

2. Methodology

This study is a qualitative-based research, through searching data and facts in the field, by observation and interview. The research is carried out in the area that is inundated and observed its impact due to the tidal flood. The interviews were conducted both with tidal flood affected communities of Tambak Lorok kampong and Tambakrejo areas and with government staff of BPBD (Local Board of Disaster Mitigation) and PU (Public Work Agency) of Semarang City in an effort to obtain information related to policies and programs to mitigate flood in Semarang city.

The survey was conducted in 2017 and involved 6 heads of households interviewed to find out the strategies undertaken in tackling the impact of the tidal flood and why they able to stay and adapt with the difficulties due to tidal flood. The explanations of the causal factors persisted were analyzed and compared with the theories and concepts related to the place of attachment [3–5]. At the end is the conclusion also with the economic, social and psychology factors of the community in the installation of tidal floods in the study area.

3. General description of study area: Tambak Lorok Kampong, Tanjung Mas and Tambakrejo Areas of Semarang City

Tambak Lorok Kampong is one of the fishing or maritime kampong in Semarang City located in the coastline of Java Sea. Tambak Lorok Kampong belongs to the administrative area of Tanjung Mas Sub-District which consists of 5 Rukun Warga (RW)s /larger neighbourhood areas.

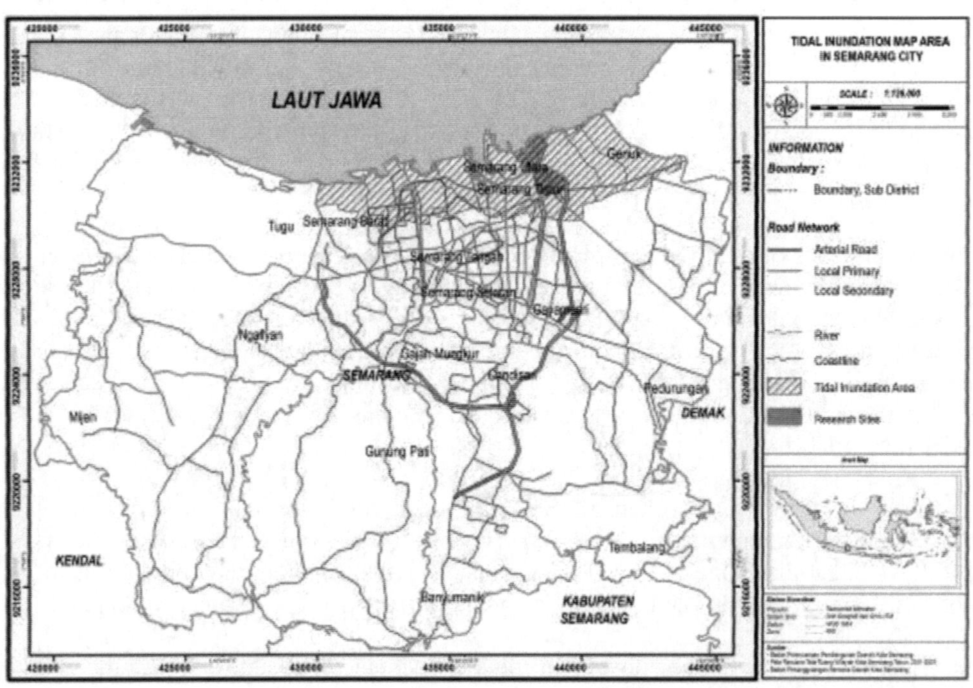

Figure 1. Tidal Inundation Map Area in Semarang City

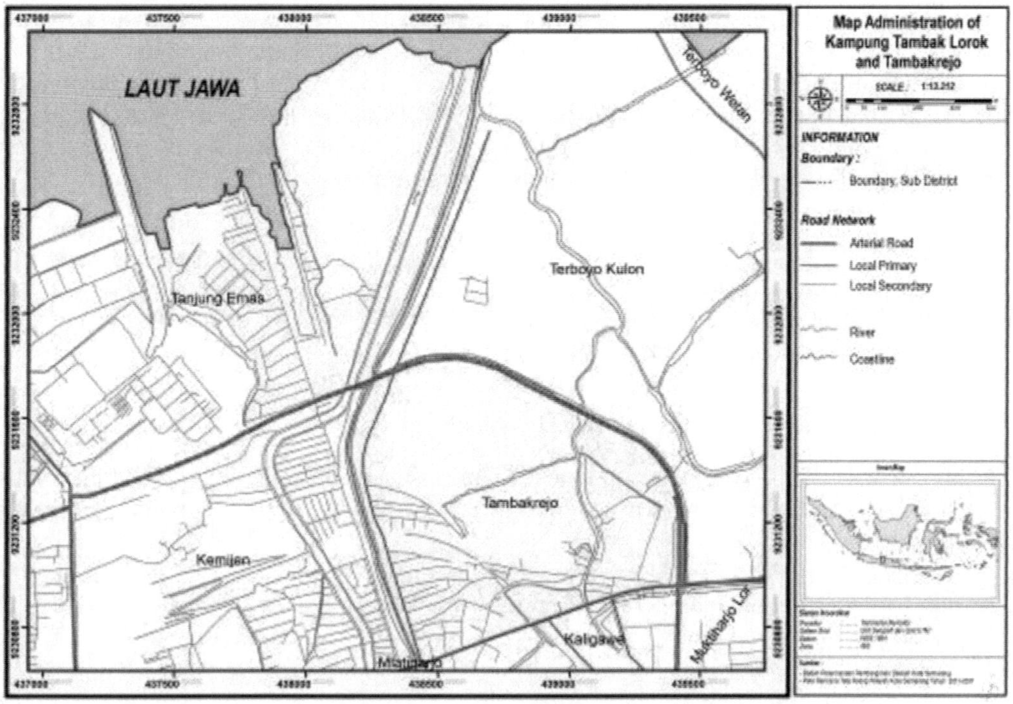

Figure 2. Map Administration of Kampung Tambak Lorok and Tambakrejo

The five RWs are RW 12, RW 13, RW 14, RW 15, and RW 16. Tambak Lorok Kampong itself has borders, namely:

a) North: Java Sea
b) East: Banger River
c) South: Arteries
d) West: Electricity generator (PLTUG)

The main function of the existence of Tambak Lorok kampong is as a residential area, fishpond and special trade for seafood [1]. It is located in the coastal city of Semarang, making the two locations become one of the areas prone to the flood disaster. Based on the results of interviews with one of the residents in Tambak Lorok Kampong, the tidal flood in Tambak Lorok Kampong has been going since 2000s.

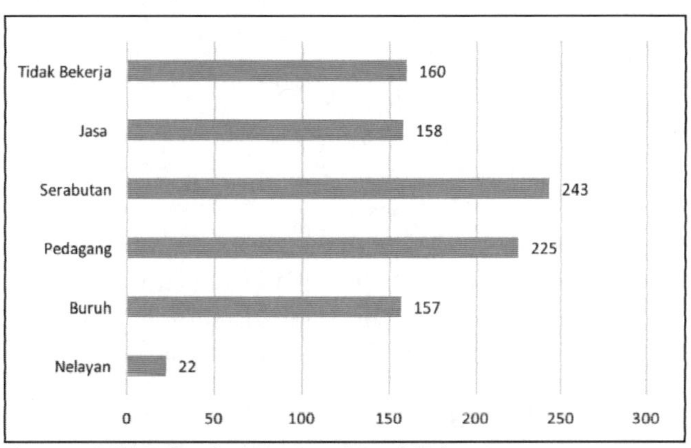

Source: PJM Kelurahan Tanjung Mas, 2014

Figure 3. Type of Jobs of Low-Incomer Family in Tambak Lorok

The occupations of local people are various, but the biggest percentage is part time worker in any jobs and some of them works as merchants, industrial workers, in service or fisherman, but some of them does not have any job (Figure 3). It seems that the local people having a low quality of life but the people of those areas actually feel comfortable to live in areas prone to tidal disasters. There are several reasons why people live in disaster-prone areas, some of which are the majority of the people who live in Tambak Lorok Kampong and Tambakrejo.

4. Affecting Factors of The Survival Ability

This section is the exploration of the affecting factors so the local people can live in the harsh condition. The factors are economic, social and psychological aspects. The discussion of the factors as part of upcoming sections.

4.1. Economic Factor Due to Close to Work-Place

Many local people have an occupation that rely on maritime based occupation, and the closeness of their house to the work place become the main factor in this economic factor. It has been explained by these:

> *"Since I was born I've lived here.... Because my father is a fisherman ...and my mother's father also lives here as well. "*
> *If for example moving house also, the distance my husband's workplace with home later also impacted".* Sunarwati,Tambak Lorok , 59

> *"But because here is already comfortable and my father and I work as a fisherman yes ..want to live here"*. Eki, Tambakrejo, 23

Although he lives at a such not in ideal and even in economic limitation condition, but as he feel comfortable and have an attachment to the place he was born and grows, he can adapt with the difficult situation. This adaptability due to the work place. Turner in his classic books in 1970s [6], has been explains that economic factor is important for low income households, so the location of house has to be close to work place [7].

> *"First because I do not have too much money so usually just buy a house here, the second is because of work factors as well. Because the distance to my workplace is pretty close "*

> *"If for example moved any, do not know where to move...., house in Semarang City and its surrounding area is expensive now. Especially now, the house price which close to the factory is expensive at the majority".* Muhsin, Tambak Lorok, 35, Port Garment Factory Worker in Port Industrial Area

The distant to the job place is also important factor of local people as it need more cost to transport.

4.2. Physical Factor

Another various efforts have been managed by the government and community of Tambak Lorok Kampong to overcome the problem of existing tidal flood. One of the efforts of the community to reduce the impact of the flood disaster is to raise the building and make the house on stage, as an interview with one of the people who live in Tambak Lorok Kampong.

> *"The effort undertaken to reduce the impact of the flood tidal at most just raise or lift-up the building, or make a sand embankment, for some people who have more money may be going to make a house on stage otherwise move out from here ... "*

Source: Survey, 2017

Figure 4. Settlement Conditions In Tambak Lorok Kampong and Tambakrejo

The Figure 4 above shows that floor or houses or even path in Tambak Lorok Kampong have been elevated to avoid inundation from tidal flood that hit the area. Even on the second photo shows that the height of the gangway from year to year increase, the house that used to still have doors and windows now is buried with soil material to elevate the road or gangway in front of the house.

Strategy and adaptation from the side of physical factor is important.

"The effort undertaken to reduce the impact of the flood tidal at most just raise the building or make a dike from the sand, if for people who have more money may be going to make a house on stage otherwise move from here. So if for example ..see many houses canneo be seen from outside, because it already closed by sand to embankment of that houses. Another effort is to repair the house, because every year the house is hit by water tidal the wall can be easily damaged. So it must be repaired continuously. " Warno, Tambak Lorok, 65

the local people put many strategies to solve the tidal flooding impacts problem. Most of them elevate their floor of house or repair the house

"... a lot of effort that I have done to overcome the impact of more severe tidal flood. From tile repairs, wall repair even up the building I have done since the first. Almost every year that I would elevate the building about 10 cm. But in a few years the water has gone up again so every year I have to raise the building so as not to be hit by more severe tidal floods. " Prihono Tambakrejo, 46

The physical intervention of local people is a routine job, although the actions are not taken place every year for every person.

"Yes, but I elevate the building is not every year, I elevate the building every two years, so fit it up I elevate 5-9 cm every two years. The cost to raise the building is much cheaper than the cost of moving house ". Muhson, Tambak Lorok, 35,

The effort to elevate the building by the local community, combine with Government efforts. The government has also made efforts to reduce the impact of tidal disasters in Tambak Lorok Kampong. Some efforts made by the government is to provide clean water from the PDAM/Clean Water local Government Owned Enterprises and make improvements to tidal floods mitigation infrastructure. The provision of clean water is very important for the people of Tambak Lorok and Tambakrejo because if people use the ground water in their settlements, the ground water in their homes is already mixed with sea water caused. An interview with one of the residents of Tambak Lorok Kampong regarding the government's efforts to overcome the problem of the flood disaster in Tambak Lorok Kampong (Figure 5).

IOP Conf. Series: Earth and Environmental Science **123** (2018) 012010 doi:10.1088/1755-1315/123/1/012010

"..the actions of the government is quite a lot of to handle the problem of tidal flood. For roads, most of the existing roads in the area here are many that are paved by the government, even if there is a road that was once stagnant is now in landfill use of land to elevate thus let the road can be passed again. In addition, there is also assistance for the provision of clean water. The clean water here is now using water from the taps, so the taste of the water here is not so salty as it used to be. "

Source: Survey, 2017

Figure 5. Government's Efforts in Reducing The Impact of The Flood Disaster

It seems that the physical condition of settlements in Tambak Lorok Kampong and Tambakrejo are quite bad, such as solid waste management that does not exist and the threat of land degradation every year. The disaster that always approaches the area to make the quality of life in Tambak Lorok Kampong and Tambakrejo very low.

In addition, the government support and also CSR from Private companies in the provision of water supply or pavement uplifting and renovation is very helpful. Based on interview to Public Work Staff (City Drainage Section and also Road and Bridge Section) The flood defence are designed infrastructure, in which formed in integration of drainage with road development and also pumping system, for example an emergency pumping in Kaligawe area and also sea defence and Toll road is part of the effort. This is part of Middle Term of Planning (RPJMD). Infrastructure development is one of the keys that make the flooded can decrease, even to eliminate. So, government policies and programme has a significant support for the physical aspect to solve the tidal flood problem. Both of the effort of local people and external from government and private sectors, has been contributed to create a better place.

4.3. Social Aspects

A mutual working among local people in the community, takes places in the area. A Gotong royong, which involves local people, for example to clean up their surrounding area, is one of cultures that also support the solid waste problems which worsen the flood in blocking the drainage system.

'If works from the kampong usually work devoted to collect garbage.... Because if the tide of sea water worsen to impact of flooding, here too a lot of garbage come into this kampong(residential areas). Sometimes, gotong royong (mutual works), are also devoted to repair the road/gangway here', Eki, Tambakrejo, 23

As part of social capital, that affect to place attachment [8,9] the culture in mutual work of gotong-royong seems contribute to support local people to adapt with their environment.

IOP Conf. Series: Earth and Environmental Science **123** (2018) 012010 doi:10.1088/1755-1315/123/1/012010

4.4. Psychological aspect: Place attachment

An emotional bonding to the place they live seems present to the local people to where they was born.

"Since I was born I've lived here .Yet even if my family and I moved.., there is also difficult adaptation, here there are many my brother who live in the area here as well. " Eki, Tambakrejo,23

Pak Eki feels comfortable, although the condition is not ideal. His family also live there, so he feels that it also contributes to the attachment of the place. A Javanese usually also has a view that 'mangan ora mangan sing penting kumpul'..which means that even they cannot afford to pay meal, the most important thing is that we can live close and together with families.

He feels comfortable to live there because, his his child's school place is also close to the area. Another local people also feels it is fine to live in the area.

"I think it's comfortable - just comfortable. My wife and children also feel comfortable living here. As far as the tidal flood will not go into my house.. feel comfortable.." Muhson, Tambak Lorok, 35

The same answers when several local people interviewed about their condition in tidal flood condition.

"But because here is comfortable, and my father and I work as fisherman.. yes want to and must stay here. Yet even if my family and I move out, there will be also difficult adaptation, here are a lot of my relatives who live in the area here as well. " Sunarwati,Tambak Lorok , 59

An emotional bonding to live even in a very limited condition area contributes an important part to make local people feel comfortable. This is what Low and Altman [2] explain that place attachment as important factor that people still stay in a place, even where considerably not ideal place to live in.

5. Conclusions

The various factors leading to the people still stay in a prone flooded areas. Economic factor become the most important factor, as they seems in a low income group, and has low capability to move from the areas. They feel that to renovate and adapt with put sacks of sands or uplifting the floor is far more easier and cheaper so that is one of their solution. And they workplace, as a fisherman or worker in Port area, is close form their house, they feel better if they live there. For the people who live or even birth at that area, the place attachment is clearer as the family member also live nearby. In limitation, they feel comfortable to live there as as far as the water not inundate house. The work together mechanism in gotong royong (mutual working) in cleaning the neighbourhood area form solid waste is also routine work, which can increase the comfort and need each other.

The place attachment is one that can explain why in such limitation, people can stick to stay in a prone disaster area. Therefore the solution regarding the people who live in prone flooded area is not as simple as to relocate them from the area. As the relocation as not the a favorite solution for government as the higher fund will be needed, also if there is any solution regarding the relocation, it seems that the relocation area should be as close as the current jobs, which are in the coastal area. The effort to solve the tidal flooding has been also need to be more rigorous. A Maritime Village Programme (Kampung Bahari) which is designed in Tambak Lorok, of President Jokowi idea, is one of a form of integrated solution, which maintain the kampong livelihood based on maritime activity. The areas then will be designed as kampong with fisherman settlement completed with fisherman port and fish market, as well as small fish industries. Not to relocate in other area which seems not compatible with the existing livelihood of the kampong, but to maintain the maritime activity. So it is one of effort to keep place attachment of the local people, therefore, the root and emotional bonding of local people is still maintained. Furthermore, not only emotional bonding, but also economically, the local people who basically in fishing activities can be increased by the fish processing small industries, and fish market also ports.

6. Acknowledgement

The author would like to thanks to Yuki Riswandha and Januarta Kusdwimayanti as researcher assistants who have contributed in this paper.

7. References

[1] Dimitra S and Yuliastuti N 2012 Potensi Kampung Nelayan Sebagai Modal Permukiman Berkelanjutan di Tambaklorok, Kelurahan Tanjung Mas *Tek. PWK (Perencanaan Wil. Kota)* **1** 11–18
[2] Low S M and Altman I 1992 Place attachment *Place attachment* (Springer) pp 1–12
[3] Hidalgo M C and Hernandez B 2001 Place attachment: Conceptual and empirical questions *J. Environ. Psychol.* **21** 273–281
[4] Morgan P 2010 Towards a developmental theory of place attachment *J. Environ. Psychol.* **30** 11–22
[5] Scannell L and Gifford R 2010 Defining place attachment: A tripartite organizing framework *J. Environ. Psychol.* **30** 1–10
[6] Turner J F C and Fichter R 1972 *Freedom to build: dweller control of the housing process* (Macmillan)
[7] Turner J F C 1977 *Housing by peopletowards autonomy in building environments*
[8] Perkins D D, Hughey J and Speer P W 2002 Community psychology perspectives on social capital theory and community development practice *Community Dev.* **33** 33–52
[9] Lewicka M 2011 Place attachment: How far have we come in the last 40 years? *J. Environ. Psychol.* **31** 207–230

The Impact of Urbanization to Forest Degradation in Metropolitan Semarang: A Preliminary Study

A W Sejati[1,2,3], I Buchori[1,2], and I Rudiarto[1,2]

[1]Department of Urban and Regional Planning, Universitas Diponegoro, Indonesia
[2]Center of Geomatic's Application for Sustainable Development, Universitas Diponegoro, Indonesia
[3]Doctoral program in Architecture and Urbanism, Universitas Diponegoro, Indonesia

Email: anangws@undip.ac.id

Abstract. This paper aims to examine the relationship of urbanization and the consequences of environmental impacts, especially the availability of forests. The analysis used remote sensing data ie Landsat 7 ETM (1990), Landsat 7 ETM (2000), and Landsat 8 OLI (2015). Analysis of forest availability used green vegetation cover parameters with NDVI techniques, while urbanization was measured by looking at the built-up area growth with NDBI techniques. The detection result in 1990 area of urban area was 74.85 km^2, in 2000 was 130.83 km^2, and in 2015 was 292.74 km^2. Given the overall size of the metropolitan area of Semarang (932.14 km^2), the proportion of the built up area in 2015 was 31% with an average of 6% per year change in 1990-2000 and 8.2% per year in 2000-2015. Urbanization in this region is characterized by forest conversion into residential and industrial buildings. With increasing area and built-up density, NDBI is found to be proportionately increasing, while NDVI decrease is significant. The results show that NDBI-NDVI correlation is negative (-0.99). It is very important for policy makers to formulate land use control policies in Metropolitan Semarang particularly in responding to the issue of forest conversion.

Keywords: Urbanization, Forest Degradation, NDBI-NDVI, Metropolitan Semarang

1. Introduction

In the era of urban century, where half of the world's population lives in urban areas [1], new challenges arise, namely the impact of urbanization on the environment, especially in Asia [2–5]. The pattern of urbanization of Asian countries like Indonesia is very different from the urbanization pattern in the Western countries in general and it is very special [6–10]. "Kotadesasi" (urbanization), urban sprawl, and economic growth on transportation routes are characteristic in which investment and industrialization can make rural areas into urban area [8,9,11–13].

After the decentralization era in Indonesia, government policies in each region became uncontrolled, especially in urbanization and industrialization control [14]. Furthermore, the increasing growth of Asean economy has further encouraged industrialization and development of built-up areas in metropolitan cities in Indonesia such as metropolitan Jakarta [15–19],Semarang [5,11,20,21], and Surabaya [22]. Like the two sides of coin, on one side of urbanization becomes an opportunity for economic growth, but on the other side, improper control policy can lead to new problems, especially environmental sustainability in the metropolitan area [23–25].

One of the environmental impacts of urbanization is the conversion of land especially on forest land. The study of land conversion by various methods has been widely conducted [26–28] but a special study of the relationship between urbanization and forest land loss especially in tropical regions like Indonesia still needs to be done. Based on this, the researchers conducted a study on urbanization relations and forest degradation with the study area in Metropolitan Semarang. The purpose of this study is to see the extent of forest conversion as a result of urbanization.

2. Data and Methods

2.1 Study Area

The study area of this paper is Metropolitan Semarang (latitudes 60 580 S and longitude 1100 250E), Semarang City and three districts affected by urbanization are: Kendal (District Kaliwungu and Boja), Demak (Sayung and Mranggen), and Semarang Regency (West Ungaran, East Ungaran, Pringapus, and Bergas) (Figure 1). Slightly different from Handayani and Rudiarto [21] that defined Metropolitan with the area affected by urban expansion in Semarang regency only in West Ungaran and East Ungaran District. Furthermore, this study area is part of Kedungsepur which is also defined as Metropolitan Semarang in the context of regional cooperation [5].

Statistical data on area studies are processed from the Center of Statistic Agency in each city and district area. The data of population growth in metropolitan Semarang used data from 1990, 2000, and 2015 (Table 1.) Statistical data was processed to provide evidence of urban growth, one of which is seen from population growth data.

Figure 1. Metropolitan Semarang

Table 1. Population data of Metropolitan Semarang year 1990, 2000, 2015

Region	Year		
	1990	2000	2015
Semarang City	1,146,931	1,368,803	1,595,187
Kaliwungu	74,991	88,025	102,574
Boja	51,567	61,898	69,219
Ungaran Barat	54,886	67,353	77,758
Ungaran Timur	44,372	58,576	69,895
Pringapus	29,763	41,297	51,772
Bergas	30,374	41,987	72361
Sayung	68,340	78,699	103,932
Mranggen	82,785	125,575	180,152
Total Population	**1,584,009**	**1,932,213**	**2,322,850**
% Growth (per year)		2.2	1.4

Source: Center of Statistic Agency and Analysis

2.2 Image Analysis

In this paper, satellite image analysis used data from Landsat with different years ie 1990, 2000, and 2015 (Table 2). Likewise, Landsat is very useful for studying large areas, especially to measure land cover such as built-up area and vegetation canopy [29–34]. The type of Landsat used in accordance with the data of the year of satellite imagery. In 1990 and 2000 the analysis used Landsat 5 TM and Landsat 7 ETM which had 7 Bands. In 2015, the analysis used Landsat 8 OLI. Special year 2015 Landsat image has experienced a lot of sharpening so that the result was better than the previous Landsat because it has 11 bands.

The combination of composite image bands for each type of analysis is critical to the interpretation of image data. The combination of bands for urban interpretation is as shown in Figure 2. To determine urban and non-urban, the false color combinations used were 7, 5, 3 for 1990 and 2000. As for 2015, the combination used 7, 6, 4. For urban classification, analysis using the NDBI formula (Eq.1) [34–36] and supervised classification [32,37,38]. NDBI used Band 5 SWIR (Short Wave Infra-Red) and Band 4 NIR (Near Infra-Red) for 1990 and 2000, while in 2015, NDBI analysis used Band 6 for SWIR and band 5 for NIR.

NDBI= (SWIR-NIR)/(SWIR+NIR).. (Eq. 1)

Furthermore, to see the condition of vegetation as the impact of urbanization, the RGB band configuration is different from urban area. The analysis used 4, 3, 2 (figure 3) band configuration, so that sharpening was more on infra-red (RED) and Near Infra-Red (NIR) sensors. The infrared sensor is getting sharper (red) when it met chlorophyll, so it can be used for green vegetation detection (Figure 3). Vegetation detection used the NDVI formula [39–42] (Eq. 2) where the RED band used band 3 and NIR bands used band 4 for Landsat 1990 and 2000, while for 2015 RED band used band 4 and NIR bands used Band 5.

NDVI = (NIR-RED)/(NIR+RED) .... (Eq.2)

Table 2. Dataset for Image Analysis

Data	Type	Year	Source
Satellite imagery	multi spectral		
	landsat 5 TM	1990	USGS
	landsat 7 ETM	2000	USGS
	landsat 8 OLI	2015	USGS

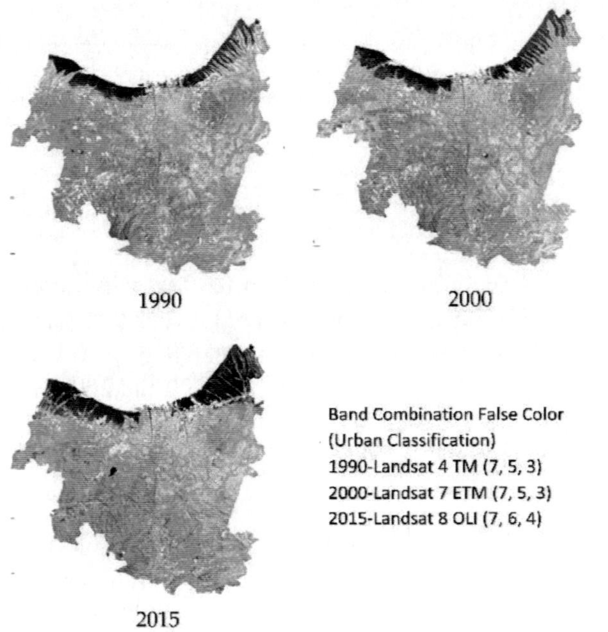

Figure 2. Landsat Band Configuration for Defining Urban Area

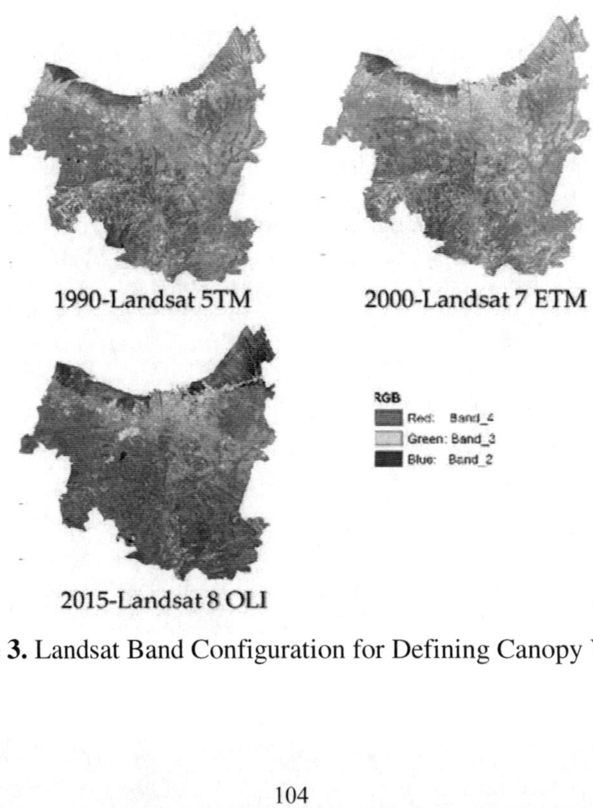

Figure 3. Landsat Band Configuration for Defining Canopy Vegetation

3 Results and Discussion

3.1 Urbanization in Metropolitan Semarang

Detection of urbanization patterns using temporal satellite images is one of the concrete steps in seeing urban development spatially. The results of image interpretation (Figure 4) show that in 1990 the urban area was 74.85 km^2, in 2000 was 130.83 km^2, and in 2015 was 292.74 km^2 (Table 3). Given the overall size of the metropolitan area of Semarang (932.14 km^2), the proportion of the built-up area in 2015 was 31% with an average of 6% per year change in 1990-2000 and 8.2% per year in 2000-2015. This change was so significant, so that the direction of urbanization can be determined from the growth direction of the built-up area.

According to Handayani and Rudiarto [21] the built-up area grows linearly with the population, so it needs a comparison with the population growth. Population growth in metropolitan Semarang from 1990-2000 were averaged 2.2% per year and in 2000-2015 an average of 1.4% per year (Table 1). Compared to the built-up area that rises up to 8.2% per year, there is a tendency that urbanization is largely not due to meeting the needs of the population but there are other factors. These factors can be seen from the growth in peri urban as an expansion area of urban Semarang.

Figure 5 shows the growth comparisons of peri-urban areas affected by urban expansion. From the figure, it is clear that Mranggen, Kaliwungu, Sayung, and Bergas are fast-growing peri-urban areas in the metropolitan urbanization of Semarang. To prove the result of NDBI calculation model and Built-up area, then field calibration is done. The calibration results show that four rapidly growing peri-urban areas are Mranggen, Kaliwungu, Sayung, and Bergas. It turns out that the distribution of the industry in the four urban areas makes the number of built-up areas grow and the growth rate is greater than the population growth rate in Metropolitan Semarang.

The phenomenon of surging areas for industry and population growth is a clear proof of the rapid urbanization process in rural areas that causes built-up land to grow rapidly and convert non-built-up land [24,43,44]. Moreover, from this growth it can be seen that the direction of metropolitan growth of Semarang tends to be evenly distributed but the dominant moment is to the east of expansion to Demak district where Mranggen and Sayung as fast-growing peri-urban triggers. This is consistent with the prediction that the urbanization of an industrialization not only gives rise to economic growth, but also gives effect to the spatial pattern of space [2,5,16]. The spatial pattern evolves due to the need for spatial development to meet the needs of industrial workers, such as residential and other supporting facilities, so that it needs to be well planned.

Table 3. Built-up area in Metropolitan Semarang (Km^2)

	1990	**2000**	**2015**
Semarang City	56.31	89.44	180.53
Kaliwungu	5.58	7.29	19.04
Boja	1.60	4.68	11.57
Ungaran Barat	2.46	5.08	10.55
Ungaran Timur	1.65	4.83	11.66
Pringapus	3.53	3.59	9.93
Bergas	1.96	4.76	12.04
Sayung	2.24	3.50	13.92
Mranggen	3.33	7.66	23.50
Total Built-up area (Km^2)	78.65	130.84	292.75

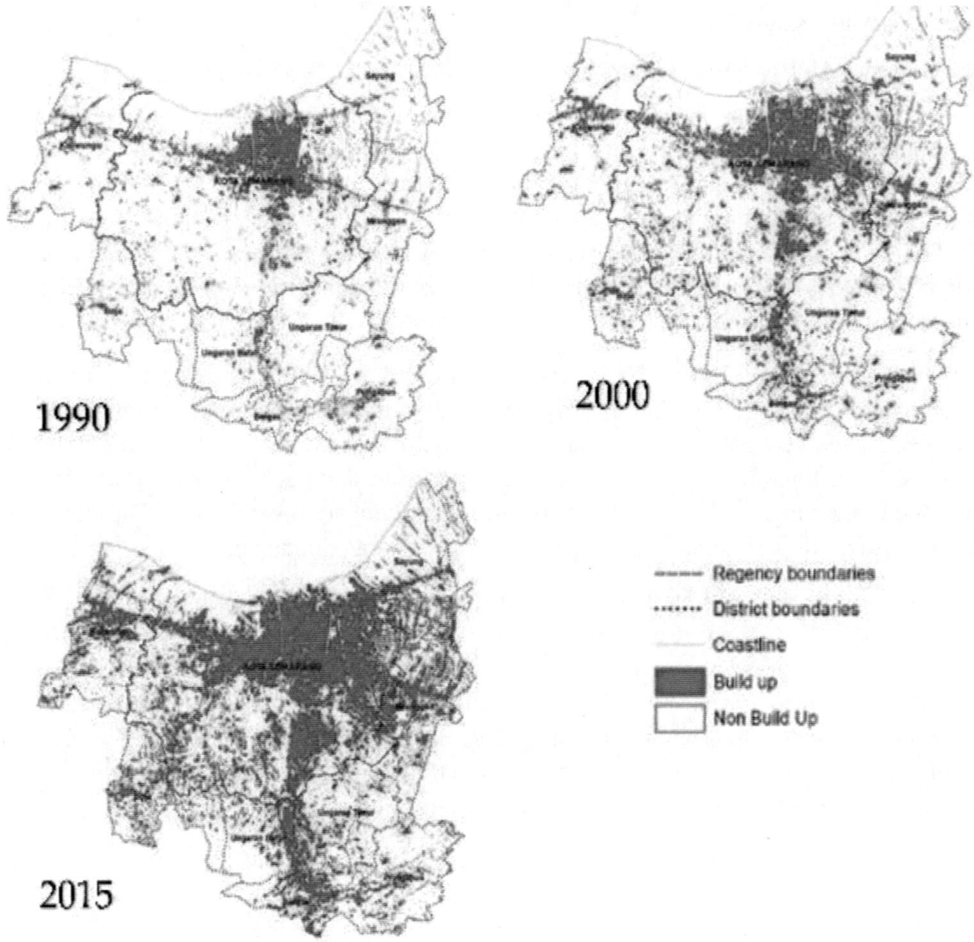

Figure 4. Built-up and non-Built-up Area 1990-2015

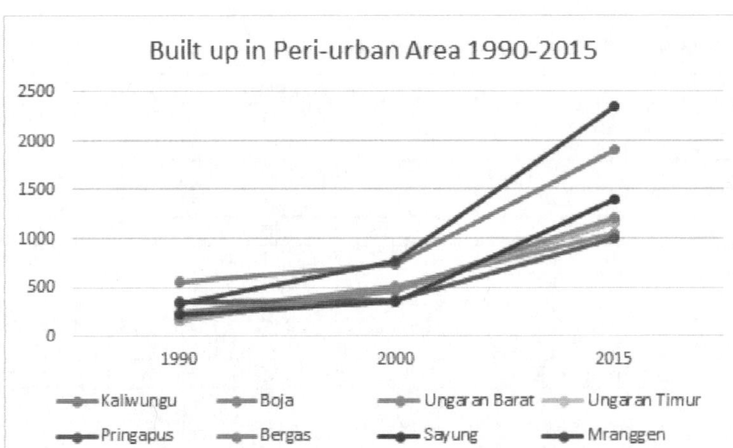

Figure 5. Comparison Built-up in peri-urban area

3.2 Forest Degradation in Metropolitan Semarang

Discussing the problem of urbanization certainly cannot be separated from land conversion. Land conversion in this paper is the reduced land cover of green vegetation that turns into a Built-up area. From the calculation with the NDVI formula (Figure 6) was the area with high vegetation density ie for the NDVI value above 0.4 decreases (Figure 7). In 1990 the number of high-density vegetation was 391.23 Km^2, in 2000 was 264.24 Km^2, and in 2015 was 242.14 Km^2. This proves that the decline is quite significant; especially in the period 1990-2000 was reduced by 32% and in 2015 decreased 8%. This characteristic was in line with the research of Sharma & Joshi [45] showing a negative relationship between built-up and vegetation. In addition the NDVI relationship with LST is negative, so it can be predicted that the reduced area of green vegetation cover will increase the UHI.

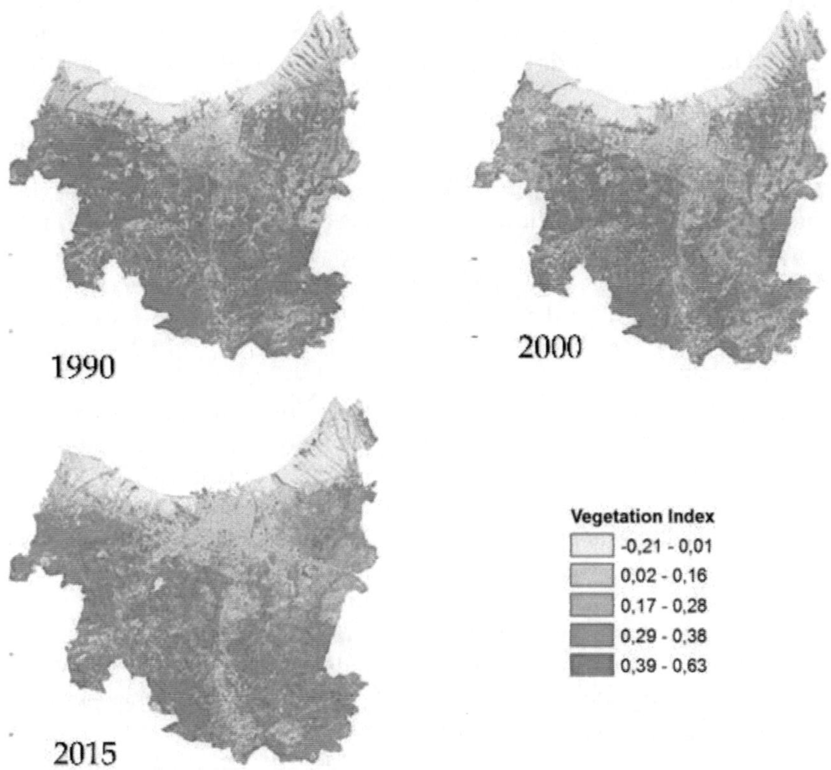

Figure 6. The NDVI result for forest detection

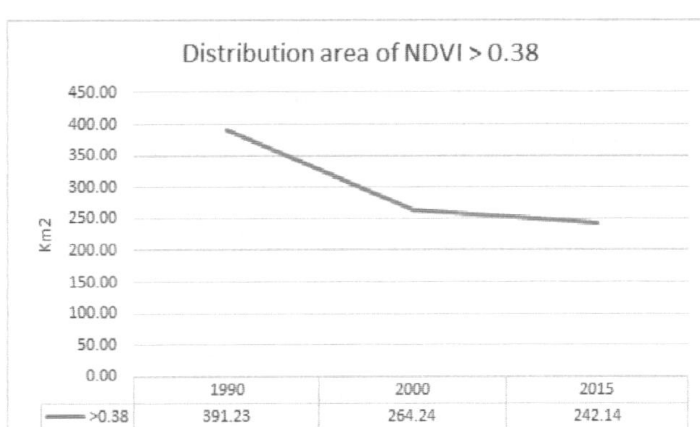

Figure 7. Degraded Green Vegetation Detection using NDVI value > 0.38

IOP Conf. Series: Earth and Environmental Science **123** (2018) 012011 doi:10.1088/1755-1315/123/1/012011

3.3 Discussion

Urban land use planning is a land use planning product that regulates land use for a period of time [46]. Planning is policy, and every policy depends on the decision maker. Policies require data and information, so that data and analysis must conform to the real conditions to support good governance systems [47]. From the analysis, it found that it was almost triple built-up growth rate for 25 years (1990-2015), it is a warning for land use planning policy makers, as the challenge of environmental issues and climate change must be responded immediately.

During that period, the population density was positively correlated with the probability of urban expansion, indicating that its development focused not only on the city core—with large population density—but also on fast growing areas in new growth centers. The relatively complete infrastructure and transportation in the city center provide a good opportunity for development, but investments such as industry also have a major effect on rural-urban transformation in the metropolitan Semarang.

Looking at the urbanization impact parameters from 1990-2015, the findings indicate that urban expansion in Metropolitan Semarang is growing as emerging new growth centers reinforce urban expansion into urban areas. In the context of urbanism, the growth of metropolitan Semarang with a new growth center, cultivated a new view of the concept of urbanization. Reduced numbers of movements to the city center have occurred. The distance to the city center does not significantly affect the growth, as facilities such as shopping centers and some downtown activity centers also grow in new growth centers. This is due to the availability of basic needs facilities in each area of urban expansion.

Furthermore, in 1990, Semarang City's CBD (center of Semarang) became the center of community activity, but by the year 2000 some community activities in fulfilling the needs of life had shifted to a new growth center such as east of Semarang City bordering on Mranggen and south of Semarang City bordering on Ungaran. The growth is widespread by 2015, so urban and peri-urban borders are merged, such as the Mranggen is no longer as rural area and Ungaran has become the new growth centers that merge with the southern part of Semarang.

Another character is industrialization. Industrial areas that originally grew around the port of Tanjung Emas in the coastal area of Semarang city (northern part) developed in a linear and also sprawl to some areas. Linearly, the industrial area expanded to Kaliwungu (Kendal) and Sayung (Demak). In 1990, the industrial area of Kendal advanced only in Kayu Lapis in Mororejo, but after 2013 the Industrial Park of Kendal quite rapidly even there is an integrated industrial area with national and international scale. Similarly with Sayung, the growth of food industry, livestock, and fertilizer from several foreign companies also expanded Sayung especially after 2010.

Returning to the main objective of this study is to show how the development of urbanization degrades the environment showing a correlation. It can be seen from the change of green vegetation cover (Forest) that disturbs the ecological balance (NDBI-NDVI correlation is negative -0.99). The vitality of environmental parameters such as NDVI analysis results in monitoring the status of forests in areas that have further implications for decision makers and policies in the municipal and regional planning sectors.

The analysis highlights that urbanization in Metropolitan Semarang coincides with a declining green status. This may explain that the environmental impact of urban expansion to peri-urban areas has greater possibilities, as new activity centers grow to form their own urban areas. On the other hand, it should be noted that the available land for expansion in downtown activity has been reduced and largely stagnant after rapid development, so growth moves more dynamically to peri-urban and poses a threat to the environment.

Urban landuse dynamics and environmental impacts require urban planners and decision makers to frame more sustainable practices than continue to promote urbanization and the rampant elite occupancy offerings in urban areas. Current spatial planning shows the focus of economic development. Control of utilization in rapidly growing areas has not been well anticipated, so the current challenge is limiting the growth of built-up areas especially in forest as protected areas and conservation.

4 Conclusion and Future Work

With increasing area and built-up density, NDBI is found to be proportionately increasing, while NDVI decrease is significant. The results show that NDBI-NDVI correlation is negative (-0.99). It is very important for policy makers to formulate land use control policies in Metropolitan Semarang particularly in responding to the issue of forest conversion. The greater the level of urbanization, the greater the conversion of forests and the extent of forest degradation. Furthermore, this phenomenon is a big challenge because controlling spatial policy is not easy but it does not mean impossible to do. This provides valuable insight for planners in intervening especially in providing recommendations to control urbanization. The future work from this paper is to measure the relationship of urbanization, degradation of forest area, and Land Surface Temperature.

5 References

[1] Bokaie M, Kheirkhah M, Daneshkar P and Hosseini A 2016 Assessment of Urban Heat Island based on the relationship between land surface temperature and Land Use / Land Cover in Tehran *Sustain. Cities Soc.* **23** 94–104

[2] Winarso H, Hudalah D and Firman T 2015 Peri-urban transformation in the Jakarta metropolitan area *Habitat Int.* **49** 221–229

[3] Szirmai A 2012 Industrialisation as an engine of growth in developing countries, 1950-2005 *Struct. Chang. Econ. Dyn.* **23** 406–420

[4] Wu Y, Shen J, Zhang X, Skitmore M and Lu W 2017 Reprint of : The impact of urbanization on carbon emissions in developing countries : a Chinese study based on the U-Kaya method *J. Clean. Prod.*

[5] Buchori I and Sugiri A 2016 An empirical examination of sustainable metropolitan development in Semarang City, Indonesia *Aust. Plan.* **53** 163–177

[6] McGee T G 1991 The emergence of desakota regions in Asia: expanding a hypothesis *Ext. Metrop. Settl. Transit. Asia* 3–25

[7] McGee T G 1989 Urbanisasi or kotadesasi? Evolving patterns of urbanization in Asia *Urban. Asia Spat. Dimens. policy issues* **108**

[8] McGee T G 2010 Building Liveable Cities in Asia in the Twenty-First Century Research and Policy Challenges for the Urban Future of Asia *Malaysian J. Environ. Manag.* **11** 14–28

[9] Setyono J S, Yunus H S and Giyarsih S R 2016 the Spatial Pattern of Urbanization and Small Cities Development in Central Java: a Case Study of Semarang-Yogyakarta-Surakarta Region *Geoplanning J. Geomatics Plan.* **3** 53–66

[10] Firman T 2004 Demographic and spatial patterns of Indonesia's recent urbanisation *Popul. Space Place* **10** 421–434

[11] Sugiri A, Buchori I and Soetomo S 2011 Sustainable Metropolitan Development: Towards an Operational Model for Semarang Metropolitan Region *Int. J. Environ. Cult. Econ. Soc. Sustain.* **7** 301–324

[12] Buchori I, Sugiri A, Maryono M, Pramitasari A and Pamungkas I T D 2017 Theorizing spatial dynamics of metropolitan regions: A preliminary study in Java and Madura Islands, Indonesia *Sustain. Cities Soc.*

[13] Handayani W 2013 Rural-urban transition in Central Java: Population and economic structural changes based on cluster analysis *Land* **2** 419–436

[14] Rukmana D 2015 The Change and Transformation of Indonesian Spatial Planning after Suharto's

New Order Regime : The Case of the Jakarta Metropolitan Area *Int. Plan. Stud.* 37–41
[15] Winarso H, Hudalah D and Firman T 2015 Peri-urban transformation in the Jakarta metropolitan area *Habitat Int.* **49** 221–229
[16] Hudalah D and Firman T 2012 Beyond property : Industrial estates and post-suburban transformation in Jakarta Metropolitan Region *Cities* **29** 40–48
[17] Firman T 2004 New town development in Jakarta Metropolitan Region: a perspective of spatial segregation *Habitat Int.* **28** 349–368
[18] Firman T 2009 The continuity and change in mega-urbanization in Indonesia: A survey of Jakarta--Bandung Region (JBR) development *Habitat Int.* **33** 327–339
[19] Elmanisa A, Kartiva A, Fernando A, Arianto R, Winarso H and Zulkaidi D 2017 Land Price Mapping of JABODETABEK, Indonesia *Geoplanning J. Geomatics Plan.* **4** 53–62
[20] Arta F and Pigawati B 2015 the Patterns and Characteristics of Peri-Urban Settlement in East Ungaran District, Semarang Regency *Geoplanning J. Geomatics Plan.* **2** 103–115
[21] Handayani W and Rudiarto I 2014 Dynamics of Urban Growth in Semarang Metropolitan – Central Java : An Examination Based on Built-Up Area and Population Change *J. Geogr. Geol.* **6** 80–87
[22] Rachmawati F, Soemitro R A A, Joko T, Adi W and Susilawati C 2015 Low-cost apartment program implementation in Surabaya metropolitan area *Procedia Eng.* **125** 75–82
[23] Surya B 2016 The Processes Analysis of Urbanization, Spatial Articulation, Social Change and Social Capital Difference in the Dynamics of New Town Development in the Fringe Area of Makassar City (Case Study: In Metro Tanjung Bunga Area, Makassar City) *Procedia - Soc. Behav. Sci.* **227** 216–231
[24] Pribadi D O and Pauleit S 2015 The dynamics of peri-urban agriculture during rapid urbanization of Jabodetabek Metropolitan Area *Land use policy* **48** 13–24
[25] Wu P C, Lay J G, Guo H R, Lin C Y, Lung S C and Su H J 2009 Higher temperature and urbanization affect the spatial patterns of dengue fever transmission in subtropical Taiwan *Sci. Total Environ.* **407** 2224–2233
[26] Phuc N Q, Van Westen A C M and Zoomers A 2014 Agricultural land for urban development: The process of land conversion in Central Vietnam *Habitat Int.* **41** 1–7
[27] Miettinen J, Hooijer A, Wang J, Shi C and Liew S C 2012 Peatland degradation and conversion sequences and interrelations in Sumatra *Reg. Environ. Chang.* **12** 729–737
[28] Zhang Y, Guo S, Liu Q, Jiang J, Wang R and Li N 2015 Responses of soil respiration to land use conversions in degraded ecosystem of the semi-arid Loess Plateau *Ecol. Eng.* **74** 196–205
[29] Wijaya N 2015 Deteksi Perubahan Penggunaan Lahan Dengan Citra Landsat Dan Sistem Informasi Geografis: Studi Kasus Di Wilayah Metropolitan Bandung, Indonesia *Geoplanning J. Geomatics Plan.* **2** 82–92
[30] Fassnacht F E, Li L and Fritz A 2015 Mapping degraded grassland on the Eastern Tibetan Plateau with multi-temporal Landsat 8 data - where do the severely degraded areas occur? *Int. J. Appl. Earth Obs. Geoinf.* **42** 115–127
[31] Wang C, Qi J and Cochrane M 2005 Assessment of tropical forest degradation with canopy fractional cover from Landsat ETM+ and IKONOS imagery *Earth Interact.* **9** 1–18
[32] Jia K, Liang S, Zhang L, Wei X, Yao Y and Xie X 2014 Forest cover classification using Landsat ETM+ data and time series MODIS NDVI data *Int. J. Appl. Earth Obs. Geoinf.* **33** 32–38
[33] Deng X, Huang J, Rozelle S, Zhang J and Li Z 2015 Impact of urbanization on cultivated land changes in China *Land use policy* **45** 1–7
[34] Varshney A 2013 Improved NDBI differencing algorithm for built-up regions change detection from remote-sensing data: an automated approach *Remote Sens. Lett.* **4** 504–512
[35] Ogashawara I and Bastos V de S B 2012 A Quantitative Approach for Analyzing the Relationship between Urban Heat Islands and Land Cover *Remote Sens.* **4** 3596–3618
[36] Zhang Y, Odeh I O A and Han C 2009 Bi-temporal characterization of land surface temperature in relation to impervious surface area, NDVI and NDBI, using a sub-pixel image analysis *Int. J. Appl. Earth Obs. Geoinf.* **11** 256–264

[37] Halder A, Ghosh A and Ghosh S 2011 Supervised and unsupervised landuse map generation from remotely sensed images using ant based systems *Appl. Soft Comput.* **11** 5770–5781
[38] Perumal K and Bhaskaran R 2010 Supervised classification performance of multispectral images *arXiv Prepr. arXiv1002.4046*
[39] Gascon M, Cirach M, Martínez D, Dadvand P, Valentín A, Plasència A and Nieuwenhuijsen M J 2016 Normalized Difference Vegetation Index (NDVI) as a marker of surrounding greenness in epidemiological studies: the case of Barcelona city *Urban For. Urban Green.* **19** 88–94
[40] Aburas M M, Abdullah S H, Ramli M F and Ash'aari Z H 2015 Measuring land cover change in Seremban, Malaysia using NDVI index *Procedia Environ. Sci.* **30** 238–243
[41] Fardani I, Darmawan S, Hakim D M, Harto A B and Wikantika K 2016 Increasing Accuracy Value in the Estimates of Carbon Stock by Using Vegetation Index from ALOS AVNIR 2 Satellite Imagery *Geoplanning J. Geomatics Plan.* **3** 1–14
[42] Gandhi G M, Parthiban S, Thummalu N and Christy A 2015 Ndvi: Vegetation Change Detection Using Remote Sensing and Gis--A Case Study of Vellore District *Procedia Comput. Sci.* **57** 1199–1210
[43] Farrell K 2017 The Rapid Urban Growth Triad : A New Conceptual Framework for Examining the Urban Transition in Developing Countries *sustainability* **9** 1–19
[44] He C, Liu Z, Xu M, Ma Q and Dou Y 2017 Urban expansion brought stress to food security in China: Evidence from decreased cropland net primary productivity *Sci. Total Environ.* **576** 660–670
[45] Sharma R and Joshi P K 2016 Urban Climate Mapping environmental impacts of rapid urbanization in the National Capital Region of India using remote sensing inputs *Urban Clim.* **15** 70–82
[46] Kaiser E, Godschalk D and Chapin F 1995 *Urban Land Use Planning* (Chicago: University of Illinois Press)
[47] Mccall M K 2003 Seeking good governance in participatory-GIS : a review of processes and governance dimensions in applying GIS to participatory spatial planning *Habitat Int.* **27** 549–573

Flood Risk Analysis in Denpasar City, Bali, Indonesia

T B Kusmiyarti[1], P P K Wiguna[1], and N K R Ratna Dewi[1]

[1]Center for Spatial Data Infrastructure Development (PPIDS), Udayana University, Bali, Indonesia

Email: tatihartanto@ymail.com

Abstract. Denpasar city is a Capital City of Bali Province and one of the leading tourist destinations in Indonesia. Denpasar area is relatively flat with high rain fall intensity with the domince of settlement. This makes Denpasar City becomes prone area of flood.
The aim of this research is to find out the spatial distribution flood hazard and the risk of population which are affected to the flood hazard. Weighting, scoring and overlaying method were used in this research. Six indicators were used to analyze the flood hazard: landuse, rainfall, type of soil, slope, altitute and drainage density. The vulnerability is analyzed per Desa or Kelurahan (Rustic/ Neighborhood) with the indicator of age, education and population density. Risk was calculated by multiplied hazard with vulnerability and divided with coping capacity. In this research, coping capacity is determined by the amount of internal budget for each *Desa* or *Kelurahan* for development purpose. Flood risk in Denpasar city is divided into five classes, very low risk, low risk, medium risk, high risk and very high risk. Total population with very high risk reached 202478 people or 13.16% of total population. The total area is 780.7 ha or 16.02% from total settlement in Denpasar city. Total population with high risk reached 202478 people or 13.16% of total population. The total area is 780.7 ha or 16.02% from total settlement in Denpasar city. The number of population with medium risk reached 202478 people or 33.51% of total population which occupied 22.95% of total settlements or 1118.18 ha. The total number of population with low risk reached 79435 people or 13.14% of total population with area of low flood risk is 716.89 ha or 14.71% of total settlements in Denpasar City. Very low flood risk with total population at risk reached 19184 people or 31.74% of total population and occupied 2003.54 areas or 41.12% of total areas of settlements.

Keywords: Flood Hazard, Flood Risk, Population

1. Introduction

Denpasar City is capital city of Bali Province, Indonesia. Beside its administrative function, Denpasar City is also the heart of tourism, education, economy, trade and services in Bali Province. Denpasar city consists of 4 districts and 43 Villages. The number of population in Denpasar always increases every year. The latest data in 2015 population of Denpasar City as much as 880,600 people [1]. This condition cause the need for land becomes increasing and eventually triggers landuse change.

Denpasar City is located in a flat until undulating topography which bordered by Lombok Strait in the south. Denpasar City is a downstream area for three major river in Bali, Ayung River, Mati River and Badung River. The physical conditions makes Denpasar City becomes prone area to flood. Not to mention that Denpasar City has high rainfall intensity. This condition and the existence of many buildings and impermeable layers due to the development of the City itself makes Denpasar City become prone area to flood.

Flood is a water runoff that exceeds the normal water level, so that the runoff from the riverbed causing a puddle on the low land on the side of the river [2]. There are two type of flooding events, the first flooding event occurring in an area that usually does not flood and both flood events occur because the flood discharge is greater than the existing drainage capacity of the river [3]. The occurrence of flooding is also caused by low soil infiltration capability, thus causing the land to be unable to absorb water [4–7].

Based on data obtained from the Regional Disaster Management Agency of Bali Province, it is said that there are some flood events occurs due to heavy rain. Several areas which at that time were submerged by the flood were Padang Sambian Village, Panjer Village, Penatih Village, Pengecutan Kelod Village and Sambian Kaja Village. In 2015 again floods occured in the area of Monang-Maning and surrounding areas, causing the death toll due to drag currents in Tukad Mati. The trigger of the flood at that time due to the amount of garbage clogging drainage channels [8]. The occurrence of floods in Denpasar City in addition to causing losses in terms of material, also makes population becomes exposed to flood hazard. The aim of this research is to study the floods risk to the population in Denpasar City.

2. Literature Review

Risk is defined as the probability of the danger or expectation of losses (death, property damage) resulting from the interaction between natural disaster and vulnerability. The purpose of risk assessment is to determine the nature and extent of risk by analysing the potential hazards and evaluating the existing conditions of vulnerability. One method of risk assessment was the qualitative method. Qualitative methods describe risks at levels such as "very high", "high", "medium", "low" and "very low". Qualitative methods were used in conditions that required rapid, precise and cost-free assessments. This method used a matching and scoring approach that emphasized the quantitative assessment of risk components. Hazards, vulnerabilities and coping capacities were the key concepts in risk assessment. Hazard as probabilities of natural events that harm and cause victims and vulnerability as the level of damage or casualties when the incident struck. While capacity was the ability of the community in tackling the dangers of natural disasters. The concept leads to an understanding of risk as the product of two concepts, namely the probability of occurrence and the intensity of the disaster event [8].

The same hazard events can result in different risk analysis due to differences in vulnerability of risk elements. For example, assessing the hazard risk of similar flood events in two different cities may result different values depending on the factors that contribute to the vulnerability and resilience of the population. Risk analysis can be used to identify locations and populations that may be affected by disaster events. This information is very important when preparing a disaster mitigation plan. A disaster mitigation plan is a dynamic process built on the practical assessment of past and current information to anticipate future hazards and provide meaningful strategies for addressing potential impacts and identified needs. The main objective of the hazard mitigation plan is to reduce vulnerability and increase community resilience, thereby reducing the potential risks [9].

3. Methodology

3.1. Area of Study

Denpasar city consists of 4 districts, and 43 sub-district. The area of Denpasar City has total area of 127.98 km^2. The area of Denpasar City is generally a relatively flat area with a slope of 0-8% and the altitude is less than 75 meters above sea level. Land use is dominated by buildings.

3.2. Tools and Materials

The tools used in this study include were: a unit of laptop with QGIS software, and writing utensils. The materials used in this study were: Denpasar City administrative map from Planning Agency of Denpasar City, Landuse Map from National Land Agency of Denpasar City, Soil Type Map from Faculty of Agriculutre of Universitas Udayana and Altitude Map, based on contour line, obtained from Planning Agency of Denpasar City, annual rainfall data in Denpasar obtained From Meteorological and Geophysical Agency, Region III of Denpasar City, and Slope Map and River Map obtained from Balai Wilayah Sungai Bali-Penida of Bali Province.

3.3. Research Implementation

The methods in this research were based on the risk equations as formulated in equation (1) [8].

$$Risk = \frac{Hazard x\ Vulnerability}{Coping\ Capacity} \quad ...(1)$$

Disaster risk is the potential loss of life, injury, or destroyed or damaged assets which could occur to a system, society or a community in a specific period of time, determined probabilistically as a function of hazard, exposure, vulnerability and capacity. Hazard is a process, phenomenon or human activity that may cause loss of life, injury or other health impacts, property damage, social and economic disruption or environmental degradation. Meanwhile vulnerability is the conditions determined by physical, social, economic and environmental factors or processes which increase the susceptibility of an individual, a community, assets or systems to the impacts of hazards. And coping capacity means the combination of all the strengths, attributes and resources available within an organization, community or society to manage and reduce disaster risks and strengthen resilience [10].

This research used the variable of flood hazard, population vulnerability to flood and institutional coping capacity which was the amount of Village's budget for development purpose. Flood hazard used matching and scoring methods, population vulnerability was analyzed by using with the indicator of age, education and population density.

3.3.1. Flood Hazard

At this stage, an analysis that requires data input from the six parameters are suspected to be the cause of flood, ie landuse, slope, altitude, soil type, rainfall and drainage density. Each parameter was given the weight and score that has the greatest influence on the flood. Table 1 listed the score and weight for each parameter [11].

Table 1. Score and Weight for Each Flood Parameter

No	Parameter	Score	Weight	Max. Score
1	Landuse	1 – 5	2	10
2	Altitude	1 – 5	1	5
3	Drainage Density	1 – 5	1	5
4	Slope	1 – 5	3	15
5	Soil type	1 – 5	2	10
6	Rainfall Intensity	1 – 5	3	15

The hazard class was divided into 5 classes. The hazard class calculation was performed by dividing the value of each parameter into equal class interval calculated according to equation (3) [11] then summed up each class of each parameter to reach the final score. The division of hazard into 5 classes resulted in 5 classes of hazard, namely very high hazard, high hazard, medium hazard, low hazard and very low hazard.

$$Class\ Interval\ = \frac{N\ maks - N\ min}{Class\ Number} \ldots (2)$$

With N max: Highest population density
N min: The lowest population density

After the weighting and scoring, then the process continued with overlay for the six parameters. Any graphical data that will be overlaid by using a UTM coordinate system to make it easier to know the unit width in meters. The purpose of doing overlay was to obtain new information that has a mapping unit of the combined graphics data. To calculate the level of flood hazard is calculated by using equation (3) as follows [8]:

Flood Potential Score = (2 x LanduseScore) + (1 x Height Score) + (1 x Drainage Density Score) + (3 x Slope Score) + (2 x Land Type Score) + (3 x Rainfall Score*) ...(3)*

3.3.2. Vulnerability

Vulnerability in this research used population's vulnerability to flood hazard by analyzing the parameter of age, education and population density with data obtained from Central Statistical Bureau of Denpasar City. Age level was calculated by grouping age level into two groups namely productive (15-64 y.o) and unproductive (0-14 y.o and >64 y.o). Level of education was calculated by dividing the population by their highest level of education, ie not attending school, finishing elementary school, finishing middle school, finishing high school, finishing vocational school, finishing university degree, meanwhile population density was calculated by dividing number of population by the area of the settlements [11].

Each indicator will be scored to create vulnerability level to each parameter then summed up to resulting a total score. Vulnerability in this research was calculated per settlements at village boundary level. And the analysis was conducted per Village.

The vulnerability class was divided into 5 classes. The vulnerability class calculation was performed by dividing the value of each parameter into equal class interval calculated according to equation (3) [11] then summed up each class of each parameter to create total vulnerable value. The division of vulnerability into was divided into5 classes that resulted in 5 classes of vulnerability, namely very high vulnerability, high vulnerability, medium vulnerability, low vulnerability and very low vulnerability.

3.3.3. Coping Capacity

The coping capacity in this research used the availability of development budget for each village to conduct disaster mitigation until post-disaster reconstruction and rehabilitation. The number of budget was then divided into equal class interval calculated according to equation (3) [11]. The division of coping capacity into 5 classes resulted in 5 classes, namely very high coping capacity, high coping capacity, medium coping capacity, low coping capacity and very low coping capacity.

3.3.4. Risk Analysis

Risk analysis was calculated by using equation (1) [8]. The division of risk was divided into 5 classes that resulted in 5 classes of risk, namely very high risk, high risk, medium risk, low risk and very low risk.Number of population at risk within a village is calculated with assumption that the population in a village was distributed evenly within village settlements using equation (4). This equation used mainly if there was more than one risk class within one village.

$$Population = \frac{Settlement\ Area}{Total\ settlement\ area} x\ total\ village\ population \ldots (4)$$

4. Result and Discussion

4.1. Flood Hazard

The area of flood hazard in Denpasar City was divided into three classes, medium hazard, high hazard and very high hazard. Denpasar City that has medium flood hazard is the area of South Denpasar and East Denpasar Sub-district, with a total area of 251.73 ha (2.09%). Based on the calculation of the total score, it can be predicted the dominant parameter that causes this area which has medium level of hazard is slope of 0% to 3% until 3%-8% with relatively flat to undulating topography. In addition, the dominant landuse is irrigated rice fields. In some areas of the water will be difficult to flow so that in some areas there will likely be inundated.

The area of Denpasar City which has high level of hazard is found in a small part of North Denpasar, West Denpasar and most of East Denpasar and South Denpasar, with total area of 4686.70 ha (38.91%). Based on the results of the calculation that cause this region has a high level of vulnerability is the average annual rainfall is quite high reaches 1982 mm. Land use is dominated by settlements and building with low to medium density. Due to the lack of natural land cover, when the condition of excess water then itcannot be absorbed by the soil.

The area of Denpasar City that has very high hazard is found in a small part of East Denpasar, partly in West Denpasar and mostly in South Denpasar, with total area of 7107,08 ha (59%). Based on the calculation of the total score, it can be predicted that the dominant parameter that cause very high hazard is the use of land dominated by settlements and building which is intended for housing, offices, trade and housing with high density. Landuse affects the amount of rainfall runoff that has exceeded the infiltration rate, resulting in the occurrence of inundation and runoff in areas that are mostly covered by buildings. The minimal natural land cover that exists in this region indicates the absence of a source of infiltration which can lead to inundation.

Denpasar City area consisting of 43 villages have varying physical conditions between villages with each other. This causes each region to have a different distribution of flood hazard. Flood hazard that occurred in Denpasar City starts from the low to very high level. Areas with low to moderate flood hazard spread over most of Peguyangan Kaja area with an area of 97.45 (33.48%) ha of total area. The village that has the widest area with high flood hazard is PemoganVillage with an area of 344.84 ha (35.76%) of the total area. While the village that has the most widespread areas with very high flood hazard is Sesetan with an area of 479.71 ha (91.1%) of the total area. Overall flood spreads with high levels of hazard are located in almost every village but with varying percentages based on the area of each village.

The secondary data of flood events of 2013-2015 floods are still frequent and spread at some point location in Denpasar City such as Padang Sambian Village, Panjer Village, Penatih Village, Pemecutan Kelod Village and Sambian Kaja Village [12,13]. There are 12 points of flood-prone locations, where 10 points are included at the result of this research. The research results are at very high flood hazard (83.3%) and the other 2 points are at high hazard (16.6%). From the data, it can be concluded that the existing research results are appropriate and can represent the conditions that occur in the field.

Seeing the occurrence of floods are still common in Denpasar City, the mitigation efforts that can be done by the government are: improving drainage channels, inviting the community to participate in maintaining environmental hygiene by not littering in the river or in other drainage channels, and making absorption wells in each house to reduce the inundation in the presence Excess water conditions. The results of the flood hazard analysis are presented in Table 2. Flood hazard spread throughout the city of Denpasar with different levels of hazard due to the physical condition of each area. Flood hazard map can be seen in Figure 1.

Table 2. Flood Hazard Class

No	Hazard Class	Area (Ha)	Percentage (%)
1	Medium	251,73	2,089
2	High	4686,70	38,91
3	Very High	7107,08	59,00

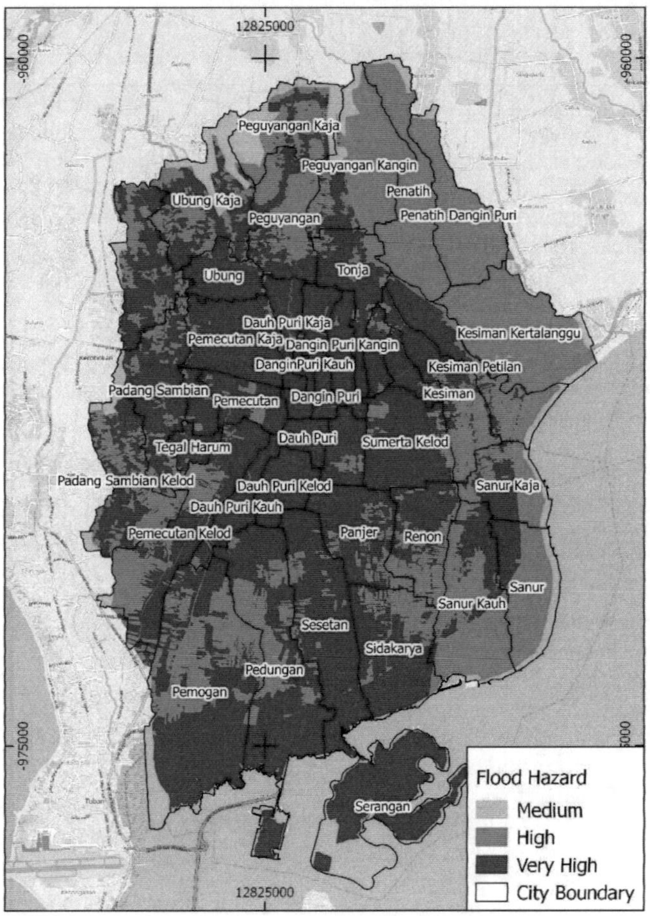

Figure 1. Flood Hazard Map.

4.2. Vulnerability

Vulnerability of population which expose to flood hazard is calculated by using age level, level of education and population density as parameter to create vulnerability class. The vulnerability of population in Denpasar city can be divided into four classes, namely very low, low, medium and high vulnerability. High vulnerability is located at high population density with low education and high unproductive age. High vulnerability with total of 19091 population located in Kelurahan Kesiman and Ubung Village. Medium vulnerability in total of 241303 population located in 19 villages in 4 Sub-district, i.e. South Denpasar, East Denpasar, North Denpasar and West Denpasar. Meanwhile, total 229043 of population are included in low vulnerability class which located in 11 villages over 3 Sub-districts, South Denpasar, North Denpasar and West Denpasar. Very low vulnerability are people located in 6 Villages over 3 Sub-districts, South Denpasar, North Denpasar and West Denpasar with total population reached 81851 people.

High vulnerability only happens on two villages of East Denpasar and North Denpasar, which is Kesiman and Ubung Village. Meanwhile medium vulnerability mainly distributed in East Denpasar Sub-district with 10 villages, meanwhile north Denpasar and West Denpasar only have 3 villages and 5 villages that categorized as medium vulnerability. Low vulnerability are distributed in South Denpasar, North Denpasar and West Denpasar sub-district with 5 villages and 6 villages each. Very low vulnerability is distributed mostly in South Denpasar with 3 villages, meanwhile North Denpasar and West Denpasar sub-district only have 2 villages and a village that categorized as very low vulnerability. Table 3 list the number of population who is vulnerable to flood hazard. Figure 2 (a) shows the vulnerability map to flood hazard in Denpasar City.

Table 3. Vulnerabilty to Flood Hazard

No	Sub-district	Village	Vulnerability	Population[1]
1	East Denpasar	Kesiman	High	9999
	North Denpasar	Ubung		9029
2	South Denpasar	Pemogan (1)	Medium	22811
	East Denpasar	Sumerta Kelod, Sumerta Kaja, DanginP uri, Kesiman Petilan, Dangin Puri Kangin, Dangin Puri Kelod, Sumerta, Sumerta Kauh, Kesiman Kertalangu, Penatih (10)		84514
	North Denpasar	Tonja, Dangin Puri Kaja, Pemecutan Kaja (3)		56931
	West Denpasar	Tegal Kertha, Dauh Puri, Dauh Puri Kelod, Dauh Puri Kauh, Pemecutan (5)		77047
3	South Denpasar	Sidakarya, Renon, Sanur Kaja, Panjer, Pedungan (5)	Low	65501
	North Denpasar	Dauh Puri Kaja, Ubung Kaja, Peguyangan Kangin, Penatih Dangin Puri, Dangin Puri Kauh (5)		66094
	West Denpasar	Dauh Puri Kangin, Dauh Puri Kauh, Padangsambian, Tegal Harum, Pemecutan Kelod, Padangsambian Kaja (6)		97448
4	South Denpasar	Sanur, Sanur Kauh, Sesetan (3)	Very Low	43062
	North Denpasar	Peguyangan, Peguyangan Kaja (2)		18624
	West Denpasar	Padangsambian Kelod (1)		20165

[1]population data was taken from center statistical bureau of Denpasar City [10]

4.3. Coping Capacity

The coping capacity for each village in Denpasar city was calculated by using the number of Village development budget which allocated for development. This budget will affect the preparation for each village if flood is coming. The availability of budget for development itself is different for each village, ranging from IDR 304,900,000.00 (USD 22,886) until IDR 9,627,400,000.00 (USD 722,663) [8]. From 43 villages, villages with the highest budget are Kesiman, Ubung Kaja,Sumerta Kaja, Padang Sambian Kelod, Kesiman Kertalangu, Sesetan, Pedungan, Dauh Puri Kaja, Penatih, Peguyangan Kangin, Pemogan and Sidakarya with total budget of more than one billion rupiah or more than 75,000 US Dollars. The highest budget that included in very high capacity is the budget from Peguyangan Kangin, Pemogan and Sidakarya Village, meanwhile the very low capacity which indicate villages with low budget for development purpose are Dangin Puri Kangin, Dangin Puri Kauh, Padangsambian, Dauh Puri Kauh, Dauh Puri, DanginPuriKaja, Penatih Dangin Puri, Tegal Harum, Panjer, and Renon.Figure 2 (b) shows the capacity for each village in Denpasar City.

4.4. Flood Risk

Flood risk analysis in this research was done by using the variable of flood hazard, population vulnerability and institutional coping capacity which was the amount of Village's budget for development purpose. Flood risk in Denpasar city was divided into five classes, very low risk, low risk, medium risk, high risk and very high risk. Very high risk areas are located in 4 villages, Dangin Puri Kangin, Dangin Puri Kaja, Dauh Puri and Dauh Puri Kauh. Total population was at risk at 50981 people or 8.43% from total population in Denpasar City with total area of 252.81 ha or occupied 5.19% of total settlements in Denpasar City. The population which risk to flood is very high is located in the middle of Denpasar City where very high hazard are located. Vulnerability in this area is medium to low with medium to low capacity as well. This indicates that the high risk is resulted by the high flood hazard but with medium to low coping capacity.

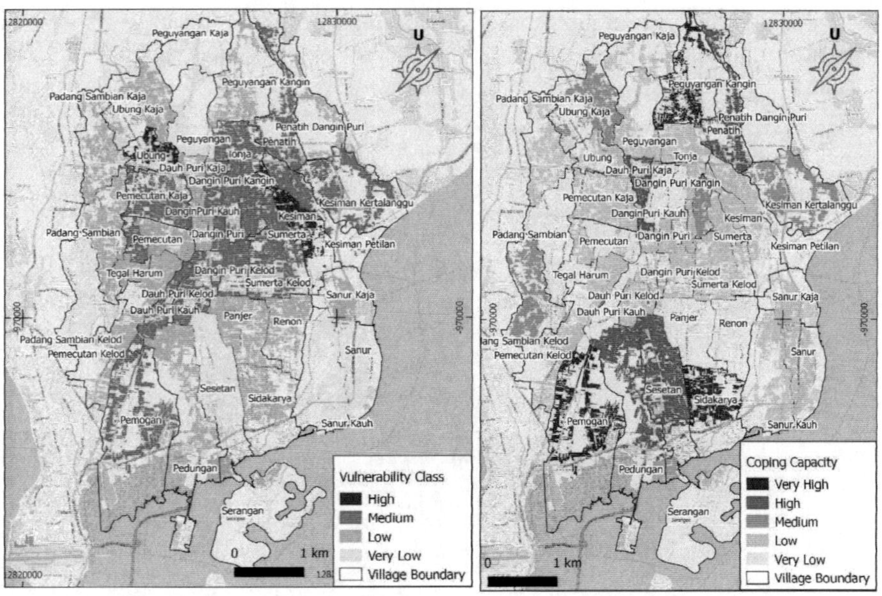

Figure 2. (a) Vulnerability Map and (b) Coping Capacity Map to Flood of Denpasar City

High flood risk areas are located in Kesiman, Ubung, Renon, Dangin Puri Kauh, Padangsambian, Panjer and Tegal Harum Village. Total population was at risk reached 202478 people or 13.16% of total population. The total area is 780.7 ha or 16.02% from total settlement in Denpasar city. The population with high flood risk located in the middle to south area of Denpasar city. The hazard in that areas are high to very high with medium to low vulnerability and very high to low coping capacity which indicates the role of hazard in building the risk value in that villages.

Medium flood risk located in 12 villages, Kesiman Village, Renon, Padangsambian, Panjer, Penatih Dangin Puri, Sumerta Kelod, Dangin Puri, Kesiman Petilan, Tonja, Dangin Puri Kelod, Sumerta, Sumerta Kauh, Tegal Kertha, Pemecutan Kaja, Dauh Puri Kelod, and Pemecutan Village. The number of population at risk is quite high, reached 202478 people or 33.51% of total population. Medium risk occupied 22.95% of total settlements which reached 1118.18 ha of total settlements in Denpasar city. Population who has medium flood risk is one of most common risk classes in Denpasar

IOP Conf. Series: Earth and Environmental Science **123** (2018) 012012 doi:10.1088/1755-1315/123/1/012012

city which occupied the center of the city area. The hazard in this area is high to very high but with medium to low vulnerability and medium to very low coping capacity.

Low flood risk is located in 10 villages in Denpasar City, namely Serangan village, Sanur Kaja, Dauh Puri Kangin, Pemecutan Kelod, Padangsambian Kaja, Sumerta Kelod, Sumerta Kaja, Kesiman Petilan, Tonja and Kesiman Kertalangu Village. The total number of population at risk reached 79435 people or 13.14% of total population with area of low flood risk is 716.89 ha or 14.71% of total settlements in Denpasar City. Population with low risk to flood is spatially distributed on the side of Denpasar City. The hazard in the area are mainly medium to high hazard with medium to very low vulnerability ad high to very low coping capacity.

Very low flood risk with total population at risk reached 19184 people or 31.74% of total population and occupied 2003.54 areas or 41.12% of total areas of settlements. Very low flood risk has the highest percentage of population at risk along with medium risk. Population with very low flood risk spatially distributed on the north side and south side of Denpasar City with high to very high hazard. The very low flood risk is result from the high coping capacity and low vulnerability which caused the risk value decreasing. Table 4 listed the flood risk class in Denpasar City

The physical condition of Denpasar City makes Denpasar has flood hazard and maximum the population is exposed to risk. Spatial distribution of very high to medium risk generally located in the center of Denpasar city. Figure 3 shows the flood risk map of Denpasar City.

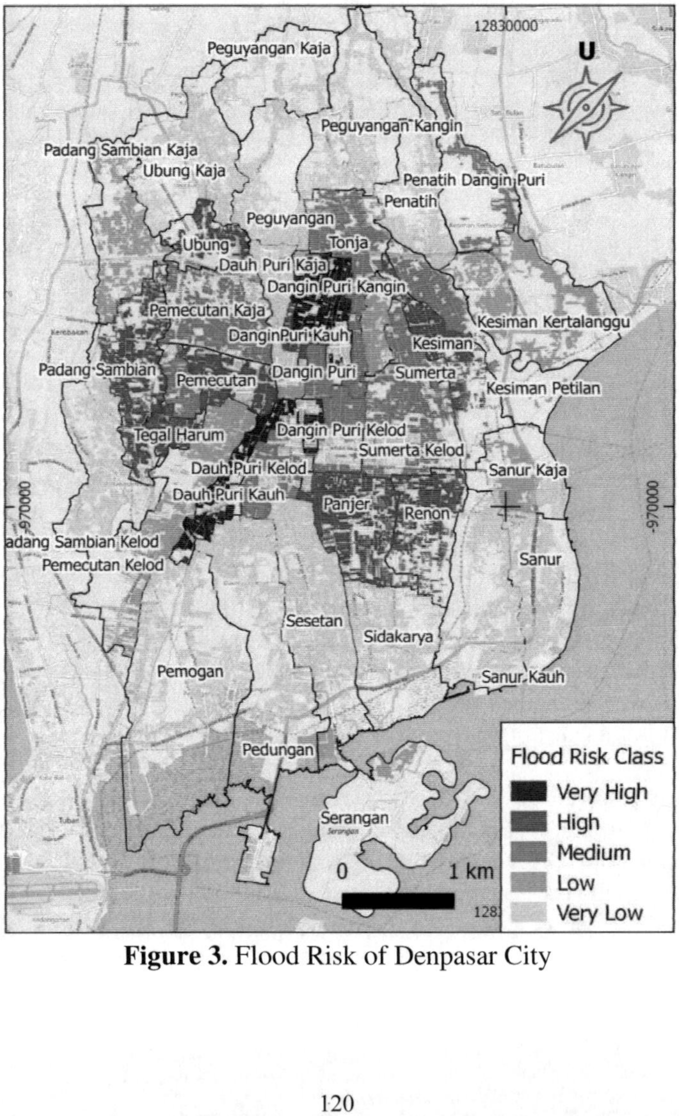

Figure 3. Flood Risk of Denpasar City

Table 4. Flood Risk in Denpasar City

Village	Settlement Area (Ha)	Population at Risk	Village	Settlement Area (Ha)	Population at Risk
Very Low Risk:			**Medium Risk (continuation):**		
Sidakarya	144.93	13590	Padangsambian	20.76	2148
Dauh Puri Kaja	68.72	18830	Panjer	23.57	2325
Pedungan	206.69	13086	Penatih Dangin Puri	72.68	5608
Ubung Kaja	157.45	24053	Sumerta Kelod	144.85	11063
Peguyangan Kangin	129.02	9951	Dangin Puri	36.71	7078
Penatih	132.66	7642	Kesiman Petilan	83.13	6508
Pemogan	218.52	22811	Tonja	127.71	13080
Sanur	146.33	9401	Dangin Puri Kelod	65.61	5843
Sanur Kauh	168.03	7724	Sumerta	56.11	7740
Peguyangan	124.18	12828	Sumerta Kauh	37.52	6932
Sesetan	270.59	25937	Tegal Kertha	50.72	15615
Padangsambian Kelod	161.83	20165	Pemecutan Kaja	164.74	29084
Peguyangan Kaja	74.59	5796	Dauh Puri Kelod	88.32	15049
Total	***2003.54***	***191814***	Pemecutan	96.25	18308
	(41.12%)	***(31.74%)***	**Total**	***1118.18***	***202478***
				(22.95%)	***(33.51%)***
Low Risk:					
Serangan	17.63	3783	**High Risk:**		
Sanur Kaja	67.88	8053	Kesiman	100.65	8218
Dauh Puri Kangin	16.69	4425	Ubung	52.71	9092
Pemecutan Kelod	193.84	27481	Renon	105.59	8450
Padangsambian Kaja	145.36	12125	DanginPuriKauh	23.38	7652
Sumerta Kelod	8.19	625	Padangsambian	170.68	17651
Sumerta Kaja	58.06	7496	Panjer	180.27	17781
Kesiman Petilan	13.15	1030	TegalHarum	147.42	10704
Tonja	18.23	1868	**Total**	***780.7***	***79548***
Kesiman Kertalangu	177.86	12549		***(16.02%)***	***(13.16%)***
Total	***716.89***	***79435***	**Very High Risk:**		
	(14.71%)	***(13.14%)***	Dangin Puri Kangin	59.72	10007
Medium Risk:			Dangin Puri Kaja	68.46	12899
Kesiman	21.81		Dauh Puri	35.26	10743
Renon	27.69	1781	Dauh Puri Kauh	89.37	17332
		2216	**Total**	***252.81***	***50981***
				(5.19%)	***(8.43%)***

[1]population data was taken from Center Statistical Bureau of Denpasar City [10]

5. Conclusions

Flood hazard in Denpasar is divided into three classes of hazard, medium hazard over an area of 251.7 ha (2.11%), high hazard of 4686.7 ha (38.9%) and very high hazard of 7107.1 ha (59%).Low to moderate hazard spatially distributed in North Denpasar Sub-district, high hazard distributed in most areas of East Denpasar Sub-district and very high hazard distributed in most areas of West Denpasar and South Denpasar Sub-district.

The vulnerability of population in Denpasar city can be divided into four classes, very low, low, medium and high vulnerability. High vulnerability with total of 19091 population located in Kelurahan Kesiman and Ubung Village. Medium vulnerability in total of 241303 population located in 19 villages, meanwhile total 229043 of population are included in low vulnerability class which located in 11 villages over 3 Sub-districts, South Denpasar, North Denpasar and West Denpasar. Very low vulnerability located in 6 Villages over 3 Sub-districts, South Denpasar, North Denpasar and West Denpasar with total population reached 81851 people.

The availability of budget for development isused as parameter for coping capacity. The availability of budget itself is different for each village, ranging from IDR 304,900,000.00 (USD 22,886) until IDR 9,627,400,000.00 (USD 722,663). Villages with the highest budget are Kesiman, Ubung Kaja, Sumerta Kaja, Padang Sambian Kelod, KesimanKertalangu, Sesetan, Pedungan, Dauh Puri Kaja, Penatih, Peguyangan Kangin, Pemogan and Sidakarya.

Flood risk in Denpasar city was divided into five classes, very low risk, low risk, medium risk, high risk and very high risk.Total population with very high risk reached 202478 people or 13.16% of total population. The total area is 780.7 ha or 16.02% from total settlement in Denpasar city.Total population with high risk reached 202478 people or 13.16% of total population. The total area is 780.7 ha or 16.02% from total settlement in Denpasar city.The number of population with medium risk reached 202478 people or 33.51% of total population which occupied 22.95% of total settlements or 1118.18 ha. The total number of population with low risk reached 79435 people or 13.14% of total population with area of low flood risk is 716.89 ha or 14.71% of total settlements in Denpasar City. Very low flood risk with total population at risk reached 19184 people or 31.74% of total population and occupied 2003.54 areas or 41.12% of total areas of settlements.

6. Acknowledgments

Authors would like to thank the Rector of Universitas Udayana, Dean of Agriculture Faculty of UniversitasUdayana, Head of Center for Spatial Data Infrastructure Development (PPIDS) UniversitasUdayana, fellow staf, researchers and professors at UniversitasUdayana for all the support, comments and suggestions that makes this paper finished.

7. References

[1] BPS Kota Denpasar 2015 *Buku Saku Statistik Denpasar*
[2] BNPB 2011 *Indeks Rawan Bencana Indonesia*
[3] Kodoatie R J and Sugianto S 2002 *Banjir: beberapa penyebab dan metode pengendaliannya dalam perspektif lingkungan* (Pustaka Pelajar)
[4] Sebastian L 2008 Pendekatan pencegahan dan penanggulangan banjir
[5] Makarov A, Mihailova A, Arefiev N, Pavlov S, Chashchina T, Terleev V and Badenko V 2015 Country Area Territory Protection from Flooding ; Construction Conditions , Problem Definition and Solution *Procedia Eng.* **117** 225–231
[6] Khan B, Iqbal M J and Yosufzai M A K 2011 Flood risk assessment of river Indus of Pakistan *Arab. J. Geosci.* **4** 115–122
[7] Marchi L, Borga M, Preciso E and Gaume E 2010 Characterisation of selected extreme flash floods in Europe and implications for flood risk management *J. Hydrol.* **394** 118–133
[8] Westen van C, Alkema D, Damen M, Kerle N and Kingma N 2009 Multi-hazard risk assessment. Distance education course, guide book *En United Nations Univ. Sch. Disaster Geo-information Manag.*
[9] Thomas D 2004 Natural Hazards Risk Assessment for the State of Colorado *University of*

Colorado and Health Science Center (University of Colorado and Health Science Center)
[10] UNISDR 2009 2009 UNISDR Terminology on Disaster Risk Reduction *Int. Strat. Disaster Reduct.* 1–30
[11] BappedaProvinsi Bali and PPLH UNUD 2006 Studi Identifikasi Potensi Bencana Alam di Provinsi Bali *Research Report*
[12] BPBD Provinsi Bali 2016 *LaporanKejadianBencana di Kabupaten/Kota Denpasar*
[13] BPS Kota Denpasar 2015 *Kota Denpasar dalam Angka 2015*

Mapping Between Bus Rapid Transit Shelter and High School Location in Semarang

D I K Dewi[1], A R Rakhmatulloh[1], and P Anggraini[1]

[1]Departmen of Urban and Regional Planning, Faculty of Engineering, Diponegoro University, Indonesia

Email: diah.intan@pwk.undip.ac.id

Abstract: The main users of public bus transport are those who have the goal to work and attend school. But the last few years there has been a decline in the use of public transport for high school students. Partly of the reason are the high use of motorcycle by student and lack of bus stop service range to high school location. This research has aim to increase the use of public transport for school students by mapping Bus Rapid Transit (BRT) shelter and school locations. The research method used are descriptive quantitative with GIs analysis tools and using spatial analysis approach

Keywords: preservation, tourism, educational religious, Kauman

1. Introduction

Transportation is an attempt to move, transport or divert an object from one place to another, where elsewhere this object is more useful or can be useful for a particular purpose [1]. Transportation will grow as the activity grows. The wider the area to be serviced, the more transportable the movement will be. Increasingly urban dwellers lead to higher activity and the number of trips of both people and goods. Problems such as congestion and other public transport problems start to emerge.

Transportation as an activity of moving goods and people from the origin to destination creates place utility and time utility, because the value of the goods becomes higher in the destination than in the place of origin. [2] Transportation is a service activity (service activities). Transportation services are needed to assist the activities of other sectors (agriculture, industry, finance, government, transmigration, land-security and others to transport goods and people in their respective activities) [2]

The transportation system has an important role for the development of a city. Implementation of the role of inappropriate transportation systems can have negative impacts that affect the development in all areas of economic, social and environmental. The development of transportation in big cities in Indonesia is increasing due to the growth and development of the city and the rate of population growth.

Besides, according to Tamin [3], one of the causes of congestion in urban areas is the increasing tendency of transportation service users to use private vehicles compared to public transport. In addition to the improving economic conditions leading to higher levels of private vehicle ownership, the declining role of public transport is also due to the low level of public transport service itself. In

IOP Conf. Series: Earth and Environmental Science **123** (2018) 012013 doi:10.1088/1755-1315/123/1/012013

essence, the low level of service involves inadequate facilities and infrastructure, long travel time, number of passengers exceeding transport capacity, low level of comfort, inadequate network system, and difficult accessibility to certain areas. In essence, the low level of service involves inadequate facilities and infrastructure like bus shelter and pedestrian path also difficult accessibility to certain areas. In addition to the significant growth in activity, travel patterns have increased considerably along with economic growth. There are 2 (two) different patterns of movement, the first pattern of movement throughout the workday or school, where the trips that occur tend to support more dominant work and school activities, while the second pattern is travel that is for the sake of leisure.

The existences of transportation facilities that serve various urban areas today generally have a lot of poor quality, thus causing various impacts to the community especially for student travel behavior. A good understanding of the factors that influence student travel behavior in general and walking to school in particular so far is not reflected in the provision of public facilities in urban areas. However this very limited understanding is reflected in current government policies based solely on the needs of parent and community preferences [4]

The operation of Bus Rapid Transit in Semarang City which is expected to increase the use of public transportation was less got public interest. But as a new program, there are certainly weaknesses such as service levels; Timeliness; Condition of facilities and infrastructure / facilities, human resources, level of user discipline, and others. There needs to be integration between the pedestrian path with the use of bus stops to increase the use of public transport [5]. Designing high-quality and safe walking and cycling environments, mixed land-use patterns and walking/cycling friendly environments are cardinal features of integrated and sustainable transport and urbanism everywhere accessible urban activities and safe, attractive walking and cycling environs – are particularly vital to the welfare of the neediest members of the world's countries [6]. Individuals walk behavior has been empirically tested for different neighborhood types; land use pattern street network, including accessibility and connectivity and pedestrian environment features. This literature provides evidence of a correlation between various aspects of the built environment and walking trips behavior. [7] By looking at these weaknesses, this research focuses on improving the condition of facilities and infrastructure, so that what is needed to improve the reach of the bus stop service with the location of the school.

The Bus Rapid Transit (BRT) system is a mass-based, road-based freight where it utilizes special and exclusive channels. The Bus Rapid Transit (BRT) system is a mass-based, road-based freight where it utilizes special and exclusive channels. In addition, the system used is a closed system where passengers can go up and down only at bus stops and of course must be equipped with a ticket system either in the form of tickets for one way or subscribe to prepaid mechanism. In order for the passengers to be comfortable on the way to and leaving the bus stop, there are reliable ferry crossing facilities, security officers at every stop, travel time schedule and also the absence of street hawkers either at bus stops or bridges except in designated places. In addition to the easy to get and leave lane bus way then from certain locations will be provided public transport route. Bus way is a special lane for bus trajectory with a view to improving the efficiency of public transport system, which is to shorten travel time and transportation cost [8]. However, the implementation of BRT in the city of Semarang currently have a special line, but preferably on the setting up and down passengers at special stops that have been provided.

2. Data and Methods

2.1. Data

2.1.1. Location of High School

In this study used data from environmental conditions around BRT and High School locations, which included a database of geographic information systems, hard copy maps, aerial photographs,

IOP Conf. Series: Earth and Environmental Science **123** (2018) 012013 doi:10.1088/1755-1315/123/1/012013

and data collected from field visits. The high school service area was first divided into a 700-square-meter grid cell and the size of the urban form was collected from the cell containing the school trip route calculated on the basis of the nearest travel route. The provision of school facilities based on SNI is only based on the range of services area without regard to the provision of other infrastructure facilities [9]. Overall, the results of the nearest route map analysis between BRT and SMA locations suggested influencing pedestrian movement patterns: density of land use and diversity, road patterns, pedestrian infrastructure, and traffic levels in the school environment. Of the attributes listed above, those with an area size, such as a road area, are counted as part of a travel route. Other attributes such as pedestrian path conditions should be available to road users based on field observations.

Table 1. The Location of Each (16) High School

	Street Network ● **High School (SMA)**	**High School**	**Built Environment Around The School**
SMA 1			SMA 1 has a strategic location because its located in the center of Semarang city, and nearby there is a provincial government center and shopping center in Simpang Lima Area
SMA 2			SMA 2 is located in a residential neighborhood along the majapahit road corridor which is the main access in the eastern region from the center to the suburbs (Pedurungan - Mranggen)
SMA 3			SMA 3 and SMA 5 are in strategic location because around Tugu Muda area and adjacent to mayor office of city of Semarang
SMA 5			
SMA 4			SMA 4 is located in neighborhood street or in residential area. The distance between the school location and the main road is about 500 meters. There is no adequate pedestrian path around the area.
SMA 6			SMA 6 is located in neighborhood street or in residential area. There is no adequate pedestrian path around the area.

IOP Conf. Series: Earth and Environmental Science **123** (2018) 012013 doi:10.1088/1755-1315/123/1/012013

	Street Network ● **High School (SMA)**	**High School**	**Built Environment Around The School**
SMA 7	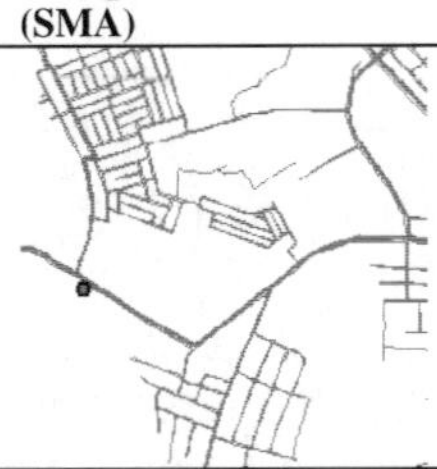		SMA 7 is located on the main road of the city that connects area A to region B. Although on the main roadside but the location of the school is not bypassed by the BRT line only local public transport.
SMA 8			SMA 8 is located on the main road of the city that connects area A to region B. Although on the main roadside but the location of the school is not bypassed by the BRT line only local public transport. There is no adequate pedestrian path along the main road, and the traffic conditions are quite crowded. This raises issues of security and inconvenience to students to walk along the road
SMA 9			SMA 9 is located on the main road of the city that connects area A to region B. Although on the main roadside but the location of the school is not bypassed by the BRT line only local public transport. There is no adequate pedestrian path along the main road, and the traffic conditions are quite crowded. This raises issues of security and inconvenience to students to walk along the road
SMA 10			SMA 10 is located on the main road of the city that connects area A to region B. Although on the main roadside but the location of the school is not bypassed by the BRT line only local public transport. There is no adequate pedestrian path along the main road, and the traffic conditions are quite crowded. This raises issues of security and inconvenience to students to walk along the road.
SMA 11			SMA 11 is located in neighborhood street or in residential area. There is no adequate pedestrian path around the area.
SMA 12			SMA 12 is located on the main road of the city that connects area A to region B. Although on the main roadside but the location of the school is not bypassed by the BRT line only local public transport. There is no adequate pedestrian path along the main road, and the traffic conditions are quite crowded. This raises issues of security and inconvenience to students to walk along the road.
SMA 13			SMA 13 is located in neighborhood street or in residential area. There is no adequate pedestrian path around the area.

	Street Network ● **High School (SMA)**	**High School**	**Built Environment Around The School**
SMA 14			SMA 14 is located on the main road of the city that connects area A to region B. Although on the main roadside but the location of the school is not bypassed by the BRT line only local public transport. There is no adequate pedestrian path along the main road, and the traffic conditions are quite crowded. This raises issues of security and inconvenience to students to walk along the road.
SMA 15	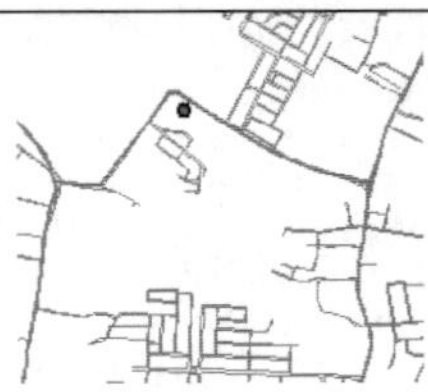		SMA 15 is located on the main road of the city that connects area A to region B. Although on the main roadside but the location of the school is not bypassed by the BRT line only local public transport. There is no adequate pedestrian path along the main road, and the traffic conditions are quite crowded. This raises issues of security and inconvenience to students to walk along the road.
SMA 16			SMA 10 is located on the main road of the city that connects area A to region B. Although on the main roadside but the location of the school is not bypassed by the BRT line only local public transport. There is no adequate pedestrian path along the main road, and the traffic conditions are quite crowded. This raises issues of security and inconvenience to students to walk along the road.

Source: Survey results, 2017

2.1.2. Locations of Bus Rapid Transit Shelter

Service coverage is one of the factors in determining the location of shelters. Site determination is determined with minimum distance to walk, i.e. 400m (according to urban housing planning / Standard of National Indonesia [10,11]. High school location within range of the bus stop services currently only around 50% covers 9 high schools (SMA1, SMA 2, SMA 3, SMA 4, SMA 5, SMA 6, SMA 8, SMA 11, SMA 13). Most high schools served by bus stops are located in the downtown area but this is not supported by good pedestrian lines. This condition will indirectly affect the interest of students to use public transportation. Because the use of public transport is not only judged by the performance of conveyance, but also affect the provision of pedestrian facilities as the link between the locations of the school to the bus stop. The description of the condition of the bus stop service to the existing school location in the city of Semarang can be clearly seen in the table and picture below.

Table 2. The Range of BRT Shelter Service to the Location of Each (16) High School

	Street Network ● **BRT Shelter**	**BRT Shelter**	**Built Environment Around the School**
SMA 1	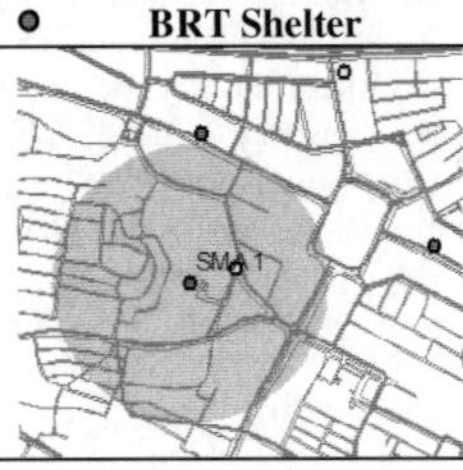		The BRT shelter in front of SMA 1 has a strategic location because its located in the center of Semarang city, and nearby there is a provincial government center and shopping center in Simpang Lima Area

	Street Network ● **BRT Shelter**	**BRT Shelter**	**Built Environment Around the School**
SMA 2	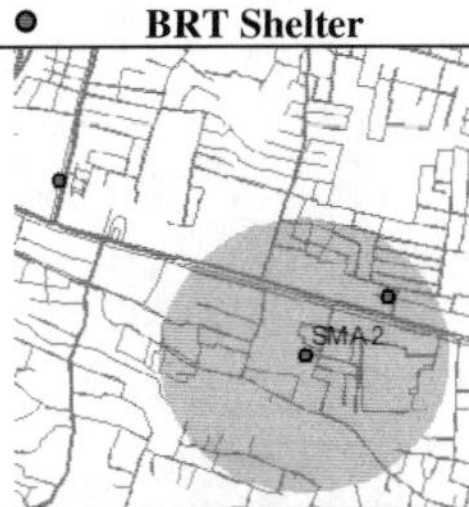		School location adjacent to the bus stop which is located beside the road Majapahit, both towards Simpang Lima (50 meters) and toward Pucanggading (200 meters). Not available pedestrian paths on both sides of the road Majapahit. Majapahit road conditions are very crowded, so it provided a bridge crossing to the bus stop towards Pucanggading.
SMA 3 SMA 5	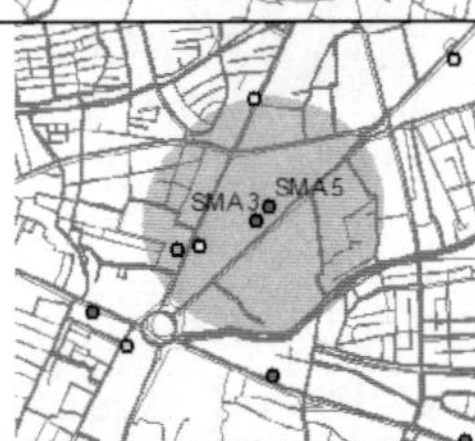		School location adjacent to the bus stop which is located beside the road Pemuda, in front of Mayor Office towards to Tugu Muda (50 meters). Because of its location in the city center has a pedestrian track that is very adequate and comfortable to use for walking
SMA 4			School location adjacent to the bus stop which is located beside the arterial road Semarang Solo, both towards Simpang Lima (50 meters) and toward Bawen (200 meters).
SMA 6		 	School location adjacent to the bus stop which is located beside the arterial road Semarang Solo, both towards Tugu Muda (500 meters) and toward Kalibanteng (500 meters).
SMA 7	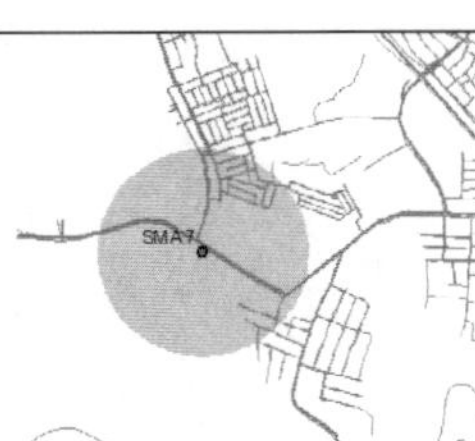		School location not reachable with the BRT shelter location services
SMA 8			School location adjacent to the bus stop which is located beside the arterial road Semarang Kendal, both towards Semarang (500 meters) and toward Kendal (500 meters).

	Street Network ● BRT Shelter	BRT Shelter	Built Environment Around the School

SMA 9	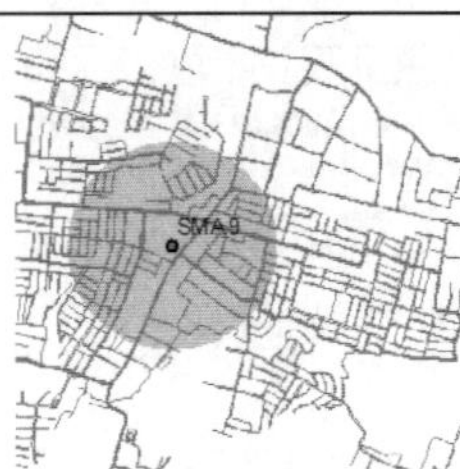		School location not reachable with the BRT shelter location services
SMA 10	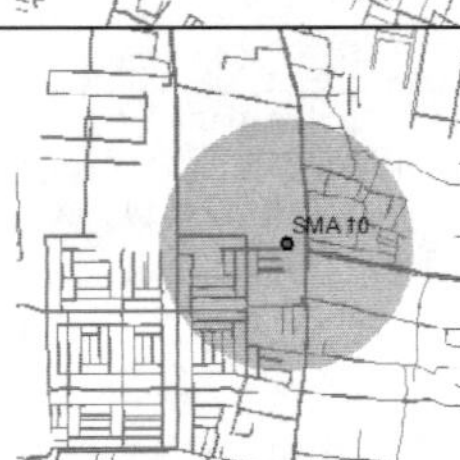		School location not reachable with the BRT shelter location services
SMA 11	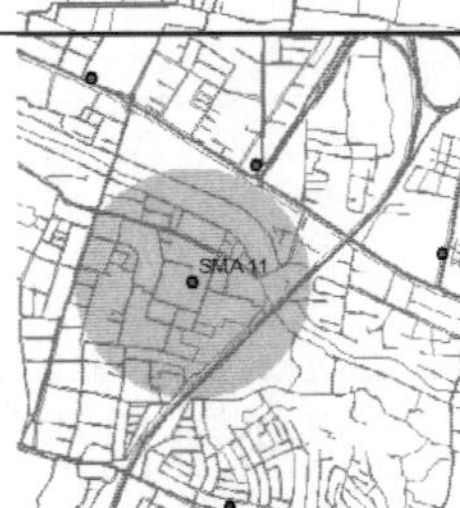		The location of the school adjacent to the bus stop which is on the road Majapahit (500meter)
SMA 12	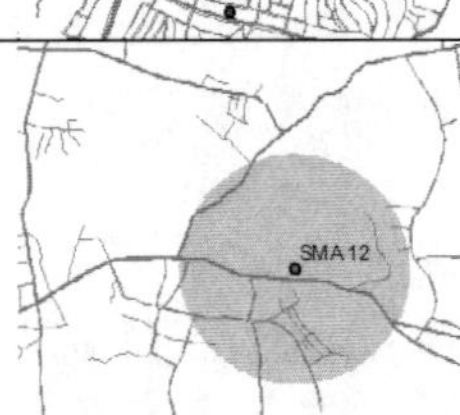		School location not reachable with the BRT shelter location services
SMA 13			The location of the school is outside the reach of the bus stop service located on the Mijen highway (> 500meter)

	Street Network ● BRT Shelter	BRT Shelter	Built Environment Around the School

SMA 14			School location not reachable with the BRT shelter location services
SMA 15			School location not reachable with the BRT shelter location services
SMA 16	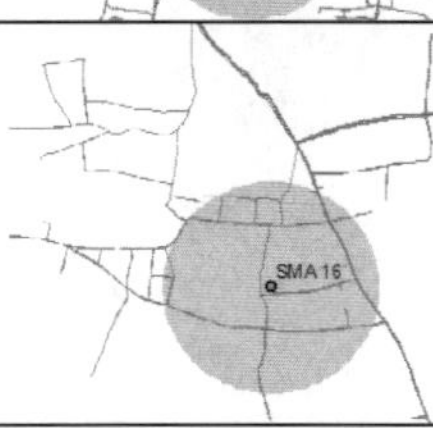 		School location not reachable with the BRT shelter location services

Source: Survey results, 2017

2.2. Methods

In this study used data from environmental conditions around BRT and High School locations, which included a database of geographic information systems, hard copy maps, aerial photographs, and data collected from field visits. The high school service area was first divided into a 700-square-meter grid cell and the size of the urban form was collected from the cell containing the school trip route calculated on the basis of the nearest travel route. Overall, the results of the nearest route map analysis between BRT shelter and high locations suggested influencing pedestrian movement patterns: density of land use and diversity, road patterns, pedestrian infrastructure, and traffic levels in the school environment. Of the attributes listed above, those with an area size, such as a road area, are counted as part of a travel route. Other attributes such as pedestrian path conditions should be available to road users based on field observations.

3. Result and Discussion

3.1. Location of High School

The location of BRT located at the location of service coverage with a radius of 400m at high school includes Tugu sub-district (SMA 8), Semarang Tengah (SMA 1, 3 and 5), Tembalang (SMA 15), Banyumanik (SMA 4), Pedurungan (SMA 2). While the coverage of services of upper secondary schools that are not served by BRT from 16 sub-districts in Semarang are in Ngaliyan sub-district (SMA 7), Mijen (SMA 13 and SMA 16), West Semarang (SMA 6), Gunung Pati (SMA 12), Genuk (SMA 10), North Semarang (SMA 14), Banyumanik (SMA 9), South Semarang (SMA 11).

IOP Conf. Series: Earth and Environmental Science **123** (2018) 012013 doi:10.1088/1755-1315/123/1/012013

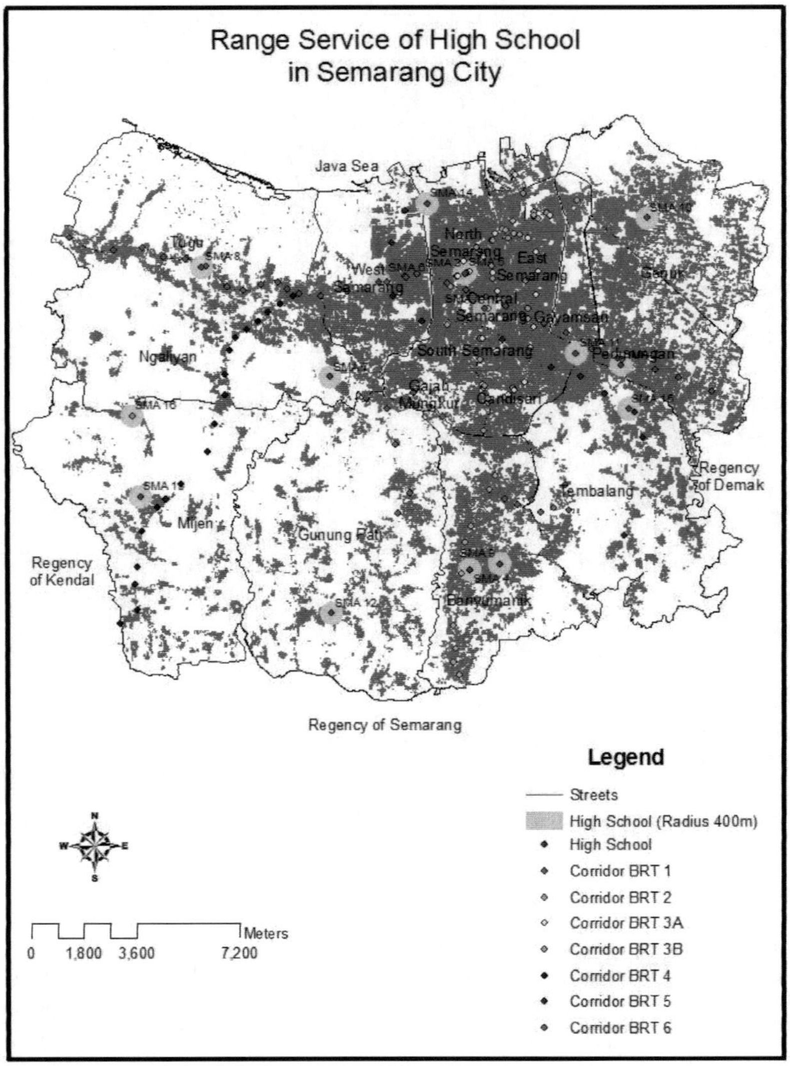

Source: Analysis, 2017

Figure 1. Mapping Range Service of High School to the Location of BRT Shelter in Semarang City

3.2. Locations of Bus Rapid Transit Shelter

Through spatial analysis, it is possible to know the range service of BRT to the reach of high school Services. Based on the analysis, it can be seen that BRT service coverage has been covered in several high school locations spread in various sub-districts in Semarang City. From figure 2, the distribution of the location of both high school and BRT facilities in Semarang City, it can be seen also, the distance / nearest location between the facilities. The proximity between facilities, providing convenience for users in utilizing existing facilities. Through network analyzes it can be seen that the location of the high school facility has the closest route to the location of the BRT facility. Based figure 3 and table 3, on the nearest distance of high school and BRT facilities in Semarang City, it is known that 44% of high school facility has the closest distance to BRT facility, while 46% of high school facility is still not reached by BRT facility.

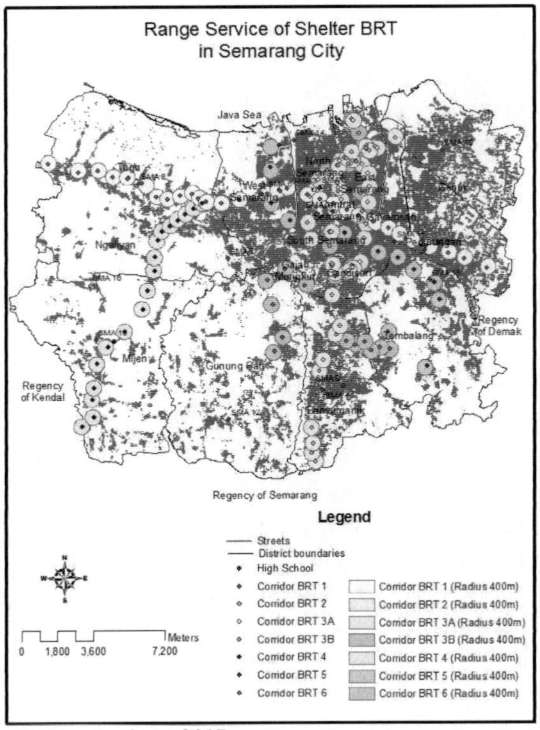

Source: Analysis, 2017

Figure 2. Mapping Range Service of Shelter BRT to the Location of High School in Semarang City

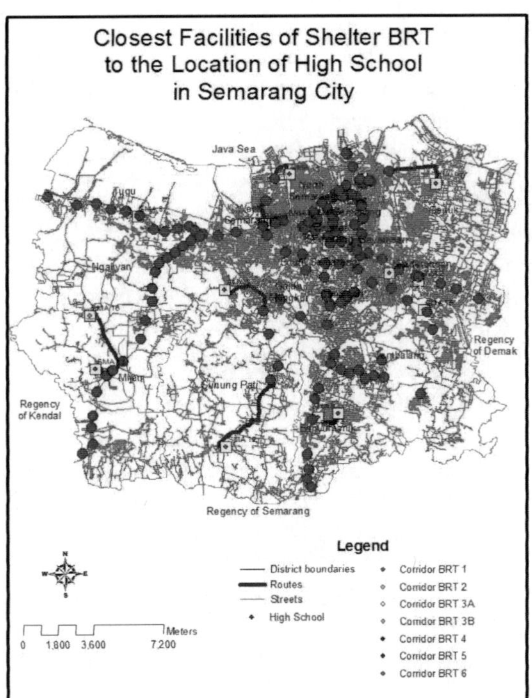

Source: Analysis, 2017

Figure 3. Closest Facilities of Shelter BRT to the Location of High School In Semarang City

Table 3. Facilities Distances (m) of Shelter BRT to the Location of High School In Semarang City

Location Facilities High School - BRT Shelter	Facilities Distances (m)
Location 1 (SMA 8) - Location 6	304
Location 2 (SMA 7) - Location 139	**3154**
Location 3 (SMA 6) - Location 129	**1433**
Location 4 (SMA 14) - Location 128	**1542**
Location 5 (SMA 1) - Location 68	62
Location 6 (SMA 10) - Location 53	**3273**
Location 7 (SMA 2) - Location 27	16
Location 8 (SMA 15) - Location 121	228
Location 9 (SMA 4) - Location 54	374
Location 10 (SMA 9) - Location 54	**1816**
Location 11 (SMA 16) - Location 119	**3366**
Location 12 (SMA 5) - Location 77	37
Location 13 (SMA 3) - Location 77	106
Location 14 (SMA 12) - Location 143	**5605**
Location 15 (SMA 11) - Location 25	**512**
Location 16 (SMA 13) - Location 105	**658**

Source: Analysis, 2017

4. Conclusion

The existence of a mass public transport in general less public interest. However, for school students using public transportation remains an option. Therefore, in improving public transport services, the city government is expected to pay attention to the placement of the bus stop location and improve or provide a safe pedestrian path. Criteria for placement of bus stops currently not noticed walking distance range for school students. Similarly, the condition of pedestrian paths connecting school locations and bus stops is not well available. Even in some schools in the suburbs of school students walking on the street, it is not safe for them. This is in contrast to the characteristics of pedestrian lanes in the downtown area such as SMA 1 (Simpang Lima) and SMA 3, SMA 5 (in Tugu Muda) connecting the school location and bus stop which is quite good and comfortable.

5. Acknowledgements

This study was supported by a grant from Department Urban and Regional, Faculty of Engineering, Diponegoro University, Semarang, Republic of Indonesia. We also grateful the board and community of Semarang City which has helped in realizing the research.

6. Reference

[1] Miro F 1997 Sistem Transportasi Kota *Bandung: Tarsito*
[2] Adisasmita R 2010 Dasar-dasar ekonomi transportasi *Makassar Graha Ilmu*
[3] Tamin O Z 2000 *Perencanaan & Pemodelan Transportasi (Second Edition)*
[4] McMillan T E 2005 Urban form and a child's trip to school: the current literature and a framework for future research *CPL Bibliogr.* **19** 440–456
[5] Herbowo N 2012 Studi Presepsi Pengguna Transjakarta Pada Koridor II (Pulogadung-Harmoni) *J. Reg. City Plan.* **23** 37–50
[6] Cervero R 2014 Transport Infrastructure and the Environment in the Global South: Sustainable Mobility and Urbanism *J. Reg. City Plan.* **25** 174–191
[7] Banerjee T, Uhm J and Bahl D 2014 Walking to school: the experience of children in inner city Los Angeles and implications for policy *J. Plan. Educ. Res.* **34** 123–140
[8] Suwandono D, Dewi D, Mussadun M and Anggraini P 2014 Optimalisasi Jangkauan Pelayanan

Halte Brt/Bus Trans Semarang *Pros. Forum Stud. Transp. Antar Perguru. Tinggi* **2**
[9] Pancarrani G and Pigawati B 2014 Evaluasi Kesesuaian Lokasi Dan Jangkauan Pelayanan Sekolah Menengah Umum Di Kecamatan Kebakkramat Kabupaten Karanganyar *Geoplanning J. Geomatics Plan.* **1**
[10] Nasional B S 2004 Tata cara perencanaan lingkungan perumahan di perkotaan *Bandung Badan Standarisasi Nas.*
[11] De Chiara J, Panero J and Zelnik M 1995 *Time-saver standards for housing and residential* development (McGraw-Hill Companies)

IOP Conf. Series: Earth and Environmental Science **123** (2018) 012014 doi:10.1088/1755-1315/123/1/012014

Global and Domestic Spheres: Impact on The Traditional Settlement of Penglipuran in Bali

G A M Suartika[1]

[1]Udayana University, Indonesia

Email: ayusuartika@unud.ac.id

Abstract: The issue addressed is how to situate the 'life-world' of a traditional Balinese village in the context of global tourism which is now worth USD$ 7.61 trillion, This problem has been brought sharply into focus, indeed amplified, by the recently adopted policy of local government in Bali to designate specific villages as tourist sites. Such action appears to be taking place in the absence of any prior studies of the social, environmental and other impacts on local communities. The following paper represents an initial attempt to sketch out the breadth of the problem from the global to the local. The paper first proceeds by examining the conceptual framework of globalisation - fundamentally an economic process, - from which global cultural identities merge. It proceeds to give an overview of policy at a global level starting with the idea of designated United Nations World Heritage sites and the inevitable commodification of 'place.' Then the issues and problems that flow from these first sections are discussed. Next, the situation in Indonesia as a whole is brought into sharp relief, focussing on policy and legislation that affect the internationally celebrated island of Bali. Finally, a case study of *Penglipuran*, a traditional *Bali Aga* village is discussed taking into account the preceding context. In conclusion, suggestions are made as to the significant research and policy issues that need to be addressed concurrently with the development of the settlement. The study documented in this paper was conducted using phenomenological approaches. It was grounded by researcher's experiences, gained from her deep involvement in community life as well as physical and site observation activities during data collection period.

Keywords: Globalization, Commodification, Tourism, *Bali Aga* Settlement, *Desa Penglipuran.*

1. Introduction

Globalisation is a historic process that has existed from immemorial time, based upon contemporary knowledge of the planet. In every age commerce and culture have been exported to the fringes of the known world. At the same time, these boundaries have been changing as explorations expanded the understanding of geography, a process allowed by the invention of new technologies – in transport, time, mathematics and other devices. Today we also understand globalisation differently, based on the development of politics, technology, economy and culture. Nevertheles, the global world we now inhabit is vastly different to any previous epoch [1]. It is no longer "known" in the traditional sense – it is now a technological construct [2].This is due to a singular invention that has compressed space and time, allowing communication to occur instantaneously wherever communications devices are located. It is called the internet.

This phenomenon, now forty years old, has affected in every level of social life. Like all prior inventions, it is politically charged, shifting power into unseen new realms frequently beyond social control as global corporations challenge all existing national boundaries. It is a process qualitatively different from all prior modes, allowing a monumental compression of wealth into fewer and fewer hands [3]. It is a concentration that has increasingly placed the wealth of corporate power ahead to many states, with the capacity to transform national policies in many countries, particularly the Global South [4]. Given the mobility of capital allowed by the internet, where vast resources can be transferred from one site to another in milliseconds, multinational corporations also have an enormous influence on public opinion, to create new forms of imperialism and monopoly practices, and to persuade governments into desired forms of cooperation and action – development policy, investment, social programs and other functions.

In this respect the European Economic Community (EEC) currently going through significant political upheaval, is exemplary. The EEC was not a mechanism devised to improve the lives of inhabitants, to enhance local culture, or to provide better social programs. It was fundamentally a method of accelerating capital formation, promoting state neo-corporatism, and advancing the interests of banks and multinational corporations. The basic mechanisms involved were the free movement of capital across national boundaries, with a similarly open system allowing labour transit-the shared distribution of migrants from developing countries. This effectively reduced the price of wages by undermining existing sanctions and increasing the labour pool. Overall, it secured greater mobility of capital, materials and populations within the EEC.

While prior imperialist practices encouraged e.g. migration from Africa, and India to the UK, and from Algeria and Vietnam to France, more recently migrants to Europe from places like Pakistan, Bangladesh, Turkey, Indonesia, Sri Lanka, the Sudan and the Middle East, have burgeoned. Within the European Economic Community, the wealthy nations have also imported populations from more impoverished states (Germany is now the fourth largest state of Turkey), as well as from poorer regions of the EEC. The overall trajectory of the process is to import labour predominantly in one direction, for work as economic migrants. The 'affluent classes' of more developed nations are then exported to these same countries for pleasure as a function of international globalised tourism [5]. The global economy and its agencies benefit from movement in both directions.

Hence, tourism can no longer be viewed as a disorganised, somewhat random product of personal choice, an extension of the 'Grand Tour" of the Victorian period. It represents a new dimension in global economic development. It is driven by private capital in allegiance with local governments who wish to improve their economic viability, reflecting the neo-liberal Agenda and its ideology of state neo-corporatism. While local economies benefit from the tourist tsunami, the overall process further expedites the concentration of global capital. The richest eight people on the planet have a net wealth of $US426 billion ($569.6 billion), equivalent to what is held by the bottom half of the world's population. In Indonesia where Penglipuran is situated, the four top billionaires possess as much capital as the bottom 100,000,000 persons, where the interest on their capital alone could displace these millions from poverty (1).

Precedents had of course existed elsewhere. In the Philippines and Indonesia for example, 'the crony capitalism' sponsored by President Marcos, was also the prevailing ideology within President Suharto's administration. Bothfused state power with private interest [6,7]. In the case of Bali, this meant that the same clique dominated the entire tourist apparatus via outright ownership, joint partnerships or shareholding investment in tourism operators and businesses, airlines bringing in tourists, ownership of the largest hotels and tourism complexes, commercial outlets and entertainment facilities for over thirty years until his demise in 1998 [8,9]. They also left certain specific forms of dereliction in their wake (Warren 2014). Similar examples are ubiquitous, and arguably exist worldwide. For example in the gambling Mecca of Macao, where a parallel scenario operates under the tutelage of Stanley Ho and other billionaires, as well as their counterparts in Hong Kong. But today, multinational penetration is vastly better organised and supported, clandestinely or otherwise, by such organisations as the United Nations and its program of designating World Heritage Sites of which Bali has five.

The whole study was conducted using phenomenological approaches, grounded by researcher's experiences gained from her deep involvement during physical and site observation. In so doing, and in the context of globalization, this paper undertakes on a case study of Penglipuran, a Bali Aga village that has been designated a tourist destination. This process is expedited through the collaboration of state, provincial and local government – focussing on Bali as a province of Indonesia, and the local government of Bangli Regency. Within this global context, our focus is one of settlement, culture and tradition [10]. So I begin at the global level by introducing the problematic of the commodification of place and the issues that flow from it in the following method (1) Introduction, (2) The commodification of place, (3) Place and the public realm, (4) *Desa Penglipuran* and government initiative of *"desa wisata"* (5). Place and space in *Penglipuran,* (6) Community administration and social structure, (7) Culture and space, (8) Touristification and cultural conservation.

2. The Commodification of Place

The totalising trajectory of the capitalist system is to commodify all use values as well as human experience [11]. It also has 'place' as a target, from Times Square in New York to *Penglipuran* in Bali. The dominating force is the eternal expansion of capital through control over media and markets such as land, as well as the all-encompassing investment funds that are necessary for the infrastructure of tourism to proceed. It does this by means of the *commodity principle* - the pivot round which all capital accumulation takes place, superficially between buyers on the one hand and sellers on the other. While the commodity can be seen merely as an object, a thing, also the social relation fundamentally underwrites global capitalist enterprise. It represents the enduring conflict between those who own the means of production, and those who do not.

Exploitation therefore takes place as an inherent part of the system due to the power imbalance between buyers and sellers and the private expropriation of collectively generated profits. Since power is concentrated at the top, the further we descend through the economic hierarchy, the less power individuals have. While trans-national corporations know exactly what they are doing, local governments are frequently naive about local impacts, since resources and background research in tourism are seriously constrained [12]. Working down the ladder, local cultures are clearly the most informationally bereft as to the workings of capital, and to government agencies and non- government agencies (NGO's) at all scales. This implies that their existence is subject to forces that are not understood; open to exploitation, and rendered culturally vulnerable to externally imposed authorities that do not necessary have their well- being at heart [13]. Or as Piketty says "The price (market) system knows no limits or morality" [3]. In this light it is critical that the people of *Penglipuran* become informed in significant detail of their legal and political environment prior to any agreements they may be offered.

While the workings of capital are somewhat over-simplified above for the purposes of explaining principles, it must be recognised that there is as much conflict between the component parts of civil society (the state, labour and the law) as there is between the various capitals (industrial, finance, commercial, property etc.)[2]. Here two points are important. First, the above processes do not take place based on friendly or mutual support, but on the basis of conflict and competition. This environment is only set to deepen and we can safely assume that as yet unforeseen consequences will arise. Second, they nonetheless take place over space and time, in places we call cities, towns and villages. Contemporary society has been categorised by Italo Calvino as 'the New Middle Ages' where there is competition between urban settlements at all scales greater share in a world of diminishing resources. The onus is then for each settlement to 'brand'–to differentiate itself from others, to ensure 'authenticity' and thus to offer a unique commodity that can be sold not only to the so-called 'creative class', but to national and international tourist audiences (2002, 2007). *Penglipuran* has just become one such commodity.

Due to the unique culture of Bali, it is clear that the process of commodifying the island's heritage begun in the Suharto era is likely to advance to levels not previously experienced [14,15].Given the inadequacies of postmodern theorising, no encompassing explanation of Bali's position within a globalised tourist universe exist, or as Samir Amin has so eloquently stated 'The critique of capitalism preceded the faddish critique now offered us by postmodern theoreticians. "I consider Post-Modernism an intellectual non-starter in the sense that it beyond its hype it offers no conceptual instruments capable of transcending the capitalist framework"[16]. So the paper will be informed more generally by political economy. So the question to be answered is 'how can this process of commodification be constrained to the benefit of local communities, and *Penglipuran* in particular, without turning the island into another Asian Disneyland'?

3. Place and The Public Realm

As part of the overarching progress of capital accumulation, trans-national corporate power is no longer satisfied with the mere sale of commodities. The commodity principle has been extended to commodifying all aspects of society. This basically encompasses social realms previously excluded from the fundamental activity of commodity production during the post- industrial period. The process now includes many functions previously allocated to the state in the service of big capital such as social services, education, the public realm and space itself. Important to this process, and therefore to this study, is the commodification of culture and cultural products [17–21].

The liberal view of globalisation is that all of this is good since the more business that is generated, the more jobs are available and a greater opportunity for employment is created, and industrialization increasingly mechanised. Supposedly, everyone benefits. While this may contain some truth, historically this has never actually happened since the economic surplus is increasingly expropriated. The fact remains that despite economic development, the structural features of the total system remain unchanged by such 'progress' given the absence of increasing social equality [3].The emerging logic suggests that the commodification of 'place' remains a prime target strategy for business in the evolution of a globalised culture that supports endless consumption of unnecessary products [1]. Once industry, commerce and all available land have been privatised, what remains? The sole arena that has not been commodified is the *public realm*, where places represent the evolution of societies, their histories, cultures, memories, and traditions. Hence the defence of culture *qua* public space becomes the last bastion of resistance to a commodified world in which *Penglipuran* is exemplary.

Over the last century, the public realm has remained 'in public hands' being administered by the state on behalf of all citizens. But in an era where the neo-corporate state has come into being, one where capital increasingly writes the rules by which it operates, the concept of *place* and all its inferences become threatened by forces never previously unleashed. From the beginning of the third millennium, there has been rapidly accelerating encroachment by the private sector into social space, for example by renting or otherwise colonising public buildings, spaces, parks, gardens and other venues for private use – many for the promotion and advertisement of commodities. Now we must briefly turn to the abstract and politically charged idea of conservation, in order to understand how the stage is being set in Indonesia and Bali for the development of tourism in small historic communities.

4. Desa Penglipuran and Government Initiative of 'Desa Wisata'

Conceptually, *Penglipuran* and the government's concept of *desa wisata* (village-based tourism development program) must be placed in a larger context. Five sites in Bali have been listed as World Heritage Association (WHA) by the United Nations Educational, Scientific and Cultural Organization

(UNESCO) as of special cultural or physical significance. These are *Jatiluwih* rice terraces, *Taman Ayun* Temple, *Ulun Danu Batur* Temple, and Lake *Batur* and *Pakerisan* Valley which is a prime example of *Subak* irrigation. Overall Indonesia has eight such sites and there are now over 1000 WHA sites worldwide. While *Penglipuran* is not one of these, the WHA is exemplary as to how *Penglipuran* might be influenced due to its special designation as an important tourism site.

In this regard, criticism of UNESCO's management and objectives must be noted - 'Indeed, the World Heritage program has garnered considerably less regard in recent years, and is seen as more concerned with economic and political influence than the stated goals of the Convention'[2]. Not only this but criticism of the management and operations of actual locations is ubiquitous 'If you go to a lot of World Heritage sites and ask for management plans with budgets and priorities, they don't exist'[3]. In addition, the search for the tourist dollar can destroy entire communities. *Kaiping* village in China is an example of this[4], and in Japan when the WHA denoted the The *Iwami Ginzan*Silver Mine special status (a huge hole in the ground) tourism numbers went from 15,000 a year to one million after designation[5]. These issues come to the fore in our conclusions about the management and community control over *Penglipuran*'s living heritage.

This living heritage is omnipresent, in that Bali has two distinct cultures, the majority belonging to *Majapahit* origins in Java and who live almost exclusively in the lowlands, and the *Bali Aga* who live almost exclusively in the highland regions. *Penglipuran*is one of more significant native settlements in Bali, belonging to the *Bali Aga* community (Reuter 2002, Suartika and Cuthbert 2017). Administratively, it is part of *Bangli* Regency of Bali Island, a Province within the Indonesian State. *Penglipuran*has a direct line of descentto another *Desa Bali Aga* of *BayungGede*, of *Bangli* Regency, both socially and ritually. Up to the present time, people of *Penglipuran* have had a special and enduring relationship with people of *Bayung Gede*. Since it is situated around 700 meters above sea level, *Penglipuran* provides a cool and comforting living environment. This quality led some to suggest that *Penglipuran* was recognised by *Bangli*'s historic Royal Family as a place to seek solace.

Since its designation as a tourist destination in 1993 (confirmed by *Bangli* Regency's Decision No. 115 Year 1993), this settlement has been recognized for its unique qualities. These regulatory mechanisms also indicate conformity with those instigated at both national (Indonesia) and provincial (Bali) levels, such as National Regulation No. 10 Year 2009 and No. 50 Year 2011on National Tourism, and Bali Provincial Regulation No. 3 Year 1991 and No. 2 Year 2012 on Cultural Tourism. Across the island, many villages have been designated, and developed under this scheme. Several *Bali Aga* settlements have been targeted and equally attract many tourists e.g. *Desa Tenganan* (of *Karangasem* Regency) and *Desa Trunyan* (of *Bangli* Regency).This is in line with Bali Province Government's policy to develop potential *desa/*s (village/s) into tourist destinations-*desa wisata*(village-based tourist destination). *Penglipuran*'s position has been officially confirmed by *Bangli* Local Government Regulation No. 10 Year 2014; No. 9 Year 2015 and No. 34 Year 2016. However *Penglipuran* is not alone. Nine other villages of *Bangli* Regency are included under the umbrella of *Desa Wisata*.

Desa wisata, in theory, is well publicised as part of a capacity building attempt aimed at achieving self-sustained communities (when most have been 'self sustaining' for records of centuries). It promotes every possible tourist interest as a potential economic benefit to the inhabitants, when in reality it must be seen as part of a profitmaking venture of larger interests. How exactly individuals and the community will benefit remains to be spelled out in some detail. In addition, community members are encouraged to participate on the basis of self- interest at every level of its development, a process seriously at odds with the communal and collective nature of the Bali Aga societies in question. Unlike the top-down strategies that were the accepted practice during the *Orde Baru* (New Order) era of the Suharto administration, the *desa wisata* program is formulated as a bottom-up scheme. Such top-down governing practices ended (in theory) following the resignation of the president in the aftermath of the 1998 economic crisis, that impacted catastrophically ont he Indonesian economy (one amongst numerous Asian countries that had to deal with devastating effects of the crisis). The idealistic role advanced by the government in this *desa wisata* scheme is intended to facilitate the whole process of developing a controlled administrative process while consciously maintaining the prominent position of the community, as both the subject and possibly the object of the whole development process.

The concept of *desa wisata* development offers the community an *idealistic* set of circumstances to enhance its own potential. In reality however, the whole process is heavily oriented to exploiting and commodifying every single unique trait the community has available for tourism related consumption. The most obvious example is the generation of significant revenues from entry fees whose function is to open up the entire community to the tourist gaze using the medium of a commodified tourist space in its entirety [10]. Many elements of built form currently regarded as sacred entities will be made accessible to visitors, and setting limits on personal movement will be almost unenforceable. Many current examples exist of domestic transformation in Bali to capture the tourist dollar. Homes become adapted by their inhabitants in their effort to sell things tourists may purchase, including snacks, gifts, or other trophies comforting to visitors. They may also be partially transformed into showrooms for many home industries, such as the production of handicrafts, woodcarving, furniture, and woven materials such as the famous *ikat* manufactured in other locations. In this case, the function of home is often further extended to accommodate the display of various products not necessarily produced on site, with the expectation of attracting purchases from potential visitors, and it is one short step to the actual rental of accommodation to tourists looking for a 'genuine' or 'authentic' Balinese experience [22].

While these practices are already widespread in tourist centres such as *Ubud* and *Sanur*, the dominant problem is how to contain economic self-interest in *Penglipuran* and other designated villages now that the practice has been fully established elsewhere. None of the above observations necessarily disagree with the specific interests of the community, nor with the idea of flourishing village life. But it is vital to ask how the *desa wisata* development program should take place here, given that cultural conservation would necessarily be more constrained than elsewhere on the island. This hopefully coincides with the government's main goals to generate revenues (somewhat inconclusively stated) –to facilitate every opportunity for the community to achieve an economically self-reliant status; as a method of reviving traditions and local culture; to support the development of a mechanism leading to community capacity building; and of course maintaining each *desa wisata*'s aptitude to provide tourists with unique experiences undiscovered elsewhere. In order to clarify these confusions and inconsistencies, it is necessary to investigate the reality of *Penglipuran* culture in greater detail below.

5. Place and Space in Penglipuran

Physically, *Desa Penglipuran* covers around 112 hectare of land. Around 15 percent is used for homes; 50 percent for dried cultivation related uses; and the rest is dedicated for bamboo forest. All members of the *krama desa* exercise their right to occupy land for domestic purposes; cultivating agricultural land; and harvesting bamboo from the forest. A strict prerequisite is applied to the latest use. Every harvest has to be approved by the *adat*, which is acquired through a *sangkep* (meeting) of *karma desa.* The most prominent physical feature of *Penglipuran* is the well aligned traditional gates of *Penglipuran* compounds. This image emblazons the front page of the official guide to the village (Figure 1).

IOP Conf. Series: Earth and Environmental Science **123** (2018) 012014 doi:10.1088/1755-1315/123/1/012014

Courtesy of Ni Made Yudantini

Figure 1. The Picturesque Facade of Homes in Penglipuran

The *Penglipuran* settlement forms a linear pattern, oriented north and south. The north is considered *hulu* (upper side), the most sacred zone, and the south is *teben* (down side) the least sacred zone. The sacred zone is a site for many temples, and the least sacred zone is where communal *setra* (cemetery) and *Dalem* Temple are located. Housing compounds are situated in between these zones, divided into two lines of homes and connected by a *rurung gede* (main community lane). The whole settlement is surrounded by a bamboo forest and dried cultivated land. An overview of the whole layout of the settlement is presented in Figure 2.

In terms of domestic architecture, the concept 'home' in *Penglipuran* has an original layout which differs radically from both Lowland Balinese and many other Bali Aga settlements. Each compound accommodates several functions represented in structures and buildings. Figure 3 demonstrates the general layout of a home munity. All housing compound have more or less have the same physical configuration. They usually consist of: Family shrine (*sanggah*), place for worships; Six columned building (*bale sakanem*), used for ritual purposes; Kitchen (*paon*), food production and sleeping; *Loji* building (*bale loji*), for sleeping and resting; Extended uses (such as toilet, garage, etc.), conditional; *Teba* (spare/unused land).

Figure 3 and 4 demonstrate the physical configuration of a typical home in *Penglipuran*. Figure 3 represents two single homes located on both east and west sides of the *rurung*/community lane.

Figure 4 illustrates a plot that has two sets of homes located on the eastern side of the *rurung gede*/main communal lane. This is a home of two *krama desa* who are related, and share one family shrine. In this case, the shrine is located in the middle of both homes. If we look closely at both figures, the family shrine is always placed in the northeastern corner. This parallels the practice of the *sanga mandala* conception in many settlements of lowland Bali where northeast is the most sacred (*utamaning utama)* zone.

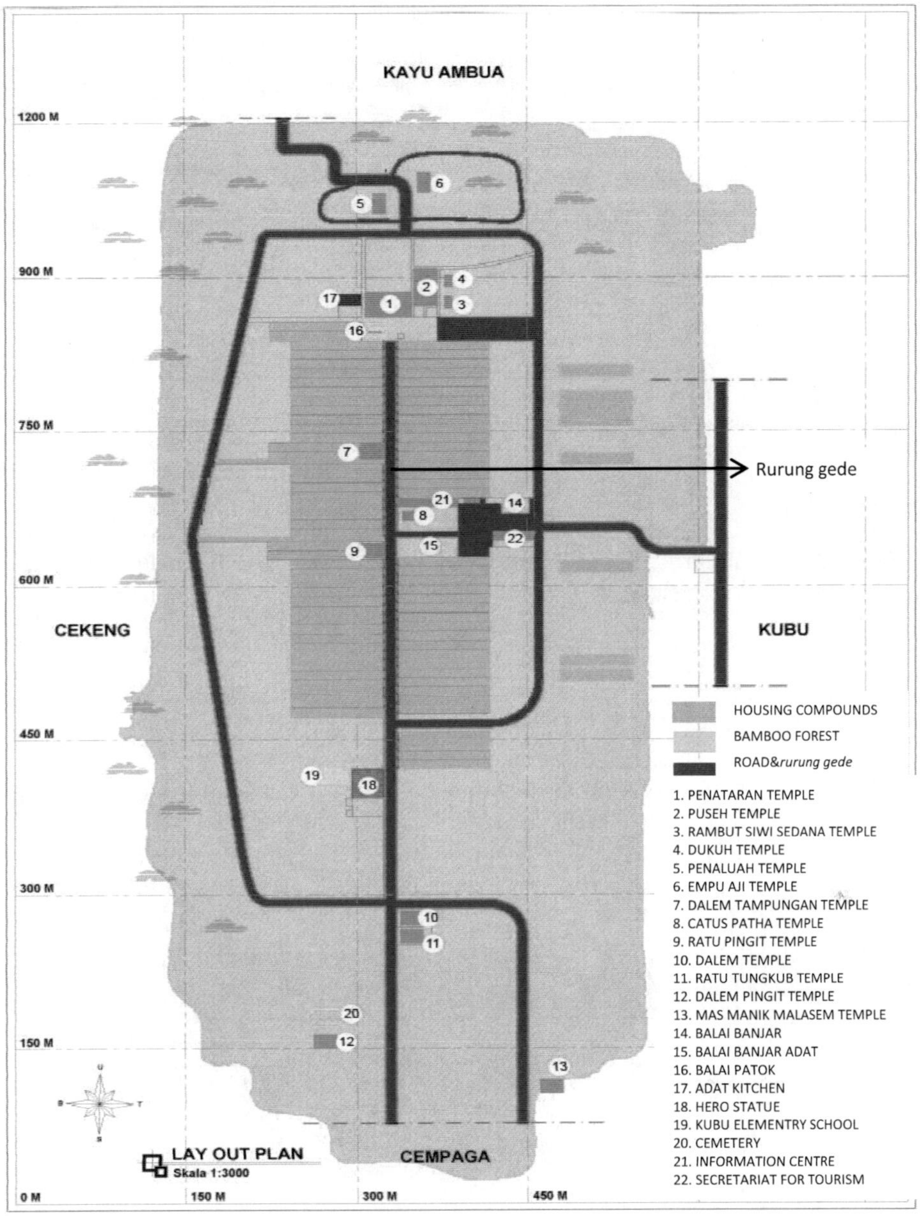

Courtesy of Agus Darma

Figure 2. Layout Plan of *Penglipuran* Settlement

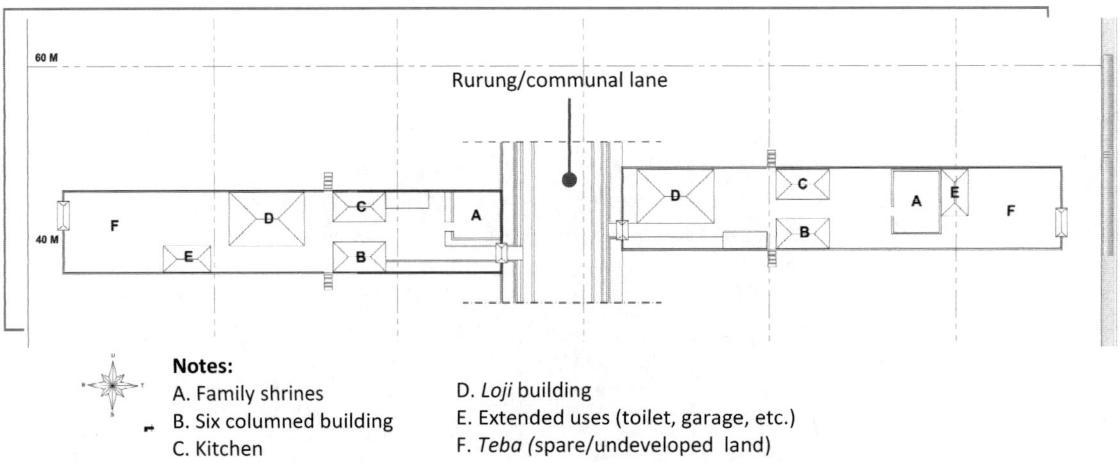

Courtesy of Agus Dharma

Figure 3. Spatial Formation of 2 Homes of Two *Krama Desa* Located on both the Eastern and Westerns Sides of the *Rurung Gede*

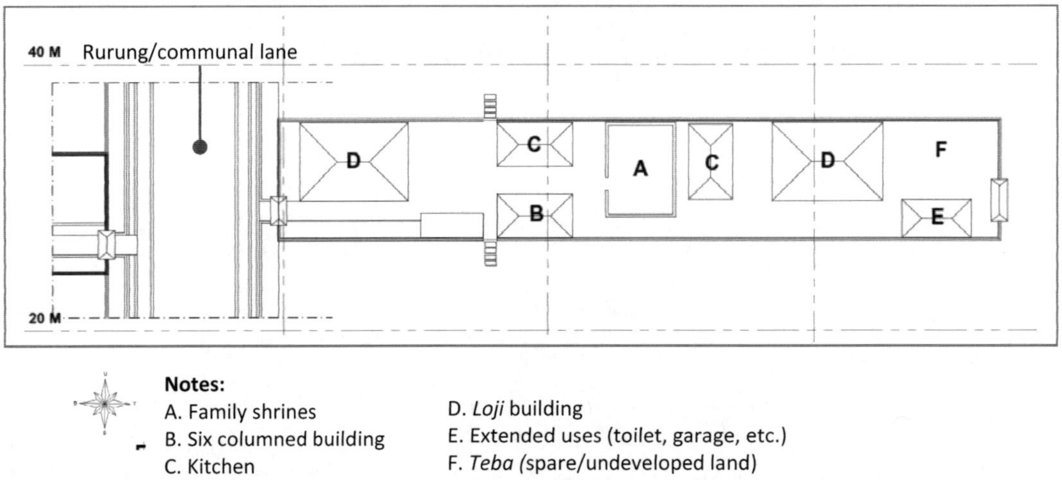

Courtesy of Agus Dharma

Figure 4. Spatial Formation of a Home of Two *Krama Desa* (Relatives) Located on the Eastern Side of the *Rurung Gede*

If we make a comparison with *Desa Tenganan* Bali Aga, another linear settlement, the sacred part of the domestic spatial formation is that located adjacent to the communal lane (*rurung*)[23]. This is a reflection of how *rurung* is viewed within different Bali Aga communities. To the *Penglipuran* conception, *rurung* is a space for circulation, while in *Tenganan* community *rurung* is conceived as part of many sacred elements within the spatial layout of the village. Thus part of the space close to the *pamesuan* (a gate to get into the house from the *rurung* or the other way around) is dedicated for structures pertained to worship. In order to situate this basic description of the physical place, it is now necessary to frame the social system that structures the spatial configuration of the village, before assessing the current and potential impact of tourism.

6. Community Administration and Social Structure

The attractiveness of *Penglipuran* as a thriving tourist destination cannot be disconnected from its status as one of Bali's native villages. Many Bali Aga villages have been assigned as tourist destinations, butare not as appealing to tourists. Since this unique socio-spatial structure will now have to accommodate government interests at all levels, as well as certain NGO's, the power relations between the state and the people remains acute [24]. *Penglipuran* has attributes that others do not possess. In order to more accurately explain its uniqueness, the main elements of the community administration and socio-spatial factors (following) need to be explained in greater detail.

As the case of other Bali Aga communities, *Penglipuran*is a patrilineal society and does not follow the four caste system widely practiced in Balinese *Majapahit* communities. Here everybody belongs to a single *sudra* caste. In this arrangement social interactions and relationships take place in a more egalitarian manner. At present, the community has 233 families. Each has its own housing compound in the centre of the settlement. Community membership is determined by one's marital status and conformance to the *adat,* community consensus and codes. Only a married couple can be considered a seat for membership (*krama desa*). Any unmarried person in the community will be included as part of an existing *krama desa* of her/his parents.

There are two types of *krama desa* exist: (i) *krama pengarep* and (ii) *krama roban*. Following the meaning of *pengarep* in Balinese which is foremost in taking responsibility, the first *krama* constitutes the governing body of the community. Members have central roles in decision making and leadership of the village. They form the core of the community and each is provided with a plot of land allocated for home and land cultivation. There are 76 *krama pengarep,* and this number is nonnegotiable. Membership is based on seniority (worked out based onthe age of one's marriage) and mastery of all written and unwritten practices and codes of the *adat*. Membership to this *krama pengarep* will be annulled when one fails to adhere to the *adat*. In this case the annulled position will be filled in by the next *krama* in line. The *krama* who has his association cancelled will then have to appoint another member as a replacement. This is usually the eldest married son in his family. *Krama roban* are the extended members of the community. *Roban* in Balinese means 'living under the same roof. Unlike the *krama pengarep,* the number of this *krama desa* continues to grow in accordance with population growth. They are the extended families of *krama pengarep*.

In terms of the governing body of the village, *Penglipuran* adheres to both the *dinas* and *adat* systems. The existence of *dinas* is a consequence of Bali integration to the Indonesian state. The whole Island is bound to implement state policy. While the *adat* is a system imbedded within and rooted to local traditions, codes, consensus and communal practices [23]. The organization of the *adat* systems in *Penglipuran* is an historic leadership scheme recognized as *Ulu Apad*. This system has 76 members (*kramadesa*) representing the whole community. The highest position is held by *Jero Kubayan* which consists of *Jero Mucuk* and *Jero Kubayan Nyoman*.

The next level below *Jero Kubayan* is *Kanca Roras* which has 12 members. Like many Bali Aga communities, one's position in this tradition is solely determined by the seniority of one's marital status which determines the pecking order. The longer one has been married, the more chance one has to be part of this *adat* leadership. Thus, where the opportunity for political manipulation and bribery is a common occurrence in the election of *dinas* leadership members, in this arrangement corruption has little chance of success. *Krama* no 13-76 will be part of *krama patilan.* The remaining community members of 167 families are classified as *krama roban*, as has been mentioned earlier.

7. Culture and Space in Penglipuran

One of the dominant considerations in the designation and allocation of cultural space is the institution of marriage. Overall, there is no restriction to marrying someone from outside the community. *Krama desa* status as well as both duties and rights are maintained. This practice differs from that of another *Desa Bali Aga* of *Tenganan* in *Karangasem* Regency. In the latter community, marrying a girl/boy who is not part oflocal *kramadesa*implies being an outcast. This means one rights to being *krama desa* will be annulled and the member will be asked to live in an area especially zoned for outcasts. Supporting equal rights for women, *Penglipuran* also permits the practice in which a woman decides to marry a man who is not part of a *krama desa* membership and live in the community. In this case, the woman will adopt the male role (*purusha*), and the man will similarly take the role of a woman (*pradana*). Since ritually the woman is now symbolically a man, she will be able to act as such in elections to the *krama pengarep*. This is similar to a widely practiced form of marriage called *nyentana* in many Balinese lowland communities.

Nonetheless, there remains significant social pressure against marrying someone living to the left, the right, or directly opposite one's home. *Penglipuran* people believe that families immediately located next to you are relatives, and one does not marry her/his relative (incest taboo). A clear spatial connection between houses can be observed from the existence of small gates (*peletasan*) connecting one family to another. This connecting *pamesuan* represents level of relationship the inhabitants ofa house have with their immediate neighbours. An open *peletasan* means a healthy relationship, while a closed one implies otherwise. This *peletasan* therefore is a social connecter as well as a symbol of continuous social ties between homes and their neighbours. Homes of *Penglipuran* often have rain water flowing freely to the adjacent neighbours loop. While this practice is considered unacceptable or even taboo to the neighbourhood life of the Balinese lowland settlements, it is a sign of closeness to the *Penglipuran* community. It is considered anti-social not to conform.

Unsurprisingly, polygamy is also taboo with the potentially serious social consequence of eviction. Those who undertake this practice will be forced from his compound and ordered to live in a special zone called *karang memadu*. Within feminism, conformance to monogamy, may be seen as some recognition and protection to women from all kind of domestic abuse. Alternatively it can also be viewed as a spatial practice, a method of social control, and a conservation strategy. As mentioned earlier, a relatively long narrowed *karang kerti* may not be suitable to contain the family of one husband; more than one wife; and a child or a couple of children. It would definitely be too crowded. If we examine the spatial formation of a *Penglipuran* home, there is only one building (called *loji*) dedicated for sleeping in a compound.

Importantly, the recognition of polygamy means one man creates more than one family. If one compound is dedicated for one family *ipso facto* this implies that one man can inhabit more than one compound. This will lessen the opportunity for his future generation to live in the *karang kerti*. Hence, the prohibition of polygamy is a sustainable practice that the community has consciously developed at all levels of the hierarchy. Further sanction simposed on members practising polygamy are prohibition to enter the main *rurung*, to enter temples, and to get involved in any *adat* activities. All of these sanctions represent social signals designed to prevent polygamy. So far, it has been highly successful. The *karang memadu* has never had any inhabitant/s. It has now been empty for centuries.

Unlike the lowland community, *ngaben*, the cremation ceremony of lowland Balinese is not practiced here. Burial is the norm rather than cremation. Every deceased person will be buried on the day of passing. During a burial procession, a male body will be laid face down, while a female body will be laid face up. The community has a cemetery located on the southern part of the settlement. As is the practice in the lowland *desa adat* (adat village), the existence of a cemetery is always accompanied with a *Dalem* Temple (see Figure 2), a sacred space to worship the God Siva-the great destroyer. There is no clear indication if the *Dalem* Temple of *Penglipuran* is also dedicated to the same worship. For many reasons – curiosity, voyeurism, novelty or the search for 'an authentic experience,' *Ngaben* ceremonies are a huge tourist magnet. For an important person, lowland ceremonies can draw thousands of people.

Krama desa rights also embody rights for living in the main settlement area; to cultivate a piece of agricultural land; and right to harvest bamboo from *Penglipuran* forest. Each *krama pengarep* is given a plot of space to build his home which is called *karangkerti*. This right to space comes with a *cecatu* (a piece of *tegalan* (dried agricultural land) and rights to harvest bamboo. The latter however can only be exercised upon community consent, which must be acquired through a community *sangkep* (meeting). So a basic rule of survival for the social system hinges on control over space – that land and social space should under no circumstances be commodified within a capitalist land market [14,23,24].

The spatial provision required for shelter and survival of *krama roban* is to be accommodated by the adat within the existing *karang kerti*. It is common to find out that many of the huge *teba* (a spare land within a housing compound) attached to a *krama pengarep*'s homes have been developed into housing compounds to shelter *krama roban*. This practice is permitted and facilitated by the *adat*. In the case of more space being required, the issue can be addressed in an *adat* meeting and extra land will be provided given community consent. Being a *krama desa* entails duties and rights. Duties include *krama desa* participation in the conduct of ritual ceremonies; adhering to the *adat* codes and practices, as well as many forms of consenting social relationships and interactions; maintaining the quality of the domestic and non-domestic living environments and every determining element associated with them.

The permit to harvest bamboo will have to meet considerations in regard to the time of harvest and the appropriate use of the material. This permit can be obtained and applied for by both *krama pengarep* and *krama roban*. *Penglipuran* community is extremely protective of their bamboo forest. They use their bamboo for all kinds of different reasons, including building materials for home construction, tools to support land cultivation, as well as bamboo appliances to facilitate their daily needs. The latest use nowadays is extended to production of bamboo-based handicrafts to be marketed to tourists and customers from other communities.

Overall, the people of *Penglipuran* are noted for a warm and welcoming demeanour. They will leave immediate tasks or preoccupations in order to give assistance to visitors. They welcome tourists to walk around their domestic compound as well as outside domestic sphere. But there is a behavioural norm not to be outside the house (at the *rurung*) after sunset (around 17:00/18:00 depending on the month). They do not object when the *Bangli* Local Government encourages the village to open their temple, bamboo forest and their homes to visitors. There is always at least one family member around when visitors come to visit the compound, since most of the *krama desa* earn their living from land cultivation, and the idea of having no one around the house for security reason is not of concern. How this unique social system interacts with tourism to date is next subject to debate.

8. 'Touristification' and Cultural Conservation

The people of *Penglipuran* seem to embrace tourism with great vigour and an eagerness to show their homes and neighbourhood to visitors. Up to the recent development, they have made their temples available to tourists and opened their *pamesuan* (home's gate) to every single person who is curious about the spatial organization of their domesticity and the built environment of their whole settlement. This includes viewing their bamboo forest; demonstrating how they produce their bamboo crafts; and indicating how their folklore informs their ritual practices, burial processions, arts, and dance. Since the introduction of the tourist industry, their *rurung gede* has been overrun on a daily basis, either by people moving to the north intending to see their *Penataran* Temple or by those strolling in the other direction to see their cemetery. In between, people walk everywhere including the parallel houses within the *karang kerti*.

While the *rurung gede* looks well maintained, many Penglipuran households have had little snack and crafts stalls attached immediately inside the *pemesuan* (gate). This is also the usual practice in the neighbouring Bali Aga village of *Tenganan*, another well known *desa wisata*. In the latter case, such a practice has long rised concerns from both within and outside the community. Some say that the accommodation of the encroaching commercial activities within *Tenganan*'s homes and public realms have colonised the intrinsic meanings and values traditionally associated with their culture [25]. On the other hand some argue that the *Tenganan* have the capacity to make choices as to the extent that tourism to intrudes into their village. The fact is, as per the case of other communities across the Bali Island, the *Tenganan* also wish to enjoy the economic benefits tourists may bring to the community.

Learning from this scenario, *Penglipuran* homes are likely to adopt the same strategy. Beginning with the sale of refreshments at the front of their homes, the *Penglipuran* people will expand their home-based business by importing tourists merchandise not necessarily produced in *Penglipuran*. This will provide additional income to that earned from land cultivation and production of bamboo crafts and appliances. This is entirely possible, since many *Penglipuran* people also nowadays work in service areas outside the community. At the same time it will weaken the originality of their culture. Over time the same people may accumulate sufficient savings to finance the enlargement of their home's *warung* (stall). Again we have the *Tenganan* precedent which is well known for their defensive attitude towards their traditions and unique cultural traits. Entering a *Tenganan* home nowadays, one seldom experiences the original 'place'. Many elements have been hidden by sarongs, statues, handicrafts, and other form of tourist merchandises. We do not need to wait for long for this to happen in *Penglipuran*, particularly when there is currently no mechanism to prevent it.

The people of *Penglipuran* conform to a strict communal schedule in which every *krama desa* has to work together on a weekly basis to clean the neighbourhood. The cleaning and maintenance of ritual places is however done on every *purnama* (full moon) and *tilem* (dark moon) day. At the same time maintaining the practice of their traditions and culture that make the whole settlement a well visited tourist destination is also on the hand of the community. These all are not cost-free duties. On the other hand, one may also argue that tourist development has created jobs for the *Penglipuran* youth, supported *Penglipuran*'s home-based bamboo production, and contributed to the whole social and ritual life of the community. These benefits have been shared by one and all.

Nonetheless, tourists who come for a brief visit, will inevitably impact on local people, bringing new perceptions, aspirations, and behaviour. This is an unavoidable part of the whole process of globalization, as when local people wish to acquire the novelties tourists bring with them, from electronic devices to clothes and other accoutrements. When basic needs are met, as they are in *Penglipuran*, desire becomes the driving force, one that has no set limits. Hence, there is also no limit to the amount of cash that can satisfy it. Every extra rupiah from the flourishing tourism industry becomes the driving force rather than the inherent cultural economy. At present, there is no prohibition to prevent this, apart from each individual's conscious aspiration to maintain the original state of her/his home and community. However, many queries arise here, including: How much longer will culture, tradition and collective social life prevail over the commodity space of the market mechanism? How can the community truly acquire total control over its own existence and the economic benefits that flow from it? How are the boundaries between community life and tourism desires to be established? Culture is a dynamic process – how are we to distinguish between culture and economy? How much change is too much in terms of tourism either contributing or being destructive to a unique community, such as *Penglipuran*?

The issue of where this whole *desa wisata* development program will be directed is yet to be outlined. Thus, if tourism is to form a sustainable element in community life, all these questions need answered. The responses should arguably be framed equitably both by government and the *krama adat* of *Penglipuran*, and not by a planning bureaucracy driven by urban politics. Such a proposal should be socialized to the targeted communities and well understood by them. Therefore, the idea of tourist development being a bottom-up and community driven program should be founded ona shared and sound knowledge rather than by financial advantage and foreign exchange. How the tangible and intangible components of the community that generate tourist interests are seen, enjoyed, shared, touched and learned without destroying it in the process, are major concerns in adopting the *desa wisata* development program.

9. Conclusion

Penglipuran is a microcosm of global forces, national heritage, state revenues and local pride. The latter however does not pay the bills, and in the case of *Penglipuran*, 60% of funds collected from entries fees contributed by visitors go to local government. A further 20% of it is for the *Sekeha Teruna Yowana Bakti* (*Penglipuran*'s Youth Association) which is jointly involved in the process of collecting these fees from tourists. *The remaining 20% share is for the community.* At this point, belief in the idea of a community based equitable system is shattered. Present development everywhere across the Island demonstrates that the *desa wisata* program is in reality a mechanism to enable local government to collect revenues upon its establishment. Furthermore, the governor has stated that he wants to have 100 villages within this designation. It may be true that this process is a community driven action rather than one imposed by government. But once the destination is set in place by the people, the government steps in and mops up the greatest share of revenue.

This general process is somewhat ubiquitous across Bali Island (and arguably the Indonesian state), one where capital reigns supreme and the mass of the people pay to support it. While the government holds the greatest percentage of collected revenues, daily maintenance and the welfare of the whole settlement and its elements are paid for by the community. In other words any surplus generated and invested by local people is expropriated by the state. In turn, the state acts as caretaker for private capital that directly contributes nothing, yet feeds of the quantum of tourism infrastructure. Abstracting principles from the UNESCO World Heritage Association indicated above, it would appear that *Penglipura*n must insist on total control over its own cultural assets and their abstraction into commodified tourist items. No government organisation can do this for them. Here we must reflect on a basic principle of capitalism, that is its capacity to destroy the sources of its own success, as it has done with nature (Suartika 2015, Cuthbert and Suartika 2014). So the less outside interference in *Penglipuran* the better it will be for everyone. This will guarantee that at least in one location, the authenticity of culture and environment will be maintained. After all is this not what tourists come to see?

10. References

[1] Sklair L 2002 *Capitalism and its alternatives* vol 65(Oxford: Oxford university press)
[2] Harvey D 2005 *Spaces of neoliberalization: towards a theory of uneven geographical development* vol 8(Franz Steiner Verlag)
[3] Piketty T 2014 Capital in the Twenty-First Century,(translated by Arthur Goldhammer), Cambridge Mass. 2014
[4] Suartika G A M 2013 The metonymic slum: home in the developing world' *J. Basic Appl. Sci. Res.* 400–406
[5] Featherstone M 1990 *Global culture: Nationalism, globalization and modernity* vol 2(Sage)
[6] Anderson B 1988 Cacique democracy and the Philippines: origins and dreams *New Left Rev.* 3
[7] Anderson B 2008 Exit Suharto *New Left Rev.* **50** 27
[8] Aditjondro G J 1995 *Bali, Jakarta's colony: Social and ecological impacts of Jakarta-based conglomerates in Bali's tourism industry* (Asia Research Centre on Social, Political, and Economic Change, Murdoch University)
[9] Picard M 2008 Balinese identity as tourist attraction: Fromcultural tourism'(pariwisata budaya) toBali erect'(ajeg Bali) *Tour. Stud.* **8** 155–173
[10] Rojek C and Urry J 1997 *Touring cultures: Transformations of travel and theory* (Psychology Press)

[11] Debord G 2012 *Society of the Spectacle* (Bread and Circuses Publishing)
[12] Britton S 1991 Tourism, capital, and place: towards a critical geography of tourism *Environ. Plan. D Soc. Sp.* **9** 451–478
[13] Urry J and Larsen J 2011 *The tourist gaze 3.0* (Sage)
[14] Warren C and Others 1993 *Adat and dinas: Balinese communities in the Indonesian state.* (Oxford University Press)
[15] Warren C 2005 Community mapping, local planning and alternative land use strategies in Bali *Geogr. Tidsskr. J. Geogr.* **105** 29–41
[16] Amin S 2014 *Capitalism in the age of globalization: The management of contemporary society* (Zed Books Ltd.)
[17] Craik J and Others 1996 The potential and limits of cultural policy strategies.-Paper presented at the Griffith University. Institute for Cultural Policy Studies. Conference (1995: Brisbane)- *Cult. Policy* **7** 177
[18] Scott A J 1997 The cultural economy of cities *Int. J. Urban Reg. Res.* **21** 323–339
[19] Miao P 2013 *Public places in Asia Pacific cities: Current issues and strategies* vol 60(Springer Science & Business Media)
[20] Hubbard P and Kitchin R 2010 *Key thinkers on space and place* (Sage)
[21] Low S M 2003 *The anthropology of space and place* (Blackwell)
[22] Putra I D G 2014 The Use of Homes for Commercial Purposes in Padangtegal Tengah Neighbourhood of Ubud *Ruang-sp. J. Built Environ.* **1** 85–102
[23] Suartika G A M 2010 *Morphing Bali: The State, Planning and Culture* (Lambert Academic Publishing)
[24] Suartika G A M 2007 Territoriality and the market system—Adat land vs. state regulations on land matters in Bali *Habitat Int.* **31** 167–176
[25] Suartika G A M 2009 Tenganan, Bali Aga and Development: A Socio-spatial Analysis of Culture *J. Archit. Plan. Stud.* **3** 44–58

Settlement Relocation Modeling: Reacting to Merapi's Eruption Incident

A Pramitasari[1] and I Buchori[2]

[1]Center of Geomatic's Application for Sustainable Development, Diponegoro University, Indonesia

[2]Department of Urban and Regional Planning, Diponegoro University, Indonesia

Email: angrenggani@gmail.com

Abstract. Merapi eruption has made severe damages in Central Java Province. Klaten was one of the most affected area, specifically in Balerante Village. This research is made to comprehend GIS model on finding alternative locations for impacted settlement in hazardous zones of eruption. The principal objective of the research study is to identify and analyze physical condition, community characteristics, and local government regulation related to settlements relocation plan for impacted area of eruption. The output is location map which classified into four categories, i.e. not available, available with low accessibility, available with medium accessibility, and available with high accessibility.

Keywords: Merapi Eruption, Settlement, Relocation, GIS Model

1. Introduction

Indonesian archipelago is one of the regions having the most active volcanoes in the world, commonly called as The Ring of Fire. That is why Indonesia has a high vulnerability in geological hazards. Located between oceans and plains which are characterized by plenty of hills and mountains, Central Java becomes prone to the volcanic eruption. Some of volcanic mountains located in Central Java are Merapi, Slamet, Sindoro, and Sumbing. Most of all, Merapi has the highest number of eruptions in the previous years and have caused a great deal of damages and losses in majority areas of Central Java. Merapi is surrounded by four districts, i.e. Magelang, Boyolali and Klaten in Central Java Province, also Sleman in D.I Yogyakarta.

Various studies have identified society's increasing vulnerability to disasters as a consequence of population expansion in hazardous areas and increasing economic and environmental strain [1–4]. Volcanic eruptions can become threats to the people living in the surrounding areas. The consequence of vulnerability from volcano eruptions can be different in each area, depending on the situation of natural events and human activities. The losses due to disasters also depend on the ability to prevent or avoid them. The impacts of disasters can be limited, depending on the type of spatial and large eruptions. The activities of Mount Merapi, for instance, can be monitored minute by minute by the use of sufficient monitoring equipment so that the casualties that may occur due to its activities can be minimized. This understanding is associated with the statement that disasters occur when hazards meet helplessness. Thus, the activity of the malicious nature will not be a natural disaster in areas without human powerlessness. As an example, volcano eruption in uninhabited areas will be less harmful than eruptions in density settlement areas.

The direct danger from volcano hazard may occur due to the throwing rocks / bombs, lava flows and hot clouds blow as an incandescent eruption, toxic gases, and thick ash fall. While the danger arising from the flow of mud mixed with rocks categorize as following hazard. This flow of flooding, either still hot or already cooled heat is often referred as "lahar". The Indonesian word "lahar" is applied as a general term for rapidly flowing, highly concentrated and poorly-sorted sediment-laden mixtures of water and rock debris from a volcano, not including normal stream flow [5,6]. Lahar following the volcanic slopes can lead to sweeping villages, rice fields or bridges.

Merapi volcano has been studied extensively on continuous research by indonesian and international scientists, government scientists, even students. Numerous projects and collaborative researches were directed at understanding the eruption impacts, management, assessment [7–10] and eruption forecasting [7,11–14].

An important lesson from this special issue is the warning outlined by several authors that Merapi volcano is capable of greater eruptions than those in the 20th century – the eruptions that may have disastrous consequences. The eruption in 4th and 5th November, 2010, was a "100-year event" [16] . It was approximately 10 times larger and more explosive than the eruptions of the past several decades, and it validated the concern about Merapi — that greater and more hazardous eruptions, like the one that took place in 1872, are a continuing threat at the volcano [1,12,16–19].

The growth of the population can also lead to disaster. A bigger population will increase the need of settlement, so it can encourage more people to live in dangerous and vulnerable areas. For example, local farmers rely for their livelihoods from the natural resources on their surrounding areas, but they often pay less attention to safety aspects. It makes many houses were established in the hazardous zone, so in the last few years, we often hear disaster incident that took many lives. Regarding this vulnerable condition, settlements are not allowed to be built in the disaster hazard so that the existed settlements built in such zone need to be relocated to more secured areas.

Relocation is the moving process from one place to another. Thus, settlement relocation can be informed as a moving process of a certain area of settlement to another place. Relocation usually takes place when the old location is subject to a natural hazard; when the old location is completely destroyed and to move the debris and to make new plotting in the old settlement is inconvenient for rapid recovery and housing purposes; also when there is a chance to relocate the settlement to land which belongs to the Government since it is generally preferred not to have to pay for the land [20]. Furthermore relocation or resettlement processes are important for reducing future hazard risk in a disaster prone area [21], especially for the settlement which existence is taken place on a restricted area, such as on the riverbanks and hazardous area of disaster. Related to those context, people who are affected by Merapi eruption need a secured place for permanent settlement due to the damaged of their settlements. Relocation process is one alternative way to provide opportunities for people who live in the hazardous area of disaster. It will give them a bigger opportunity to carry on their lives in a better and secured condition.

The discussion about Merapi's risk assessment, mitigation and other geothermal studies are commonly discussed, but the research that is intended to analyze about GIS relocation model based on physical condition, governmental and community aspects might be a new thing. This paper elaborates three aspects: physical, government and community aspects as an input in making GIS model for relocating the settlements existing in hazard zones area. The object of the study is communities in Kemalang Sub-district, who lives in Balerante Village, Klaten District. Balerante Village is considered as the most vulnerable because settlements damages caused by Merapi's eruption in 2010 are mostly happened in this place.

IOP Conf. Series: Earth and Environmental Science **123** (2018) 012015 doi:10.1088/1755-1315/123/1/012015

2. Methods

This research uses quantitative approach, it is divided into several analyses; analysis based on physical aspect, government aspect and community aspect. Scoring and buffer analyses are applied in the process. The next step is analyzing the model simulation using model builder. While the last is making a descriptive analysis on the final output of the model. This research also uses descriptive quantitative method to explain and explore more about the phenomena. In descriptive ways, relocation model for impacted settlements can be formulated. Each indicator in natural condition of a scope area is converted in value and score. The conversion is needed to simplify numeric analysis process, so the land suitability degree of settlements on relocation model can be obtained. Variables in the physical aspect are obtained from the literature, while variables in the government and community aspect are obtained from direct interview of government officials and local village leader.

2.1. Physical Aspect

Physical condition of nature plays a big role; it helps define the most suitable land for a new settlement location. Land suitability for development of the city should consider several aspects, i.e. physical, socio-economic conditions, accessibility, environment and ecology, the potential of local resources as well as political factors [22].

Natural physical aspect is determined as one of the main analyses by setting four variables as a measurement. It consists of topographical land slopes, geology condition, hydrology condition, and soil type condition. The maps of physical variables are overlaid based on each weight. This step is also known as weighted overlay. Weighting is done to show the influence of each variable to its related aspect. Weighted overlay is expressed in percent (%) where the total amount is 100%. Influence level on each variable is varied, so it also gives different proportion on the weight. Land slopes variable has the biggest weight in comparison to other variables. It has 40% weight while other variable only has 20%. The reason is related to the natural condition of study area that tends to be sloping in most of the area. This sloping condition gives bigger influence on the risk of natural disasters such as eruption, earthquake, and landslide. Thus the weight of land slopes variables is twice bigger than others.

Table 1. Score and Classification of Physical Aspect Variables

Variable	Characteristic	Score	Weight
(1) Land slopes	Very steep (>45%)	1	40%
	Steep (25 – 45%)	2	
	Low steep (15 – 25%)	3	
	Sloping (2 – 15 %)	4	
	Flat (0 – 2 %)	5	
(2) Geology condition	Dacite	1	20%
	Merapi volcanic rocks	3	
	Old merapi volcanic rocks	5	
(3) Hydrology condition	Scarce Groundwater	1	20%
	Local productive aquifer	2	
	Widespread productive aquifer	3	
	Widespread with medium productivity	4	
	Widespread aquifer with high productivity	5	
(4) Soil type condition	Litosol - grey regosol	1	20%
	Grey brown regosol	3	
	Grey regosol	5	

Source: Minister of Forestry Decree No.873/UM/II/1980 and No.683/KPTS/UM/1981; Verhoef 1994; modified by researcher 2014

On the other hand, scoring is done on each variable with a value of 1 to 5 where the higher the score, means the higher suitability level of the land. For an example, areas with a high degree of slope will earn a small score because the higher the slope, the smaller the score. Meanwhile, the scoring on the hydrological conditions depends on the availability of the sources of water. This method is also applied for geology and soil type condition. The score is given according to the type of rock and the soil.

2.2. Government Aspect

One of the most important aspects to be considered in relocation action is regulatory factors. The policy about spatial planning has been regulated by the central government. However, it needs to be in line with the relevant local authorities. The criteria chosen as variables in the government aspects have been adapted to land use regulation based on government spatial plan of Klaten. The government aspect variables are also inferred as elimination criteria. The reason is mainly based on the regulation that these areas are not allowed or permitted to be relocation referrals. Score is also given on each variable of governmental aspect. The score in the government aspect is ranging from 0 to 1. The scoring method is given to identify restricted zones so it could be eliminated. A zero (0) score is given to the restricted area while score of one (1) is given for unrestricted area.

Table 2. Score and Classification of Government Aspect Variables

No.	Variables	Area Classification	Score
1	Conservation Area	Merapi's National Park & Riverbanks Area	0
		Other Area	1
2	Riverbanks Area	0-50 Meter Buffer Area	0
		>50 Meter Buffer Area	1
3	Existing Settlement Area	Build Up Area	0
		Non Build Up Area	1
4	Volcano Hazard Area	Hazard Zone 1,2,3	0
		Non Hazard Zone	1

Source: Law No. 26 of 2007; Presidential Decree No. 32 of 1990; Presidential Decree no. 57 of 1989; Regulation of the Minister of Agriculture and 837/KPTS/UM/11/1980 No.683/KPTS/UM/8/1981; modified by researcher, 2014

2.3. Community Aspect

The analysis in community aspect is emphasizing on the range of accessibility to roads, evacuation routes, shelter, high voltage wires, and other public facilities such as the range of educational facilities, health facilities as well as worship facilities. The determination of an ideal access to reach facilities was done by creating a buffer on roads, high voltage wires and evacuation centers. The buffer distances on each aspect were different from one to another. It depends on the rules and assumptions adapted based on the public's ability to access the facility. The more varied the facilities and infrastructures of the residential means the higher the possibility it is chosen as the residential location.

2.4. The Existing Road

The road used as a parameter in this study is a collector road. In this variable, there are some classifications which depend on the buffer distance. It measured from how far people can walk within a period of 10-15 minutes. The classifications criteria are <500m, 500-1000m, and > 1000m.

Table 3. Score and Buffer Distance of Existing Road Variables

No.	Distance	Walking Time Estimation	Score
1	<500 m	<10 minute	5
2	500-1000 m	10-15 minute	3
3	>1000 m	>15minute	1

Source: Modified by Researcher, 2014

2.5. Evacuation Route

Based on the field observation, evacuation route on Kemalang districts do not use special track but use the available collector raods. Therefore, the determination of buffer distance on evacuation routes uses the same lassification as the existing road variables.

Table 4. Score and Buffer Distance of Evacuation Route Variables

.No.	Distance	Walking Time Estimation	Score
1	<500 m	<10 minute	5
2	500-1000 m	10-15 minute	3
3	>1000 m	>15 minute	1

Source: Modified by Researcher, 2014

2.6. Shelter Point

The buffer distance on shelter point is 500 meter. This distance is considered as the safest distance, so a score of 5 is given as the highest score. Meanwhile a score of 1, as the lowest score, is given to the settlement location with over 500 meters' distance.

Table 5. Score and Buffer Distance of Shelter Point Variables

No.	Distance	Walking Time Estimation	Score
1	<500 m	<10 minute	5
2	>500 m	>10 minute	1

Source: Modified by Researcher, 2014

2.7. High-Voltage Tower (SUTET)

The unavailability of electricity network map is the main reason why the researcher chose high voltage wires as a reference. The buffer distance of voltage wires is SNI 04-6918-2002 which refers to "free space and minimum free distance of high-voltage wires." The rules are described as follows:

a) The minimum vertical clearance of conductors to the building is 9 meters.
b) The minimum clearance from the horizontal axis to the tower is 22 meter for a single circuit of 500 kilovolt high-voltage wires and 17 meter for double circuit of 500 kilovolt high-voltage wires.

The high-voltage tower found the study area is 1000 kilovolt high-voltage wires. However, there is no explanation about the double or single circuit description. Based on the information, a safe buffer distance is assumed as far as 50 meters.

Table 6. Score and Buffer Distance of High-Voltage Tower Variables

No.	Distance	SNI Regulation	Score
1	<50 m	Not Recommended	1
2	>50 m	Recommended	5

Source: SNI 04-6918-2002, Modified by Researcher, 2014

2.8. Public Facilities Availability

Access to the health, educational and worship facilities are determined using scalogram diagram method. Statistical data of facilities on each village are obtained from statistical bureau. The output of scalogram diagram is a facility access hierarchy from low, medium to high. The next step is making the public facility map through spatial process. The specific information about each facility could be seen in the following table.

Table 7. Scalogram Diagram, The Availability Of Education, Health & Worship Facilities

No.	Village	EDUCATION FACILITIES			HEALTH FACILITIES				WORSHIP FACILITIES				Total Score	Criteria
		Kindergarten	*Elementary School*	*Junior High School*	*Health Center*	*Sub Health Center*	*Maternity Clinic*	*Clinic*	*Mosque*	*Musholla*	*Catholic Church*	*Christian Church*		
1	Balerante	1	2	0	0	0	0	0	4	2	0	0	4	Middle
2	Bawukan	1	2	0	0	1	0	0	4	1	1	1	7	Middle
3	Bumiarjo	1	1	1	0	0	0	0	2	21	0	0	5	Middle
4	Dompol	1	2	0	0	0	0	0	3	6	0	0	4	Middle
5	Kemalang	1	2	2	0	0	0	0	3	6	0	0	5	Middle
6	Kendalsari	2	2	0	0	0	0	0	3	9	0	0	4	Middle
7	Kaputran	2	3	2	1	0	1	0	5	7	1	1	9	High
8	Panggang	1	1	0	0	0	0	0	3	0	0	0	3	Low
9	Sidorejo	2	2	0	0	1	0	0	2	11	0	0	5	Middle
10	Tahun	2	2	0	0	0	0	0	1	5	0	0	4	Middle
11	Tangkil	3	2	0	0	1	0	0	6	10	1	1	7	Middle
12	Tegalmulyo	1	2	0	0	0	0	0	11	3	0	0	4	Middle
13	Tlogowatu	1	3	0	0	0	0	0	2	14	0	2	5	Middle

Source: Klaten in figures; Modified by Researcher, 2014

3. Results

3.1. GIS Model and Output Classification

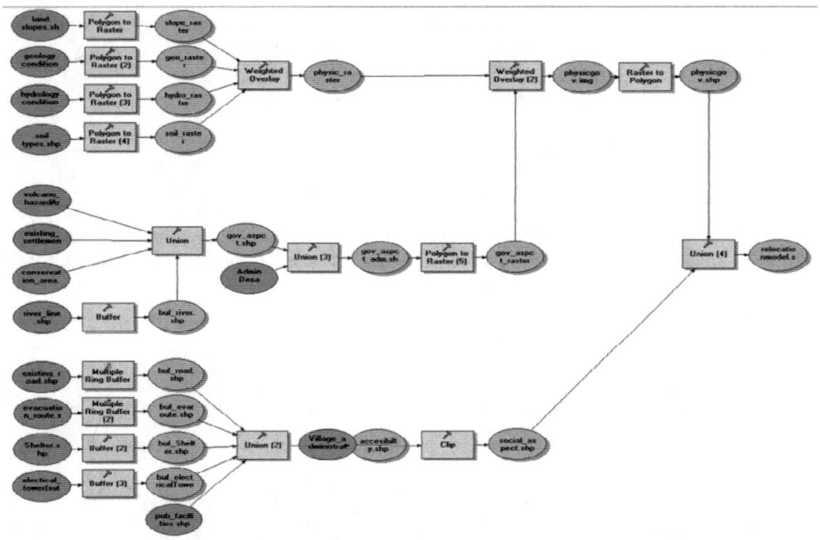

Source: Analysis of researcher, 2014

Figure 1. Relocation Area Scheme on GIS Model Builder

The GIS functions help in managing the spatial data and visualizing the results [23]. Analysis of relocation model is split into three sub-models; physical aspect, government aspect, and community aspect. Splitting model method is intended to make a clear explanation in each step so it will be more understandable. The complete relocation model is shown in the figure below:

The output of GIS relocation model is a table and map which classified location based on its availability and accessibility status. Relocation map as the model output was divided into four classifications, i.e. not available, available with low accessibility, available with medium accessibility, and available with high accessibility. The criteria of relocation area are shown on the following table.

Table 8. Relocation Area Based on Criteria in Kemalang Sub District

No.	Criteria	Area (m^2)	Percentage (%)
1	Not available	35.43748056	68%
2	Available – low accessibility	10.88708415	21%
3	Available – medium accessibility	3.382845381	6%
4	Available – high accessibility	2.722800762	5%
	Total	52.4302107	100,00

Source: Analysis of researcher, 2014

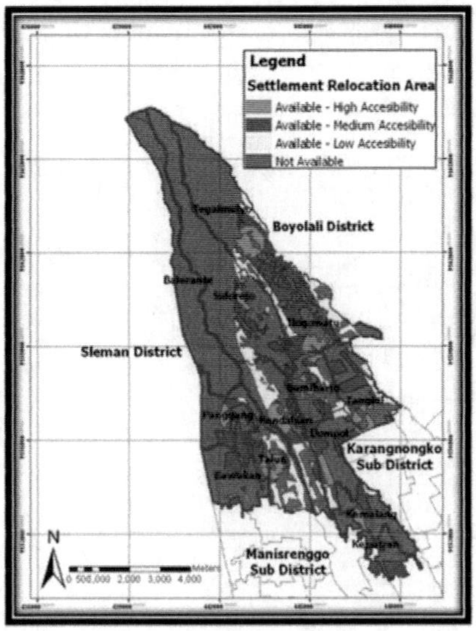

Source: Analysis of researcher, 2014

Figure 2. Settlement Relocation Area Map

On relocation model, governments have planned to provide 150m2 land to the society for each household. The provision is 100m2 for the building and 50m2 for open space. The numbers of household that need to be relocated are 165. It means that government needs to provide 2.5 ha of land for relocation area. The total available areas with each level of accessibility that approved by government aspect in every village are elaborated as follows:

a. Available with low accessibility: Bawukan (67.59 ha), Bumiarjo (56.70 ha), Dompol (76.33 ha), Kemalang (33.45 ha), Kendalsari (118.71 ha), Panggang (60.90 ha), Sidorejo (110.30 ha), Talun (102.74 ha), Tangkil (103.04 ha), Tegalmulyo (125.02 ha), and Tlogowatu (233.93 ha);
b. Available with medium accessibility: Bawukan (47.32 ha), Bumiarjo (7.24 ha), Dompol (5.39 ha), Kemalang (40.37 ha), Kendalsari (9.05 ha), Kaputran (41.17 ha), Panggang (72.49 ha), Talun (11.71 ha), Tangkil (97.95 ha), and Tegalmulyo (3.64 ha);
c. Available with high accessibility: Bawukan (6.65 ha), Bumiarjo (21.76 ha), Dompol (21.94 ha), Kemalang (16.43 ha), Kendalsari (39.13 ha), Panggang (14.90 ha), Sidorejo (15.59 ha), Talun (32.82 ha), Tangkil (21.77 ha), Tegalmulyo (53.66 ha), and Tlogowatu (26.22 ha).

In contrary with government's point of view, society prefers to move communally and demands a settlement area as large as the previous neighborhood. On the interview, the chief of the residents said that every household owns about 1000m2 of land in the current area. It means mathematically the overall land requirement is 16.5 ha. The total available area with each level of accessibility that approved by community aspect in every village are elaborates as follows:

a. Available with low accessibility: Bawukan (67.59 ha), Bumiarjo (56.70 ha), Dompol (76.33 ha), Kemalang (33.45 ha), Kendalsari (118.71 ha), Panggang (60.90 ha), Sidorejo (110.30 ha), Talun (102.74 ha), Tangkil (103.04 ha), Tegalmulyo (125.02 ha), and Tlogowatu (233.93 ha);
b. Available with medium accessibility: Bawukan (47.32 ha), Kemalang (40.37 ha), Kaputran (41.17 ha), Panggang (72.49 ha), and Tangkil (97.95 ha);
c. Available with high accessibility: Bumiarjo (21.76 ha), Dompol (21.94 ha), Kendalsari (39.13 ha), Talun (32.82 ha), Tangkil (21.77 ha), Tegalmulyo (53.66 ha), and Tlogowatu (26.22 ha).

In comparison to the land requirement on government aspect, the choice becomes more limited. However, this requirement can still be met even in relatively fewer options. The relocation area available according to the physical, government, and community aspect is shown on the following table and map:

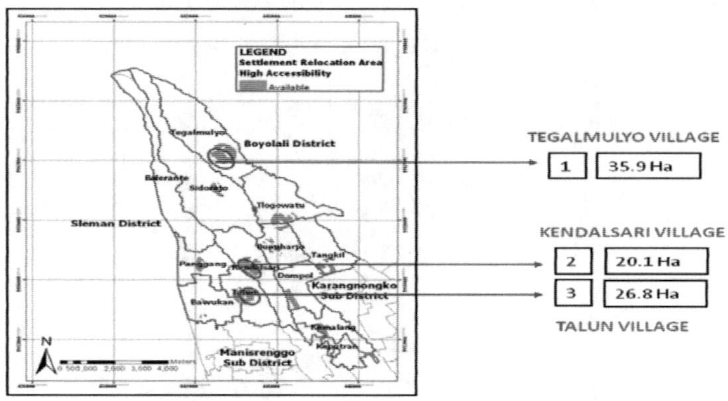

Source: Analysis of researcher, 2014

Figure 3. Priority Relocation Area Map

Table 9. Relocation Area Priority Based on Physical, Government & Community Aspect

Village	Area (ha)	Total Area (ha)	Relocation Area Priority
Bawukan	4.16	4.16	
Bumiarjo	2.69	9.61	
	2.91		
	4.01		
Dompol	2.84	19.56	
	4.12		
	12.6		
Kemalang	3.69	8.97	
	5.28		
Kendalsari	4.8	38.3	
	13.4		
	20.1		2
Panggang	14.9	14.9	
Sidorejo	13.6	13.6	
Talun	5.59	32.39	
	26.8		3
Tangkil	2.5	20.6	
	8.28		
	9.82		
Tlogowatu	5.67	23.09	
	7.42		
	10		
Tegalmulyo	3.8	50.3	
	10.6		
	35.9		1

Source: Analysis of researcher, 2014

IOP Conf. Series: Earth and Environmental Science **123** (2018) 012015 doi:10.1088/1755-1315/123/1/012015

The final result founds that the priority relocation areas approved by the physical, government and community aspect are Tegalmulyo as the first priority with 35.9 ha area, Kendalsari as the second priority with 20.1 ha, and Talun as the third priority with 26.8 ha.

4. Conclusion

Good settlement relocation model should integrate at least three aspects: governmental, community and natural physical aspects. Those aspects have strong relations one and another, without it relocation model becomes less accurate because in reality the process will involve government as the regulator and community as the object. Moreover, natural condition is a basic factor because it is related to the conditions of an area. Government aspect becomes a priority in relocation model because it is related to the policies and regulations. Policies and regulations were made by the government through a complicated process of assessment with various considerations. Regulation about restriction is made because in fact, inadequate conditions have been found in the field. Analysis on community aspect is also very important. However, the object of relocation is communities, so the fulfillment of their needs, comforts and securities on the new settlements or environments should get more attention.

5. References

[1] Sparks R S J, Aspinall W P, Crosweller H S and Hincks T K 2012 Risk and uncertainty assessment of volcanic hazards *Risk and Uncertainty Assessment for Natural Hazards* ed J Rougier, S Sparks and L Hill (Cambridge: Cambridge University Press) pp 364–397

[2] Buchori I and Tanjung K 2013 Developing a Simulation Model for Predicting Inundated Areas Affected by Land Use Change: A Case Study of Keduang Sub-watershed *Int. J. Environtmental Sustain.* **9** 79–108

[3] Buchori I, Sugiri A, Maryono M, Pramitasari A and Pamungkas I T D 2017 Theorizing spatial dynamics of metropolitan regions: A preliminary study in Java and Madura Islands, Indonesia *Sustain. Cities Soc.*

[4] Buchori I and Sugiri A 2016 An empirical examination of sustainable metropolitan development in Semarang City, Indonesia *Aust. Plan.* **53** 163–177

[5] Smith G A and Fritz W J 1989 Volcanic influences on terrestrial sedimentation *Geology* **17** 375–376

[6] Vallance J W 2000 Lahars *Encycl. volcanoes*

[7] Lavigne F, Thouret J-C, Voight B, Young K, LaHusen R, Marso J, Suwa H, Sumaryono a, Sayudi D S and Dejean M 2000 Instrumental lahar monitoring at Merapi Volcano, Central Java, Indonesia *J. Volcanol. Geotherm. Res.* **100** 457–478

[8] Jenkins S F, Spence R J S, Fonseca J, Solidum R U and Wilson T M 2014 Volcanic risk assessment: Quantifying physical vulnerability in the built environment *J. Volcanol. Geotherm. Res.* **276** 105–120

[9] Wilson G, Wilson T M, Deligne N I and Cole J W 2014 Volcanic hazard impacts to critical infrastructure: A review *J. Volcanol. Geotherm. Res.* **286** 148–182

[10] Thouret J-C, Lavigne F, Kelfoun K and Bronto S 2000 Toward a revised hazard assessment at Merapi volcano, Central Java *J. Volcanol. Geotherm. Res.* **100** 479–502

[11] Hidayat D, Voight B, Langston C, Ratdomopurbo A and Ebeling C 2000 Broadband seismic experiment at Merapi Volcano, Java, Indonesia: very-long-period pulses embedded in multiphase earthquakes *J. Volcanol. Geotherm. Res.* **100** 215–231

[12] Voight B, Sparks R S J, Hammond J, Shalev E, Malin P, Kenedi C, Minshull T A, Paulatto M, Mattioli G, Hidayat D and Widiwijayanti C 2010 Active source seismic experiment peers under soufrire hills volcano *Eos (Washington. DC).* **91** 245–247

[13] Wassermann J and Ohrnberger M 2001 Automatic hypocenter determination of volcano induced seismic transients based on wavefield coherence—an application to the 1998 eruption of Mt. Merapi, Indonesia *J. Volcanol. Geotherm. Res.* **110** 57–77

[14] Zlotnicki J, Bof M, Perdereau L, Yvetot P, Tjetjep W, Sukhyar R, Purbawinata M A and others 2000 Magnetic monitoring at Merapi volcano, Indonesia *J. Volcanol. Geotherm. Res.* **100** 321–336

[15] Surono, Jousset P, Pallister J, Boichu M, Buongiorno M F, Budisantoso A, Costa F, Andreastuti S, Prata F, Schneider D, Clarisse L, Humaida H, Sumarti S, Bignami C, Griswold J, Carn S, Oppenheimer C and Lavigne F 2012 The 2010 explosive eruption of Java Merapi volcano — A " 100-year " event *J. Volcanol. Geotherm. Res.* **241–242** 121–135

[16] Jousset P, Pallister J, Boichu M, Buongiorno M F, Budisantoso A, Costa F, Andreastuti S, Prata F, Schneider D, Clarisse L, Humaida H, Sumarti S, Bignami C, Griswold J, Carn S, Oppenheimer C and Lavigne F 2012 The 2010 explosive eruption of Java's Merapi volcano—A "100-year" event *J. Volcanol. Geotherm. Res.* **241–242** 121–135

[17] Jousset P J P S 2013 The 2010 eruption of Merapi volcano *J. Volcanol. Geotherm. Res.* **261** 1–6

[18] Marhaento H and Kurnia A 2015 Refleksi 5 Tahun Paska Erupsi Gunung Merapi 2010: Menaksir Kerugian Ekologis Di Kawasan Taman Nasional Gunung Merapi *Geoplanning J. Geomatics Plan.* **2**

[19] Marhaento H 2016 GIS-based analysis for assessing landslide and drought hazard in the corridor of mt. Merapi and mt. Merbabu national park, Indonesia *Geoplanning J. Geomatics Plan.* **3** 15–22

[20] Bayulke N 1983 Lessons to be learned from relocation of villages and use of prefabricated houses in post-disaster housing in Turkey *Socio-Architectural Asp. Hous. Earthquake-prone areas Turkey* 17–25

[21] Hepperle E, Dixon-Gough R, Mansberger R, Paulsson J, Reuter F and Yilmaz M 2015 *Challenges for Governance Structures in Urban and Regional Development/Fragen zur Steuerung von Stadt-, Land-und Regionalentwicklung* (vdf Hochschulverlag AG)

[22] Golany G 1976 *New-town planning: Principles and practice* (John Wiley & Sons)

[23] Kalogirou S 2002 Expert systems and GIS: an application of land suitability evaluation *Comput. Environ. Urban Syst.* **26** 89–112

IOP Conf. Series: Earth and Environmental Science **123** (2018) 012016 doi:10.1088/1755-1315/123/1/012016

Understanding Student Travel Behaviour in Semarang City

O R Manullang[1], W P Tyas[1], N Anas[1], and F N Aji[1]

[1]Urban and Regional Planning Department, Diponegoro University, Tembalang, Semarang 50275, Indonesia

Email: oktomanullang73@gmail.com

Abstract. The highest movement in Semarang City is dominated by motorcycles, which reached 79% of the number of vehicles. Highest percentage movement use motorcycle caused the highest percentage accident by motorcycle users, which reached 66% and 9% involving high school students. This happens because of the dependence of motorcycles usage in fulfilling the needs of movement in the city of Semarang. Understanding student travel behavior based on their activities is used to know travel needs and the cause of dependence on motorcycle usage. Analysis method in this study use network analysis to compare the potential accessibility and actual accessibility **to** known why motorcycle chosen by students as the main mode. In addition, phenomenology analysis is used to explain the intent and reasons the data produced by network analysis. The analysis result indicates that the high use of motorcycles by high school students in the Semarang city due to the absence of other effective and efficient modes in fulfilling the movement needs. Even, the student which can potentially use public transport preferred to use a motorcycle. This mode is more effective and efficient because of its flexibility and lower costs.

Keyword: Motorcycle, Network analysis, Student, Travel Behavior

1. Introduction

In doing their activities, human has to travel, in other words, the need or desire to engage in an activity in different locations will illustrate travel pattern [1]. In addition to work activities, school activity is one of the highest contributors movement in the city [2]. To fulfill the needs of the school, students need to move from origin (home) to the destination (school) both motorized and non-motorized. Motorized movement in Semarang City is dominated by motorcycles, it can be seen from the population which reaches 79% of the total vehicles in Semarang City [3]. The dominance of motorcycle use is in line with the percentage of accident numbers involving this mode, which reaches 66% of 5,378 incidents in 2012-2014. These incidents have involved 9% of Senior High School students [4].

Students tend to use motorcycle because of the limitation in their movement [5]. The dependence of the motorcycle occurred on students based on three main points [6], are (1) the shape and structure of the city accompanied by the inability of the public transportation system provides the needs of the movement, and causing the dependence of private vehicles, (2) motorcycle has the highest utility than other modes because of more effective and more efficient for students, so that it is suitable for student's activities, and (3) the contract of transportation culture implemented by providing parking infrastructure in school tend to attract new motorcycle users. The dependency will increase while teenager considered motorcycle usage as a lifestyle for traveling in a term to fulfill their social status [7]. Safety driving has not been a concern for motorcycle users, especially for students rider in Semarang City. Students use motorcycles with the permission of their parents, even though the student does not have a driver's license. All these factors aggravate transit culture. The lack of service of public transportation with existing travel demand cause motorcycles chosen as primary mode [6]. This study aims to understand travel pattern, travel behavior and causes of the dependence of high school students in motorcycle users. This study is expected can be the basis for recommending urban transportation policies that match with characteristics and needs of student, so it can be an effective and efficient solution and eliminate dependence on motorcycle usage.

2. Literature

At first, transportation approach was based on four stages modeling, but it was considered to be deficient in describing demand because it considered each trip was not related to other trip conducted by individuals because of the aggregate assumption [1]. Four stages modeling only accumulate the demand, while the movement is very complex where there is a trip chain from the origin to the destination [1]. According to Kitamura [8], there are some measurable indicators in travel behavior, there are:

a. Travel Distance
b. Travel Time
c. Travel Cost
d. Travel Frequency
e. Mode Use

The activity-based concept is a new development to analyze travel behavior, this concept will have a long series to understand the participation of a set activity. Indicators used in this concept such as area based measures, distance based measures, count based measures and duration based measures have all been used to derive travel patterns [9,10]. This research is focused on motorcycle users so the mode use is not considered as an indicator. Measurable indicators that are used in this study is travel distance, travel time, and travel frequency.

In addition to work activity, educational activity is one of the highest trip generators [2]. Students' trip pattern has a fundamental difference between "heading towards" trip and "return" trip from school [7]. "Heading to school" trip has a simple pattern with fixed destination and arrival time while "return" trip has various complex patterns because of the various activities after school. The return trip from school has three trip patterns:

a. Immediate Return, the return trip from school after the last class straight to go home. Students from the suburbs are likely to have this pattern because of the public transport in their home requires them to go home during the peak hours. Besides, first year students are also likely to have this pattern.
b. Stay and Return, the return trip that is made after a certain period of time of staying at school for in school-activities. Besides immediate return pattern, first year students are also likely to have this pattern.
c. Indirect Return, the return trip made any time after the last class but have another destination before coming back home. This pattern is purposed to fulfill recreation, shopping, and extra lesson needs. Students in downtown are likely to have this pattern because they have more alternative modes and better operational time. Generally, senior year students will have this pattern because of the needs of an extra lesson or part time working.

3. Method

The method of data analysis used in this study was network analysis using ArcGIS and phenomenology analysis. Network Analysis (NA) collectively is a macro transportation modeling to see the relationship between objects connected by the transportation that forming travel pattern. Measures of accessibility from these transport models can be classified into potential accessibility and actual accessibility [10]. Network analysis used to analyze potential accessibility of various modes. Potential accessibility in this study is limited to walk, public transports using, and motorcycles using potential. The result of potential accessibility will be compared to actual accessibility so that the public transport route potential will come as the result.

Motorcycle potential accessibility was analyzed by choosing the fastest route with closest facility tools. Network dataset used in this study is the minimum speed in every road distinguished by the hierarchy. The minimum speed standard is according to SNI (Indonesian National Standard) 03-6967-2003 about general requirements of network system and street geometric assuming that each road in the same hierarchy will have the same speed.

Table 1. Minimum Speed Based on Roads Hierarchy

Hierarchy	Minimum Speed (Km/hour)
Primer Artery	60
Primer Collector	40
Primer Local	20
Secondary Artery	30
Secondary Collector	20
Secondary Local	10

Source: SNI 03-6967-2003 tentang Persyaratan Umum Sistem Jaringan dan Geometrik Jalan Perumahan

Potential of walking and -of transit accessibility were analyzed by using service area tools measured from the school location and BRT Trans Semarang's Shelter distribution. According to SNI 03-6967-2003 about general requirements of network system and street geometric, people in Indonesia only made it walk 400 meters in a hot weather, for shopping activities and carrying some goods they only made it walk 300 meters. According to that, service area distance from school to shelter is about 300 to 400 meters. Actual accessibility obtained by visualizing students trip chain based on their travel diary. Phenomenology approach as qualitative method used to understand the complexity of the travel behavior and the reason of students dependency of motorcycle usage in Semarang City.

4. Result and Discussion

4.1. Potential Accessibility

Potential accessibility on foot and public transport was analyzed by using service area tools measured from the point of the school location and the distribution of shelters with walk distance about 300 m and 400 m from that point. Potential accessibility by walking to school and public transportation based on normal walking distance can be seen in Figure 1. Based on the results of the service area analysis, only 1% of respondents have the potential to walk with common mileage. 19.6% of respondents have the potential to use public transportation BRT Trans Semarang. While 79,4% others have to use private vehicle to reach school. Potential accessibility of motor vehicle users is estimated based on the shortest route to the school, the distribution of the route can be seen in Figure 2.

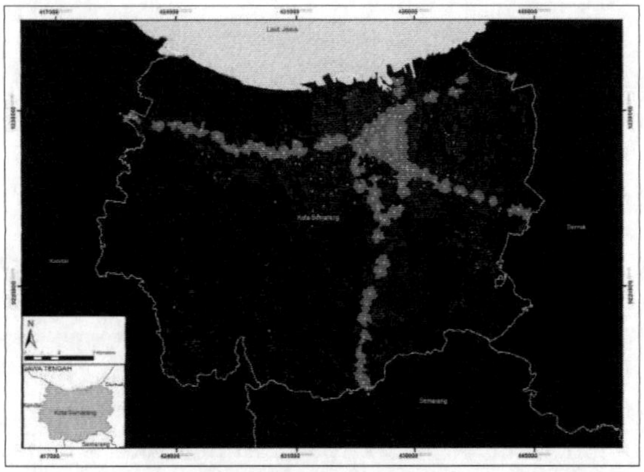

Figure 1. Walkability Potential Access of school and BRT shelter

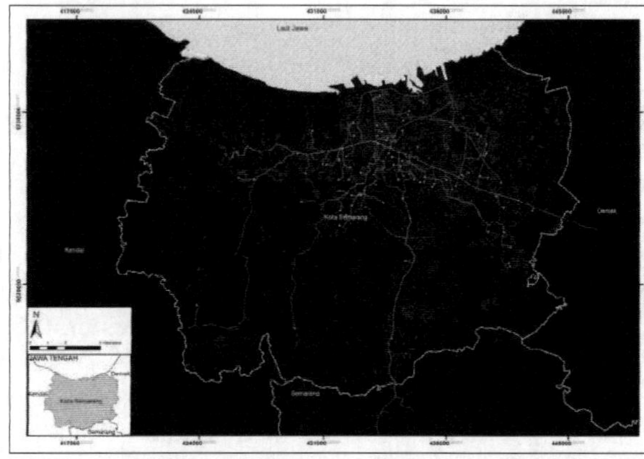

Figure 2. Potential accessibility motorcycle

4.2. Actual Accessibility

Actual accessibility of motorcycles in this study can be classified into three types of travel patterns, namely immediate return, stay and return and indirect return. Based on the visualization of trip chain data, the student's travel pattern has a fundamental difference between " to " and 'back' 'trips from' 'school. The " toward " school trip has a simple pattern with the goal and the arrival fixed time, this pattern is always through the shortest route to get to school. Meanwhile, the journey back from school has a variety of complex patterns because more activities take place after school.

Immediate Return and Stay and Return travel patterns have the same frequency, distance, travel time and travel route. The difference between these patterns is the type of activity and duration activity. 49.5% of students have immediate return pattern, in general, these students are first-year students or students who are not actively participating in the organization or extracurricular. 15.5% of students have stayed and return patterns, in general, these students actively participate in school organization or extracurricular that conducted in schools. Actual accessibility immediate return and stay and return pattern have the same route with potential accessibility because it has one travel destination so the route is the shortest route.

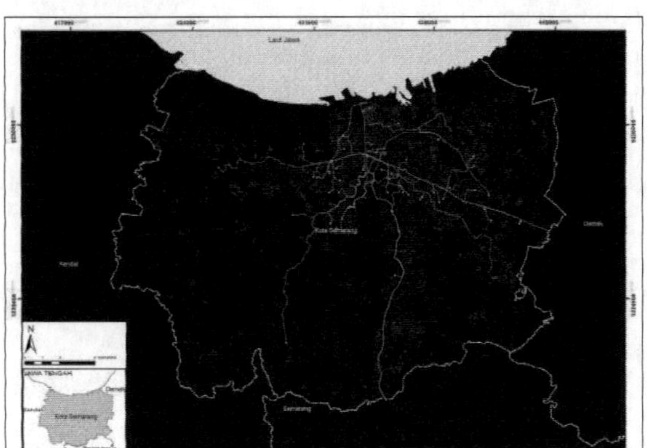

Figure 3. Actual Accessibility immediate return and stay and return pattern

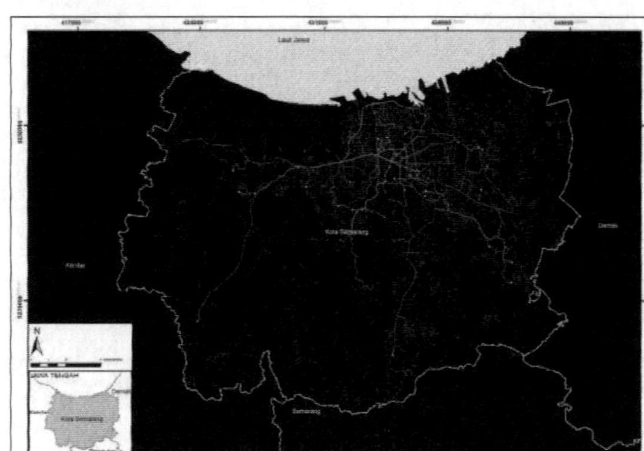

Figure 4. Actual accessibility indirect return pattern

Based on the results of travel diary questionnaire, 35% of students have Indirect Return patterns. Indirect return patterns occur on the way home, while the journey to school has a simple pattern that is the fastest and shortest route because it has only one destination. Indirect patterns occur due to the need for course activities, part-time work, extracurricular activities, organizational activities, hang out, shopping and other activities after school.

Potential accessibility and actual accessibility of high school student's motorcycle users in Semarang have differences. Actual accessibility has a more complex route than potential accessibility, the difference in routes can be seen in Figure 6. Potential accessibility only predicts the route of the student trip directly to the destination so that this route only has a match with the travel pattern immediate return and stay and return.

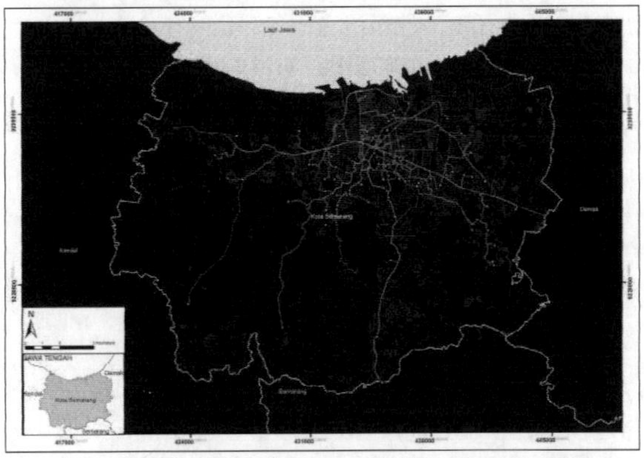

Figure 5. Actual accessibility motorcycle

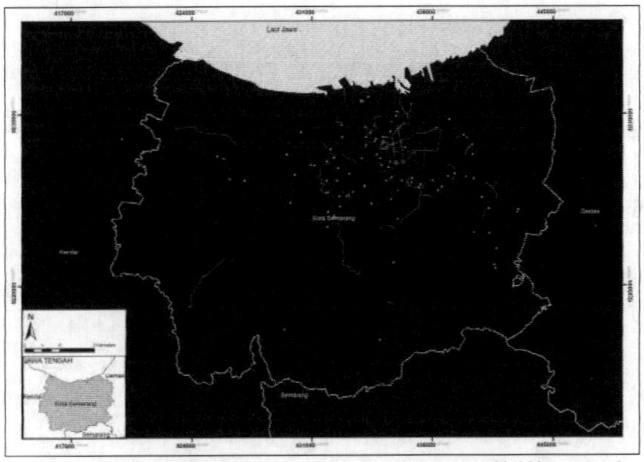

Figure 6. Different potential accessibility and actual accessibility motorcycle

Based on the results of questionnaires and depth interviews of students motorcycle users in Semarang City, students decided to use motorcycles because motorcycles are considered more flexible in determining routes and travel time and more efficient in cost. In addition, the absence of public transportation and family members who can lead the students makes the reason for parents to initiate their children to use a motorcycle even though they do not have not a driver's license, even 88.9% of respondents do not have a driver's license. Most students learn driving techniques at the age of 12-15 years when the physical conditions are met. Although the student's origin location has another potential accessibility mode, the student still keeps the motorcycle as the main mode of movement. Another mode of use occurs when compulsion can not use a motorcycle as it has not been allowed by parents to carry motorcycles, motorcycles are damaged, and when the school establishes policies only two- and three-years student that are allowed to take the motor to school. The average daily travel frequency conducted by the learner is 3.55 or about 4 trips per day with average travel cost is Rp.18.762,89 per week or approximately Rp.3.752.58 per day. When compared to travel expenses using BRT Trans Semarang then the required daily cost is Rp.4.000,00 per day or Rp.20.000,00 per week.

5. Conclusion

The high use of motorcycles by high school students in Semarang City due to the absence of other modes of choice are considered more effective and efficient in traveling. In fact, students who have the potential for accessibility to use public transport modes still make the main choice motorcycles in movement. Motorcycles become the most effective and efficient mode because it is more flexible in determining the route and travel time. Even not significant, the costs incurred will be cheaper than public transport. Hence, the provision of transportation system should be based on user activity, not only on the main activity but also consider the supporting activities. The provision of public transport must be matched with the needs of users and become the first choice to trip.

6. Acknowledgement

The authors are grateful for the funding of DIPA FT Undip 2017 from Engineering Faculty of Diponegoro University. It is a support and would enrich research of travel behavior in the developing country.

7. References

[1] Philip M 2013 Activity Based Travel Behavioural Study and Mode Choice Modelling **2** 181–190
[2] Dharmowijoyo D B E, Susilo Y O and Karlström A 2016 Relationships among discretionary activity duration , its travel time spent and activity space indices in the Jakarta Metropolitan Area , Indonesia *JTRG* **54** 148–160
[3] BPS 2015 *Kota Semarang dalam Angka 2015* (Semarang)
[4] Polrestabes 2015 *Laporan Laka Lalin Kota Semarang Tahun 2014* (Semarang)
[5] Kamargianni M, Polydoropoulou A and Goulias K G 2012 " Teenagers ' Travel Patterns for School and After-School Activities ." *Procedia - Soc. Behav. Sci.* **48** 3635–3650
[6] Setiaji R W 2011 Pembentukan Ketergantungan Sepeda Motor Siswa SMA/SMK Kota Ungaran
[7] Alemu D D and Tsutsumi J G 2011 Determinants and spatial variability of after-school travel by teenagers *J. Transp. Geogr.* **19** 876–881
[8] Kitamura R 2010 Life-style and travel demand 679–710
[9] Kamruzzaman M, Hine J, Gunay B and Blair N 2011 Using GIS to visualise and evaluate student travel behaviour *J. Transp. Geogr.* **19** 13–32
[10] Ardiansyah S and Rakhmatullah A 2014 Model Spasial Statistik Kepemilikan Sepeda Motor di Kecamatan Banyumanik, Kota Semarang *Geoplanning J. Geomatics Plan.* **1** 56–64
[11] Becker U and Gerike R 2008 Background-Needs as the focus of transport planning. In How to Define and Measure Access and Need Satisfaction in Transport 5–24

Street Lighting Infrastructure Assessment Using Discriminant and GIS Method on Mount Merapi Evacuation Road

R P Izdihar[1], M Maryono[1], W Widjonarko[1], and S Rahayu[1]

[1]Department of Urban and Regional Planning, Diponegoro University, Semarang 50275,Indonesia

Email: rakanpramoe@gmail.com

Abstract. This research aims to assess street lighting infrastructure in rural-urban of Mount Merapi Evacuation road. Three evacuation road/corridor; Mriyan-Boyolali, Wonodoyo-Boyolali and Samiran-Boyolali are selected as case study. By using discriminant this study examine 6 variables namely type of lamp, physical component, height, time, power and cons consumption. In addition this study also using GIS method to assessing geographical feature as of previous result. According to the discriminant analysis, the characteristic of street lighting could be distinguished as two characteristic, while from the GIS assessment, the study found three characteristic of geographical street lighting feature.

Keywords: street lighting infrastructure, Mt. Merapi, evacuation road corridor

1. Introduction

Study concerning to street lighting infrastructure have been conducted worldwide. Many theme have been researched such as concerning to crime [1,2], accident [3,4] energy and efficiency [5,6], smart and intelligent system [7,8]. However, research associated toward evacuation and characteristic of corridor very limited. This study aims to assess the quality street lighting infrastructure in associated with preparedness of evacuation process in mount merapi hazard. Three corridor surrounding mount Merapi namely Mriyan-Boyolali, Wonodoyo-Boyolali and Samiran-Boyolali are selected as case study. In addition this study will explore the physical and management aspect. The physical variable such as type of lamp, physical component, height, time, power and cons consumption, while the management variable such as providing and maintenance system

2. Methods and Study Area

This research utilized discriminant method to analysis street lighting characteristic in the three corridors. This method is one of the powerful method to classify the characteristic of infrastructure, since many study have been conduct by using this method at least for 27 years. For example, study of

IOP Conf. Series: Earth and Environmental Science **123** (2018) 012017 doi:10.1088/1755-1315/123/1/012017

infrastructure impact [9], study to define significant urban rail infrastructure [10], Moreover, this research utilized spatial method such as GIS to explore the spatial aspect between corridors. This study also assess the management aspect namely, providing and maintenance system, in associated with community concern in preparedness. Different with previous, for example, community concern in disaster waste [11], lava flow [12], vulnerability of infrastructure [13], this study utilized interview, simple FGD and observation method during data collection. Data characteristic classified by 6 variables for discriminant analysis and classification as three group for final assessment such as good, medium, and bad.

The study area covers 3 villages spread in 3 different sub-districts, namely Mriyan - Boyolali, wonodoyo - Boyolali, and Samiran- Boyolali. The three villages have geographical characteristic such as geographical slopes, classification of disaster prone area. The length of each corridor as follow (figure 1):

a. Mriyan Village - Boyolali City corridor has length : 12,672 km;
b. Wonodoyo Village - Boyolali City corridor has a length : 14,616 km;
c. Samiran Village - Boyolali City corridor has a length : 17,724 km

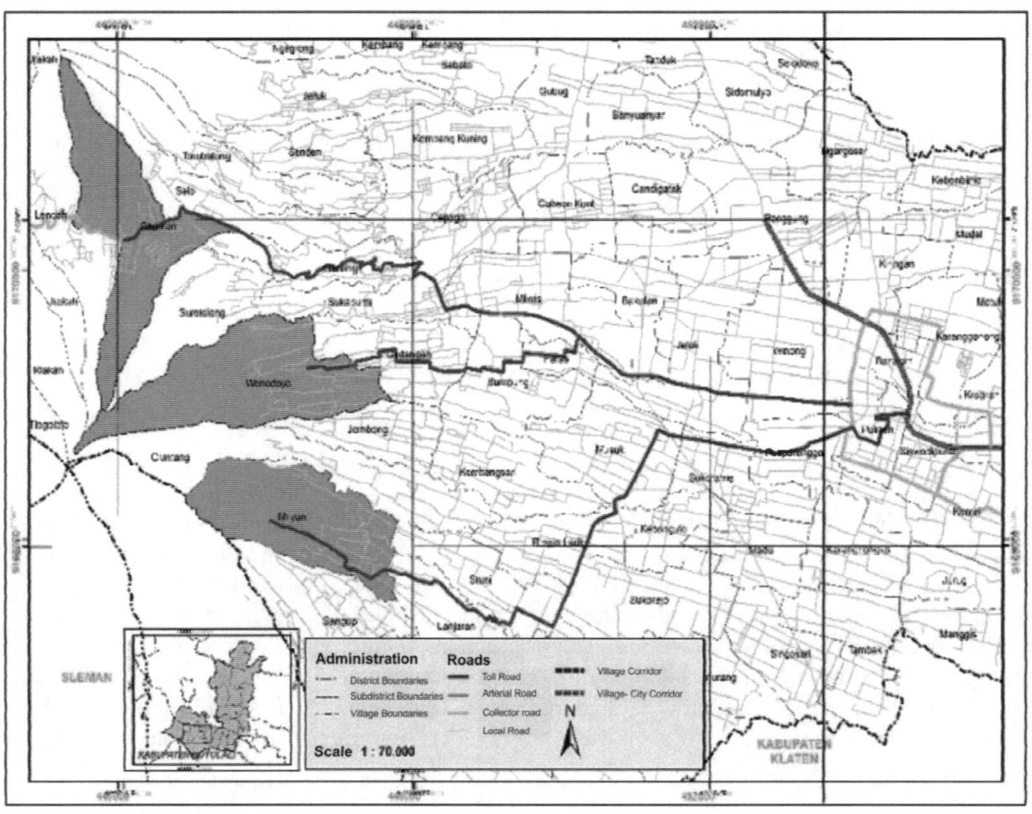

Figure 1. Study Area, Mriyan, Wonodoyo, Samiran Corridor

3. Result and Discussion

3.1. Physical Data and Characteristic of Street Lighting

Table 1. Description of Corridor, Sub Corridor, Number of Street Lighting, Average of Power and Time of Lighting in Study Area

Corridor	Sub corridors	Zone and classification of prone Area	Number of street lighting	Average of Power	Time of Lighting
Mriyan – Boyolali	Mriyan	Disaster Prone II	27 point	15 watt	12 hours
	Sruni	Disaster Prone II	13 point	15 watt	10 hours
	Ringinlarik	Disaster Prone III	16 point	20 watt	10 hours
	Musuk	Disaster Prone III	25 point	50 watt	12 hours
	Sukorame	Disaster Prone III	10 point	50 watt	10 hours
	Pusporenggo	Non – Disaster Prone	6 point	150 watt	10 hours
		Lenght of corridor : 12, 672 km			
Wonodoyo – Boyolali	Wonodoyo	Disaster Prone II	17 point	15 watt	12 hours
	Gedanagan	Disaster Prone II	15 point	15 watt	12 hours
	Sambung	Disaster Prone III	20 point	20 watt	12 hours
	Paras	Disaster Prone III	9 point	50 watt	12 hours
	Jelok	Disaster Prone III	7 point	50 watt	10 hours
	Winong	Non – Disaster Prone	10 point	150 watt	10 hours
		Lenght of corridor : 14, 616 km			
Samiran – Boyolali	Samiran	Disaster Prone II	10 point	15 watt	12 hours
	Selo	Disaster Prone II	10 point	15 watt	12 hours
	Genting	Disaster Prone III	15 point	20 watt	10 hours
	Mliwis	Disaster Prone III	10 point	50 watt	10 hours
	Winong	Non – Disaster Prone	10 point	150 watt	10 hours
		Lenght of corridor : 17,724 km			

Source : Authors, 2017

As depicted in table 1, totally there are 17 sub corridor in with represent rural-urban region. The length of corridors are 17,724 km. The upstream corridors are dominated with area as classify for disaster prone II at while the downstream corridor are disaster prone III. Sub corridor of Mriyan and Wonodoyo are classify as rural zone, sub corridor of Sruni, Gedangan are classify as rural-urban, while Ringinlarik, Musuk Sukorame, Purporenggo, Sambung, Paras, Jelok, Winong, Samiran, Selo, Genting, Mliwis, Winong are classify urban zone [14]. The physical condition such as power of street lighting is diverse among 15-150 what. The highest power of street lighting in corridor of study area is 150 what while the highest is 150 what. The time of street lighting on tend to similar between 10-12 our per day.

Table 2. Description of Physical Characteristic of Street Lighting in Study Area

Physic			**Sub Corridor and Area**
Height	**Physical Component**	**Type of Lamp**	
Height of street lighting no more than 5 meters	In this areas, the physical material used for street lighting is still reasonable, for example made of wood / bamboo so that not good enough resilience	Flourescent Tube Lamp	Mriyan , Wonodoyo Samiran, Gedangan , Sambung Selo and Genting (sub corridor 1)
Quite varied, there street lighting which has a height of less than 5 meters, there is also 5 - 10 meters	In this area l-urban areas, the physical material of street lighting is good enough. Made of metal, has a lamp shield, so good enough resilience	SOX	Ringinlarik, Pusporenggo, Jelok, Mliwis, and Winong (sub corridor 2)
Cost & Energy			**Area**
Time of Lighting	**Power**	**Cost Consumption**	
The average time of street lighting lighting in the village area is for 12 hours / day.	P day : Average Power of lamp x Time of Lighting x street lighting Exist : 15 watt x 10 hours/ day x 177 street lighting : 26,550 watt/hour P month : P day x 30 days : 26,550 watt/hour x 30 : 796,500 watt/hour : 796. 5 kWh	Cost Analysis : P month x Electricity Rates : 796. 5 kWh x Rp 1,385.00 : Rp 1,103,152. 50	Mriyan, Wonodoyo Samiran, Gedangan, Sambung, Selo dan Genting (sub corridor 1)
The average time of street lighting lighting in the village area is for 10 hours / day.	P day : Average Power of lamp x Time of Lighting x street lighting Exist : 150 watt x 12 hours/ day x 53 street lighting : 95.400 watt/hour P month: P day x 30 daysi : 95,400 watt/hour x 30 : 2,862,000 watt/hour : 2,862 kWh	Cost Analysis : Cost Incurred : P month x Electricity Rates : 1,800 kWh x Rp 1,385.00 : Rp 3,963,870.00	Ringinlarik, Pusporenggo Village, Jelok, Mliwis, Winong (sub corridor 2)

Source : Authors, 2017

As depicted in table 2, Fluorescent Tube Lamp, is used in sub corridor of Mriyan, Wonodoyo, Samiran, Gedangan, Sambung, Selo and Genting. This is the sub corridors 1. In this areas, the physical material used for street lighting is still reasonable, for example made of wood / bamboo so that not good enough resilience. The height of the lamp in this area is no more than 5 meters. The average time of street lighting lighting in the village area is for 12 hours / day. The consumption of power estimated 796, 5 kWh per month with the cost of energy is IDR 1,103,152.50 per month.

In sub corridor 2 namely; Ringinlarik, Pusporenggo, Jelok, Mliwis, and Winong, type of lampt that used is SOX. In this area l-urban areas, the physical material of street lighting is good enough. Made of metal, has a lamp shield, so good enough resilience. The height of lamp is Quite varied, there PJU which has a height of less than 5 meters, there is also 5 – 10 meters. The average time of street lighting lighting in the village area is for 10 hours / day. Energy consumption per month estimated 2.862 kWh and cost of energy per month estimated IDR 3,963,870.00

According to the characteristic of physical of street lighting could be shown that there is a significant different between sub corridor 1 group of corridors Mriyan, Mriyan, Wonodoyo, Samiran, Gedangan, Sambung , Selo and Genting comparing to the physical characteristic of street lighting in sub corridors 2 group of corridors Ringinlarik, Pusporenggo, Jelok, Mliwis, and Winong.

3.2. Characteristic of Management System of Street Lighting

As depicted in table 3, In the sub corridor 1 as group of Mriyan, Wonodoyo, Samiran, Gedangan, Sambung, Selo and Genting, the providing system of street lighting is Community base. However the level of understanding toward energy efficiency is very limited. In sub corridor 2 as group of corridors Ringinlarik, Pusporenggo, Jelok, Mliwis, and Winong, they received support from the government. So in this sub corridors the role of community in providing and management system not dominant.

Table 3. Description of Management Characteristic of Street Lighting in Study Area

Provide And Maintenance		**Area**
Provide System	Management System	
- **Community based** - **The ability of the community is limited, so they need an assistance from goverment**	- Community based (People who pay their own electricity bills, and for lamp replacement by their self)	Mriyan, Wonodoyo, Samiran, Gedangan , Sambung, Selo dan Genting (sub corridor 1)
- Received a direct assistance from the goverment - The role of the community not dominant	- Rely on government assistance (both for electricity bills and cost maintenance)	Ringinlarik, Pusporenggo, Jelok , Mliwis, dan Winong (sub corridor 2)

Source : Authors, 2017

3.3. Discriminant Analysis

The discriminant analyis is used to understanding dominant faktor or variable in the clasification. The result of discriminant analysis is that the physical variable of lamp type is the most influential. The discriminant function according to sub corridor 1 and sub corridor 2 of this study are as follow:

Sub corridor 1 $= -11{,}988 - 8{,}673\ X1 + 0{,}332\ X2$,
Sub corridor 2 $= -4{,}895 - 2{,}525\ X1 + 0{,}139\ X2$

Tabel 4. Variables in the Analysis

Step		Tolerance	Sig. of F to Remove
1	Type of Lamp	1,000	,000

Source : Authors, 2017

Tabel 5. Eigenvalues

Function	Eigenvalue	% of Variance	Cumulativ e %	Canonical Correlation
1	3,694[a]	100,0	100,0	,887

Source : Authors, 2017

Tabel 6. Structure Matrix

	Function
	1
type_lamp	1,000
height_street lighting[a]	,594
cost[a]	,456
power_consumption[a]	,456
time_lighting[a]	-,382
ketersediaan[a]	,071

Source : Authors, 2017

3.4. GIS analysis of Street Lighting

The GIS analysis the overall quality of street lighting. The result of assessment by using scoring system. The street lighting quality of sub corridor of Musuk Pusporenggo Sukorame Wonodoyo, Paras, Sumbung, and Gedangan is less quality. Sub corridor of Mriyan, Sruni, Lanjaran, and Ringinlarik is moderate quality. And the sub corridor of Samiran, Selo Genting, Mliwis, Manyaran, Sukabumi, Bakulan, Bolo, Jelok Village, Winong the street lighting quality is good.

Tabel 7. Street Lighting Quality

	Red Zone	*Yellow Zone*	*Green Zone*
Quality	That is a corridor which has Public Street Lighting with less quality	That is a corridor which has Public Street Lighting with medium quality	That is a corridor which has Public Street Lighting with good quality
Sub corridors	Musuk Pusporenggo Sukorame Wonodoyo Paras, Sumbung, and Gedangan	Mriyan, Sruni, Lanjaran , and Ringinlarik	Samiran, Selo Genting, Mliwis, Manyaran, Sukabumi, Bakulan, Bolo, Jelok Village, Winong

Source : Authors, 2017

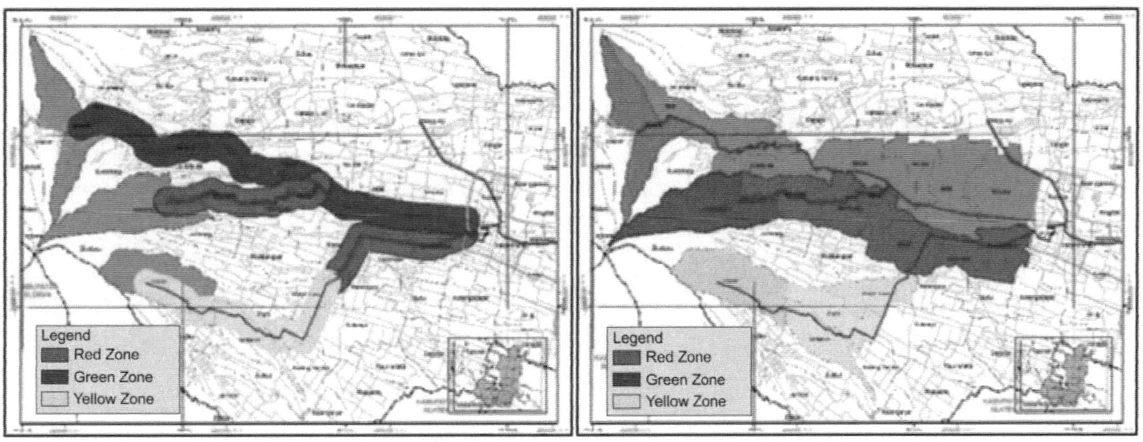

Source : Authors, 2017

Figure 2. Assesment Based on Discriminant and GIS Aplication

4. Conclusions

The study have been Analysis Street lighting characteristic as follow:

a. Two characteristic of street lighting with the model could be distinguished by using discriminant parametric model:
   1) Sub corridor 1 = -11,988 – 8,673 X1 + 0,332 X2
   2) Sub corridor 2 = - 4,895 – 2,525 X1 + 0,139 X2

   With the Group of sub corridor 1 are Mriyan, Wonodoyo, Samiran, Gedangan, Sambung, Selo and Genting and Group of sub corridor 2 are Ringinlarik, Pusporenggo, Jelok, Mliwis, and Winong.

b. Three characteristic of geographical feature of street lighting are
   1) Sub corridor with less quality of street lighting quality at Musuk Pusporenggo Sukorame Wonodoyo, Paras, Sumbung, and Gedangan is less quality.

2) Sub corridor with moderate of street lighting quality at Mriyan, Sruni, Lanjaran , and Ringinlarik is.
3) sub corridor with good quality of street lighting at Samiran, Selo Genting, Mliwis, Manyaran, Sukabumi, Bakulan, Bolo, Jelok Village, Winong

5. Acknowledgment

This research was funded by Anual year Budget of Faculty Engineering Research Schema for year 2017, under contract No. 434/UN7.P/HK/2016. I would like to thank my partners during discussion and help in this study. The Government of Boyolali for the permission, Pak Sukardi, Ibu Suparmi who help for motorcycle. The Leader of Mriyan, Wonodoyo and Samiran Village for many information during FGD and interview, Laras Kun Rahmanti and Intan Hapsari Hasmatika for the survey and collecting data, Mr. Darno for providing the motorcycle during survey, my friend Hafiz Satria in helping me to understand the framework of this study; and also Anindya Putri Tamara for helped me in editorial writing.

6. References

[1] Ramsay M and Newton R 1991 The effect of better street lighting on crime and fear: A review
[2] Farrington D P and Welsh B C 2002 *Effects of improved street lighting on crime: a systematic review* (Home Office London)
[3] Beyer F R and Ker K 2009 Street lighting for preventing road traffic injuries *Cochrane Libr.*
[4] Wanvik P O 2009 Effects of road lighting: an analysis based on Dutch accident statistics 1987--2006 *Accid. Anal. Prev.* **41** 123–128
[5] Kostic M and Djokic L 2009 Recommendations for energy efficient and visually acceptable street lighting *Energy* **34** 1565–1572
[6] Müllner R and Riener A 2011 An energy efficient pedestrian aware Smart Street Lighting system *Int. J. Pervasive Comput. Commun.* **7** 147–161
[7] Popa M and Cepi$\$csc$\$ua C 2011 Energy consumption saving solutions based on intelligent street lighting control system *UPB Sci. Bull., Ser. C* **73** 297–308
[8] Wojnicki I, Ernst S, Kotulski L, Se A and Others 2014 Advanced street lighting control *Expert Syst. Appl.* **41** 999–1005
[9] Looney R and Frederiksen P 1981 The regional impact of infrastructure investment in Mexico *Reg. Stud.* **15** 285–296
[10] Lane B W 2008 Significant characteristics of the urban rail renaissance in the United States: A discriminant analysis *Transp. Res. Part A Policy Pract.* **42** 279–295
[11] Nakayama H, Shimaoka T and Others 2015 Identification of Factors Affecting Stakeholders' Intentions to Promote Preparedness in Disaster Waste Management: A Structural Equation Modeling Approach 九州大学工学紀要 **74** 79–98
[12] Gregg C E, Houghton B F, Paton D, Swanson D A and Johnston D M 2004 Community preparedness for lava flows from Mauna Loa and Hual{$\$=a}lai volcanoes, Kona, Hawai 'i *Bull. Volcanol.* **66** 531–540
[13] Wilson T M, Stewart C, Wardman J B, Wilson G, Johnston D M, Hill D, Hampton S J, Villemure M, McBride S, Leonard G and Others 2014 Volcanic ashfall preparedness poster series: a collaborative process for reducing the vulnerability of critical infrastructure *J. Appl. Volcanol.* **3** 10
[14] Bappeda Boyolali 2016 *Study of Rural-Urban classification in Boyolali*

Transportation Limitation Access to the Small Islands (Case Study: Banggai Laut Regency)

S Sunarti[1]

[1] Urban and Regional Planning Department, Faculty of Engineering, Diponegoro University

Email: n4rti08@gmail.com

Abstract. Indonesia is as an archipelago and maritime country, the large number of Islands owned and scattered in all directions makes a challenge for the Government in equitable development. Development in Indonesia has not been spread evenly and tends to focus on the big island, while the smaller islands are still far behind and lack of government attention. One of them is the lack of infrastructure especially the access to the small islands. Among the small islands in Indonesia with minimal maritime infrastructure or transportation is Banggai Laut Regency, Central Sulawesi Province. This Regency is a new regency that separate itself from its previous regency that is Banggai Kepulauan Regency in about 4 years ago. For the development of the Banggai Laut Regency, access to reach that regency is quite difficult. Therefore, the aim of this research is to find infrastructure development strategy to support the development of Banggai Laut Regency. The research method used was the concurrent model mixed method. Data collection method was done with primary data through field observation and interview, secondary data through literature and document review. Analytical techniques used are qualitative descriptive and Map Overlay techniques using GIS to describe the characteristics of study areas and spatial relationships between islands. The results of this research conclude that the Banggai Laut Regency requires infrastructure development particularly maritime transportation to enhance accessibility of the community headed to Banggi Laut Regency or headed to another island from the Banggai Laut Regency.

Keywords: Accessibility, Transportation, Small Islands, Development

1. Introduction

Equity of development is still as a homework for governments in various countries, especially for developing countries. In Indonesia, equity of development is a challenge for the government considering Indonesia's vast territory. Indonesia as an archipelagic country and a maritime country has approximately 17,504 islands that consist of big and small islands [1]. Indonesia's small islands have diverse roles and potentials such as marine tourism potential, energy and defense, environmental services and others. The contribution of small islands in Indonesia has not been matched by government attention in various sectors. One of them is the government's lack of support for the Small Islands management. Development allocations undertaken by the government are still centered on large islands and still prioritize on land area counts rather than on water areas [2]. Infrastructure has an important role in regional development, such as sea transportation to facilitate the movement of goods, services, logistics, and people from one island to another in an effort to improve the welfare. In the current condition, there are still many small islands in Indonesia that do not have adequate transportation infrastructure so that to reach small islands is quite difficult, whereas without transportation, it will be difficult for the community to get benefit from production activity [3]. These transport limitations have an impact on economic development on small islands where they often fall behind compared to large islands with adequate infrastructure.

One of the small islands that do not have adequate infrastructure is Banggai Laut Regency in Central Sulawesi Province. Banggai Laut Regency is a new regency, which separated itself from Banggai Kepulauan Regency and only about 4 years old. Access to Banggai Laut Regency is quite difficult, so it becomes an obstacle to develop the potential of the region. To reach the Banggai Laut Regency from the Sulawesi Island takes a long time and requires several changes of transportation type. In addition, transportation infrastructure within Banggai Laut Regency itself is not adequate, whereas the district has many different potentials for each island. If there is good accessibility in the movement of goods, services and people and the availability of adequate transportation infrastructure, then the potential can be well developed to improve the welfare of society and the regional economy. Based on the problem, the purpose of this research is to find infrastructure development strategy to support the development of Banggai Laut Regency.

The steps taken to achieve the research objectives include: 1.) identification of geographic conditions, population, potentials and problems in Banggai Laut Regency, 2.) identification of transportation facilities, infrastructure and accessibility in Banggai Laut Regency; 3.) analyze of accessibility development strategy in Banggai Laut Regency to develop the region.

2. Research Method

The research approach used in this research was the concurrent model mix method. This model was chosen because researchers combined qualitative and quantitative data to conduct a comprehensive analysis of existing research problems [4]. The quantitative method used was descriptive, through the frequency distribution of the criteria and variables used, while the qualitative method was used to explore and synthesize various information. The method of data collection was divided into two namely: primary and secondary data. Primary data were gained through field observations related to transportation infrastructure conditions and interviews with agencies and communities on transportation conditions between small islands in Banggai Laut Regency. Secondary data were obtained literature review and documents obtained from related institutions. Analytical techniques used are qualitative descriptive and Map Overlay techniques using GIS to describe the characteristics of study areas and spatial relationships between islands.

3. Literature Review

Small Island is an island with an area smaller than or equal to 2,000 km^2 along with its ecosystem unity. Coastal Resources and Small Islands are biological resources, non-biological resources, artificial resources, and environmental services. Biological resources include fish, coral reefs, seagrass beds, mangroves and other marine biota. Non-living resources include sand, seawater and seabed minerals. Artificial resources include marine and fishery-linked marine infrastructure, and environmental services in the form of natural beauty, seabed surface underwater installations associated with marine and fisheries and ocean wave energy contained in the Coastal Zone [5].

While Transportation is the activity of moving goods (cargo) and passengers from one place to another [3]. Transportation has several important roles for society, economic and socio-political development of a country [3] namely 1.) Transportation and Community Life, ie transportation is beneficial for people's life in terms of transporting raw materials and products produced by a region and distributed to other areas, 2.) Geographical Specialization, ie each region has different specialties, such as Tasikmalaya, Kalimanyan, Yogyakarta handicraft will be different in each region. With adequate transport infrastructure, such goods can be sold and marketed outside their specialist area, 3.) Economical Production, ie a production will be useful and economical if there is enough transportation mode, in the absence of transportation, the community will not get benefit of production, and 4.) National and Hankamnas development, ie the role of transportation for national development, development of the entire territory of Indonesia, equality of development and national defense of the Indonesian nation. There is a lot of understanding of accessibility, such as accessibility is a measure of potential or ease of people to achieve goals in doing a journey [6]. In line with the previous understanding, accessibility is a concept that combines land use systems geographically with the transport network system that connects them [7]. To express accessibility, accessibility levels can be meassured through two things: distance and condition of the infrastructure, with the following classifications [8]:

Table 1. Accessibility Classification According to Black [8]

Distance	Far	Low Accessibility	Mid Accessibility
	Near	Mid Accessibility	High Accessibility
Infrastructure Condition		Very bad	Excellent

Accessibility is very influential factor on the development of a region, where the infrastructure sector is one of the sectors that have an important role in spurring economic growth, because basically the infrastructure sector acts as a liaison of various kinds of economic activity. One of which is Road construction, it has accessibility function to open less developed area and open mobility function to spur the developed area [9]. Transportation is not only the process of transporting people and goods from one location to another but having an impact on the economic condition of the region [10]. The existence of transportation improvement projects has an impact on the benefit of the poor and road users, namely to increase access to trade, education, health and so on, then decrease vehicle operating costs, and to save time or shorten travel time and increase in capital value [11].

4. Results and Discussion

Banggai Laut Regency is an archipelagic regency consisting of four major islands namely Banggai Island, Labobo Island, Bangkurung Island and Bokan Islands. Banggai Laut owns 7 districts, 4 districts in Banggai Island that is North Banggai, Banggai, Central Banggai and South Banggai Districts. While the three other islands each consisting of one district that is Labobo District, Bangkurung District and Bokan Islands District. The islands of Banggai Laut Regency are lowland areas where the entire island is <500 meters above sea level but has varying slopes, ranging from 0-2%, 2-15%, 15-40% to 40%. Here is the administrative boundary map of Banggai Laut Regency.

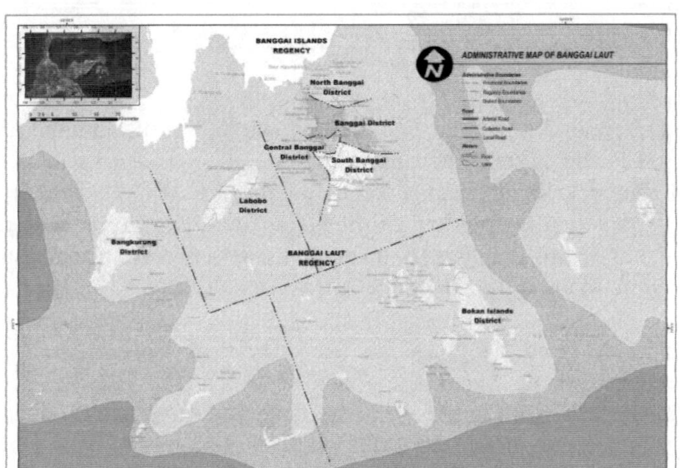

Source: Development Planning Agency at Sub-National Level of Banggai Laut Regency, 2015

Figure 1. Administrative Boundary Map of Banggai Laut Regency

Its form as archipelago and separated from the mainland of Central Sulawesi Province, makes Banggai Laut Regency quite remote and has the least population compared to other regencies or cities in Central Sulawesi Province. While in Banggai Laut Regency the population is mostly concentrated in the largest island, population and population density can be seen in following figure and table from each island:

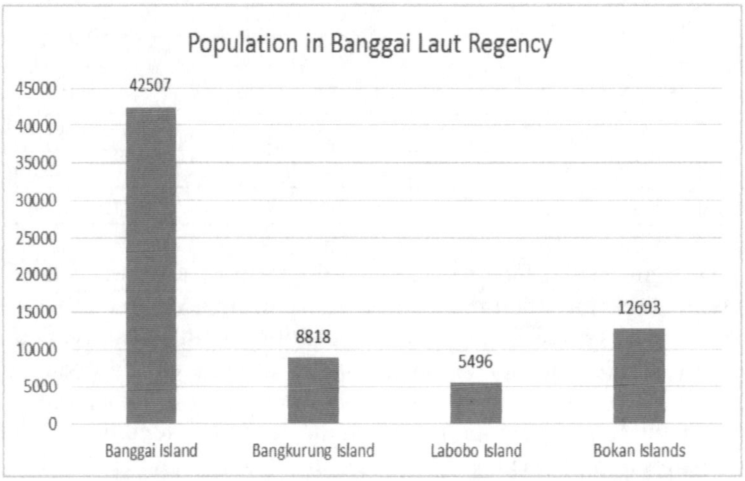

Figure 2. Population in Banggai Laut Regency, 2015 [12]

Table 2. Population Density in Banggai Laut Regency 2015 [12]

Island	Population	Area (Km^2)	Population Density
Banggai Island	42.507	294,39	144
Bangkurung Island	8.818	116,55	76
Labobo Island	5.496	85,65	64
Bokan Islands	12.693	229,08	55
Jumlah Total	69.514	725,67	96

From **Figure. 2** and **Table. 2** it can be seen that the largest population and the highest population density exist in Banggai Island. This is because Banggai Island is the main island located Banggai District as the Capital Regency. Banggai District is as the center of government and acts as a Center for Local Activities in the space structure of Banggai Laut Regency. Because on that issue all activities are more concentrated in Banggai Island, so that the facilities and infrastructure of this distric is more complete compared to the other islands.

Characteristics of residents in Banggai Laut Regency occupying each island is quite diverse, among others, inhabited by various tribes such as Bajo tribes, Banggai tribes, Bugis tribes, Buton tribes and Javanese tribes. Each of these tribes has characteristics and traditions that are different from each other. The Bajo tribe has a tendency not to be out of their community, and in its history the Bajo tribe is quite difficult to merge with the mainland people, they never interact and even avoid other tribes because their activity is mostly done at sea. However, nowdays the value and customs of the Bajo tribe begin to shift and begin to open up to interact with the mainland people because of the demanding needs of the economy [13]. Although the cultural values of the Bajo tribe have begun to change and begin to open up to interact with people outside the Bajo tribe, the intensity of their interactions will be different when compared to other tribes such as the Bugis tribe. Bugis tribe has a culture of wander since the 17th century and able to blend and adapt easily in places they stop [14].

The livelihoods of most people in Banggai Laut Regency are in agriculture, plantation, forestry, hunting and fishery as well as financial institution, real estate, leasing and corporate services. Both have the same percentage of 38%. Geographical condition which is an archipelagic regency and culture of several tribes such as Bajo tribe who live in the sea is also a driving factor of most people engaged in fisheries, namely as fishermen and cultivation of marine resources such as seaweed and pearl one of them. The livelihood of residents in Banggai Laut Regency can be seen in the following figure:

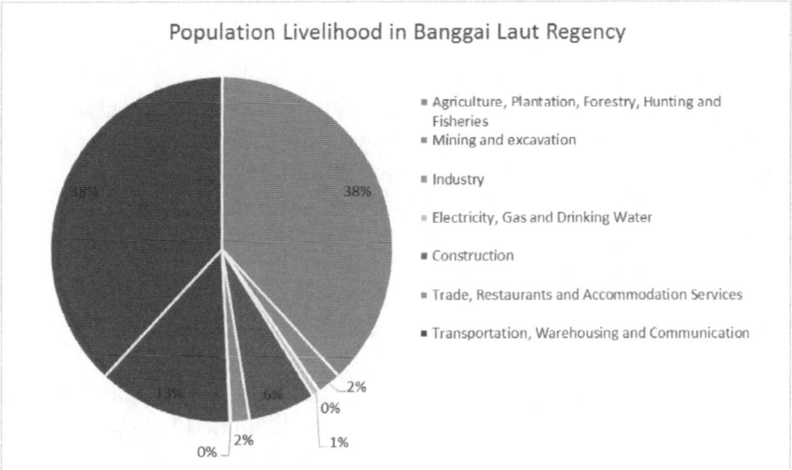

Figure 3. Population Livelihood in Banggai Laut Regency [12]

The potential of Banggai Laut Regency in addition to human resources, is the nature potential that is very beautiful and rich. Potentials are owned by the leading sectors are the fisheries and marine sector, that is seaweed in Bangkurung Island, Banggai Island and Bokan Islands Island. In addition to seaweed, there is also a pearl-producing island. Banggai Laut Regency also has potential in tourism sectors, especially marine tourism because of the beauty of the sea that is owned, among others, jellyfish lake in Bokan Islands, Kelapa Lima beaches, White Sands Beach and other marine tourism. Another potential of Banggai Laut Regency is mining where mining potential in Banggai Laut Regency is granite and quartz sand. Potential granite stone mines themselves are found in north of Banggai Island, Bangkurung Island and Labobo Island.

In addition to the potential, Banggai Laut Regency has various problems that could hamper the development of areas such as the threat of earthquake, landslide and tsunami prone disaster in all islands in Banggai Laut Regency and low accessibility to Banggai Laut Regency and within the regency itself.

The condition of land transportation infrastructure in Banggai Laut Regency can be seen from road condition in Banggai Laut Regency and can be seen in the following diagram:

IOP Conf. Series: Earth and Environmental Science **123** (2018) 012018 doi:10.1088/1755-1315/123/1/012018

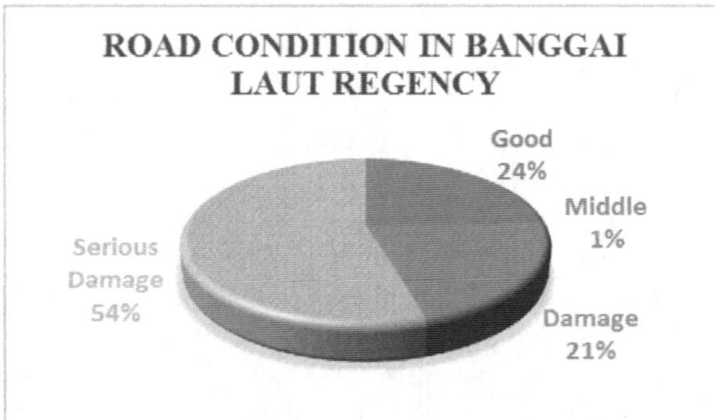

Figure 4. Road Condition in Banggai Laut Regency, 2015 [12]

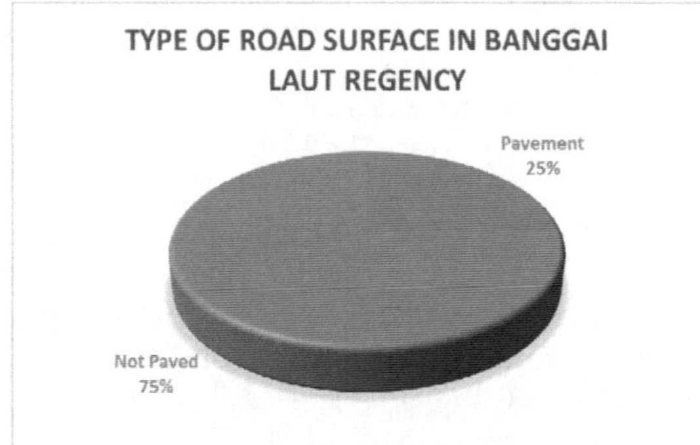

Figure 5. Type of Road Surface in Banggai Laut Regency, 2015 [12]

The **Figure 4 and 5** above show that more than half of all roads in Banggai Laut Regency are 54% having condition with serious damage level. The remaining 21% of roads are damaged, 1% are in moderate condition and only 24% of roads are in good condition. As for the type of road surface that has been paved only 25%, the remaining 75% still not paved. Road conditions that are not feasible to cause inter-regional relationships are less developed quickly, thus requiring further handling, especially in relation to regional development efforts. With the condition of land transportation infrastructure such as it will inhibit the movement and slow the travel time to finally affect the important aspects in the development of one region is the economic aspect.

In addition to road conditions, transportation infrastructure in Banggai Laut Regency has not spread evenly in all villages. Where from the statistical data obtained that there are still villages that have transportation infrastructure that can not be passed by four-wheeled vehicles at a certain time, when it rains and not even passable by four-wheeled vehicles throughout the year. Here is a graph of the number of villages that use land transportation infrastructure according to the existence of roads that can be passed by four-wheeled vehicles:

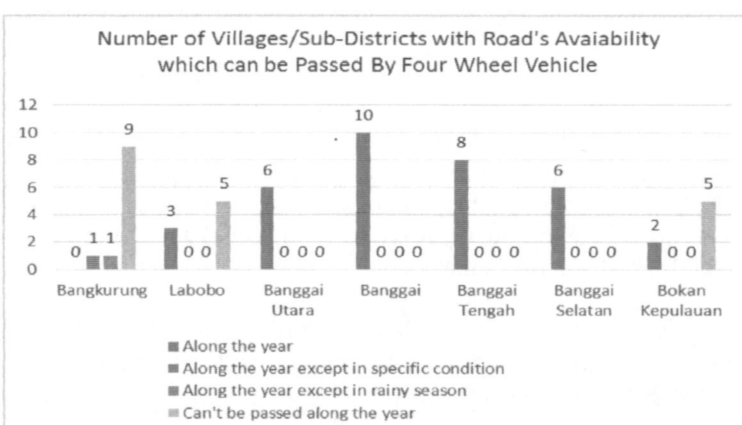

Figure 6. Number of Villages/Sub-Districts with Road's Avaiability which can be passed by four wheel Vehicle [15]

From the **Figure 6** above can be seen that, some villages in Banggai Laut Regency still has a road that can not be passed by four-wheeled vehicles. From seven districts, Bangkurung district is in poor condition, because there is 1 village which the road inside cannot be passed by a four-wheeled vehicle at a certain time, 1 village which the road can be passed by four-wheeled vehicles but not in rainy season, and even 9 Villages which road infrastructure cannot be passed by four-wheel vehicles throughout the year. Unlike districts in Banggai Island as the main islands of Banggai Laut Regency, namely North Banggai, Banggai, Central Banggai and South Banggai districts, where all villages in that four districts can be accessed by four-wheeled vehicles throughout the year. This shows that the development of transportation infrastructure has not been evenly distributed throughout the island, but still concentrated in Banggai Island as the center of activity in Banggai Laut Regency.

Banggai Laut Regency as an archipelago-shaped regency, sea transportation becomes the most crucial transportation to help people movement both to Banggai Laut Regency from other regency as well as the movement within the regency itself. Sea transportation in Banggai Laut Regency is boats without motor, outboard motor boats and inboard motor boats. Following figure is the number of boat in Banggai Laut Regency:

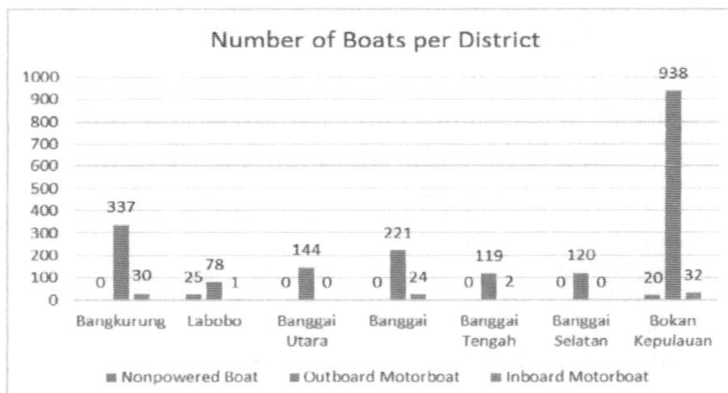

Figure 7. Number of boats per district, 2015 [16]

From the **figure. 7** it can be seen that the most sea transportation in Banggai Laut Regency is outboard motor boat and it is available in all districts. However, the problem is the existence of sea transportation in the form of boats that are used both for fishing activities and for crossing activities between islands has not been matched by the number of docks are adequate. The docks only exist on the big islands of Banggai Island, Labobo Island, Bangkurung Island and large islands in Bokan Islands, whereas for small islands such as those scattered around the main island of Bokan Islands do not have docks, so boats and motor boats are only anchored at the beach's border. The location of the dock can be seen in the following figure 8

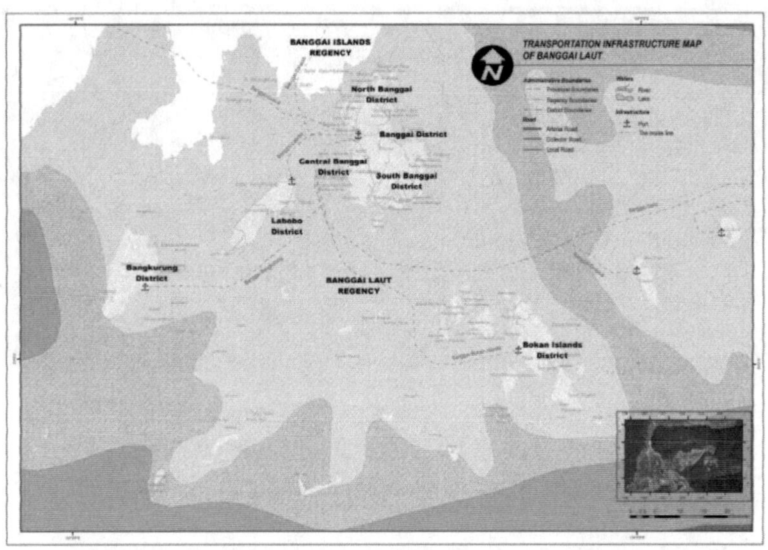

Source: Development Planning Agency at Sub-National Level of Banggai Laut Regency, 2015

Figure 8. Dock Location in Banggai Laut Regency

IOP Conf. Series: Earth and Environmental Science **123** (2018) 012018 doi:10.1088/1755-1315/123/1/012018

Moreover, since sea transportation is the main transport in Banggai Laut Regency and considering its topography as an archipelagic district, the presence or absence of travel is highly dependent on weather conditions and sea waves. These factors can also affect accessibility in Banggai District because in addition to slow travel time, weather conditions and bad sea waves can increase the risk of travel until the cancellation of travel between islands.

The journey to Banggai Laut Regency can be reached by a limited mode of transportation where the journey to Banggai Laut Regency can not be directly reached by plane because it has no airport. Air transportation only arrived in Luwuk City, Banggai Regency. After using the air transport proceed with 2 alternative choices that are directly by sea or combination between sea and land transportation. The sea trip directly to Banggai Laut is for 9 hours by wooden ship, while for combination with sea and land transportation that is reached by speedboat to Salakan, Banggai Kepulauan Regency then continued by land transportation to the port at south end of Banggai Kepulauan Regency and continued again by using a small speedboat that can only accommodate 5 people to the only port on Banggai Island. Overall travel time from Luwuk to Banggai Laut Regency by using the combination of land and sea route is for 8 hours. The trip length to Banggai Laut Regency is also influenced by weather factors such as high wave or sea storm. Here is a route map to Banggai Laut Regency through a combination of sea and land transportation, white line describes the route, yellow dot describes the transit location, the rhombus symbol shows the use of sea transportation mode and the triangle symbol shows the use of land transportation mode:

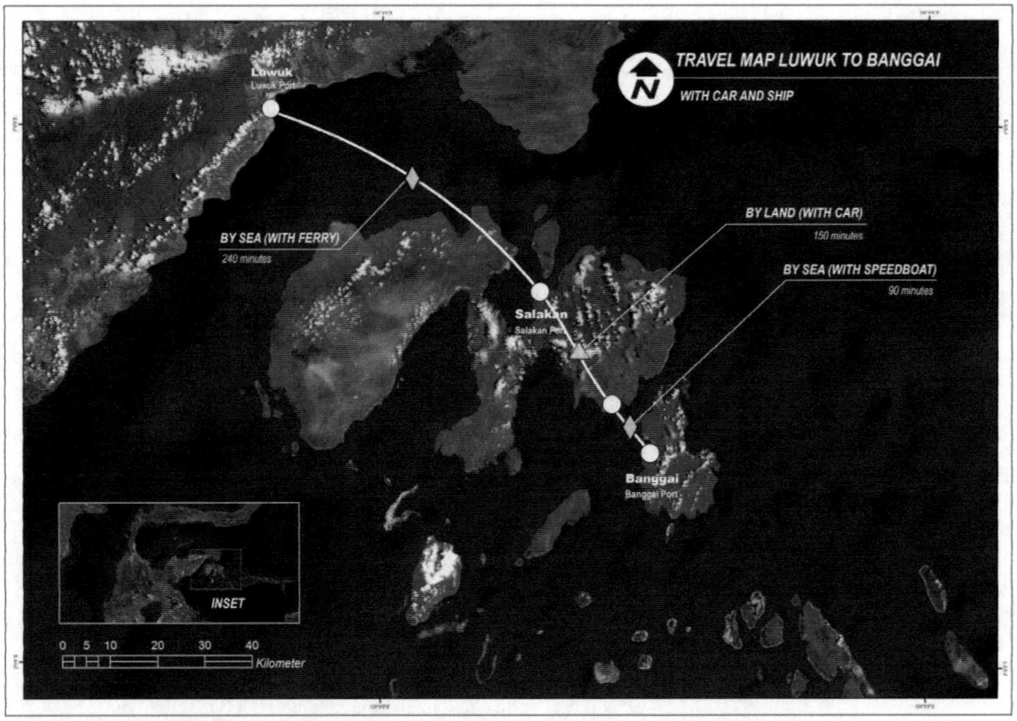

Source: Development Planning Agency at Sub-National Level of Banggai Laut Regency, 2015

Figure 9. Travel Map from Luwuk to Banggai Laut by a combination of land and sea transportation

Banggai Laut district consists of several islands, so the inter-subway trips can only be reached by sea and with varying distance. The distance from the Capital Regency with the most distant district is Bokan Islands, which is about 34.5 miles of sea travel that can be reached by using wooden boats or speed boats. Here are the distance and choice of transportation modes that can be used to go to other District from Banggai district as a Capital Regency:

Table 3. Distance between districts and transportation that can be used [17]

Regency/Capital City	District/Capital City	Distance		Can be reach by vehicle
		Land (km)	Sea (mil)	
Banggai Laut/ Banggai/ Lompio	Bangkurung/ Lantibung	-	31,7	Wooden ship / Speedboat
	Labobo/ Mansalean	-	9	Wooden ship / Speedboat
	North Banggai/ Lokotoy	15	-	Two-wheeled and four-wheeled vehicle
	Banggai/ Lompio	0	-	On foot
	Central Banggai/ Adean	5	-	Two-wheeled and four-wheeled vehicle
	South Banggai/ Matanga	28	-	Two-wheeled and four-wheeled vehicle
	Bokan Kepulauan/ Bungin	-	34,5	Wooden ship / Speedboat

Based on the data mentioned in the above points can be measured accessibility level as in Table 1 by grouping the data into two things, the distance and condition of facilities and infrastructure [8] as follows:

Table 4. Data Classification by Variable

Variable	Existing Condition of Banggai Laut Regency
Distance	• The distance between Banggai Regency and other surrounding areas: - With Banggai Regency 130 Km - With Banggai Kepulauan Regency 78 Km - With Makassar City 602 Km - With Palu City 418 Km • Distance between the Capital District with Other Districts - With Bangkurung District 31,7 mil - With Labobo District mil - With North Banggai District 15 Km - With Central Banggai District 5 Km - With South Banggai District 28 Km - With Bokan Kepulauan District 34,5 mil • Travel time from Banggai Regency using Wooden ship directly to Banggai Laut Regency for 9 Hours, and if using combination of land and sea route takes time for 8 hours • To reach Banggai Laut Regency, it is necessary to change transportation modes both from air transportation and land transportation to sea transportation
Facilities and Infrastructure Condition	• 54% of roads are heavily damaged and 75% of roads have not been paved • There are still villages where road conditions can not be passed by 4 wheeled vehicles either at certain times, during the rainy season or not passable throughout the year • Sea transport (boats and motorboats) have not been matched by adequate dock quantities • Docks are only available on large islands, while small islands do not have docks as ship berths • Do not have Airport yet, so the only way to get to Banggai Laut Regency is by sea • Travel duration and travel time depend on weather and sea waves

IOP Conf. Series: Earth and Environmental Science **123** (2018) 012018 doi:10.1088/1755-1315/123/1/012018

From the table it can be concluded that from the distance side, Banggai Laut Regency is classified as Far, seen from distance with other regency / city around it, distance between islands, long time travel and required change of various kinds of transportation to reach Banggai Laut Regency. In terms of infrastructure condition, Banggai Laut Regency is categorized as very bad condition, judging from the poor road condition, port and dock with insufficient number, and limited choice of transportation mode to Banggai Laut Regency. From both sides, it is found that the accessibility level in Banggai Laut Regency is included in the low classification or can be seen in the following table:

Table 5. Position of Banggai District accessibility level based on Black measurement [8]

Distance	Far	Low Accessibility	Middle Accessibility
	Near	Middle Accessibility	High Accessibility
Infrastructure Condition		Very Bad	Excellent

Explanation:
Existing condition in Banggai Laut Regency
Accessibility Level in Banggai Laut Regency

Interaction between regions with Banggai Laut Regency can be known by using the calculation of gravity model. This model can be used as a reference to estimate the attractiveness of a location with the location around it. It can also be applied to estimate the number of trips between the two places based on their attractiveness. The variables used to know the interaction between regions with the gravity model is the total population of Banggai Laut Regency, population in other area and the distance between Banggai Laut Regency and other area [18]. Then obtained the following results:

Table 6. Interaction between regions and inter-districts in Banggai Laut Regency

Interaction Between Banggai Laut Regency with Other Regency/City

Regency A	Regency B	Population Regency A	Population Regency B	Distance (km)	Interaction	Interaction Criteria
Banggai Laut	Banggai	69,514	354,402	130	2,915,491	Low
	Banggai Kepulauan	69,514	114,980	78	2,627,456	Low
	Makassar	69,514	1,469,601	602	563,779	Low
	Palu	69,514	367,342	418	292,294	Low

Interaction Between Capital Regency with Other District

District A	District B	Population District A	Population District B	Distance (km)	Interaction	Interaction Criteria
Banggai	Bangkurung	23,214	8,818	31.7	407,410	Low
	Labobo	23,214	5,496	9	3,150,226	Low
	North Banggai	23,214	6,517	15	1,344,761	Low
	Central Banggai	23,214	7,265	5	13,491,977	Low
	South Banggai	23,214	5,511	28	326,358	Low
	Bokan Islands	23,214	12,693	34.5	495,115	Low

Criteria :
High: Interaction > 4 (x 10^9) *Mid: Interaction 2–4 (x 10^9)* *Low: Interaction 0-2 (x 10^9)*

IOP Conf. Series: Earth and Environmental Science **123** (2018) 012018 doi:10.1088/1755-1315/123/1/012018

From the **Table 6** can be seen that the interaction between Banggai Laut Regency with other Regency/Cities around it as well as between Capital Regency with other districts included in the criteria low. This is caused by the population is much less than the surrounding districts and other cities and the distance is quite far. In addition, based on observations, the low interaction between Banggai Laut Regency and surrounding Cities or District is influenced by distance and population, also supported by internal factors that is community activity and community habit, also transportation infrastructure to Banggai Laut Regency and infrastructure in the region as described in the previous section, namely the limited choice of transportation modes, inadequate conditions of transportation facilities and transportation infrastructure and also relatively longer travel times. Here is a map of inter-regional interaction in Banggai Laut Regency:

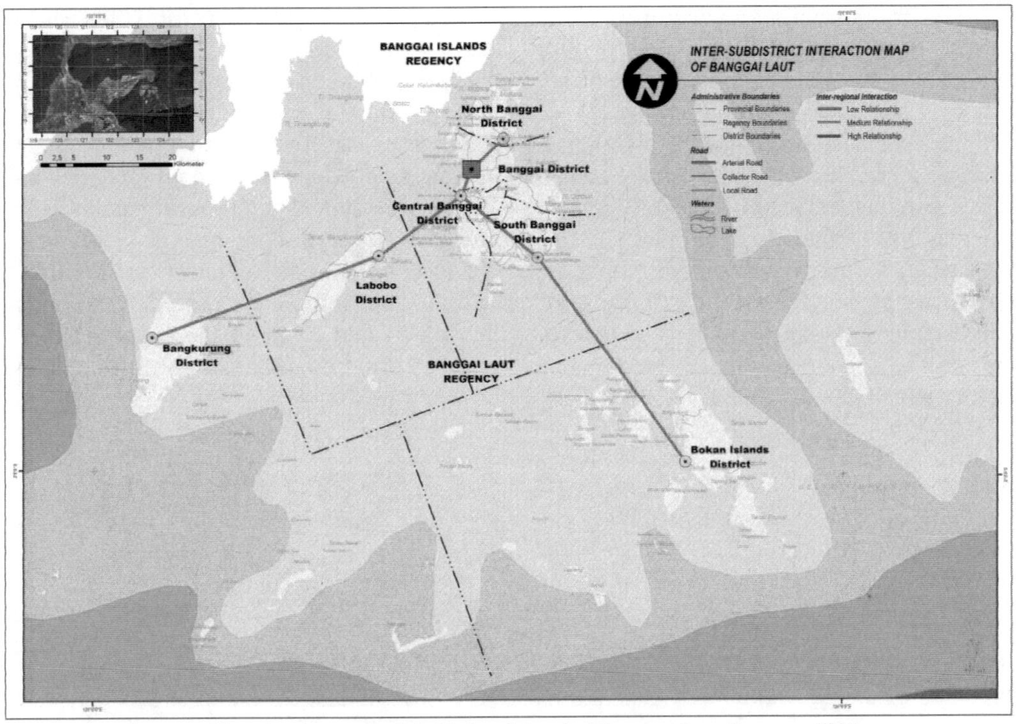

Source: Development Planning Agency at Sub-National Level of Banggai Laut Regency, 2015

Figure 10. Map of District interaction in Banggai Laut Regency

Banggai Laut Regency has some problems, one of them is the accessibility with low accessibility level namely limited choice of transportation modes, inadequate transportation facilities condition, the distance is relatively far from other regencies and cities around it and also travel time is relatively longer. In addition, inter-regional and inter-island interactions lie in the low criteria because of far distance and the population is small. Whereas in the development of a region, infrastructure has an important role in its success, where infrastructure is one of the sectors that have an important role in spurring economic growth, because basically the infrastructure sector acts as a liaison of various kinds of economic activity [9]. In addition, the transportation itself has several benefits, one of which is a geographical specialization in which each region has different specialties or potentials, with the existence of adequate transportation then goods can be sold and marketed outside the specialist area [3]. Therefore, several strategies are needed to develop transportation accessibility to achieve the development of Banggai Laut Regency.

IOP Conf. Series: Earth and Environmental Science **123** (2018) 012018 doi:10.1088/1755-1315/123/1/012018

The problem of road quality in Banggai Laut Regency is inadequate, where 54% of roads are severely damaged and 75% of roads have not been paved and there are still villages that can not be passed by four-wheeled vehicles not only in certain conditions but can not be passed by four-wheeled vehicles throughout the year. It also shows the uneven distribution of infrastructure in Banggai Laut Regency. Therefore, a strategy is needed to improve the quality of roads, repairing damaged roads, paving roads and widening roads for areas that cannot be passed by four-wheeled vehicles to open and improve access in Banggai Laut Regency. Besides, it is also necessary equitable distribution of transportation infrastructure development throughout the island in Banggai Laut Regency so that adequate infrastructure condition is not only centered on Banggai Island and the potential and economy of other islands can be developed by the existence of better transportation access.

Banggai Laut Regency has only one port on each of the major islands, while in other smaller islands it has no ports such as Toropot Island, Kokudang Island and others, so boats and outboard motorboats only lean on the beach. Therefore, the strategy is needed to increase the number of docks/ports not only on the big islands but also in other small islands in Banggai Laut Regency.

The transportation mode choice to Banggai Laut Regency is still very limited, where Banggai Laut Regency does not have an airport so the route that can be reached is only two alternatives, namely by sea or sea and land combination which both have long travel time. Therefore, a strategy is needed to increase the type of transportation mode in the form of additional airports and flight route to Banggai Laut Regency. Airport development in Banggai Laut Regency to open connections to other regency and provinces, so that accessibility to Banggai Laut Regency can be reached easily and quickly. With easy accessibility among regions, it can develop the region and improve the economy and develop other sectors to support regional development. Strategies to add docks/ports and airports can be seen on the following map:

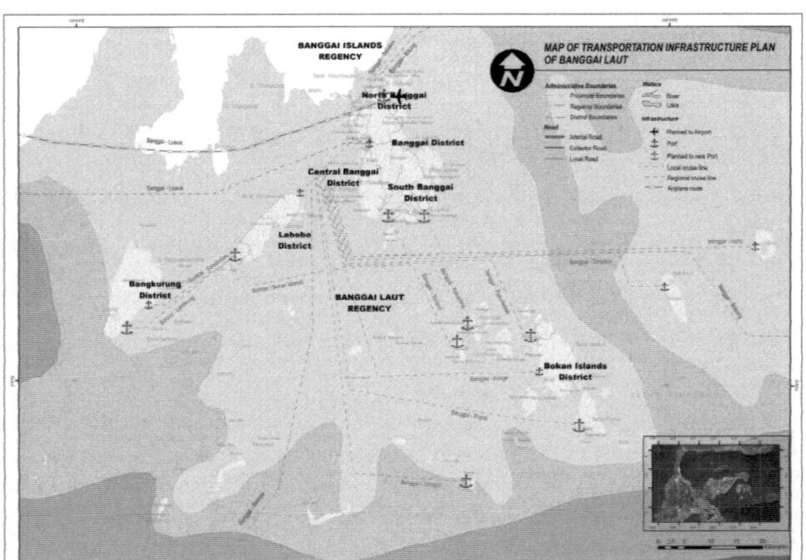

Source: Development Planning Agency at Sub-National Level of Banggai Laut Regency, 2015

Figure 11. Transportation Infrastructure Plan in Banggai Laut Regency

5. Conclusion

Banggai Laut Regency is an archipelagic district which still has some problems, namely, the low accessibility to Banggai Laut Regency and low accessibility within the district itself, that can be seen from the condition and availability of facilities and infrastructure that have not been sufficient which then impact on the length of travel time required. For the development of Banggai Laut Regency as a new regency, transportation infrastructure is needed that can open and facilitate access between islands and to other regencies / cities around it. The development or strategy undertaken is to improve the quality of roads, to increase the types of transportation modes in the form of air transportation namely airport and flight route to Banggai Laut Regency, increasing the number of docks and equitable infrastructure development throughout the island.

6. Acknowledgement

Thank you to the Local Government of Banggai Laut Regency as the party giving the fund and PT. Bina Karya Tiyasa as a consultant conducting the preparation of RP3KP of Banggai Laut Regency.

7. References

[1] Kurniawan D A 2014 Pengembangan Transportasi Laut dalam Mendukung Kebijakan Pengentasan Kemiskinan di Kawasan Timur Indonesia *The 17th FSTPT International Symposium*
[2] Central Burau of Statistic 2015 Luas Daerah dan Jumlah Pulau Menurut Provinsi 2002-2015 *www.bps.go.id*
[3] Martudji T 2017 Pemerintah Harus Serius Benahi Pulau Kecil Terluar dan Masyarakatnya *Nusantara.News*
[4] Salim A 2006 *Manajemen Transportasi* (Jakarta: Raja Grafindo Persada)
[5] Creswell J W 2003 *Research design: Quantitative, qualitative, and mixed methods approaches (2nd ed.)* (Thousand Oaks, CA: Sage)
[6] Republik Indonesia 2014 *Undang-Undang No. 1 Tahun 2014 tentang Perubahan atas Undang-Undang No. 27 Tahun 2007 tentang Pengelolaan Wilayah Pesisir dan Pulau-Pulau Kecil* (Indonesia)
[7] Susantono B 2004 *1-2-3 Langkah Kecil Yang Kita Lakukan Menuju Transportasi Yang Berkelanjutan, Referensi Ringkas Bagi Proses Advokasi Pembangunan Transportasi* (Jakarta: MTI)
[8] Tamin O Z 1997 *Perencanaan dan Pemodelan Transportasi* (Bandung: Institut Teknologi Bandung)
[9] Black J A 1981 *Urban Transport Planning: Theory and Practice* (London: Cromm Helm)
[10] Kurniawan D A and Purwoto H 2009 Kajian Dampak Infrastruktur Jalan terhadap Pembangunan Ekonomi dan Pengembangan Wilayah *Simposium XII FSTPT* (Petra Surabaya University, Surabaya)
[11] Alex R and Bathiche T 2006 Poverty Reduction and Transportation Infrastructure in Timor-Leste
[12] Central Burau of Statistic Bangai Laut Regency 2016 *Banggai Laut in Figures 2016*
[13] Suryanegara E, Suprajaka and Nahib I 2015 Social Change on Bajo Tribe: Case Study in Wakatobi Islands, Southeast Sulawesi) *Maj. Globe* ***17*** *No.1* 067–78
[14] Kompas 2017 Budaya Merantau Suku Bugis *Pressreader.com*
[15] Central Burau of Statistic Banggai Laut Regency 2014 *Village Potential Statistics of Banggai Laur Regency 2014*
[16] Central Burau of Statistic Banggai Laut Regency 2015 Number of Fishing Boats by Subdistrict and Type of Boat in Banggai Laut Regency, 2015 *banggailautkab.bps.go.id*
[17] Central Burau of Statistic Banggai Laut Regency 2015 Distance between Regency Capital to Sub District Capital *banggailautkab.bps.go.id*
[18] Reilly W J 1929 Methods for the study of retail relationships

IOP Conf. Series: Earth and Environmental Science **123** (2018) 012019 doi:10.1088/1755-1315/123/1/012019

Modeling the Dynamic Interrelations between Mobility, Utility, and Land Asking Price

E Hidayat[1], I Rudiarto[2], F Siegert[3], and W D Vries[3]

[1]Institute of Road Engineering, Indonesia
[2]Department of Urban and Regional Planning-Diponegoro University, Indonesia
[3]Technische Universitat Munchen, Germany

Email: edwin.hidayat@pusjatan.pu.go.id

Abstract. Limited and insufficient information about the dynamic interrelation among mobility, utility, and land price is the main reason to conduct this research. Several studies, with several approaches, and several variables have been conducted so far in order to model the land price. However, most of these models appear to generate primarily static land prices. Thus, a research is required to compare, design, and validate different models which calculate and/or compare the inter-relational changes of mobility, utility, and land price. The applied method is a combination of analysis of literature review, expert interview, and statistical analysis. The result is newly improved mathematical model which have been validated and is suitable for the case study location. This improved model consists of 12 appropriate variables. This model can be implemented in the Salatiga city as the case study location in order to arrange better land use planning to mitigate the uncontrolled urban growth.

Keywords: Modeling, Land asking price, Urban growth, Salatiga City

1. Introduction

Regarding the future interest, planning a city should consider sustainable development. The challenge is to linking economic activities, social issues, and environmental impacts. One way to achieve sustainable development is by using land use planning to control economic and social activities, so as a result, allocation of land should fit particular uses. Land use is very important as a benchmark for parceling mechanisms, also as a benchmark for zoning the urban environments. Zoning has a function to evaluate the environmental consequences, as well as to support the future decision making in order to mitigate the negative effects of urban development [1].

Urban development is influenced by population growth and rapid urbanization process. These cause urban sprawl which leads to unsustainable practices that cause ecological, social and environmental problems [2]. Furthermore, population growth triggers a rapid land use change particularly converting farmland into housing areas. The increase of population is equal to the increasing demand for housing. While, a high density of housing areas evokes new issues, like the probability of hazards, such as fire hazards, an increasing need of water, and the high impacts of houses on ecosystem function [3]. Moreover, high population density and the establishment of new settlements may entail serious problems in water supply, energy provision, and utilities [4]. Similar findings by Mohammady [5] show that the expansion of urban areas results in a lack of infrastructure, increase of environmental pollution, and limits urban services. Eventually, unplanned urban growth will lead to hampering the sustainable development and lessen the socio-economic quality of the inhabitant itself.

Urban sprawl and population growth trigger dynamic socio-economic quality which leads to dynamic land and property price. The new population needs more space to live so that the demand for land and housing also rises. One example to provide additional housing option is establishing the road redevelopment program. The purpose of this program is to gentrify old neighborhoods which have low price become a new housing alternative for the new population. However, this program also gives impact on the transaction prices of nearby housing, because it pushes up property and rental prices near redevelopment sites [6]. Another research finding indicates the real estate price is connected to transportation facilities. Empirical results demonstrate that rail transit facilities can markedly elevate real estate prices [7]. In addition, cities with a larger scale of urban rail transit will achieve explosive growth in land value, and there will be a very large gap in land value between cities with different scale of urban rail [8].

From the environmental perspective, environmental quality affects the prices of land and real estate properties. For instance, a land on a location near to the industrial plant usually prone to noise and air pollution produced by industries [9], buyers usually tend to ignore these type of land and resulted in low land value. In addition, the availability of water which included as utility becomes a consideration. The property sale prices will increase or decrease in response to water quality. Then, a long-term water quality monitoring is needed to identify the effectiveness of alternative land development [10]. Furthermore, people also consider sanitation and sewerage as a part of flooding risks, and the absence or presence of sanitation and sewerage become influential factors in determining the market value [11].

Based on the previous paragraphs, road development and transportation facilities are factors relatively close to people's mobility. Meanwhile, water and electricity, as well as sanitation and sewerage, are factors related to utility. Furthermore, several researchers used a specific term such as land value, market value, property price, and real estate price. These terms, in general, are commonly purposed to explain the impact of mobility and utility. Hereafter, the terms used in this research is land asking price. Understanding the interrelation between mobility, utility, and land asking price is important. The presence or absence of mobility will influence the utility and land asking price. On the other hand, the presence or absence of utility will also influence the mobility and land asking price. Therefore, the reciprocal relation between mobility, utility, and land asking price is important to enhance the land use planning. The aim of this research is to provide alternative information to forecast the land asking price which has reciprocal relation with mobility and utility. The final results of this research are the instruments which can be used to determine the price of land objectively to manage the uncontrolled urban growth.

2. Data and Methods

2.1. Research Design

This research is a deductive research. Based on the existing theory, the author has hypotheses that mobility, utility, and land asking price have dynamic interrelationship. Table 1 shows the steps and procedures to answer the hypotheses in a consecutive way.

Principally, there are 2 data collection methods used in this research for gathering information, namely secondary data collection and primary data collection. Ensuring the accuracy and appropriateness of the data is needed to achieve a decent conclusion. Secondary data collections in this research are collected by studying the literature related to land valuation modeling or concerned field using the online procedure. In addition, the supplemental data is needed by visiting the related offices. The land market data is used as the land asking price whereas land asking price is set up by the seller with an additional fee for the property agent. In addition, Salatiga city has 23 sub-districts and each sub-district is collected 3 land asking prices [12]. The total number of land asking price data is 69 data for a whole city. Primary data in this research are required in order to validate the step 1 in research design. The data is based on the expert opinion. The expert is a respondent who has graduated from university level and has experience more than 5 years in the related field.

Prior to data processing, the data are classified based on their purposes and types. Validation is required when there are peculiar data found. The researcher then is converting the data into the data type. For example, data for the statistic approach is converted into numeric data. If the data are not quantitative, then dummy method is applied.

a. For determining the Influence Variables, the existing models are compared. If a variable is accommodated by minimum 3 researchers, this variable will consider as the influence variable. Then validated using expert opinion.
b. The Most Suitable Variables is generated using statistical analysis method. The selected existing models are calculated using actual data from the case study location. The result is decided based on the comparison of statistical analysis test.
c. Construction of newly improved model is used statistical analysis that is the linear regression. Then, validated using several statistical tests.

Table 1. Research Design

Step	Purpose	Activities
1	The influence variables which is suitable for the case study location	The mobility and utility aspects are broken down to generate the influence variables towards land asking price. 9 existing models are compared. Then, validation using expert opinion
2	The most appropriate models	Selected existing models are compared to choose the most appropriate one in appraising land asking price in case study location. 3 existing models are calculated using multiple linear regression with the actual data from the case study location
3	The improved new models	Model A is developed using variables from step 1. New improved model (model B) is constructed with variables of model A and additional variables of the most appropriate existing model (step 2).

Source: Author, 2017

2.2. Case Study: The Salatiga City

Salatiga is one of the cities in the Central Java province of Indonesia. Compared to other cities in central java, Salatiga city is typically a small city because it only has area 56,78 km^2, meanwhile, the average size of the city in Central Java is about 1358 km^2. Salatiga consists of 13,75% agriculture land and 86,25% non-agriculture land. Salatiga city is divided into 4 districts (Kecamatan) and 23 sub-districts (Kelurahan). Total population in 2015 is 183.827 people [12]. Salatiga is located about 47 km south of Semarang, the capital city of Central Java Province. It is also located about 55 km from Surakarta, the culture and tourism city in Central Java. Therewith, Salatiga is located on the arterial road network connecting Semarang to Surakarta. Thus, Salatiga has very high traffic intensity. The public transportation in Salatiga according to Marsudi [13] follows the radial pattern. It means the CBD (central business district) as the central of the city. In general, people are moved from the housing area to the city center. It causes the accumulation of the private car and public transportation mainly passes the main roads, and leads to the bad use of the existing road network. Moreover, Salatiga is included in the emerging extended metropolitan region (EMR) Kedungsepur, which is going to be an activity center in Central Java. Therefore, Salatiga is a very strategic city to be developed.

The city of Salatiga was chosen as the case study location. It is comprised of 4 districts (Kecamatan) and each district has several sub-districts (Kelurahan). The total numbers of sub-districts are 23 for a whole city. As much as 3 samples of land asking price data are taken in each sub-district in order to represent the land asking price distribution in particular sub-district.

The land price which is used as the dependent variables is the land asking price/offer price instead of transaction/deal price. Land asking price is chosen due to the nature of the price itself. Land asking price is much reasonable because it is included in the real value type. Additionally, the seller usually set the land asking price by using comparative and cost method. Thus, land asking price can represent market price because transaction price/deal price is same as the land asking price or less than land asking price. Land asking prices in this research are taken from the property agent in Salatiga City. It means the price consists of expected price from the seller with an additional fee from the agent. In terms of mobility, the variables represent this aspect are variables related to people movement and accessibility facilities. In this research, the accessibility infrastructure and transportation facilities are also included as the mobility. While for the utility are variables that related to urban services or supporting infrastructure. Both mobility and utility in Salatiga city are elaborated.

3. Results and Discussion

3.1. Determining the Influence Variables

The mobility aspect and the utility aspect are extracted using the literature review method to determine the influence variables toward land asking price. 9 previous journal articles and master theses are selected. The previous study by Ai [14] or R1 which discussed the modeling of land value focusing on residential land. A similar topic of residential land value is also discussed by Lehner [15] or R2. Meanwhile, Chiarazzo et. al. [9] or R3 conducted a research which emphasized on the environmental aspect towards real estate price, while Bohman and Nilsson [16] or R4 conducted a research about the impact of commuter trains toward property value. Subsequently, Nazir, Othman, & Nawawi [17] or R5 was employed hedonic pricing model in determining housing value. Furthermore, Tutuko and Shen [18] or R6 studied the effects of land use zoning on housing development, while the impacts of living conditions on housing development were studied by Tao [19] or R7. Meanwhile, Sue and Wong [20] or R8 conducted a research focused on the political economy of housing price. The last, Topçu [21] or R9 elaborated the accessibility effects on urban land value. R (researcher) is term to help in comparison of existing models. From these 9 articles, each article has its own method and variables. However, these articles still have similarities and are relevant to the discussion about mobility aspect and utility aspect in the dynamic interrelations with the land asking price, property price or housing price. Variables generated from previous researches are classified into mobility aspect or utility aspect. Table 2 presents variables which influence the land and property price.

Table 2. Variables influence Land and Property Price

Variables	R1	R2	R3	R4	R5	R6	R7	R8	R9	Aspect
Distance/travel time to the CBD	X	X	X	X	X	-	-	-	X	Mobility
Access/Distance to bus stop / train station	X	X	X	X	X	-	-	X	X	
Distance to the nearest the main road/highway	X	X	-	X	X	X	-	-	-	
Access to Hospital/Health Center	X	-	-	-	-	X	-	-	-	Utility
Access to Nearest park/open space	-	-	-	-	X	X	X	-	X	
Access to nearest school	-	-	-	-	X	X	-	X	X	
The presence of good sanitation	-	-	-	-	-	X	X	-	X	
Availability of Electricity	-	-	-	-	-	X	X	-	-	
Availability of garbage disposal system	-	-	-	-	-	X	-	-	-	
Availability of Clean water system	-	-	-	-	-	X	X	-	-	

Source: Author, 2017

IOP Conf. Series: Earth and Environmental Science **123** (2018) 012019 doi:10.1088/1755-1315/123/1/012019

Table 2 shows some variables included in several studies. However, some of them are applied only by 1 or 2 researchers, variables of distance/travel time to central business district (CBD) for example, are applied by 6 researchers. Meanwhile, the availability of the garbage disposal system is used only by 1 researcher. Thus, on determining the influence variables, only variables included in minimum 3 studies will be considered as the influential variables. These variables will be used for testing the dynamic interrelation among mobility, utility, and land asking price.

Moreover, to make sure the suitability of variables generated by the literature review, validation needs to be done in order to find out the compatibility of the real condition of mobility and utility in the case study area. Interview method has been employed. In this research, the respondents are persons who have knowledge and experience in the related field in Salatiga City. A minimum requirement for the respondent is graduated from university level and has work experience more than 5 years in related field. The respondent is chosen based on the disposition of each intended organization and must fulfill the minimum requirements. Table 3 presents the summary of interview results. It is only showed the respondents opinion related to the validation of influential variables related to land price.

Table 3. Summary of Interview results for Influential Variables

Respondent	Opinion
Mr. Jadi Amali, Head of Sub-division for Urban Planning at Bappeda	New development of CBD area will increase the land price since the nearby property will be converted into a commercial area. Infrastructure has a strong influence in the property market.
Ms. Sri Kusrini Martuti, Head of General Affair Section at BPN	Method for determining land tax or property tax based on the regulation of Financial Minister number 150/PMK.03/2010 about classification and designation of object value. BPN as the regulator in maintaining the land administration does not have the responsibility to control the land value. The land value or property price depends on the market price BPN provides a land value map or zone land price (ZNT), which can be accessed at http://peta.bpn.go.id/. This zone land price is based on market price
Mr. Robert A, Lending Officer at Bank of Stated Owned Company	The valuation procedure is divided into 3 categories, namely, physical, tenure, and fair market value. The physical category consists of Land use, risk towards flood, high voltage electricity path, prone to landslides, land contour, and soil fertility (for agriculture land). The physical category also points of access to the road (road width), access to electricity, access to landline (telephone), clean water facility, school, market, gas station, access to mosque/church, and distance to the cemetery. Tenure category depends on the freehold, building use right, right to use only. Fair market value category uses the comparison method of nearby property.
Mr. Lodewyk Pattiwael, CEO at Rumah Salatiga (Property agents)	There are many factors in determining the land value. However, in general, land value depends on the supply and demand.

Source: Interview result, 2017

Based on the interview result, it can be seen that respondents from BPN and property agent do not declare specifically about variables which influence land value. Meanwhile, respondents from Bappeda and Bank analyst emphasize that infrastructure gives strong influence in determining the land market price or property price.

Variables generated from the mobility aspect consist of 3 variables, namely, Distance/travel time to the CBD, access/distance to bus stop, and distance to the nearest main road. Access or distance to the CBD is classified as mobility because in transportation theory, CBD is included in the trip attraction. People move from housing (generation) area to attractive area, whereas CBD always attracts a lot of people. Studies about trip attraction to CBD/commercial area/shopping center have been done by many researchers, for example D'Andrea and Francoso [22] and Uddin, Hasan, Ahmed, Das, Uddin, & Hasan [23]. The variable of distance to bus stop and distance to the nearest main road are classified as mobility due to a close relationship with the accessibility. Meanwhile, the variables considered for analysis from utility aspect are 3 variables, namely access to nearest park/public space, and access to the nearest school, and the present of good sanitation. These variables are classified as utility because public space, school, and drainage system can be described as public services, the Government has responsibility to provide those utilities because support the people daily activities. The access to nearest public space and access to the nearest school are also attracting people. However, those variables are classified as utility instead of mobility because of public space and schools are attracting only a particular user for a particular time. In addition, Ortuzar and Willumsen [24] is stated that the commercial land use is the highest trip attraction. While, public space and school are areas included in the recreational and educational land use, so, much suitable included in the utility.

Therefore, the 6 final variables generated from the literature review and validated using expert opinion is used to construct the model A. However, on the one hand, the bus stop variable and main road variable in some existing studies are classified as transportation factor or accessibility aspect. These variables are not something new for land valuation. Similarly, school variable or public space variable in some existing researches are classified as infrastructure factor or environment aspect. These variables are not new things in determining the land or property price. On the other hand, these 6 final variables adopted from related literature and validated using interview methods are appropriate variables in the case study location, Salatiga city. In addition, these final variables representing mobility and utility are a new alternative for land valuation method in Salatiga for specifically, and in Indonesia for generally.

3.2. The most appropriate models

Some of the existing models are compared in order to understand the variable compatibility in the case study area, Salatiga City. The existing models are chosen based on the variety of variables employed in each model. The selected existing models are also chosen based on the similarity characteristic of mobility behavior. The chosen models are models which constructed using data from the Asian region. The proposed models are Singapore model by Lehner [15], Hankou-China model by Ai [14] and Labuan-Malaysia model by Nazir, Othman, and Nawawi [17]. Table 4 presents the statistical analysis of the existing models.

Table 4. Statistical Analysis of Existing Model

Existing Model	Rsquare (>0,5)	Multicollinearity (>0,1)	F-test (<0,05)	Durbin-watson (-2 until 2)
Singapore Model	0,267	Free	0,006	1,003
Hankou – China model	0,194	Free	0,058	1,004
Labuan – Malaysia model	0,647	free	0,0000	1,847

Source: Author, 2017

The Singapore model, the Hankou-China model, and the Labuan-Malaysia model have been validated using several statistical tests such as t-test, f-test, multicollinearity, R-Square, and durbin-watson test. The result indicates that the Labuan-Malaysia model is the most appropriate model. This model can be adapted and implemented in the case study location, Salatiga city. The Labuan-Malaysia model also provides the most accurate variables in forecasting the land asking price.

IOP Conf. Series: Earth and Environmental Science **123** (2018) 012019 doi:10.1088/1755-1315/123/1/012019

The most appropriate model which has the most suitable variables is Labuan-Malaysia model. This model has been validated using several statistical tests. This model provides the most accurate variables in forecasting the land asking price because this model has variables which are different from the other model such as distance to the nearest religious place. This variable is mostly related to culture instead of mobility or utility. In general, the characteristic of people who live in Salatiga city is religious people, thus, distance to the nearest religious place is a variable that is suitable for the Salatiga city culture. Another variable is the distance to nearest police station. This variable should be included in the safety aspects instead of in mobility or utility. The feel of safety belongs to everyone. People want to live safely, so, this variable also proofs that the Labuan-Malaysia model has the most appropriate variables.

3.3. The new improved models

The new models are mathematical models. These models consist of model A and model B. Model A is a model constructed with 6 variables based on the literature review and validated with interview method. Model B is a model constructed with 6 variable of model A and additional variables of Labuan – Malaysia model. Table 5 presents statistical analysis of new models.

Table 5. Statistical Analysis of New Models

Mathematical Model	Rsquare (>0,5)	Multicollinearity (>0,1)	F-test (<0,05)	Durbin-watson (-2 until 2)
Model A	0,225	Free	0,006	0,969
Model B	0,679	free	0,0000	1,636

The model A is constructed using statistical analysis and validated using several tests. The variables generated from the mobility aspect consist of 3 variables and 3 variables for utility. However, a variable from utility aspect namely the presence of good sanitation is ignored because all data have a same value. In term of statistical analysis, this variable cannot be calculated. It does not mean the presence of good sanitation does not influence the land asking price rather this variable cannot be applied only for this particular data. The mathematical model A is illustrated in the equation as follow.

$$Y = 2724367.976 - 793.228x1 - 55.731x2 - 122.434x3 - 116.133x4 - 103.439x5$$

Whereas:
Y = land asking price (rupiah/square meter), x1 = distance to CBD (meter), x2 = distance to nearest bus stop (meter), x3 = distance to main road (meter), x4 = distance to nearest green/public space (meter), x5 = distance to nearest school (meter)

Model B is constructed using statistical analysis and validated using several tests. Model B variables are an adaptation of model A variables and Labuan-Malaysia Model variables. Model A consist of 6 variables and 10 variables from Labuan-Malaysia model is added. However, 3 variables of both models are similar, namely the distance to a bus stop, access to nearest park/public space and access to nearest school. Thus, the same variables are removed, so that there are no redundant variables. In addition, a variable of good sanitation is also removed because the data are all the same. All location in the case study have a drainage/sanitation and the condition of this utility is also in good condition in flowing the rain water, so, all location in the case study have low risk to urban flood. However, for other case study or another area which have different conditions, the variable of good sanitation should be applied. Therefore, model B is comprised of 12 variables, adapting variables of model A and Labuan-Malaysia model. The mathematical model B is illustrated in the equation as follow.

$$y = 3999524.155 + 673653.495x1 - 195790.872x2 + 249435.115x3 + 155582.405x4 - 2166410.369x5 - 423282.769x6 - 443791.287x7 + 463726.491x8 + 381131.834x9 + 324238.643x10 - 737.260x11 - 171.333x12$$

Whereas:
Y = Land asking price/offered price by the seller (rupiah), x1 = access to local market / commercial place (1 if under 500 meter; 0 if other), x2 = access to religious place (1 if under 500 meter; 0 if other), x3 = the land parcel location/hook position (1 if corner lot; 0 if other), x4 = road access (1 if more than 4 meter and paved; 0 if other), x5 = gap of sales price compare to zone land price/BPN map (1 if the gap under, 1 million rupiah; 0 if other), x6 = access to bus stop (1 if under 1000 meter; 0 if other), x7 = access to public space (1 if under 2000 meter; 0 if other), x8 = access to the nearest school (1 if under 1000 meter; 0 if other), x9 = access to nearest river (1 if under 1000 meter; 0 if other), x10 = access to police station (1 if under 1000 meter; 0 if other), x11 = distance to main road (meter), x12 = distance to cbd (meter).

After comparing the model A and model B using statistical test, it can be concluded that model B has better performance compare to model A. Model B has higher R-square and less negative variables. Therefore, model B is a newly improved model which fits the case study location, Salatiga city. Model B consist of 5 mobility variables and 2 utility variables also 5 supporting variables.

Finally, after calculating and comparing the model A and model B, the result is model B as the newly improved model. The variables of model B consists of mobility variables, utility variables, and supporting variables. Mobility variables are access to local market/commercial place, road access/width, access to bus stop, distance to the main road, and distance to CBD. Utility variables are access to public space and access to the nearest school. Moreover, supporting variables are access to the religious place, comparison of sales price and zone land price, the hook land parcel location, access to nearest river, and access to the police station. In addition, the variable of the presence of good sanitation is cannot be calculated because the data are all the same. This variable is not yet confirmed influence land asking price or not.

4. Conclusion

The main results of this research are instruments in a form of mathematical model which can be used to assess the land price objectively. The newly improved mathematical model can be used by the planner and transport engineer in order to arrange better land use planning. The model B can be used to anticipate the uncontrolled urban growth in Salatiga city. The new improved model can be used as an instrument in the preliminary land acquisition process to develop new collector road, planning the bus stop location. In addition, the mathematical model B also can be used by the land-appraiser. This model is useful to predict the land asking price for property investment. For the Government, the mathematical model can be used as an early prediction for land taxation particularly for the industry, and for planning appropriate distance of new public/green space to the industrial area, commercial area, etc.

5. Acknowledgments

This research is funded by the Ministry of Public Works-Indonesia as part of the scholarship for the master program.

6. References

[1] Gerundo R and Grimaldi M 2011 The Measure of Land Consumption Caused by Urban Planning *Procedia Eng.* **21** 1152–1160
[2] Du R 2016 Urban growth: Changes, management, and problems in large cities of Southeast China *Front. Archit. Res.* **5** 290–300

[3] Vukomanovic J, Doumas S, Osterkamp W and Orr B 2013 Housing Density and Ecosystem Function: Comparing the~Impacts of Rural, Exurban, and Suburban Densities on Fire~Hazard, Water Availability, and House and Road Distance~Effects *Land* **2** 656–678
[4] Afandiyev V, Eminov Z and Nagiyev S 2014 Economic and geographic factors affecting the development of Greater Baku *J. Urban Reg. Anal.* **6** 203
[5] Mohammady S 2014 A Swarm Optimization Based Method for Urban Growth Modelling *Environ. Res. Eng. Manag.* **69**
[6] Yau Y 2011 Does comprehensive redevelopment change the housing price gradient? A case study in Mongkok, Hong Kong *Urbani izziv* **22** 98–106
[7] Zhang X, Liu X, Hang J, Yao D and Shi G 2016 Do Urban Rail Transit Facilities Affect Housing Prices? Evidence from China *Sustainability* **8** 380
[8] Yang Y, Zhang P and Ni S 2014 Assessment of the Impacts of Urban Rail Transit on Metropolitan Regions Using System Dynamics Model *Transp. Res. Procedia* **4** 521–534
[9] Chiarazzo V, Caggiani L, Marinelli M and Ottomanelli M 2014 A Neural Network based Model for Real Estate Price Estimation Considering Environmental Quality of Property Location *Transp. Res. Procedia* **3** 810–817
[10] Chang H, Thiers P, Netusil N R, Yeakley J A, Rollwagen-Bollens G, Bollens S M and Singh S 2013 Relationships between environmental governance and water quality in growing metropolitan areas: a synthetic view through the coupled natural and human system lens *Hydrol. Earth Syst. Sci. Discuss.* **10** 7395–7424
[11] Cupal M 2015 Flood Risk as a Price-setting Factor in the Market Value of Real Property *Procedia Econ. Financ.* **23** 658–664
[12] BPS Kota Salatiga 2016 *Kota Salatiga dalam Angka*
[13] Marsudi M 2006 *Analisis Kinerja Mobil Penumpang Umum (MPU) dan Sistem Jaringan Trayek di Kota Salatiga* (Magister Teknik Sipil)
[14] Ai P 2005 Residential land value modelling
[15] Lehner M 2011 *Modelling housing prices in Singapore applying spatial hedonic regression* (IVT, ETH Zürich)
[16] Bohman H and Nilsson D 2016 The impact of regional commuter trains on property values: Price segments and income *J. Transp. Geogr.* **56** 102–109
[17] Nazir N N M, Othman N and Nawawi A H 2015 Role of Green Infrastructure in Determining House Value in Labuan Using Hedonic Pricing Model *Procedia - Soc. Behav. Sci.* **170** 484–493
[18] Tutuko P and Shen Z 2016 The Effect of Land Use Zonings on Housing Development: The Introduction of Cdl Approach in the Border Area of Surabaya and Sidoarjo Regency, Indonesia *Procedia-Social Behav. Sci.* **227** 107–114
[19] Tao L W 2015 Living conditions—The key issue of housing development in Beijing Fengtai District *HBRC J.* **11** 136–142
[20] Sue E D W and Wong W-K 2010 The political economy of housing prices: Hedonic pricing with regression discontinuity *J. Hous. Econ.* **19** 133–144
[21] Topcu M and Kubat A S 2009 The analysis of urban features that affect land values in residential areas *Proceedings of the 7th International Space Syntax Symposium, Stockholm, Sweden* pp 8–11
[22] D'Andrea A F and Francoso M T 2015 Application of Cluster Analysisin Trip Attraction Models of Campinas Metropolitan Region *Int. J. Eng. Technol.* **7** 377–381
[23] Uddin M, Hasan M R, Ahmed I, Das P, Uddin M A and Hasan T 2012 A comprehensive study on trip attraction rates of shopping centers in dhanmondi area *Int. J. Civ. Environ. Eng.* **12** 12–16
[24] de Dios Ortúzar J and Willumsen L G 2011 *Modelling Transport* (John Wiley & Sons, Ltd)

IOP Conf. Series: Earth and Environmental Science **123** (2018) 012020 doi:10.1088/1755-1315/123/1/012020

Lifestyle as an Influential Factor to Urban Mobility Transport: a Case Study of Semarang City, Indonesia

I Ismiyati[1] and F Hermawan[1]

[1] Department of Civil Engineering, Diponegoro University

Email: ismiyati_hs@yahoo.com

Abstract. Most of urban spatial structures in developing countries apparently face a typical phenomenon, as well as in Indonesia. The development of the urban spatial structure has the effects, namely to create polycentric pattern (sprawl). Moreover, communication technology believes that the factors of distance and density are highly considered in the organization of the urban structure. In other words, a distance problem is overcome by communication technology, in terms of interaction among people; in running their activities, mobility or distance is not a problem at all. Urban structure as path which is dependent is unable to intervene for an optimum form of urban structure because of dynamic of development objectives. In facts, lifestyle of inhabitant particularly concerning residential and vehicle ownership influences the mobility transport on the tremendous changes in developing countries. On the contrary, this research points out that mobility transport contributes to transportation problems as it becomes increasingly inefficient. Therefore, a sporadic traffic jam and increasing carbon emission issues have risen on the urban phenomenon. It is important to investigate the lifestyle, in terms of residential choice and vehicle ownership to reshape the urban spatial structure. The research aims to draw the urban spatial growth which extends to the phenomenon process toward polycentric pattern and inefficient transport mobility patterns triggering transportation problems in the context of Indonesia. The results confirm that lifestyle regarding residential choices to suburban area and vehicle ownership preference are unable to create the efficient mobility transport, either by cost, density consequences or vehicle ownership as orientation. This research recommends the local authority from multi-disciplinary sector, in particular public policy making to issue permission for authority of land use; residential area and transport agencies for reconciliation with regard to life style aspects in urban spatial planning.

Keywords: Urban Mobility Transportation, Urban Lifestyle, Semarang City

1. Introduction

The rapid development of urban population forces the city expansion to move towards the suburbs. Such situation is what both developing and developed countries are currently facing. The development is due to the rising urbanization which leads to more demand for land in the periphery [1,2]. In developing countries like Indonesia, the expansion is unexpectedly sprawl which extends towards the outskirt. This condition is as the impact of continuous urbanization that is translated into increasing needs for land in the periphery. In addition to great expansion of the city, urban and rural integration also appears, while technology continues to evolve; then a shift in social structure is inevitable [3]. The high growth of metropolitan and megacity on every continent is considered the consequence of economic growth. The economic growth is unfortunately often responded with the changing lifestyle, such as private ownership of residential and vehicle [4]. The indication can be seen from the high growthof population living in the suburbs, where the more than 5 km daily commuting from home to the work place does not become a constraint. Figure 1 presents the population density in the suburban area of Semarang.

Bertaud [5] has pointed out that big cities in the world have far been shifting their pattern from the monocentric to the polycentric (many centres). Moreover, the polycentric cities have more efficient transportation than the monocentric because of denser population with more orderly and well managed infrastructure. Research by Hermawan [6] on Isoprice in densely populated residences in three different areas of study in Semarang city suggests that the densely populated outskirts show more efficient mobility if the public transport is designed to have a big capacity, but if they employ small-capacity public transport, the mobility in the area turns to be more inefficient.

Indonesia is a developing country where almost all the cities plan development towards the periphery, while traffic jam always appears on almost the entire road network up to the city centre. The transformation from the monocentric to the polycentric image can also be identified in several urban areas in Indonesia especially in the areas of study in Semarang city as described in Figure 1, showing the population density which shifts towards the suburbs.

The research findings of Ismiyati [4] explain that in terms of transportation mobility, 70% inhabitants of Semarang are likely to be consumerist and conduct long commuting from the suburb to the city centre to reach work place and school. This condition displays that despite the shifting pattern of the city from the monocentric to the polycentric, the mobility and transportation infrastructure is deemed inefficient, because the mobility of the inhabitans is done using private transports [7].The purpose of this study is to analyze and describe the spread patterns of residential density which moves towards the periphery, as well as analyzing and exploring how the lifestyle leads to the inefficiency in mobility and urban transport infrastructure.

2. Research Methods

The research was undertaken using the quantitative method followed with field observation to obtain the data on land value on each subdistrict in Semarang City, the distance to the city center, as well as the spread of residential density. A qualitative method was also implemented through in-depth interviews to obtain information on the characteristics of the population in the suburbs. The purpose of combining these two methods is to explore the social phenomenon that is unable to explain solely through a quantitative method.

3. Results and Analysis

3.1. Population Distribution Patterns of Semarang City

The result of the analysis implies that the population distribution moves to the city periphery as seen in Figure 1, as well as on the analysis of the relationship between density and the distance to the center as described in Figure 2. The urban development tends to show the establishment of many centres in the suburbs that lead to the polycentric pattern. The development of the polycentric cities in Indonesia occurs as the impact of less integration and coordination between spatial policy and the urban transport policy. Meanwhile, the polycentric pattern or sprawl tends to become less beneficial for advancement strategies in public transport due to the irregular population mobility from the origin to the destination that will also lead to irregular settlement pattern. Hence, the cost for development on transport infrastructure is more expensive compared to those in monocentric cities. Monocentric cities are believed to enjoy more benefits because this monocentric pattern provides better access yet cheap for accommodating mobility to work. These facts indicate inefficient control systems for urban development in Indonesia, as the transportation problems are hitting some urban areas in the country.

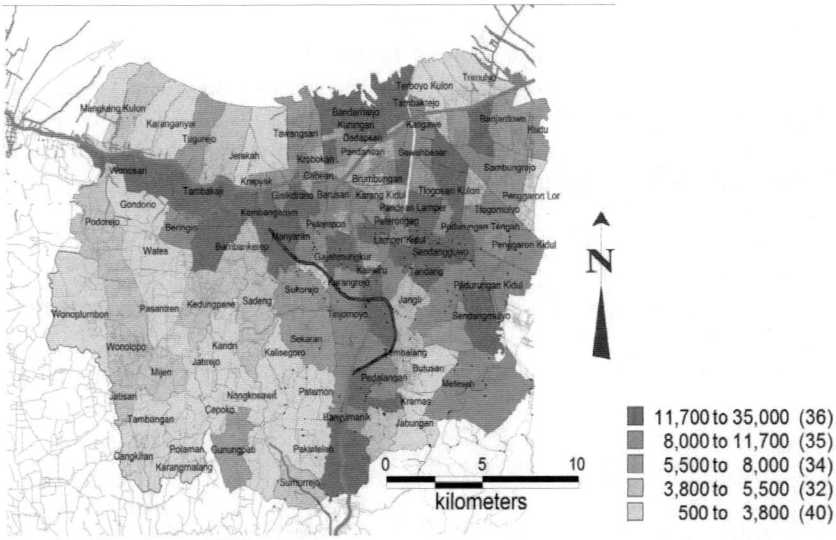

Source: Hermawan, 2011

Figure 1. Population Density Distribution in Suburban of Semarang

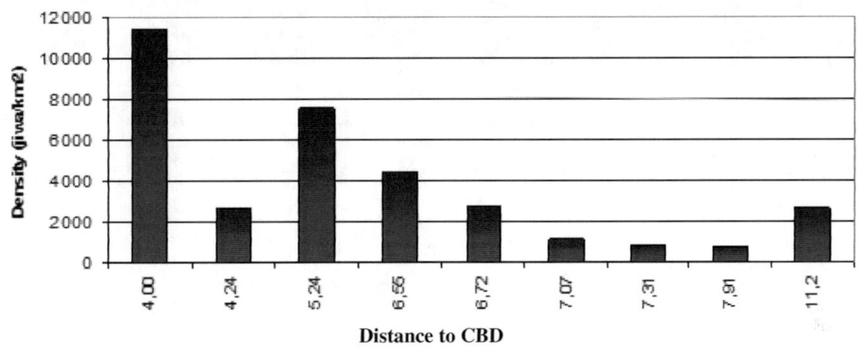

Figure 2. Population Density in the Suburban of Semarang City and the Distance to CBD

According to Bertaud [5] in his analysis on metropolitan cities in several developing countries, these cities are now experiencing changes in patterns from the monocentric to the polycentric. His analysis also tells that the transportation infrastructure in polycentric cities is more efficient because of higher population density underserved. Research by Hermawan [6] on Isoprice for suburbs with greater volume of population density, the public transport services are more efficient if having large capacity, while in the current situation, most suburbs are still served by small capacity public transport. Figure 3 shows the results of in-depth interviews in several subdistricts in the periphery and presents the potential population density and use of private and public transport mode. The figure also depicts that the population distribution in the outskirts of Semarang City and mobility patterns of the inhabitants using private cars are 50% > on average.

3.2. Analysis of the Accessibility, Land Value and the Distance to CBD

The influence of accessibility to population distribution pattern and land value is shown in Figure 3 and Figure 4. These figures explain that variation in land value is as the impact of different level of accessibility and the activity of transportation system that is stated with the intensity of land use (IGL), in addition to other factors.

IOP Conf. Series: Earth and Environmental Science **123** (2018) 012020 doi:10.1088/1755-1315/123/1/012020

Differences on topography in the existing suburbs also impact to the varying land prices, thus varying land prices can be found in some locations despite the same distance from the city square (Simpang Lima Semarang). The distribution of land value in Figure 4 indicates population distribution is concentrated on the suburban areas. This is not because of sheer distance considerations and cheaper land value as what is described by Alonso in his theory. Yet, it is the difference in land slope and social shifts from the technological advancement that influences the land value.

The theory by Alonso [8,9] about land lease and transport costs explains that a place far from the center will have lower land value but higher and more expensive transportation costs. However, in Semarang, the reality shows the contrary, since regardless the far distance to the city centre (10 km to 30 km), a residential area built near public transportation and superb settlements would be valued higher and the cost of transportation would be cheaper as it is more flexible compared to using public transport services.

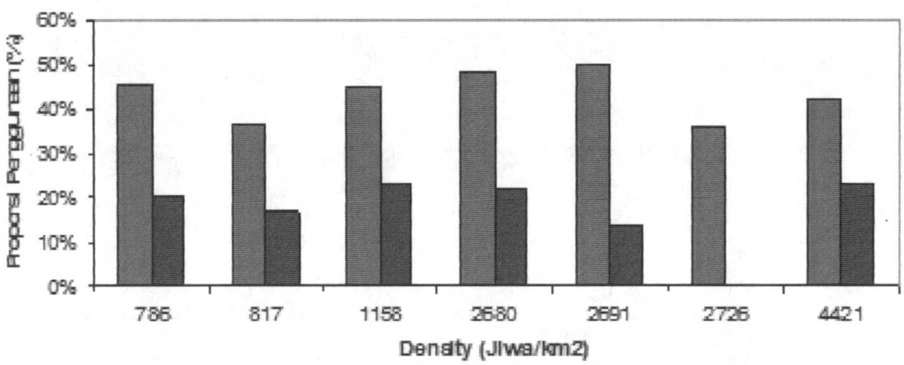

Figure 3. Population Density of Semarang Vs Potential of Transportation Mode Ownership

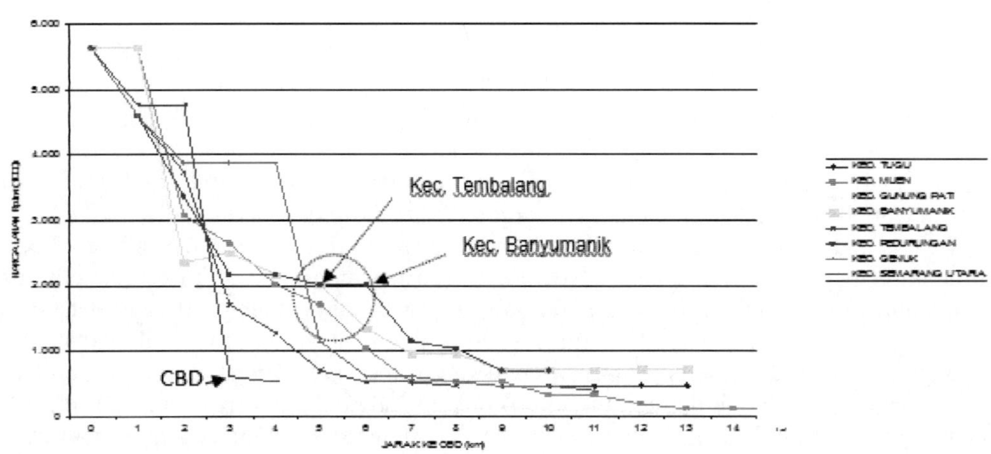

Source: Ismiyati, 2014

Figure 4. Land Value of Each Subdistrict in Semarang City and the Distance to CBD

3.3. The Influence of Lifestyle to Changes in Mobility

The explanation above shows that mobility influences lifestyle as the impact of advancement in information technology. The results of the study found social development towards new life style which affects transport mobility. In this case, Indonesian community is having social awareness that creates a lifestyle and eventually triggers to the rise of transportation issues.

4. Conclusions and Suggestions

4.1. Conclusions

The differences in topography and access cause different city development plans as in Semarang City. Suburban areas connected with the highway such as Tembalang and Banyumanik are reported to show the highest density. The difference in the land value in the Semarang periphery is influenced by the ease of access and environmental factors, as depicted in the data that land value in the suburban areas of Semarang namely Tembalang and Banyumanik is the highest, though the areas are considered highland.

The influence of lifestyle is seen from the high land value at Banyumanik and Tembalang subdistricts and with the number of residentials exceeding the standard density. On the other hand, in terms of mobility, more than 50% population took more than 5 km journey using their private cars. Mobility is influenced by lifestyle, thus making the polycentric cities less efficient. In fact, such situation is happening in most developing countries including Indonesia.

4.2. Suggestions

The cities in the developing countries, such as cities for students that are undergoing expansive development are expected to conduct behavior study prior to addressing the needs of urban transport because each city possesses different characters and accomodates different population.

5. References

[1] Webster D and Muller L 2009 Peri-urbanization: Zones of rural-urban transition *Hum. Settl. Dev. 1* 280
[2] Catanese A J and Snyder J C 1989 *Perencanaan Kota* (Penerbit Erlangga)
[3] Bogardus E S 1925 Measuring social distance *J. Appl. Sociol.* **9** 299–308
[4] Ismiyati I, Soetomo S and Riyanto B 2011 *Mobilitas Transportasi Dikaitkan Dengan Pemilihan Tempat Tinggal Di Kawasan Pinggiran Kota Semarang* (Universitas Diponegoro)
[5] Bertaud A 2004 The spatial organization of cities: Deliberate outcome or unforeseen consequence?
[6] Hermawan F, Riyanto B and Ismiyati I 2008 Konsep Pengembangan Angkutan Umum Yang Humanis Di Daerah Suburban Berbasis Karakteristik Wilayah (Studi Kasus Kecamatan Banyumanik-Semarang) *Prosiding Seminar Nasional Transportasi USM*
[7] Kim D-S, Mizuno K and Kobayashi S 2003 Modeling urbanization by accessibility in rapid-growth areas *J. urban Plan. Dev.* **129** 45–63
[8] Alonso W and others 1964 Location and land use. Toward a general theory of land rent. *Locat. L. use. Towar. a Gen. theory L. rent.*
[9] Yunus H S 2000 *Struktur tata ruang kota* (Pustaka Pelajar)

Assessing green waste route by using Network Analysis

I H Hasmantika[1] and M Maryono[1]

[1]Department of Urban and Regional Planning, Diponegoro University, Semarang 50275, Indonesia

Email: intan.hapsari17@pwk.undip.ac.id

Abstract. Green waste, such as waste from park need treat proper. One of the main problems of green waste management is how to design optimum collection. This research aims to determine the optimum green waste collection by determining optimum route among park. The route optimum was assessed by using network analysis method. And the region five of Semarang city's park within 20 parks in chose as case study. To enhancing recycle of green waste, three scenarios of treatment are proposed. Scenario 1 used one integrated treatment facility as terminal for enhancing recycle of green waste, Scenario 2 used two sites and scenario 3 used three sites. According to the assessment, the length of route of scenario 1 is 36.126 km and the time for collection estimated is 46 minutes. In scenario 2, the length of route is 36.471 km with a travel time is 47 minutes. The length of scenario three is 46.934 km and the time of collection is 60 minutes.

Keywords: Green waste, GIS, Network Analysis

1. Introduction

Green waste should be treated properly to ensuring the resilience of urban green space [1] Planning for green waste management is needed to ensure that urban green park sustainability is in line with many perspectives of function, such as ecology, aesthetic, recreation and leisure in city [2] Green waste management is important to maintenance the function of urban green space to mitigate high temperatures in urban landscapes [3]. Urban green waste management is also important for mitigating urban flood waste management that usually dominant in flood waste component such in flood waste in Bangkok Thailand, during huge flood in 2011[4]. Moreover, green waste optimum management, could be utilized to enhance biomass base-energy [5].

One of the critical factors to achieve the proper green waste management is the optimum condition of transportation system [6]. This study aims to examine the route of the green waste collection from park. Moreover, there are two sub systems on the green waste transportation that will be examined; the first is route and the second is time of green waste collection for enhancing recycling system.

2. Data and methods

2.1. Study area

Several studies concerning on the sustainability issues of Semarang city have been conducted. For example the challenging of sustainability in Semarang [7], inundated issues of Semarang [8,9]. Study

IOP Conf. Series: Earth and Environmental Science **123** (2018) 012021 doi:10.1088/1755-1315/123/1/012021

concerning on the leisure level in Semarang city that have been initiated in 2002 by Guntur Suworo [10], however an information concerning to the maintenance of urban green open space, such as park maintenance not yet discuss. This indicator is very important to analyze in growing cities such as Semarang.

The study area is park in Candisari and Tembalang district at which the two districts are as representative of Semarang Growth in recent year. Candisari has a unique characteristic housing while Tembalang district is one of the education centre zone in Semarang. 20 parks at which 13 in Candisari district and seven parks in Tembalang district (Fig 1) were selected to examine the model of green waste collection optimum route. The initially and preliminary study shown that there is no regular schedule of green waste colleting, sweeping. The transport of green waste is done around 10:00 to 12:00 pm, however there no route for enhancing the recycling process. The green waste directly transport to Jatibarang landfill that have been over capacity.

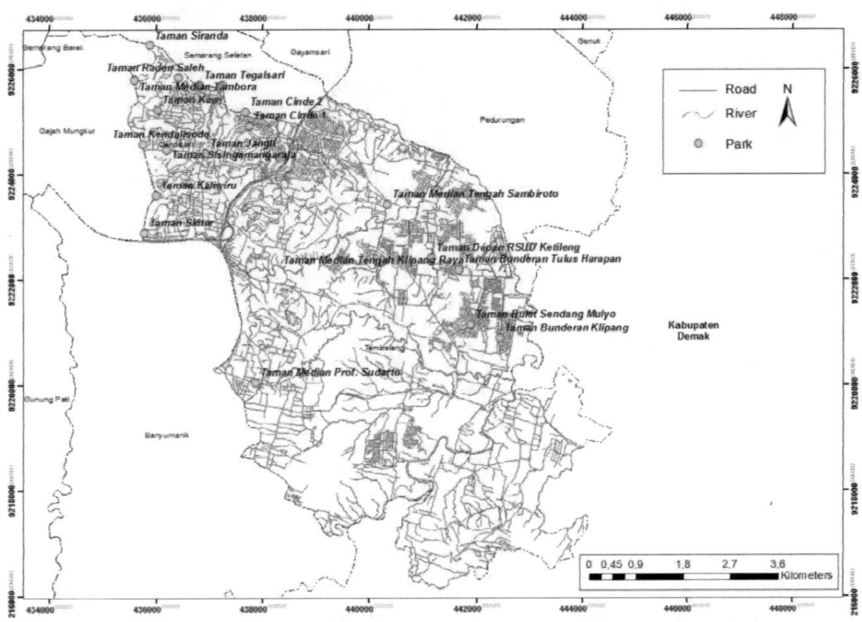

Figure 1. Study area and distribution of park in Candisari and Tembalang District Semarang

2.2. Method

The approach of the study is spatial assessment by utilizing network analysis method to examine the optimum route of green waste collection and transport. Buchori at all, 2015 have initiated to analyze spatial change and growth in Semarang [11], This study used Network Analysis in GIS tools at district level to explore more detail of about spatial attribute such the length and distance of optimum route. The method of assessment run as follow: (1) determining integrated waste treatment unit location as the starting point of departure, (2) determining the transit point, (3) determining the end point or destination, and (4) determining the non-viable path.

3. Results and discussion

3.1. Determination of Integrated Waste Treatment Unit Locations

Spatial approach to analyze the location of integrated waste treatment unit for sorting, composting and recycling have been initiated by Teixeira et all [12]. It is one of the platforms to reduce the waste generation entering the landfill. Refers to the conception, recently, in Candisari and Tembalang districts have only one, call "Compost House" in Karanganyar Gunung Village. According to the

IOP Conf. Series: Earth and Environmental Science **123** (2018) 012021 doi:10.1088/1755-1315/123/1/012021

survey and measurement, the Compost House is over capacity to treat 105.785 kg of organic waste and 15.466 kg of inorganic in. According to the assessment of the 41 temporary sites, two sites were selected to be used as integrated waste treatment unit. Therefore, there are three units and three locations of integrated waste treatment facility (TPST) that were examined for the building the model, namely to Karanganyar Gunung, Tegalsari, Bulusan (Figure 2).

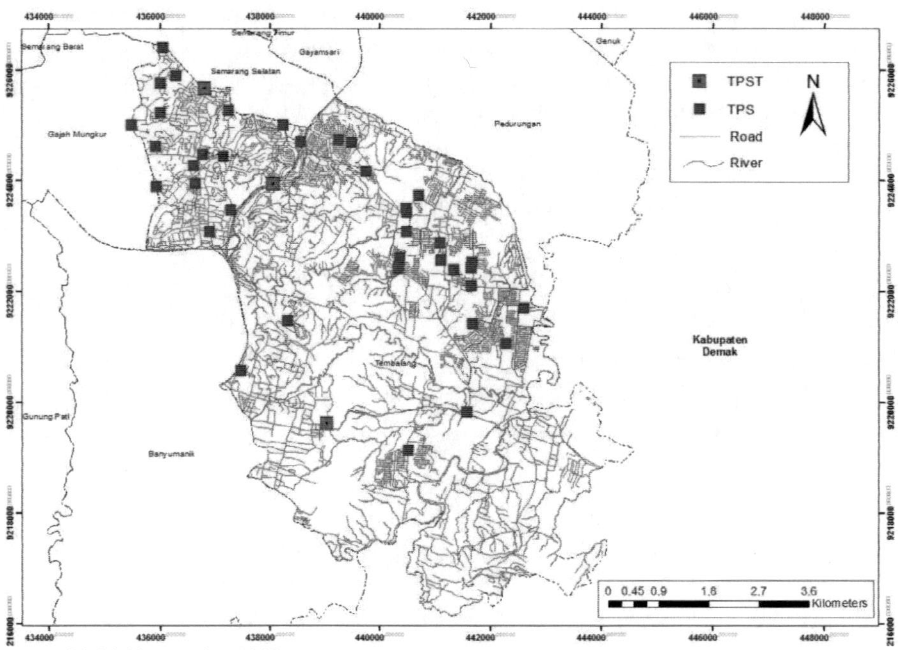

Source : Field Observation,2017

Figure 2. Distribution and location of selected integrated waste treatment facility unit (TPST) in Candisari and Tembalang District

3.2. Optimal Route of Green Waste Disposal

The analysis began with the determination of the starting point of the transport route, i.e. from the pool truck, then continued with the transit points in each of park and ends at the integrated waste treatment facility unit. The assessment of optimal route is conducted by selecting the road characteristic, park and the site of integrated unit facility. For example the "toll road and the road less than 3 meters is inserted into the restriction attribute of GIS tools, as code that road cannot be used to transfer green waste. According to the three temporary collection sites of integrated facility, three transport route scenarios were examined in this study. Scenario 1 used one site, Scenario 2 used two sites and scenario 3 used 3 sites, with advantages and disadvantages in each scenario.

The route on scenario 1 shows that the route for green waste disposal starts from pool truck, parks and temporary collection site in Tembalang. All of park waste which has collected is sorting before. Then, sorted inorganic waste is sold to collectors for recycling, while organic waste is brought to the landfill or to the composting site. The length of route is 36.126 meters and takes time about 46 minutes. The route on scenario 2 is similar to scenario 1. The difference is the route through two sites, starting from pool truck – parks – temporary collection site in Tegal Sari – parks – temporary collection site in Tembalang. The function of two sites is to divide that park waste, so it does not accumulate in one place. This route has a length of 36.471 meters with a travel time of about 47 minutes. The route on scenario 3 uses 3 sites, one of them is "Compost House". This site only accepts organic waste that has been sorted, so it is used as the end point of the waste transport route. The length of route is 46.934 meters and takes time about 60 minutes. The result of the analisys is depicted in figure 3.

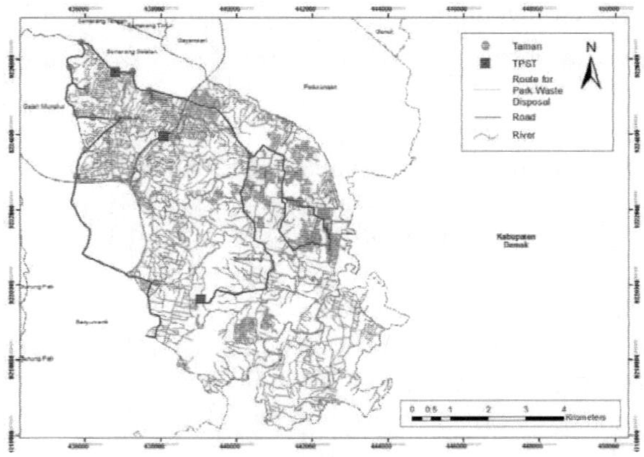

Scenario 1

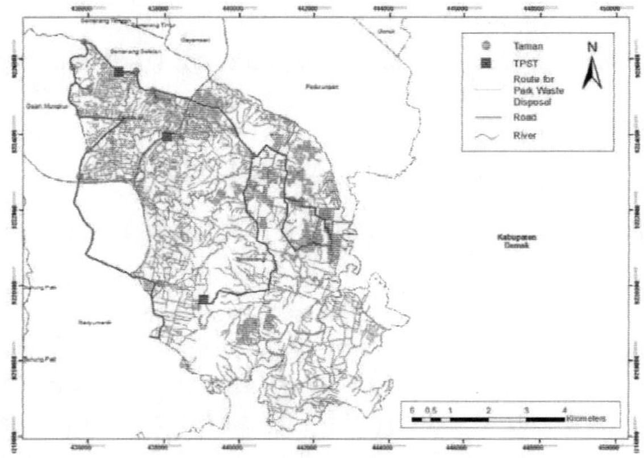

Scenario 2

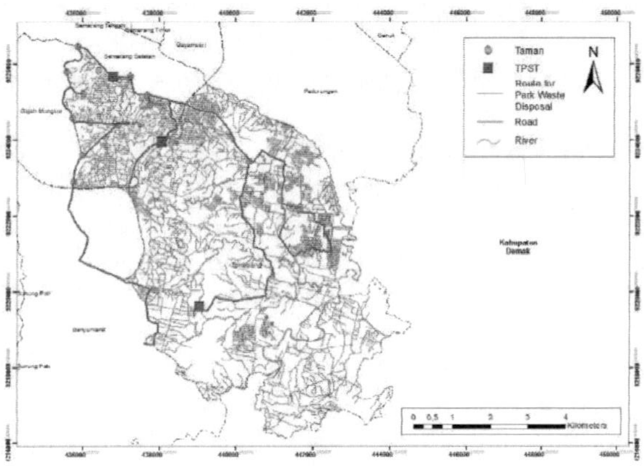

Scenario 3

Figure 3. Result of three Green waste route scenario by using network analysis in Candisari and Tembalang, District, Semarang Indonesia

4. Conclusions

The optimum route of green waste collection can be determined by using Network Analysis method. According the study of green waste collection from park in Candisari and Tembalang, the method shown the good result. According to the optimum integrated green waste recycling scenario 2 is the best result comparison to scenario 1 and 3. In the case of green waste transport route, this analysis is able to show the shortest route and the fastest travel time, however understanding of the real situation such as toll roads or roads that are less than 3 meters wide, is needed to verify manually.

5. Acknowledgment

This research is funded by the Directorate of Research and Community Services, Ministry of Research, Technology and Higher Education for year 2017 with contract number 344-42/UN7.5.1/PP/2017. Author would like to thank to Mr. Dalkam concerning to many information and data for completing this research. Author would like to thank to Andi Hadigy Nur Alamsyah, Dhita Mey Diana, Dhafina Almas, Aji Uhfatun Muzdalifah, and Dea Chintantya, Laras Kun Rahmanti Putri and Rakan Pramoe Izdihar for helping in collecting data, measuring and survey of green waste generation, and thanks to Bayu Rizqi who has helped Arc GIS assessment and utilization.

IOP Conf. Series: Earth and Environmental Science **123** (2018) 012021 doi:10.1088/1755-1315/123/1/012021

6. References

[1] Tzoulas K, Korpela K, Venn S, Yli-Pelkonen V, Kaźmierczak A, Niemela J and James P 2007 Promoting ecosystem and human health in urban areas using Green Infrastructure: A literature review *Landsc. Urban Plan.* **81** 167–178

[2] Mukhtar S, Wafa W, Halimzai H and Shams A K 2016 Planning for the Solid Waste Management of Central Park in New Capital Development of Afghanistan *J. Environ. Prot. (Irvine,. Calif).* **7** 805

[3] Norton B A, Coutts A M, Livesley S J, Harris R J, Hunter A M and Williams N S G 2015 Planning for cooler cities: A framework to prioritise green infrastructure to mitigate high temperatures in urban landscapes *Landsc. Urban Plan.* **134** 127–138

[4] Nakayama H, Shimaoka T, Omine K, Patsaraporn P, Siriratpiriya O and others 2013 Solid waste management in Bangkok at 2011 Thailand floods *J. Disaster Res.* **8** 456–464

[5] Raud M, Mitt M, Oja T, Olt J, Orupõld K and Kikas T 2017 The utilisation potential of urban greening waste: Tartu case study *Urban For. Urban Green.* **21** 96–101

[6] Meisel F and Thiele N 2014 Where to dispose of urban green waste? Transportation planning for the maintenance of public green spaces *Transp. Res. Part A Policy Pract.* **64** 147–162

[7] Sugiri A, Buchori I and Soetomo S 2011 Sustainable metropolitan development: Towards an operational model for Semarang Metropolitan Region *Int. J. Environ. Cult. Econ. Soc. Sustain.* **7** 301–323

[8] Marfai M A and King L 2008 Potential vulnerability implications of coastal inundation due to sea level rise for the coastal zone of Semarang city, Indonesia *Environ. Geol.* **54** 1235–1245

[9] Khadiyanto P, Soetomo S and Hadi S P 2015 Settlement adaptation on a seawater tide overflow area at the north part of Semarang, Indonesia *J. Flood Risk Manag.*

[10] Guntur Suworo A D E 2002 *Leisure Center di Semarang* (Jurusan Arsitektur Fakultas Teknik Undip)

[11] Buchori I, Sugiri A, Hadi S P, Wadley D and Liu Y 2015 Developing a geographic information system-based assessment model for sustainable metropolitan development: The case of the semarang metropolitan region, Indonesia *Am. J. Environ. Sci.* **11** 62–75

[12] Teixeira J, Antunes A P and de Sousa J P 2004 Recyclable waste collection planning----a case study *Eur. J. Oper. Res.* **158** 543–554

Monitoring the Restored Mangrove Condition at Perancak Estuary, Jembrana, Bali, Indonesia from 2001 to 2015

R Ruslisan[1], M Kamal[1,2], and F Sidik[3]

[1]Department of Geographic Information Science, Faculty of Geography, Universitas Gadjah Mada, Bulaksumur, Yogyakarta, Indonesia
[2]PUSPICS Faculty of Geography, Universitas Gadjah Mada, Bulaksumur, Yogyakarta, Indonesia
[3]Institute for Marine Research and Observation, The Ministry of Maritime Affairs and Fisheries, Jembrana, Bali, Indonesia

Email: m.kamal@ugm.ac.id

Abstract. Mangrove is unique vegetation that lives in tidal areas around the tropical and subtropical coasts. It has important physical, biological, and chemical roles for balancing the ecosystem, as well as serving as carbon pool. Therefore, monitoring the mangrove condition is very important step prior to any management and conservation actions in this area. This study aims to map and monitor the condition of restored mangroves in Perancak Estuary, Jembrana, Bali, Indonesia from 2001 to 2015. We used IKONOS-2, WorldView-2 and WorldView-3 image data to map the extent and canopy cover density of mangroves using visual delineation and semi-empirical modelling through Enhanced Vegetation Index (EVI) as a proxy. The results show that there was a significant increase in mangrove extent from 78.08 hectares in 2001 to 122.54 hectares in 2015. In term of mangrove canopy density, the percentage of high and very-high canopy density classes has increased from 32% in 2001 to 57% in 2015. On the other hand, there were slight changes in low and medium canopy density classes during the observation period. Overall, the result figures from both area extent and canopy density indicates the successful implementation of mangrove restoration effort in Perancak Estuary during the last 14 years.

Keywords: Mangroves, Monitoring, Perancak Estuary, SAVI

1. Introduction

Mapping and monitoring mangrove forest is essential to provide a baseline of mangrove status and support for management and conservation policy. Mangroves have been recognized among the most important objects in coastal areas which has many ecological functions. It is a breeding ground for juveniles of aquatic organism, providing habitat and food chain for coastal flora and fauna, protecting the coastline from tsunami and sedimentation, and attracting tourists [1,2]. According to Donatoa et al. [3], mangrove forest has important function as carbon sink in tropical forests and has the higher average of carbon storage compared to other forest type. Despite this ecological importance of mangrove, FAO [4] reported that 36% of the world mangrove forest is damaged. The decreasing mangrove ecosystem either from natural or anthropogenic disturbances cause loss of many ecological functions of mangroves. These including the high intensity of coastal abrasion, higher rate of sea water intrusion, loss of fish habitat, decreasing nutrition supply for aquatic organism, and thus causing imbalance of the ecosystem interface between land and sea ecosystem [5]. These threats of mangrove forest corroborate the need to provide efficient tools for mapping and monitoring the status of mangroves both their extent and conditions.

Remote sensing has been acknowledged as one of the most efficient tools for mapping and monitoring environmental changes such as vegetation and other non-renewable resources [6]. Among of the key advantages in using remote sensing for mangrove monitoring are the provisions of synoptic and repeated coverage of mangrove sites [7]. Remote sensing images provide sources for monitoring mangrove forest through the change detection procedure. The changes between mangrove forest extent and status from one date to the other can be monitored using multi-temporal images. The rapid development of Geographic Information System (GIS) also contributes to the development of monitoring methods for environmental changes. Remote sensing and GIS can be integrated in this context; GIS-based change detection requires multi-temporal images as data source, and overlay analysis in GIS is a powerful tool to perform the change analysis [8].

The Perancak estuary in Jembrana Bali is covered by mangroves that experienced an anthropogenic pressure. Most of mangroves forest in this area were altered to aquaculture for economic purposes, which threatening the ecosystem. The Institute of Marine Research and Observation [9] explained that these changes took place in 1980s and mostly from mangrove forest to shrimp ponds. Since 2000s the restoration efforts have been done to the abandoned shrimp ponds to restore the ecological functions of the ecosystem [10]. Therefore, it is important to monitor the growth rate of Perancak mangrove forest to assess the progression of forest restoration process. One of the key parameters of forest physical characteristics observable from remote sensing images is the forest canopy cover density. It is defined as the percent forest area occupied by the vertical projection of tree crowns [11]. It shows an aggregate of leaves from tree crown at a forest stand which can be used to indicate the health of forest and wild life habitat, predict wood plant composition and measure leaf area index (LAI) [11,12]. This study aims to map and monitor the condition of restored mangroves in Perancak Estuary, Jembrana, Bali, Indonesia from 2001 to 2015 by means of mangrove canopy cover density. Several specific objectives are to (1) use high-spatial resolution remote sensing data to estimate mangrove canopy cover, (2) assess the accuracy of canopy cover estimation result, and (3) monitor the condition of mangroves at Perancak Estuary from 2001 to 2015.

2. Data and Methods

2.1. *Study site*

Perancak Estuary is a mangrove preservation site located in Jembrana Sub-district, Jembrana Regency, Bali (114°37'11.7" - 114°37'12.2" E and dan 08°24'33.5" - 08°24'36.1" S, Figure 1). The study site approximately 876 ha, consists of more than 390 ha of the area as aquaculture (both active and inactive fishpond) and 78.6 ha as mangrove forest. This area of Perancak Estuary includes six villages, which are Perancak, Pengambengan, Air Kuning, Budeng, Yeh Kuning, and Sangkar Agung [13].

Perancak Estuary has a strategic functions as a producer of biological natural resources (source of fisheries, agricultural and timber products, germplasm sources), as transportation facilities, as a port location for ships, fishing sites, as aquaculture and recreation or tourism facilities[13]. Mangrove forests in Perancak Estuary with area of about 177.09 ha are the remaining forest after the land conversion to aquaculture in 1980s [9]. According to Susiana [14], the condition of mangrove ecosystem in this site is divided into natural and rehabilitated mangrove areas, and each have similar biodiversity status with tree stand density more than 1,500 tree/ha. There are nine mangrove species found in the area of Perancak and Tuwed with *Sonneratia alba* as the most dominant species. Based on the magnitude of the important vegetative value index, the nine mangrove species are listed respectively from the largest to the smallest; *Sonneratia alba*, *Rhizophora apiculata*, *Avicennia marina*, *Bruguiera gymnorrhiza*, *Rhizophora stylosa*, *Rhizophora mucronata*, *Ceriops decandra*, *Ceriops tagal* and *Excoecaria agallocha* [15].

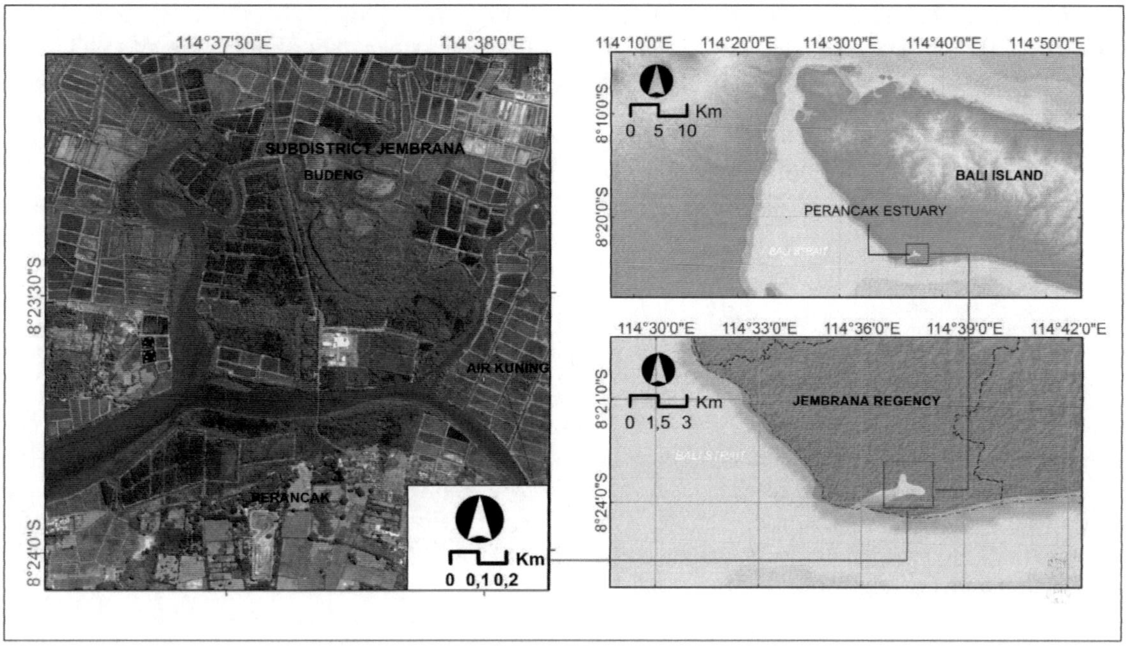

Figure 1. Study Site at Mangroves of Perancak Estuary, Bali, Indonesia

2.2. Image data

The images used in this study are a high resolution satellite imagery possessed by BPOL (Institute for Marine Research and Observation) Jembrana, Bali. The images used in this study are WorldView-3, WorldView-2, and IKONOS-2 (Table 1). The World View images have been orthorectified and radiometrically corrected at standard level. While for IKONOS-2, it has been geometrically corrected at standard level. There are four multispectral bands used in this study, which are red, green, blue and near infrared bands. These four bands are used for vegetation index calculation to estimate the multi-temporal mangrove canopy density and to monitor changes in mangrove condition resulting from restoration efforts in Perancak Estuary.

Table 1. Remote Sensing Image Specification Used in This Study

Images	Acquisition date	Spasial resolution	Spectral attributes (nm)
Worldview-3	16 April 2015	1,31 m (multi) 0,31m (Pan)	Coastal blue (400-450), blue (450-510), green (510-580), yellow (585-625), red (630-690), red edge (705-745), NIR1 (770-895), NIR2 (860-1040), Panchromatic (450-800).
Worldview-2	16 August 2010	1,84 m (multi) 0,48 m (Pan)	Coastal blue (400-450), blue (450-510), green (510-580), yellow (585-625), red (630-690), red edge (705-745), NIR1 (770-895), NIR2 (860-1040), Panchromatic (450-800).
IKONOS-2	2 August 2001	3,2 m (multi) 1 m Pan	Blue (450-520 nm), Green (520-600 nm), red (625-695 nm), NIR (760 – 900 nm), Panchromatic (450-800).

Level 2A of Worldview-2 and Worldview-3 images are standard geometric and radiometric-corrected image that have position shifts at nadir <3.5 meters. Worldview-3 image was recorded on April 16, 2015 at 09:42:20 and has cloud cover of 2.3% with a recording angle of 59.3°, and Worldview-2 was recorded August 16, 2010 at 09:51:47 with an angle of 57.8° and 0.7% of cloud cover. While IKONOS-2 image was recorded on August, 2 2001 at 09:58:00 with cloud cover of 38% and the angle of recording 56.12°. A cloud cover of 38% caused the IKONOS-2 image condition being slightly darker. It indicates that the image quality of IKONOS-2 is less than the other images which have very little cloud cover of less than 5%. This is because the acquisition time of IKONOS-2 was in the rainy season. In terms of spatial resolution, the Worldview-3 and Worldview-2 pixel sizes have been resample by Digital Globe to 2 meters and for IKONOS-2 to 4 meters.

2.3. Image Preparation

Remote sensing image processing for change detection requires that all images have the same geometry and radiometric characteristics [8]. Image processing was carried out in this study for geometric, radiometric and atmospheric correction between image, and extraction of vegetation index. Geometric correction between images was done using WV-3 as reference image because it is the latest sensor which geometric aspects are stable. The radiometric correction was conducted up to top of atmosphere (TOA) reflectance level, and then continued with relative atmospheric correction using DOS (Dark Object Subtraction).

Vegetation information extraction was done by applying the vegetation index formula on each image. The vegetation index used included Normalized Difference Vegetation Index (NDVI), Soil-adjusted Vegetation Index (SAVI), and Enhanced Vegetation Index (EVI) (Table 2). NDVI is classified as a generic vegetation index because it uses red, green and NIR bands for vegetation index extraction [16]. SAVI is a vegetation index used to reduce the influence of soil backgrounds. Coefficient value of L = 1 is used to reduce soil variance at low vegetation density used, while L = 0.5 and L = 0-0.25 are used for medium and high vegetation density, respectively. The effects of soil almost all disappear with higher vegetation densities that have smaller L factor [17]. While, the EVI is a vegetation index used to reduce atmospheric disturbances and reflections of soil backgrounds all together in vegetated area so that the index results will be more sensitive to biomass and canopy cover density [18].

Table 2. Vegetation Indices Used in This Study.

Vegetation index	Formula		Reference
NDVI	$\frac{\rho NIR-\rho red}{\rho NIR+\rho red}$	(1)	[20]
SAVI	$\frac{\rho NIR-\rho red}{\rho NIR+\rho red+L} x\ (1+\mathrm{L})$	(2)	[18]
EVI	$G\frac{\rho NIR-\rho red}{\rho NIR+C1*\rho red-C2*\rho blue+L}$	(3)	[19]
Note: For SAVI, the L value range between 0 (high vegetation density) to 1 (low vegetation density). For EVI, the coefficient used is as follow G= 2.5, C1= 6, C2= 7.5, and L= 1.			

2.4. Canopy Density Estimation

Digital photography using hemispherical lenses is a fairly accurate approach in estimating the density of forest canopy cover [19]. The use of fisheye lenses is aimed at obtaining a 180° viewing angle, thus being able to see leaf cover more flexibly on each layer [20]. Measurements with hemispherical photographs were performed with upward taking with a shooting height of approximately 1.5 meters from the ground surface and covering an area of approximately 2.3 m^2 [21]. An example of an upward hemispherical photo acquisition is shown in Figure 2a.

The sampling scheme used in this study was purposive sampling which is based on variation of mangrove canopy cover and sample accessibility. The advantage of this approach is its ability to obtain similar samples representing an object at geographically difficult sites from accessible point sampling [22]. The determination of the sample location is done visually from the image by considering the qualitative density of the mangrove forest. Assessment of mangrove conditions in the field was conducted with 10x10 meter plot for tree level (Figure 2b). The selection of this size is because the mangrove area is relatively heterogeneous and mixed for each size, so the size of the stake and seedlings are covered by the tree level when projected from the image. Assessment of the condition is done only at the tree level by considering the tides as the basis for shooting with a height of 1.5 meters sensor/camera on the ground.

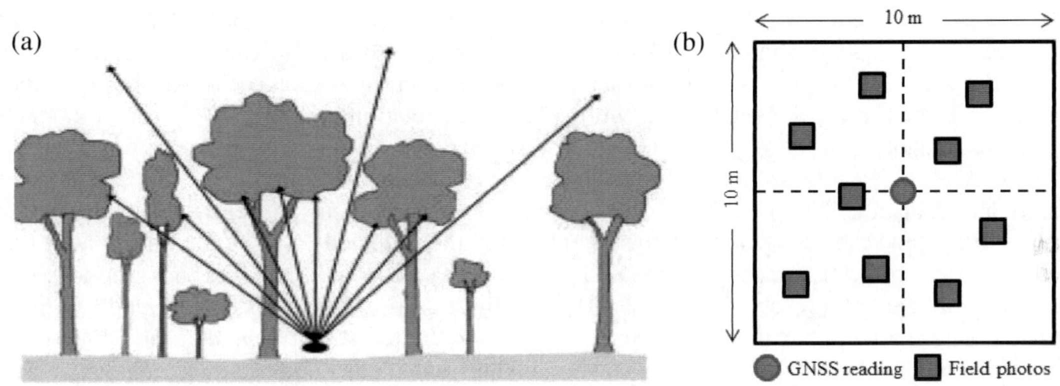

Figure 2. Hemispherical field photos: (a) upward projection of the sensor field of view [23], and (b) field plot and measurement points.

The field sampling was conducted between 1 and 9 May 2016, and was collecting 70 field samples which consist of 36 samples for model development and 34 samples for model validation. On each sampling plot, we recorded the position of the centre of sampling plot using Global Navigation Satellite System (GNSS) receiver and nine random upward facing field photos of the mangrove canopy using a hemispherical camera system. The canopy cover values were then calculated by the CAN-EYE software (http://www6.paca.inra.fr/can-eye) based on the samples collected in the field. The development of canopy cover density map was conducted by correlating field canopy density value with of the corresponding pixels of NDVI, SAVI, and EVI vegetation index values using linear regression analysis. In this case, the relationship analysis was done between vegetation index images as predictor (independent variable (x)) and field data as the response (dependent variable (y)). From the regression analysis, vegetation index that has the highest coefficient of determination (R^2) will be used as the basis for estimating and monitoring of mangrove canopy cover density in the research area.

2.5. Accuracy assessment

The canopy density accuracy assessment was performed using a 1:1 plot of goodness of fit. The aim of this accuracy method is to evaluate the accuracy of the mangrove canopy density estimation from the image (y-axis) based on the ideal value plot (1:1 diagonal line) with the reference value, in this case was the validation field sample (x-axis). From the relationship result, if the data position is above the 1:1 diagonal line, the estimate is over-estimated, and *vice versa* [24]. To obtain a numerical accuracy value, in additoin to the plot of goodness of fit 1: 1, we also performed an accuracy test with Standard Error (SE) (Equation 4) as the following:

IOP Conf. Series: Earth and Environmental Science **123** (2018) 012022 doi:10.1088/1755-1315/123/1/012022

$$\textbf{SE}\ (\textit{Standard Error}) = \sqrt{\frac{\Sigma(Y-Y')}{n-2}} \quad (4)$$

where Y = reference value, Y' = estimation value, n = number of sample. The smaller the value of SE, the smaller the error and the more accurate the result of the canopy cover density model.

3. Results and Discussion

3.1. Canopy cover density modelling

The development of canopy cover density map was done by modelling the relationship between field canopy density and vegetation indices value (NDVI, SAVI, and EVI). This is a type of parametric inferential statistics because the data type used has ratio scale (i.e. it has absolute 0 values). The statistical process used to draw conclusions requires the data to be normally distributed. From data analysis, the result of data normality calculation using Kolmogorov-Smirnov method shows that field canopy cover density data is normally distributed with value 0.06 (has significance value > 0.05). The result of the model relationship from 36 field samples is shown in Figure 3a. The three selected indices have a positive correlation with field data with Pearson correlation close to 0.7, or have moderate strength of correlation.

Based on Figure 3a, EVI is able to explain a better data relationship because it considers the atmosphere factor and the background of the soil affecting the reflected value of the image. SAVI has a better coefficient of determination than NDVI, because SAVI considers the soil background aspect, where in the study area with low canopy density conditions tend to have soil background that is not very wet but has an effect on over-estimation at low tide and under-estimation at high tide. By looking at the coefficient of determination, the EVI has the highest value ($R^2 = 0.557$), therefore we use this index for further processing. We applied the EVI regression model to estimate the mangrove canopy cover density, where Field Canopy Density = 1.6052 EVI – 0.3604, and the mapping result is presented in Figure 3b. There is no specific pattern found from the canopy density map. However, most of the high canopy density values are located at the abandoned fishponds and within the nature mangrove forest. While the low canopy density values are scattered across the study site.

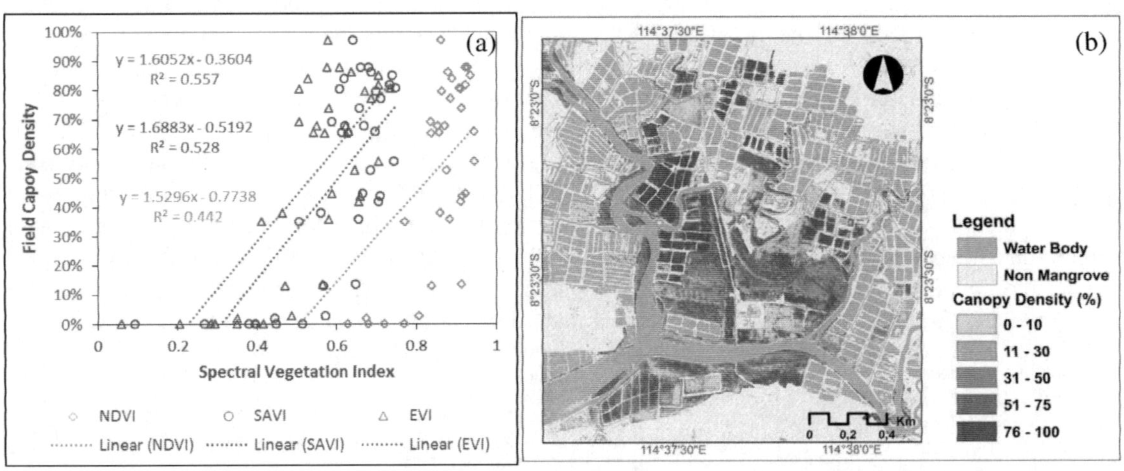

Figure 3. Canopy cover density model: (a) scatter graph between field data and vegetation indices, and (b) canopy density map from EVI model

3.2. Accuracy assessment results

The accuracy assessment used in this research is the plot of goodness of fit 1: 1 and SE. The results in Figure 4 show the distribution of the estimated value of canopy density mangrove value to the corresponding field canopy density value. The linear regression line indicates that there is under-estimation at low canopy density, and over-estimation at high canopy density value. This indicates that the background soil has significant influence on the estimation of the density of the canopy from the image. Because the soil that exists at low densities is not very wet, thus contributing a high reflection due to understory covers. The low accuracy is affected by several factors such as low GNSS accuracy due to positioning problem caused the incompatibility between the sampling positions in the field with the image. In addition, higher image spatial resolution result in more variation within a single object on the image and thus decrease the estimation accuracy [21].

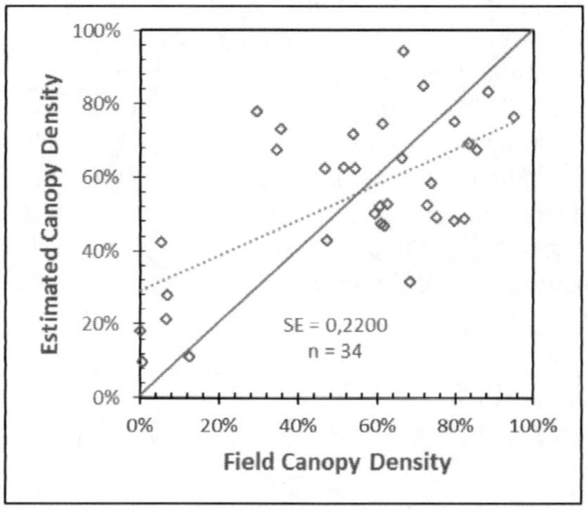

Figure 4. Plot Goodness of Fit 1:1 of EVI

The over-estimation at low density (0-10%) tends to occur due to the reflectance of seedlings, stakes or shrubs on the mangrove understorey. The low density mangrove forest makes understorey vegetation observable from remote sensing images. On the other hand, there is an under-estimation result at 50-80% canopy density, where the field measurements were conducted at the tree level at 1.5 meters above the soil surface. Hence, the density of shrubs and mangrove seedlings is not recorded, but these features are identifiable from the image and increase the value of the canopy cover density.

3.3. Multi-temporal analysis

This research is aimed to monitor the condition of canopy density by using vegetation index as a proxy. The canopy density condition in 2015 was developed from the relationship between field data and vegetation index data using linear regression. EVI was used because it has the best coefficient of determination value. The distribution of mangrove canopy cover density in 2010 and 2001 were developed using field samples collected in 2016. We selected samples where the forest condition did not change much from 2001-2015. Regression functions generated each year of observation is presented in Table 3. In 2001 the coefficient of determination only reached 0.527, meaning the image data was only able to explain about 53% of the field canopy density distribution. While, in the year 2010 the coefficient of determination is 0,816 or about 82% of image data able to explain the field data. The low coefficient of determination in 2001 was mainly attributed to the low quality of IKONOS-2 image mentioned in section 2.2. Other reason might be related to the time difference between image acquisition and field campaign.

Table 3. Regression Functions at Different Years

r	R^2	Regression function	Year
0.746	0.557	Y = 1.605 x – 0.3604	2015
0.903	0.816	Y = 1,419 x – 0.1216	2010
0.726	0.527	Y = 0.980 x – 0.4713	2001

The results of mangrove canopy cover density mapping in 2001, 2010 and 2015 are presented in Figure 5a-c. Overall, we can see the change pattern of Perancak Estuary mangroves; the canopy density of is increasing throughout the observation years. It is indicated by an increase of area extent of dark green features on the maps. From change detection point of view, there are two major changes happen in this area. First, there were land cover changes of non-mangrove area to mangrove area. This pattern is observable from the 2001 – 2015 canopy density maps, especially in the fish ponds area. This pattern indicates that the efforts of restoring mangroves by replanting in abandoned fish pond were successful. We can also see this pattern form the graph in Figure 5d and Table 4, where the total area extent of mangroves increased from 2001 to 2015. Second, there were significant canopy cover density increases during the fourteen years period. This type of change is identified in the natural mangrove forest patch in the middle of the research site, and within some abandoned fish ponds. This pattern indicated a significant growth of mangrove forest in this area. Figure 5d also confirms this pattern; the dark green bar (very high canopy density class) is increasing from 2001 to 2015. This finding is supported by Rahmania et al. [25] where they also found a significant canopy cover increase in this area from 2001 to 2014.

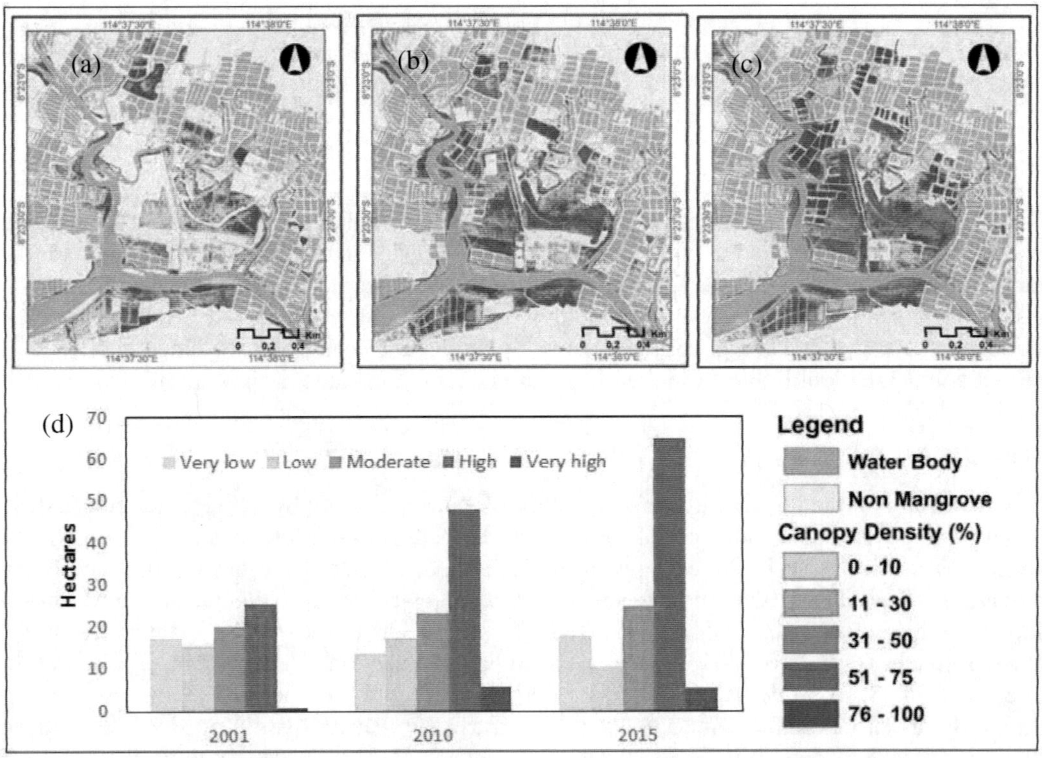

Figure 5. Mangrove canopy cover density in (a) 2001, (b) 2010, (c) 2015, and (d) the canopy density area comparison between years

Tabel 4. Area extent of mangrove canopy cover (in hectares).

Category	Canopy cover (%)	2001	2010	2015
Very low	0-10	29.68	27.02	30.59
Low	11-30	10.52	12.10	9.54
Moderate	31-50	11.31	15.10	17.56
High	51-75	16.17	27.98	34.94
Very high	76-100	10.42	25.00	29.91
Total (ha)		78.08	107.19	122.54

3.4. Mangrove Change at Perancak Estuary

Mangrove monitoring in this study conducted from 2001, 2010 and 2015 to observe changes in the area of mangrove forest and its canopy density conditions. The result of visual interpretation indicates that mangrove at Perancak Estuary shows improvement from year to year, in terms of area extent and density of the canopy. In 2010, mangroves expanded by 29.1 ha from 2001 (Table 5). However, there were about 6.51 ha mangrove areas disappeared in some locations; hence, the area extent of mangroves in 2010 was 107.19 ha. From the maps, there was an increase in mangrove extent in 2015 at about 15.35 ha, and the total mangrove extent was 122.54 ha. However, in 2015 there was an area reduction of 1.37 ha. During the period of 14 years there were 70.19 ha of mangrove forest that remains stable. This result shows that in terms of extent of mangrove Perancak from 2001-2015 is improved significantly as seen from Figure 6a. In terms of mangrove canopy density value, there was a significant increase from 2001 to 2010, which amounted to 70.07 ha. Then from 2010 to 2015 there was an increase in mangrove canopy density as much as 48.78 ha. This shows improvement in the quality of mangrove canopy cover in the study area (Figure 6b).

Table 5. Mangrove changes recorded in 2001, 2010 and 2015.

Changes	Mangrove extent			Mangrove canopy cover		
	2001	2010	2015	2001	2010	2015
Increase	-	29.1	15.35	-	70.07	48.78
Decrease	-	6.51	1.37	78.08	24.83	25.44
No change	78.08	71.57	70.19	-	18.82	49.71
Total (ha)	78.08	107.19	122.54	78.08	113.71	123.92

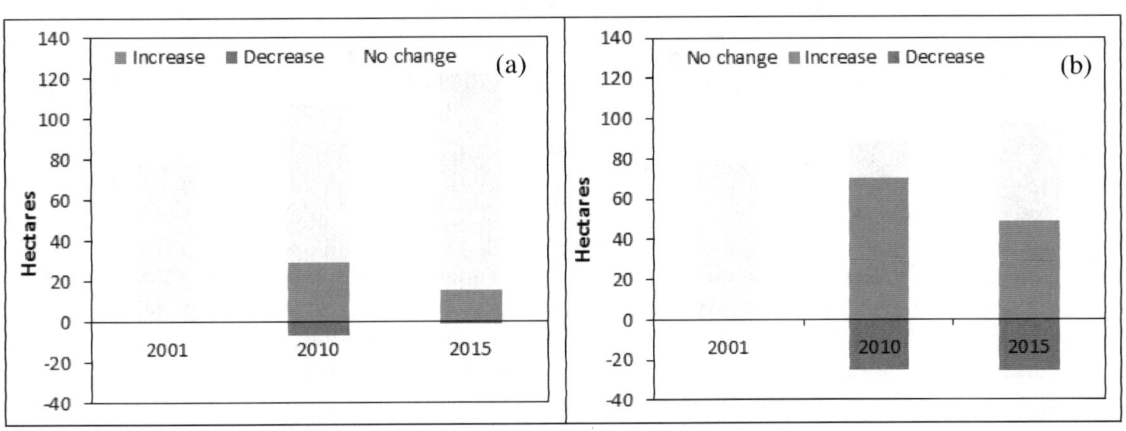

Figure 6. Changes in Perancak Estuary mangroves; (a) mangrove extent, and (b) mangrove canopy cover density.

The pattern of canopy density development from 2001-2010 and 2010-2015 has changes as shown in Figure 7a and b. The pattern of density development with green colour indicates mangrove area that has increased the density of the canopy, with the dominant distribution being in the planting area. Mangroves basically have a slow development [26], but the pattern does not apply to mangrove planting. Distance of mangrove planting was arranged about 1-2 meters, so within 1-2 years it canopy covers the mangrove gaps. The natural mangrove area tends to decrease. This pattern is in line with the research of Rahmania et al. [25] at Perancak, which monitor open and closed canopy cover. Their result shows that open canopy was increasing in the natural mangrove area in 2014, while for the closed canopy area did have a significant increase in the area of planting.

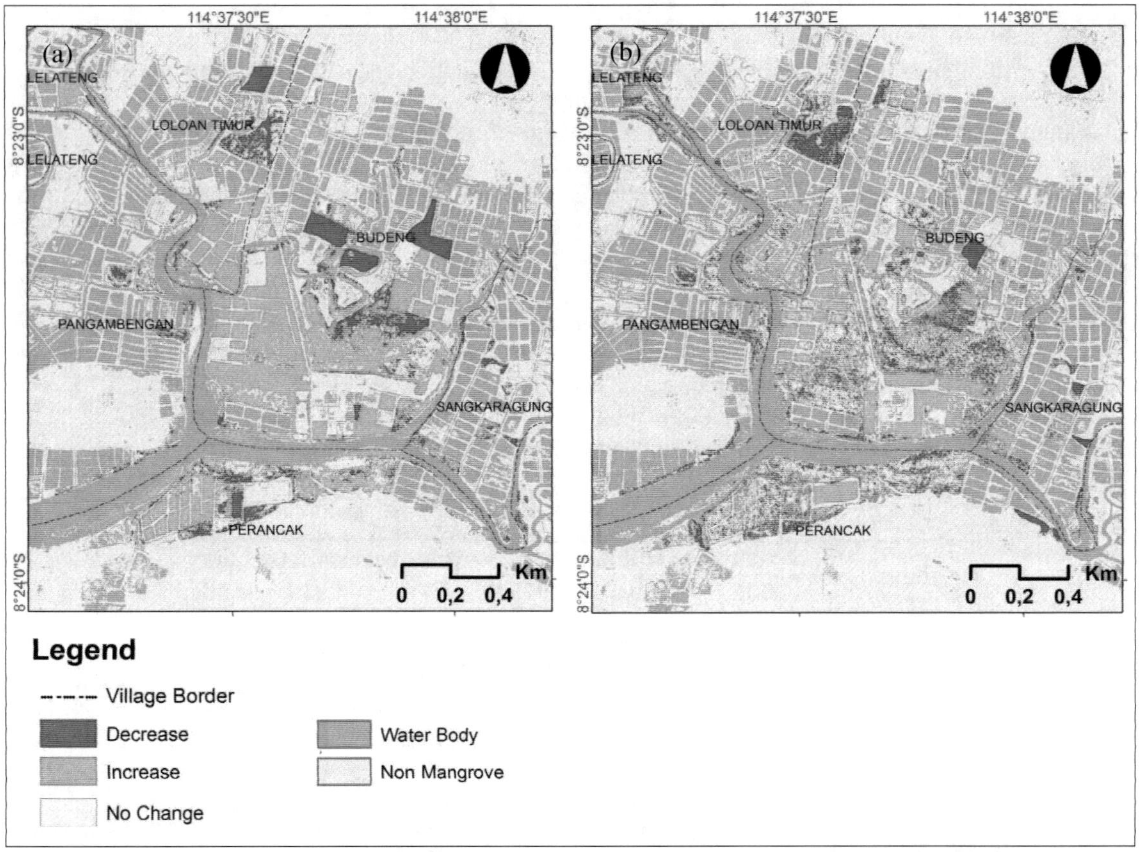

Figure 7. Changes in Perancak Estuary mangroves; (a) from 2001 to 2010, and (b) from 2010 to 2015

Planting activities produce new mangroves plants with relatively rapid vegetation growth. In addition, the pattern of adjacent planting triggers the closure of the canopy and almost all areas of planting have areas that meet closely. According to Rahmania et al. [25], planting in the fish pond area made mangrove propagules locked in the pond and prevent it to spread across the site. It is also one of the new individual developments that led to the growth of the relatively fast growing mangrove area. However, some mangrove planting experiencing leaf decay and dieback, especially *Rhizophora sp.* This condition was due to the competition of photosynthesis among individual mangrove which grows close to each other. Decreased mangrove areas are more affected by the human factor, for example the opening of mangrove areas for shrimp ponds since the 1980s has triggered the changes in mangrove conditions in Perancak Estuary [10]. However, organic litter in mangrove canopies with very high density (75-100%) are more affected by natural factors such as sea level rise or human disturbance.

4. Conclusion and Future Work

This study employed remote sensing data to estimate and monitor mangrove canopy cover density at Perancak Estuary in 2001, 2010 and 2015. Several vegetation indices were used as proxy to estimate this information from semi-empirical relationship between the indices and field data. Overall, remote sensing data are successfully map mangrove extent and canopy cover density in the study site. From the estimation result, the density of the canopy in Perancak Estuary in 2015 is dominated by high density (50-75%). The resulting accuracy assessment of the estimation is moderate, but enough to distinguish the density level of mangrove canopy. From the statistical analysis we found that the best vegetation index is EVI with SE value of 0.220. 2. The extent of Perancak Estuary mangroves was increased from 78.08 ha in 2001, to 107.13 and 122.54 ha in 2010 and 2015, respectively. Similar pattern also found in mangrove canopy density where the condition was tend to improve at planting area marked by increasing of canopy density of about 48.78 ha of mangrove. However, we also noticed some canopy density decrease of about 25.44 ha in natural mangrove. Some future directions proposed as the follow up of this study are (1) to use real time image or image that is obtained not too far from field campaign, (2) to use similar image dataset throughout the year of observation, and (3) to evaluate the mangrove changes using at a shorter period of observation (i.e. yearly or each six months).

5. Acknowledgement

The authors are grateful for the accomplishment of this paper and express their gratitude for the support of BPOL who has provided image data and licensing for research and those who have assisted in this research.

6. References

[1] Lugo A E and Snedaker S C 1974 The ecology of mangroves *Annu. Rev. Ecol. Syst.* **5** 39–64
[2] Hutchings P and Saenger P 1987 Ecology of mangroves *ISBN 0 7022 2015 9*
[3] Donato D, Kauffman J B, Murdiyarso D, Kurnianto S, Stidham M and Kanninen M 2012 *Mangrove adalah salah satu hutan terkaya karbon di kawasan tropis*
[4] FAO 2007 *The World's Mangrove 1980-2005: A Thematic Study in The Framework of The Global Forest Assestment 2005*
[5] Kordi M 2012 Ekosistem Mangrove: potensi, fungsi, dan pengelolaan *Rineka Cipta. Jakarta* **255**
[6] Jamalabad M S 2004 Forest canopy density monitoring using satellite images *Geo-Imagery Bridging Continents XXth ISPRS Congress, Istanbul, Turkey, 2004*
[7] Giri C, Ochieng E, Tieszen L L, Zhu Z, Singh A, Loveland T, Masek J and Duke N 2011 Status and distribution of mangrove forests of the world using earth observation satellite data *Glob. Ecol. Biogeogr.* **20** 154–159
[8] Lu D, Mausel P, Brondizio E and Moran E 2004 Change detection techniques *Int. J. Remote Sens.* **25** 2365–2401
[9] BROK 2004 *Laporan Riset Observasi dan Kajian Pemanfaatan Kawasan Konservasi Laut*
[10] Proisy C, Rahmani R, Viennois G, Andayani A, Baudel S, Fahran R, Gusmawati N F, Germain O, Lemonnier H, Mbay N and others 2014 Monitoring changes on mangroves coasts using high resolution satellite images. A case study in the Perancak estuary, Bali
[11] Gill S J, Biging G S and Murphy E C 2000 Modeling conifer tree crown radius and estimating canopy cover *For. Ecol. Manage.* **126** 405–416
[12] Paletto A and Tosi V 2009 Forest canopy cover and canopy closure: comparison of assessment techniques *Eur. J. For. Res.* **128** 265–272
[13] BROK 2009 *Laporan Riset Observasi dan Kajian Pemanfaatan Kawasan Konservasi Laut*
[14] Susiana S Diversitas Dan Kerapatan Mangrove, Gastropoda, dan Bivalvia Di Estuari Perancak, Bali
[15] DKPK 2014 *Profil Dinas Pemerintah Kabupaten Jembrana*
[16] Danoedoro P 2012 Pengantar penginderaan jauh digital *Yogyakarta Andi*

[17] Huete A R 1988 A soil-adjusted vegetation index (SAVI) *Remote Sens. Environ.* **25** 295–309

[18] Huete A, Didan K, Miura T, Rodriguez E P, Gao X and Ferreira L G 2002 Overview of the radiometric and biophysical performance of the MODIS vegetation indices *Remote Sens. Environ.* **83** 195–213

[19] Chianucci F and Cutini A 2012 Digital hemispherical photography for estimating forest canopy properties: current controversies and opportunities. iForest, 5 (6): 290-295 *Commun. by Roberto Tognetti*

[20] Weiss M and Baret F 2010 can-eye version 6.1 user manual *EMMAH (Mediterranean Environ. Agro-Hydro Syst. Model. Lab. French Natl. Inst. Agric. Res. (INRA), Avignon, 47p*

[21] Kamal M, Phinn S and Johansen K 2016 Assessment of multi-resolution image data for mangrove leaf area index mapping *Remote Sens. Environ.* **176** 242–254

[22] McCoy R M 2005 *Field methods in remote sensing* (Guilford Press)

[23] Jennings S B, Brown N D and Sheil D 1999 Assessing forest canopies and understorey illumination: canopy closure, canopy cover and other measures *For. An Int. J. For. Res.* **72** 59–74

[24] Kamal M, Hartono H, Wicaksono P, Adi N S and Arjasakusuma S 2016 Assessment of Mangrove Forest Degradation Through Canopy Fractional Cover in Karimunjawa Island, Central Java, Indonesia *Geoplanning J. Geomatics Plan.* **3** 107

[25] Rahmania R, Proisy C, Germain O, Gaspar P, Viennois G, Prosperi J, Andayani A, Subki B, Farhan A R, Sidik F and others 2015 13 Years of changes in the extent and physiognomy of mangroves after shrimp farming abandonment, Bali *Analysis of Multitemporal Remote Sensing Images (Multi-Temp), 2015 8th International Workshop on the* pp 1–4

[26] Twilley R R, Rivera-Monroy V H, Chen R and Botero L 1999 Adapting an ecological mangrove model to simulate trajectories in restoration ecology *Mar. Pollut. Bull.* **37** 404–419

IOP Conf. Series: Earth and Environmental Science **123** (2018) 012023 doi:10.1088/1755-1315/123/1/012023

Measuring Carbon Emissions from Deforestation at Donggala Regency, Central Sulawesi Province, Indonesia

I Nahib[1] and J Suryanta[1]

[1]Geospatial Information Agency, Cibinong, Jawa Barat, Indonesia

Email: irmnahib@gmail.com, irmadi.nahib@big.go.id

Abstract. Forest is a natural resource that is very important and beneficial for the livelihood either directly or indirectly. Forest has a variety of ecological functions. One of forest functions is to maintain the amount of stored carbon. The forest area changes into non-forest area resulted in reducing forest functions as a provider of environmental services. This study aims: 1) to determine the deforestation during the period of 2000-2011, 2) to make model of the landcover change using logistic regression model, 3) to measurecarbon emissions and valuation based on impact of deforestation. The materials used in this study are : a) Indonesian Topographic Map at Scale 1: 50,000, Geospatial Information Agency (BIG) , b) landcover map (year of 2000 and 2011), scale 1 : 250,000, produced by director general of forestry planning, ministry of environment and forestry, 3) environmental variables (dependent variable) such as : distance from roads, distance from streams, elevation and slope. The spatial analysis is done by land change modeler which is module in Idrisi Terrset. Meanwhile calculations of carbon storage and economic value which are done by ecosystem service modelers also as a Idrisi Terrset. The results show that the rate of deforestation during the period of 2000-2011 at Donggala as high as13,448.07 ha or about 1,222.55 ha per year. The impact of the forest cover changes resulted in the decrease of carbon storage up to 3.66 million tons or equivalent to 13.42 million tons of carbon emissions. Economic losses caused carbon emission in period 2000-2011 up to US$ 38,188,465 (net present value, *NPV*)

Keywords: NPV model, Carbon Emission, GIS, Land use Change Modeler

1. Introduction

Forest is natural resources that are very important and useful for life and living either directly or indirectly. Direct benefits from the existence of the forest are timber, non-timber products and wildlife. While the indirect benefits are environmental services, such as watersheds, aesthetic function, an oxygen supplier and carbon sink.

Forest destruction, climate change and global warming reduce the indirect benefit of forest because forest is the largest carbon sink and play very important role in global carbon cycle and can hold carbon at least 10 times greater than other vegetations prairie grass, crops and tundra [1].

In 2008, the United Nations launched REDD (United Nations Collaborative Programme on Reducing Emissions from Deforestation and Forest Degradation in Developing Countries) provides a mechanism to mitigate climate change by sequestering forest carbon. REDD also promotes the secondary ecosystem service benefits associated withthis forest conservation, including protection of biodiversity and water quality [2].

The average annual deforestation in Indonesia for the period 2000-2012 was 671,420 hectares, accounts for 525,516 ha of deforestation in mineral land and 145,904 ha of deforestation in peat land. During this period,9 percent of deforestation occurred in Sulawesi Sulawesi [3]. Meanwhile, refers [4] the rate of deforestation in period 2000-2009 for Sulawesi Island was 1,667,840.59 Ha (15.58 %)

The Central Sulawesi province has forest about 4.2 million ha, so as to have a strategic role in the implementation of REDD +. The deforestation occurred in central Sulawesi Province 432,111.50 Ha (10.15 %). Donggala regency is one of regency in central sulawesi province. Meanwhile [5], in years 2000 that has forest 401.930 ha. The deforestation the period 2000-2011 was 14.790 ha (12.59 %).

Application of remote sensing and geographical information system was used to estimate land cover changes from multi temporally information. Integration from carbon factor was taken from secondary data (previous studies) and land cover changes data highly expected to give information about carbon stocks changes [6].

Managing landscapes for carbon storage and sequestration requires information about how much and where carbon is stored, how much carbon is sequestered or lost over time, and how shifts in land use affect the amount of carbon stored and sequestered over time. Valuation is applied to sequestration, not storage, because market prices relate only to carbon sequestration. Discount rates are multipliers that typically reduce the value of carbon sequestration over time [7]

How much carbon changes lost in previous studies have not been presented in spatial data (map) and based only on statistical figures, this study is based on spatial data and numbers so it will be easier to understand the form of distribution. The study aimed: 1) to determine the deforestation during the period of 2000- 2011, 2) to make modeling the forest cover change using logistic regression model, 3) to measure carbon emissions and valuation base on impact of deforestation.

2. Materials and Method

Tool use in this study are ArcGis and TerrSet Software. Data Used were Indonesian Topographic Map, Scale 1: 50,000, Geospatial Information Agency (BIG); Land Use Land Cover Map of Donggala Regency, Scale 1: 250,000 (2000 and 2011), Directorate General of Forestry Planning Ministry of Enviromental and Forestry (**Figure 1**). Included in the data are four enviromental variable (dependent variable) such as distance from streams, distance from streams, elevaton and slopes, Indonesian Topographic Map, Scale 1: 50,000, Geospatial Information Agency (**Figure 2**). Meanwhile economic data area carbon price (US$ 15) and discount rates (7 %). A table of land use cover (LULC) classes, containing data on carbon stored in each of the four fundamental pools for each LULC class. This study only calculate carbon above (carbon density in aboveground mass (Mg/Ha) (**Table 1**).

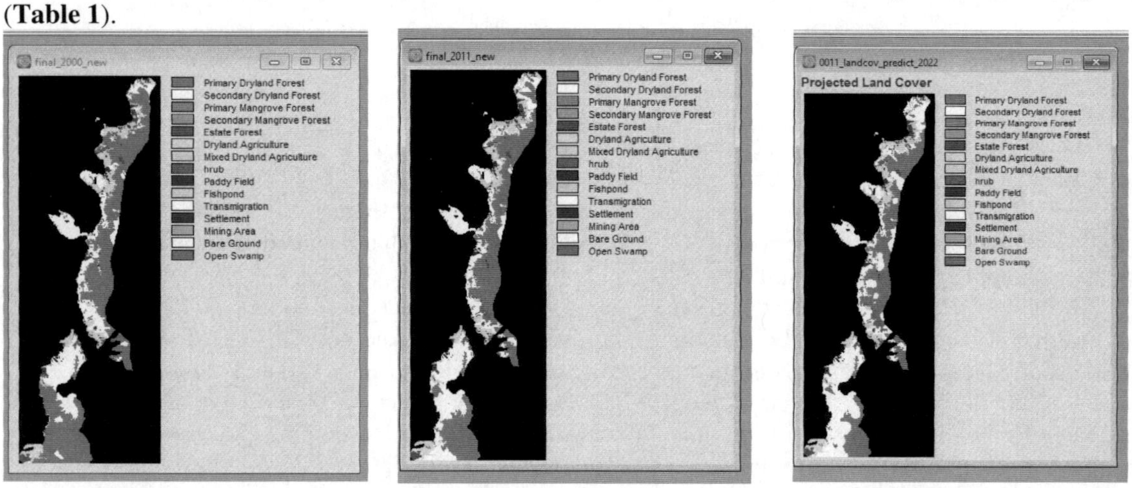

Landcover 2000 Landcover 2011 Landcover 2022 (Projected)

Figure 1. Donggala's Landcover

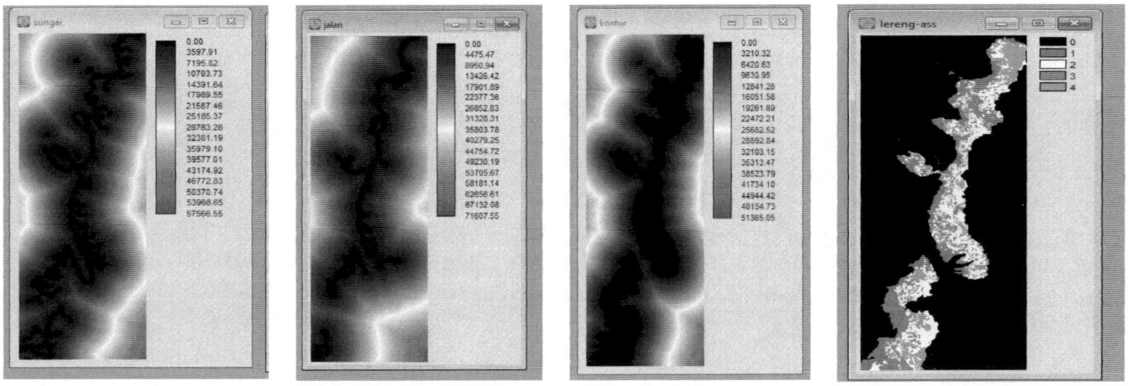

a. Distance from streams (m) b. Distance from road (m) c. Elevation (meter) d. Slopes (degrees)

Figure 2. Environmental variable (dependent variable)

Table 1. Emission Factor for Baplan land cover class in Central Sulawesi

No	Land Cover	C Above
1	Primary dryland forest (PF)	195.4
2	Secondary dryland forest (SF)	169.7
3	Primary mangrove forest(PMF)	170
4	Secondary mangrove forest (SMF)	120
5	Estate crop (EP)	63
6	Pure dry agriculture (AUA)	8
7	Mixed dry agriculture (MxUA)	10
8	Shrub (Sr)	15
9	Paddy Field (Rc)	5
10	Fish pond/aquaculture (Po)	0
11	Transmigration areas (Tr)	10
12	Settlement areas (Se)	1
13	Mining areas (Mn)	0
14	Bare ground (Br)	0
15	Open Swamp (WB)	0

Source: Refer 8

Logistic regression model (LRM) was used to model and analyze the lancover change in IDRISI TerrSet. The objective of the present study was to assess the importance of the explanatory variables on landcover change from 1990 to 2000 and predicting the probability of change by 2011. The binary presence or absence is the dependent variable for the periods 1990–2000. The predicted landcover of 2011 was validated using ROC / AUC (Relative Operating Characteristic/Area Under Curve) module of IDRISI TerrSet. The ROC module is comparing a suitability image depicting the likelihood of that class occurring (the input image) and a boolean image showing where that class actually exists (the reference image). The ROC curve is the true positive fraction vs false positive fraction and the AUC is a measure of overall performance [8].

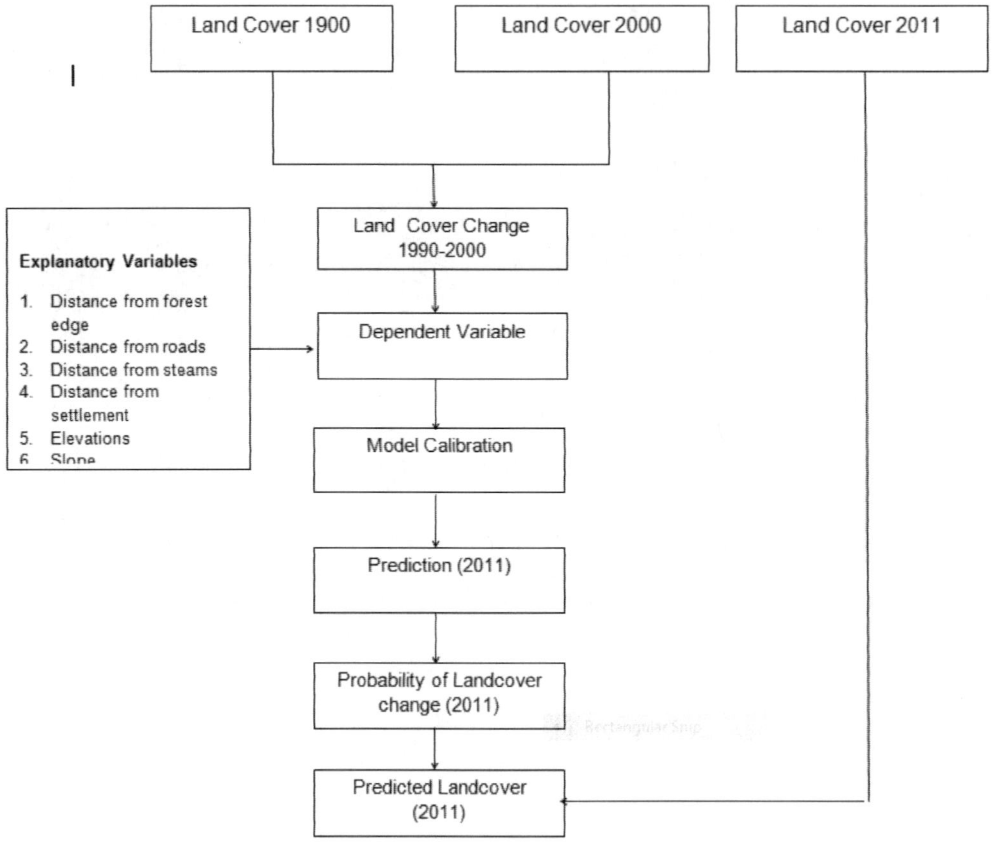

Figure 3. Flowchart stage research activities

2.1. Carbon storage/ Sequestration Model and Valuation

Calculation of emissions using the stock changes approach (*stock difference*) which was measured at two different time points using two factors, namely: activity data and emission factors. The main data of carbon stocks changes was derived from the data of the land cover change [6].

Carbon strorage/ Sequestration model needs the map and data tables, including the economic data. Land use/land cover (LULC) map must be A raster format, with a LULC code for each cell. The dataset should be projected in meters (UTM 50 S). The year (2000, 2011 and 2022) depicted by the LULC map, for use in calculating sequestration and economic values [7].

The carbon prices observed in these instruments vary significantly, from less than US$1/tCO2e to $US130/tCO2e. Majority of emissions (85 percent) are priced at less than US$10/tCO2e (Kossoy, et.al, 2016). In this study use assumptions the social cost of carbon (SCC) US$15 per metric ton of C and market discount rate (r = 7% per year). The value of carbon sequestration over time for a given parcel x is:

$$value\_seq_x = V\frac{sequest_x}{yr\_fut - yr\_cur}\sum_{t=0}^{yr\_fut-yr\_cur-1}\frac{1}{\left(1+\frac{r}{100}\right)^t\left(1+\frac{c}{100}\right)^t}$$

3. Results and Discussion

3.1. Landcover Change

Landcover change analysis was done for Donggala compared time series data from 2000 until 2011. **Table 2** show the changes of land cover Donggala Regency in from 1990 to 2022*.

Table 2. Recapitulation of Donggala Regency's Landcover from 2000, 2011and 2022*

No	Type of Landcover	Years 2000		Years 2011		Years 2022 *	
		Ha	Percent	Ha	Percent	Ha	Percent
1	Primary Dryland Forest	259,717.59	54.52	206,347.59	43.31	165,387.60	34.72
2	Secondary Dryland Forest	129,316.23	27.14	169,246.62	35.53	209,263.41	43.93
3	Primary Mangrove Forest	562.5	0.12	506.88	0.11	506.88	0.11
4	Secondary Mangrove Forest	300.15	0.06	347.31	0.07	347.31	0.07
	Sub Total Forest	389,896.47	81.84	76,448.40	79.02	75,505.20	78.82
5	Estate Forest	7,704.72	1.62	7,704.72	1.62	7,704.72	1.62
6	Dryland Agriculture	43,060.23	9.04	44,007.12	9.24	44,599.50	9.36
7	Mixed Dryland Agriculture	10,820.07	2.27	15,131.16	3.18	15,131.16	3.18
8	Shrub	11,324.88	2.38	18,744.57	3.93	19,095.39	4.01
9	Paddy Field	7,905.87	1.66	8,666.82	1.82	8,666.82	1.82
10	Fishpond	1,336.95	0.28	1,345.41	0.28	1,345.41	0.28
11	Transmigration	151.11	0.03	151.11	0.03	151.11	0.03
12	Settlement	3,721.50	0.78	3,721.95	0.78	3,721.95	0.78
13	Mining Area	218.52	0.05	218.52	0.05	218.52	0.05
14	Bare Ground	216.72	0.05	216.72	0.05	216.72	0.05
15	Open Swamp	47.79	0.01	47.79	0.01	47.79	0.01
	Sub Total Non Forest	86,508.36	18.16	99,955.89	20.98	100,899.09	21.18
	Total	476,404.83	100	476,404.83	100	476,404.83	100

Source : Result of Analysis of Landcover Map in year 2000, 2011 and 2022 (Projected)*

The forest cover area in Donggala in 2011 was 376,448.40 ha or approximately 79.02 % of the total area. It has been decreased by 13,448.07 ha (16.15%) compared to 2000. The forest degradation was 40.960 ha or 4.096 ha year$^{-1.}$

Primary dryland forest conditions in Donggala Regency in 2000 covered 259,717 ha and in 2011 reduced into 206,347.59 ha. A reduction was 53,370.00 ha or approximately 20.553 % over the 11 years. The average deforestation of primary dryland forest occurred in Donggala was 1.86 % per year or about -4,851.82 ha per year. The reduction was caused by deforestation which has changed primary dryland forest into a secondary dryland forest. The reduction of primary dryland forest caused an additional extensive secondary dry forestdirectly, because deforestation in Indonesia is the selective cutting of trees which had 50 cmand up diameter of the tress.

In 2000 the secondary dry forest area in Donggala was 129,616.38 ha and in 2011 decreased to 169,593.93 ha. The replenishment of secondary dryland forest area was 39,977.55 ha, or approximately 46.59 % over the 11 years. The increasing average of secondary dryland forest was 4.23 % per year, or about 3,634.32 ha per year. The replenishment of secondry dryland forest was caused by the degradation, which has changed the primary dryland forest into secondary dryland forest. Tabel 3 shows lancover changes (increasing and decreasing).

Table 3. Landcover changes of Donggala Regency peroid 2000-2011 and 2011-2022*

No	Type of Lancover	2000 -2011		2011- 2022*	
		Ha	Percent	Ha	Percent
1	Primary Dryland Forest	-53,370.00	-20.55	-40,959.99	-19.85
2	Secondary Dryland Forest	39,930.39	30.88	40,016.79	23.64
3	Primary Mangrove Forest	-55.62	-9.89	0.00	0.00
4	Secondary Mangrove Forest	47.16	15.71	0.00	0.00
	Total Forest	-13,448.07	16.15	-943.20	3.79
5	Estate Forest				
6	Dryland Agriculture	0.00	0.00	0.00	0.00
7	Mixed Dryland Agriculture	946.89	2.20	592.38	1.35
8	Shrub	4,311.09	39.84	0.00	0.00
9	Paddy Field	7,419.69	65.52	350.82	1.87
10	Fishpond	760.95	9.63	0.00	0.00
11	Transmigration	8.46	0.63	0.00	0.00
12	Settlement	0.00	0.00	0.00	0.00
13	Mining Area	0.45	0.01	0.00	0.00
14	Bare Ground	0.00	0.00	0.00	0.00
15	Open Swamp	0.00	0.00	0.00	0.00
	Total Non Forest	*13,447.53*	*15.54*	*943.20*	*0.94*

Over the 11 years (2000-2011) the increase of non-forest area was13,448,07 ha or 15.54 % compared to the condition of the forest cover in 2000. The rate of change was 1,222.50 ha per year or the reduction was 1.41 % per year compared to non-forest areas condition in 2000. The increase of non-forest areas was caused by the activity of forest land conversion into non-forest areas (other uses). The next figure 4 shows the trend of forest changes in Donggala regency.

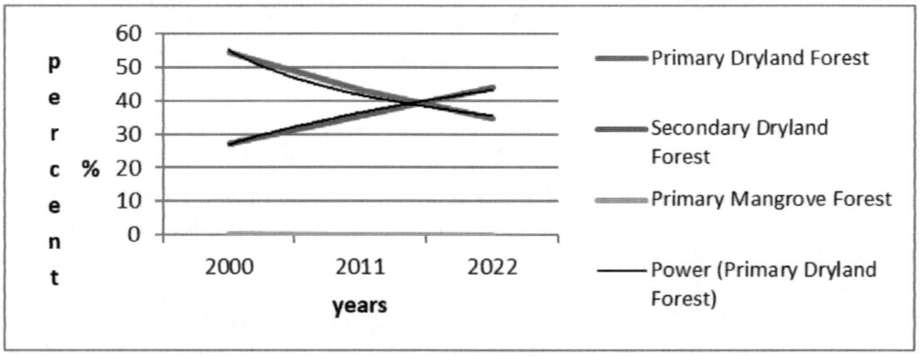

Figure 4. Farest Changes from 2000 to 2022

Referring to the above graph can be explained the primary dry forest decline is inversely proportional to the rise in the number of secondary dry forest that reaches 15% between 2012 and 2022. Donggala District is one of district in Central Sulawesi Province, wich contribution of 12.59 % total deforestation in Central Sulawesi Province in period 2000-2011 [5]. The forest cover in Donggala district in 2011 was 387,140 ha or approximately 81,26 % of the total area. It has been decreased by 14.790 ha (3.68 %) compared to 2000. The rate of dorestation in period 2000-2011 is 344.55 ha year^{-1}. Motivated by Sumargo et al. [4] the rate of deforestation in Donggala distric is lower than deforestation in central sulawesi 43.211,15 ha year^{-1} .

The changes of forest cover was caused by the deforestation, either planned or not. Planned deforestation is usually in the form of changes planned by the government for the benefit of forest land for plantations, agricultural or residential development, which is carried out lawfully in accordance with the legislation. Unplanned deforestation is a deforestation through illegal activities. The forest degradation can be caused by illegal or unauthorized activities, such as harvesting and illegal logging.

3.2. Carbon storage/ Sequestration model and valuation

The output from model simulation is raster maps, produced by carbon stroage and sequestration modul are 1) Map of current carbon storage, Map of future carbon storage, Carbon sequestration map, Map of economic value of carbon **(Figure 5)** and 2) a summary of storage and sequestration and net present values of sequestration (**Table 4**).

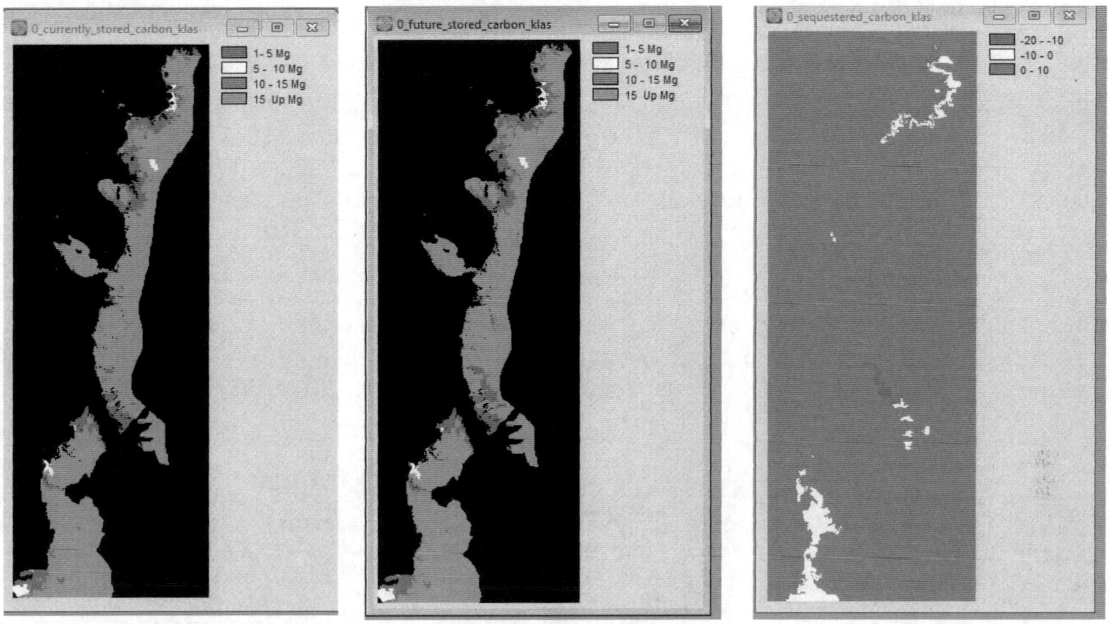

a. Current carbon storage b. future carbon storage c. Carbon sequestration

Figure 5. Map representing Carbon strorage and Carbon sequestration in period 2000-2011

Table 4. Carbon storage, Sequestration and Net prensent value in Donggala Regency

No	Parameter	Period 2000-2011	Period 2011-2022*
1	Total current carbon (Mg)	73,978,140	70,487,819
2	Total value of currently stored carbon (US $)	110,967,2117	10,57,317,306
3	Total scenario carbon (Mg)	70,487,819	69,285,088
4	Total sequestered carbon (Mg)	-3,490,321	-1,202,732
5	Total value of sequestered carbon (US $)	-38,188,465	-13,159,386

Total current carbon (year 2000) is 73.980 Mton. Meanwhile in the future, total carbon (year 2011) is 70.49 million ton. Hence, total sequestered is -3.49 million on.Negative values indicate carbon lost to the atmosphere. Areas with large negative e values will have the largest changes in land cover. In the study, Donggala has been decreased by 13,448.07 ha (16.15%) compared to 2000. The forest degradation was 40.960 ha or 4.096 ha year$^{-1.}$

By measuring the amount of C storage in the bodies of living plants (biomass) in the landscape, can calculate the amount of CO_2 in the atmosphere is absorbed by plants. Donggala area almost as large an area with carbon strotage content more than 15 up Mg per cell (166,67 Mg per ha).

The impact of land cover changes caused decreased carbon stocks, especially in the forest area. In 2000 the forest area is 389,896 ha, equivalent to the carbon storage of 72.83 million ton. In 2011, decline in forest area of by 13.448 ha, resulted in a decrease of carbon 3.66 million tons.

The reduction of CO_2 in the air by the plants was called seprocess (C ssequestration).This C sequestration process occurs for the survival of plants which need sunlight, carbon dioxide gas (CO_2) is absorbed from the air and water as well as nutrients absorbed from the soil. Through the photosynthesis process, the CO_2 in the air is absorbed by plants and converted into carbohydrates, afterwards they are distributed throughout the body of the plants and eventually are dumped throughout the plant body. Thus, measuring the amount of C stored in the body of living plants (biomass) in a field can describe the amount of CO_2 in the atmosphere absorbed by plants [9].

The map of the Net Present Value (NPV), is the economic value of total sequestration in Donggala Regency area. It could be observed from 2000 to 2011 (and 2011 to 2022 *) there was a value of US$ -187.48 to 25. 269 per metric ton per pixel in areas. Meanwhile in period 2011- 2022 * there was a value of US$ 0 to - 177,66.The output map are the changes in land cover and its effects in sequestration. It is possible to see that the raster has a large variety of classes and that more areas are close to becoming emmiters while others have already made the transition [10].

Base on Table 4, the Net Present Value (NPV) over time of total sequestration (between the current and the future) in Donggala Regency is US$ -38.19 million caused by the carbon emissions, base on prioce of carbon is US$ 15 and discount market is 7 %. Negative values indicate carbon lost to the atmosphere (emissions of CO_2). Costs required to restore the initial conditions, so that the forest come back to normal and can absorb CO_2.

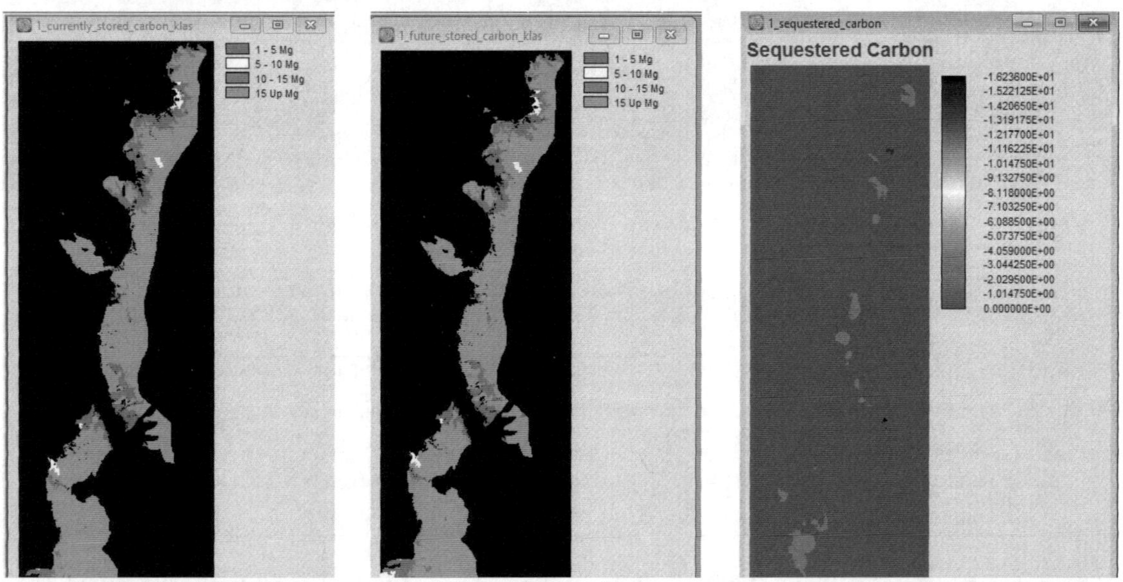

a. Current Carbon Storage b. Future Carbon Storage c. Carbon Sequestration

Figure 6. Map representing Carbon strorage and Carbon sequestration in period 2011-2022

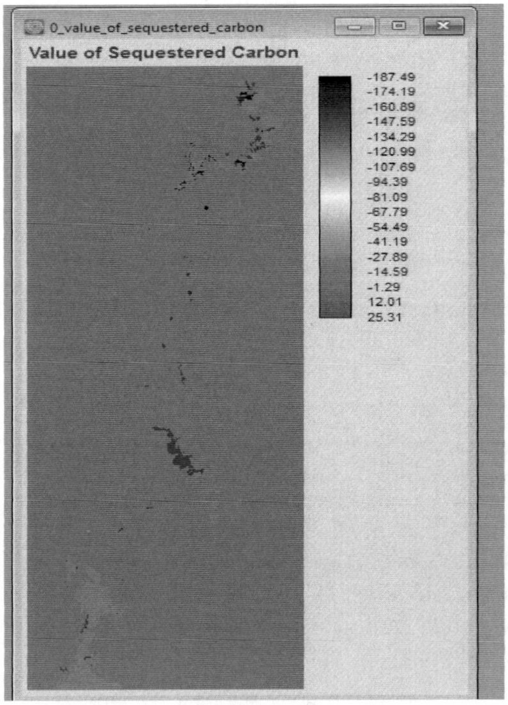

a) NPV (2000-2011)

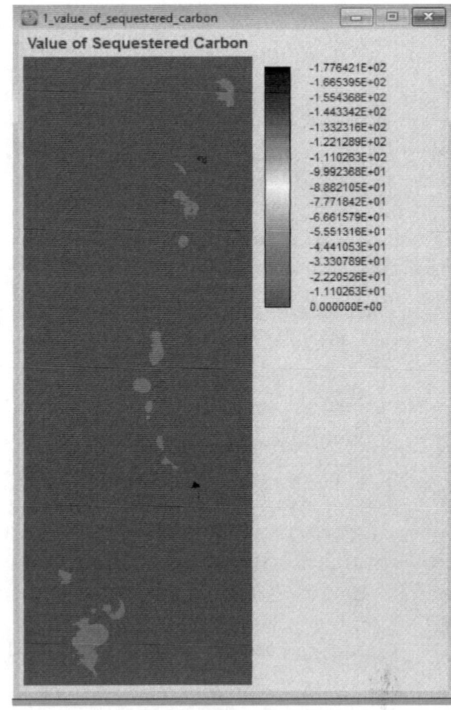

NPV (2011-2022*)

Figure 7. Map representing the carbon Net present value of Sequestration 2000-2011 and 2011-2022 (prediction)

4. Conclusion

The forest cover area in Donggala in 2011 was 376,448.40 ha or approximately 79.02 % of the total area. It has been decreased by 13,448.07 ha (16.15 %) compared to 2000. The forest degradation was 40.960 ha or 4.096 ha year^{-1}. Total of carbon storage (only carbon above) throughout the land cover in Donggala Regency in 2011 reached 3.49 million tons C. The carbon storage had been decreased by 317,301.91 ton C or the average was 317,301.91 tons C year^{-1} compared to the carbon storage in 2000.

The impact of land cover changes caused decreased carbon stocks, especially in the forest area. In 2000forest area is 389,896 ha, equivalent to the carbon storage of 72.83 million ton. In 2011, decline in forest area of by 13.448 ha, resulted in a decrease of carbon 3.66 million tons. The net present values of sequestration (between the current and the future) US$ -38.19 million caused by the carbon emissions, based on price of carbon is US$ 15 and discount market is 7 %.

5. Acknowledgments

We are grateful to Centers for Research, Promotion and Cooperation, Geospasial Information Agency (BIG) for the data and financial support.

6. References

[1] Adiriono T 2009 *Method of Measuring Carbon (Carbon Stock) on Industrial Plantation Forest Crassicarpa Acasia Type* (Thesis. Faculty of Forestry Programs Graduate. University of Gadjah Mada. Yogyakarta. not published)

[2] Gibson L, Lynam A J, Bradshaw C J A, He F, Bickford D P, Woodruff D S, Bumrungsri S and Laurance W F 2013 Near-complete extinction of native small mammal fauna 25 years after forest fragmentation *Science (80-.).* **341** 1508–1510

[3] UNFCCC 2014 *National Forest Reference Emission Le,el for Deforestation and Forest Degradation in the Context of the Acti,ities Referred to Decision 1/CP.16*
[4] Sumargo W, Nanggara S, Nainggolan F and Apriani I 2011 *Potret Keadaan Hutan Indonesia 2000-2009 (The state of Indonesia's Forests 2000-2009)*
[5] Turmudi and Nahib I 2015 *Potret Hutan Sulawesi Tengah Berdasarkan Data Geospasial* (IPB press. Bogor)
[6] Nahib I and Widjojo S 2016 The Impact of Landcover Changes on Carbon Stock: A Study Case In Central Kalimantan Forest
[7] Sharp R, Tallis H T, Ricketts T, Guerry A D, Wood S A, Chaplin-Kramer R, Nelson E, Ennaanay D, Wolny S, Olwero N, Vigerstol K, Pennington D, Mendoza G, Aukema J, Foster J, Forrest J, Cameron D, Arkema K, Lonsdorf E, Kennedy C, Verutes G, Kim C K, Guannel G, Papenfus M, Toft J, Marsik M, Bernhardt J, Griffin R, Glowinski K, Chaumont N, Perelman A, Lacayo M, Mandle L, Hamel P, Vogl A L, Rogers L, Bierbower W, Denu D and Douglass J 2016 *InVEST +VERSION+ User's Guide* (The Natural Capital Project, Stanford University, University of Minnesota, The Nature Conservancy, and World Wildlife Fund)
[8] Kumar R, Nandy S, Agarwal R and Kushwaha S P S 2014 Forest cover dynamics analysis and prediction modeling using logistic regression model *Ecol. Indic.* **45** 444–455
[9] Hairiah K, Sitompul S M, van Noordwijk M and Palm C 2001 *Methods for sampling carbon stocks above and below ground* (ICRAF Bogor, Indonesia)
[10] Klass A B and Wilson E J 2008 Climate change and carbon sequestration: Assessing a liability regime for long-term storage of carbon dioxide

IOP Conf. Series: Earth and Environmental Science **123** (2018) 012024 doi:10.1088/1755-1315/123/1/012024

Spatial Distribution of Socioeconomic Characteristics in Rural's Java: A Case from Three Different Rural Areas in Central Java

I Rudiarto [1], W Handayani [1], H B Wijaya [1], and T D Insani [1]

[1] Department of Urban and Regional Planning, Diponegoro University, Semarang, Indonesia

Email: iwan.rudiarto@undip.ac.id

Abstract. The intention of this paper is to combine the socio economic development into spatial aspect with the fact that the divided sub study areas are differed in its socio-economic characteristic. The research was conducted in three different rural areas, i.e.; coastal area in Sayung sub district – Demak Regency, plain area in Delanggu sub district – Klaten Regency, and mountain area in Kledung sub district – Temanggung regency. Spatial interpolation technique has been applied in order to identify the spatial distribution of socioeconomic data. The results show that socioeconomic characteristic in plain area and coastal area is more varied and regularly distributed as compared to the mountain area. Educated people are less found in the plain area while in coastal and mountain area the condition is better. Coastal area is identified as the prone area to the disaster issues and therefore socioeconomically vulnerable. The result of this research is very important to the development policies that need to undertake regarding to socioeconomic development in each associated location.

Keywords: Socioeconomic, Spatial Interpolation, Household, Rural Area, Central Java

1. Introduction

Socioeconomic characteristics in a rural area is very much related to the resource use owned by the farming families as well as the location where the family lives [1–4]. In the socioeconomic development, the location of the rural's family is indicated by the settlement distribution in a specific place. Generally, rural settlements are in high pressure of the urbanization and have been significantly transformed into a more urbanized area [5]. The existence of settlements in rural area is a central unit as it may reveal the relationships between land and the people, the historical background of a community, as well as the socio-political connections [6,7]. Therefore, the settlement distribution in rural area will also show the socioeconomic conditions of the people. On the other hand, resource use in the rural area indicates the relationship between economic and environmental aspect. Income gain by the farming family represents the economic aspect while land resource utilisation represents the environmental aspect. Farming families are the main player between those two aspects where the decision on how to utilise the resource very much depends on the family [8].

Rural area is mostly dominated by the agricultural land where income of rural people is majorly generated from this sector. Usually, the location of agricultural land owned by the farming family is located nearby the settlement area where the family lives. Even, most of the houses of farm families is located within their farm field which made the land maintenance is much easier. Therefore, the assessment of socioeconomic characteristics of rural area becomes more attractive since it may reveal the capability of rural people in maintaining their land resource which implicates to the family income. In addition, rural areas have a great diversification in terms of physical environment, ecology, and pattern of land use as well as their socioeconomic characteristics [9]. Due to their great diversification, each socioeconomic characteristics of rural area has it owns specification either in coastal, plain, or mountain area where spatial concern is the major aspect. Knowing the socioeconomic characteristics of each typical rural area would certainly useful in understanding the socioeconomic development and what type of characteristic is found more in one region as compared to others.

To understand the socioeconomic development, a set of socioeconomic characteristics data need to be elaborated, tabularly and spatially. The assessment of spatial distribution of socioeconomic characteristics in typical rural area by using Geographical Information System (GIS) and Remote Sensing (RS) has become widely concerned to describe more clear understanding on the specific issue [10–12]. The concept of spatial distribution is derived from the interaction of resource use and farming activities indicated by the location of each farming family. Spatial description on socioeconomic characteristic therefore is able to define the situation more clearly instead of a table format which the researchers normally did. Tables succeed in organizing information for easy comparison and for analysis in spreadsheets or statistical packages, but they do so by largely ignoring one particular property of each object: its geographic location [13]. By distributing and analyzing table data into its geographical location, a better understanding on how the data correlated can be spatially presented. This paper is amed to describe the socioeconomic characteristis of farming family by assesing the socioeconomic development of three different rural areas in Central Java.

2. Data and Methods

2.1 Study Area

Three different rural areas were selected as the study area, i.e.; rual coastal area, rural plain area, and rural mountain area. The reason of selecting those rural areas is to compare and to show how the socioeconomic characteristices distributed in each rural area. Those three rural areas are Sayung sub district in Demak Regency for coastal rural area, Delanggu sub district in Klaten Regency for plain rural area, and Kledung sub district in Temanggung Regency for mountain rural area. Concerning the elevation level, the study areas located on the different altitudes level following a gradient line from coastal to mountain area. Sayung sub district is located on 0 - 3 meters above sea level (asl); Delanggu from 100 up to 200 meters asl, and Kledung varied from 1138 m up to 1500 m (asl). The location of study area are shown in Figure 1.

2.2 Data Needs

To provide basic data on socio economic characteritics, randomly selected farm family was applied in study areas by handing out standardized questionnaires to 83 farm families (see Figure 2) distributed in study area. Those 83 farm families were distributed more less equal in three study areas; 23 samples in Sayung subdistrict, 30 samples in Delanggu subdistrict, dan 30 samples in Kledung subdistrict. Specifically in Sayung subdistrict, less samples number were applied due to less farm families found in that area. Selection of farm family surveyed is based on their geographical location where the even distribution is preferable. Socioeconomic data that are being captured were related to farm income, off farm income, family income, education level, and farm yield in 2017.

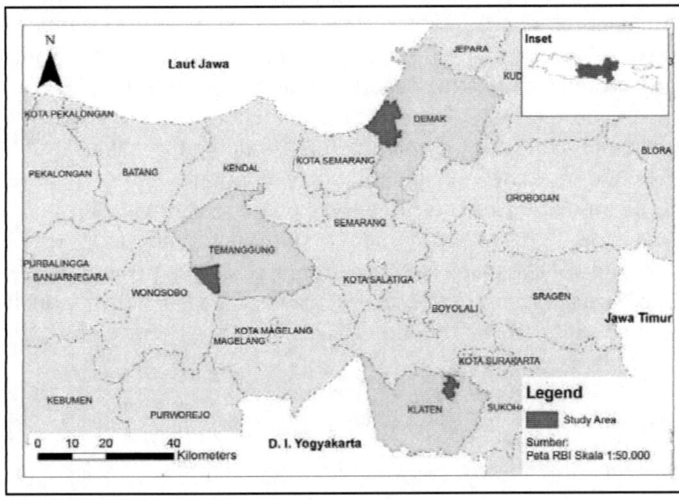

Figure 1. Study Area

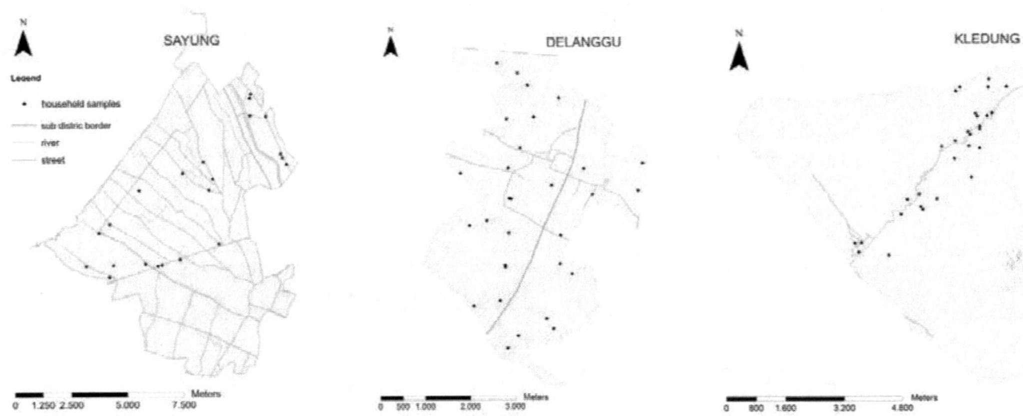

Figure 2. Samples Distribution in Study Areas

2.3 Method

A spatial calculations were carried out with interpolation technique to generate data for the whole study areas. Prior to the interpolation technique, each location of surveyed families was recorded using the Global Position System (GPS) to locate them into their geographical location. This step is very important to be performed as the location of each sampling family will be the link between socio economic data and its spatial aspect. By linking all related data and information to its spatial location, spatial analysis in GIS can be carried out.

IOP Conf. Series: Earth and Environmental Science **123** (2018) 012024 doi:10.1088/1755-1315/123/1/012024

The interpolation technicque was done from point data collected throuh the micro level survey. Spatial interpolation is the procedure of estimating the value of a field variable at un-sampled sites within the area covered by sample locations or in simple words, given a number of whose locations and values are known. Kriging method was applied as the interpolation technique to generate socioeconomic data for each stuy area. Krigging method involves several steps in its process and more appropriate used if the user has already known the spatially correlated distance or directional bias in the data [4].

Interpolating socioeconomic data from sampled points is intented to identify all the unknown value of unsampled area within the study area where specific socioeconomic data can be generated to the whole study area. However, the interpolation results may over or underestimate the conditions at the edge of surface. The interpolation, particularly at the border of study area and at the less sample point area, is often continued with unrealistic values. Once the last sampling point is passed, the derived trend continues with the same gradient as before the sampling point and makes the values rise or decline inappropriately in some cases. To avoid these possible errors, the interpolated grid layers were classified, thus only the class bandwidth is readable from the maps [3].

3. Results and Discussion

3.1 Farm İncome Distribution

Farm income shows the economic ability of a farm in a year to provide an economic surplus to the farming family. Farm income in Sayung sub district is dominated by fish farming activity that is commonly found in coastal rural area. Delanggu sub district as a plain rural area is one of the main rice producer in Central Java and it dominates its household farming activity. On the other hand, Kledung sub district as a mountain rural area shows a dominant farming activity for tobacco and vegetables. Table 1 below shows the amount of average farm income from each study area. The highest average farm income comes from Kledung sub district with Rp 164.213.000,- for each household, followed by Sayung sub district with Rp 36.388.000,- and the lowest is Delanggu sub district with Rp 11.404.000,-. These outcomes can give the raw insight on the value of the product from each area type.

Table 1. Average Farm Income

Sub District	Area Type	Avg. Farm Income (in 1000 rupiah)
Sayung	Coastal Rural	36.388
Delanggu	Plain Rural	11.404
Kledung	Mountain Rural	164.213

Spatial calculation was applied to describe further differentiation of farm income between area types. Through interpolation technique that has been explained before based on information from household interview, Figure 3 shows the distribution of farm income of each study type. The darker colour portray the higher income. Kledung sub district is dominated by the darker green while Delanggu sub district has its darker green mainly in the central area. Although Sayung sub district has higher average farm income than Delanggu, the distribution of its farm income is generally on the lower side as shown in the lighter green colour that dominate it. However, the spatial differentiation in study areas is only to show the distribution of farm income among the interviewed household but not to take into account as the representative for the whole area. The more samples will influence the spatial distribution and accordingly change the spatial differentiation as well.

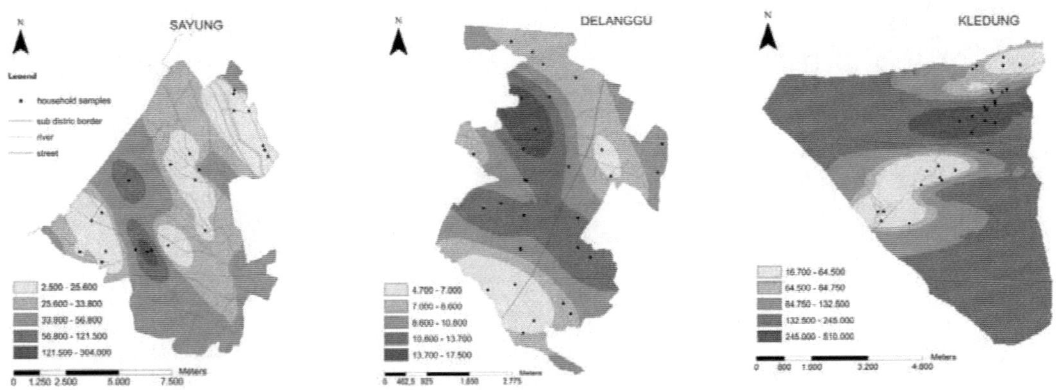

Figure 3. Farm Income Distribution (in 1000 Rupiahs) in Study Areas

3.2 Off-Farm İncome Distribution

Off-farm income derived from off-farm activities per year that take place outside the family's own responsibility of farm business. As farming becomes the main activity for almost all the household in study area, the existence off-farm income for some households is not as vital as farm income in supporting family livelihood. The involvement of family members in off-farm activities can be classified into two categories in general: permanent and seasonal jobs. Permanent job of the interviewed households are government official, private employer, and trader. Seasonal job is related to all types of jobs that family members irregularly involved such as field worker, building worker, and driver.

Table 2. Average Off-Farm Income

Sub District	Area Type	Avg. Off-Farm Income (in 1000 rupiah)	Household with Off-Farm Activities
Sayung	Coastal Rural	17.346	7
Delanggu	Plain Rural	17.928	13
Kledung	Mountain Rural	7.613	15

In the study area, there are 42% households from total interviewed households who are engaged in off-farm activities. Kledung sub district has a highest total of 15 households or 50% of composition followed by Delanggu sub district with 13 households or 43% and the last is Sayung sub district with 7 households or equal to around 30%. Eventhough there is not always a household member that engage in off-farm activities, for the households who do, a member who engage in off-farm activity can has more than one off-farm jobs.

As shown in Table 2, the highest off-farm income is Delanggu sub district with the average of Rp 17.928.000,- followed closely by Sayung sub district with Rp 17.346.000,- and Kledung sub district with Rp 7.613.000,-. Spatial distribution of off-farm income (Figure 4) shows clearly the spatial differentiation of off-farm income from high level to the low level income. It can be inferred as well that the amount level of family who gained off-farm income had been increased from coastal rural area which is Sayung

IOP Conf. Series: Earth and Environmental Science **123** (2018) 012024 doi:10.1088/1755-1315/123/1/012024

sub district to mountain rural area type which is Kledung sub district. More households are engaged in off-farm activities the higher the area.

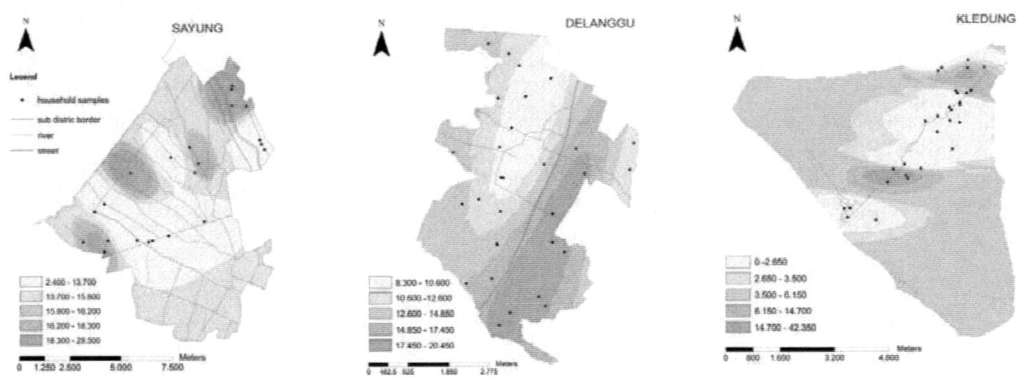

Figure 4. Off-Farm Income Distribution (in 1000 rupiahs) in Study Areas

3.3 Family İncome Distribution

Family income is the main criteria in economic ability of family to support family needs and expenses. Family income is generated in one year by integrating farm and off-farm incomes. Usually, family income of the rural area households are largely contributed by farm income, but among the three study areas, Delanggu sub district shows different condition where off-farm income makes up 61% of the average family income while farm income only contributes 39%. Although in average the contribution of off-farm income is higher to the family income in this sub district, there might be some households that their farm incomes make up their family income more. For Sayung sub district, average family income is made up by farm income at 68% while for Kledung, the average family income dominantly comes from farm income at 96%.

Table 3. Average Family Income

Sub District	Area Type	Avg. Family Income (in 1000 rupiah)	Compotition	
			Farm income	Off-farm income
Sayung	Coastal Rural	53.735	68%	32%
Delanggu	Plain Rural	29.332	39%	61%
Kledung	Mountain Rural	171.826	96%	4%

Table 3 shows that the average family income in all three study areas is alined with the amount of average farm income. Kledung sub district has the highest average family income with Rp 171.826.000,- followed by Sayung, that barely has a third of Kledung's, with Rp 53.735.000,- and Delanggu sub district with Rp 29.332.000,-. Spatial analysis using interpolation is also used to see the distribution of family income in three study area (Figure 5). The darker orange colour covers most of high family income level in study area. Lighter orange colour shows low family income level that is distributed more in Sayung sub district and some part of Delanggu subdistrict. The interval of family incomes in each study area can also indicate the income gap of the households. From the picture we can see that although Delanggu sub district has the lowest family income, the income gap is nogt as severe as Sayung sub district and Kledung sub district.

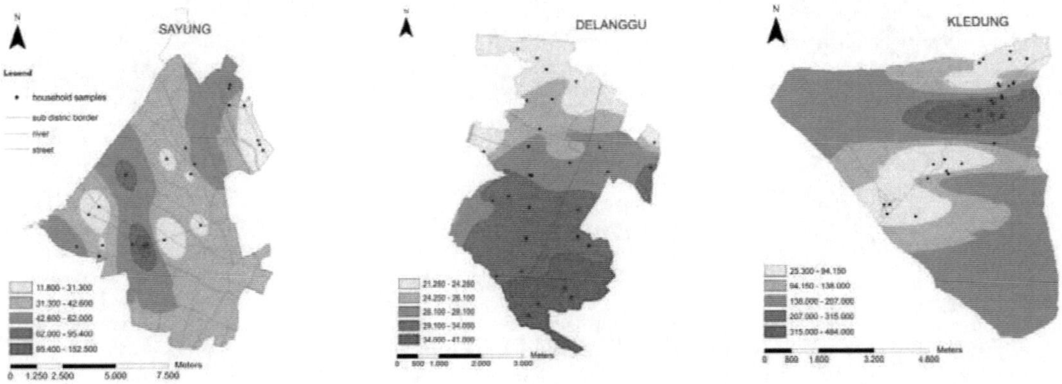

Figure 5. Family Income Distribution (in 1000 rupiahs) in Study Areas

3.4 Education Level

Educational background of family members determines family's perception in society and how they make decission surrounding their lives. The higher education level the wider knowledge and information can be gained which is very useful in managing and developing the ability of various family resources. In general, analysis from interview data shows that level of education in study area is noticeably medium where only a little number of people found illiterate as shown in table 4.

Table 4. Education Level

Sub District	Area Type	Education Level			Weighted Education Index
		Illiterate	Elementary - High School	Higher Education	
Sayung	Coastal Rural	2	60	16	0,82
Delanggu	Plain Rural	1	83	13	0,77
Kledung	Mountain Rural	2	102	5	0,82

Weighted education index shows that there are more higher-educated people in Sayung sub district (coastal rural area) and Kledung sub district (mountain rural area) with average index of 0.82 compared to Delanggu sub district which has average index of 0.77. However, from the number of total people from each education level, Sayung sub district and Delanggu sub district have more people with higher education level than Kledung sub district. It can explains that the level of high education following altitudinal gradient and increasingly from higher altitude area to lower altitude area.

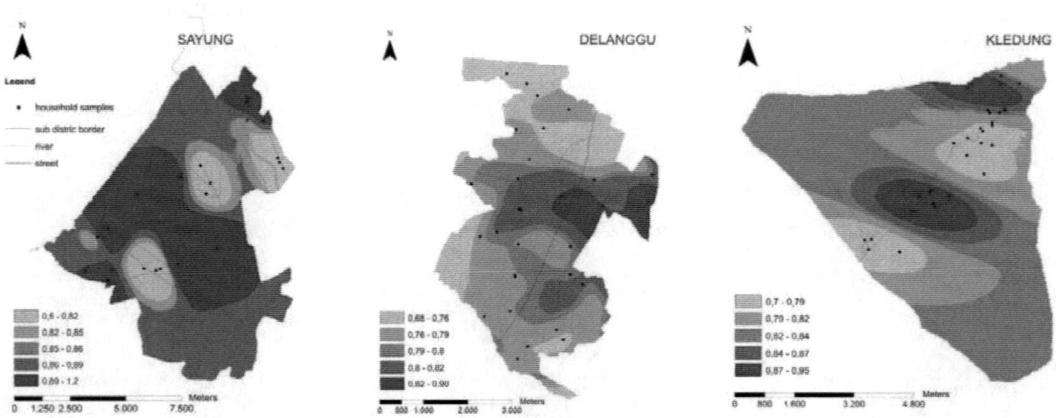

Figure 6. Weighted Education Index

Spatial distribution of weighted education index was built by using interpolation technique that allows measurement of surrounded data where there is no data available. As shown in Figure 6, the distribution of educated people is expressed by violet colour where the darker the colour indicate a higher index. Sayung sub district overall shows indication that higher-eductaed people are spread mostly in its central area and they have higher index than the other two study area. Delanggu sub district shows more lighter colour which mean that it has less higher-educated people. Lastly, Kledung sub district shows that it is placed in the middle level with more area covers by moderate amount of darker colour yet less with lighter colour. This empirical result may arise due to the possibilities of different level of educational facilities and infrastructures that exist.

3.5 Farm Yield

Farm yield shows the amount of a commodity that was harvested per unit of land which in this reasearch is measured in kilograms per hectare (kg/ha). It shows the productivity of farm activity. The size of land is not the main factor that can make a higher level of productivity. The opposite condition of the corelation between land size and land productivity is shown in Table 5. The area with biggest number of leverage and size has the lowest number of farm production.

Table 5. Average Farm Yield

Sub District	Area Type	Avg. Land Size (ha)	Avg. Farm Yield (kg/ha)
Sayung	Coastal Rural	2,21	3.233
Delanggu	Plain Rural	0,54	11.214
Kledung	Mountain Rural	1,15	9.832

Sayung sub district has the highest average land size of 2,21 ha but it can only produce fish product as much as 3.233 kg/ha. Kledung sub district has average land size of 1,15 ha with farm yield of 9.832 kg/ha while Delanggu sub district with only average land size of 0,54 ha can produce 11.214 kg/ha of rice. This opposite condition between the land size and farm yield can also be caused by the difference land size it is needed to produce each kind of commodity. The spatial distribution of farm yield is as shown in Figure 6 where darkest red colour dominates central area of Delanggu sub district while the same dark colour only appears in the smallest part of both Sayung sub district and Kledung sub district. But at the same time, Delanggu sub disrict also has the most area covered in lighter red colour than the other study areas which indicate that a lot of its area has a tendency of producing smaller amount of product.

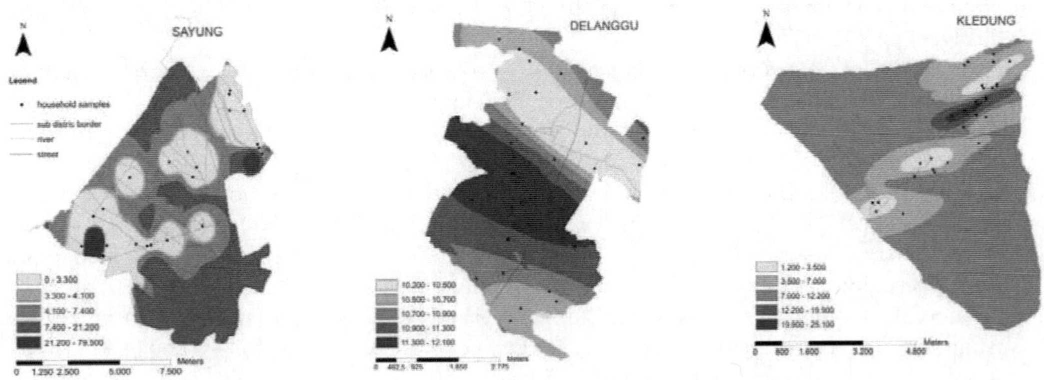

Figure 7. Farm Yield Distribution (kg/ha) in Study Area

4. Conclusions

Spatial assessment of socioeconomic characteristics in three study areas had shown how the socioeconomic development were distributed and what kind of characteristics found more dominant as well as which rural area has a better condition as compared to other rural area. In general, the spatial pattern of socioeconomic distribution was found more regular with more less equal distributed in coastal area - Sayung sub district and in plain area - Delanggu sub district as well. While in mountain area - Kledung sub district the spatial pattern was quite irregular where most of the socioeconomic characteristics distribution was found centred. This condition occurred due to the location of the household samples mostly located along the main road.

In terms of socioeconomic characteristics, mountain area – Kledung sub district was found more advance in family income where most of the family gain the income from the tobacco farming. It is also shown that the closer the area to the urban area, the higher the family income as indicated in plain area – Delangu sub district. As the expanded urban area of Solo city, Delanggu sub district has more opportunity to have more income from different sources. Surprisingly, the education level of plain area which is close to the urban area of Solo was not linearly related to the education condition where more educated people is usually found in urban area as the access to education facilities is more possible. On the other hand, plain area – Delanggu sub district has more productivity, indicated by the farm yield per hectare as compared to mountain and coastal area. It looks like the farmer in plain area has more capability in optimizing or even maximizing the input concerning the average land size.

5. Acknowledgement

This paper is part of the research on "Spatial Planning Modelling for Sustainable Rural based on Resources and Biodiversity" funded by Diponegoro University, Indonesia.

6. References

[1] Bhatta G D 2010 *Socio-economic and spatial assessment of smallholder peri-urban farming in the middle mountains of Nepal* (Margraf Publishers)

[2] Doppler W and Bahadur K C K 2013 Impact of Natural Resources and Infrastructure on Future Livelihood in Mountain Farming in the Himalaya Region *The Future of Mountain Agriculture* (Springer) pp 145–170

[3] Lentes P 2003 *The Contribution of GIS and remote sensing to farming systems research on micro- and regional scale in North west Vietnam* vol 52(Margraf Verlag)

[4] Rudiarto I 2010 *Spatial Assessment of Rural Resources and Livelihood Development in Mountain Area of Java: A Case from Central Java-Indonesia* (Margraf)

[5] Dumreicher H 2008 Chinese villages and their sustainable future: the European Union-China-*Research* Project "SUCCESS" *J. Environ. Manage.* **87** 204–215

[6] Jones E E 2010 An analysis of factors influencing sixteenth and seventeenth century Haudenosaunee (Iroquois) settlement locations *J. Anthropol. Archaeol.* **29** 1–14

[7] Palmisano G O, Govindan K, Loisi R V, Dal Sasso P and Roma R 2016 Greenways for rural sustainable development: An integration between geographic information systems and group analytic hierarchy process *Land use policy* **50** 429–440

[8] Doppler W, Praneetvatakul S, Munkung N, Sattarasart A, Kitchaicharoen J, Thongthap C, Lentes P, Tai D, Grueninger M and Weber K 2006 *Resources and livelihood in mountain areas of South East Asia: Farming and rural systems in a changing environment* (Margraf)

[9] Huby M, Cinderby S, Crowe A M, Gillings S, McClean C J, Moran D, Owen A and White P C L 2006 The association of natural, social and economic factors with bird species richness in rural England *J. Agric. Econ.* **57** 295–312

[10] Liu Y S, Yang R and Li Y H 2013 Potential of land consolidation of hollowed villages under different urbanization scenarios in China *J. Geogr. Sci.* **23** 503–512

[11] Walsh S J, Evans T P, Welsh W F, Entwisle B and Rindfuss R R 1999 Scale-dependent relationships between population and environment in northeastern Thailand *Photogramm. Eng. Remote Sensing* **65** 97

[12] Fox J 2003 *People and the environment: Approaches for linking household and community surveys to remote sensing and GIS* vol 1(Springer Science & Business Media)

[13] Goodchild M F and Janelle D G 2004 Thinking spatially in the social sciences *Spat. Integr. Soc. Sci.* 3–22

Erosion Prediction Analysis and Landuse Planning in Gunggung Watershed, Bali, Indonesia

N M Trigunasih[1], T Kusmawati[1], and N W Yuli Lestari[1]

[1]Center for Spatial Data Infrastructure Development (PPIDS), Udayana University, Bali, Indonesia

Email: tri5963@yahoo.com

Abstract. The purpose of this research is to predict the erosion that occurs in Gunggung watershed and sustainable landuse management plan. This research used the USLE (Universal Soil Loss Equation) methodology. The method used observation / field survey and soil analysis at the Soil Laboratory of Faculty of Agriculture, Udayana University. This research is divided into 5 stages, (1) land unit determination, (2) Field observation and soil sampling, (3) Laboratory analysis and data collection, (4) Prediction of erosion using USLE (Universal Soil Loss Equation) method, (5) The permissible erosion determination (EDP) then (6) determines the level of erosion hazard based on the depth of the soil, as well as the soil conservation plan if the erosion is greater than the allowed erosion, and (7) determining landuse management plan for sustainable agriculture.
Erosion which value is smaller than soil loss tolerance can be exploited in a sustainable manner, while erosion exceeds allowable erosion will be conservation measures. Conservation action is the improvement of vegetation and land management. Land management like improvements the terrace, addition of organic matter, increase plant density, planting ground cover and planting layered header system will increase the land capability classes. Land use recommended after management is mixed plantation high density with forest plants, mix plantation high density with patio bench construction, seasonal cultivation and perennial crops, cultivation of perennial crops and cultivation of seasonal crops.

Key words: erosion prediction; land use planning; watershed

1. Introduction

Watershed according to Law No. 7 Year 2004 on the Watershed is a land area that is a unity of ecosystem with river and its tributaries, which functions to accommodate, store, and drain water from rainfall to the lake or naturally occurring sea, whose boundaries on land constitute topographical separators and boundaries at sea up to waters areas that are still affected by land activities. Watersheds consist of upstream, middle and downstream. Upper river basin ecosystems are an important part because they have a protection function for all parts of the basin. This protection in terms of water system functions, upstream and downstream areas have biophysical linkages through the hydrological cycle [1]. Watershed has specific characteristics and is closely related to its physical conditions such as soil type, land use and topography. Land use change in the upstream will also affect the downstream area in the form of decreasing the reservoir storage capacity or silting of river and irrigation channels which in turn may increase the potential of flooding.

Optimal and well-managed watershed conditions will support all activities and businesses in the region. Conversely, poor management will cause damage to the watershed. The rapid deterioration of watersheds affected by the land use change and poor land use planning will increase the danger of erosion that occurs in the watershed. The occurrence of floods, landslides, droughts and high erosion is a sign of damage to the watershed. Erosion causes the loss of fertile soil layers for plant growth and reduced soil's ability to absorb and retain water. Land carried by erosion is deposited in some places such as reservoirs, lakes or irrigation channels. This sediment will cause silting in river bodies, reservoirs or irrigation canals. This condition will increase overland flow and flood hazard.

Damage to the watershed includes damage to the biophysical or water quality aspects. Watershed damage will occur faster if the community is not aware of the importance of maintaining and preserving the surrounding environment. Besides the role of the very large community, the government's role in the formation of regulations that pay attention to the preservation of natural environment will also affect the rapid slowdown of the watershed that occurred. The rapid degradation of the watershed is influenced by the number of buildings that do not pay attention to river borders, cutting trees without a selective cropping system, planting seasonal crops on sloped lands above 45% of these measures will increase the erosion hazard occurring in the watershed. The occurrence of floods, landslides, droughts and high erosion is a sign of damage to the watershed. Erosion is the event of moving or transporting soil or parts of the land from one place to another by natural media [1]. Factors that affect erosion are climate, topography, plants, soil and humans. Factors that can be controlled by humans are plants while climate and topography are directly not controlled by humans and for land can be controlled indirectly by certain processing [2].

Watershed damage caused by erosion can be minimized by taking conservation action. Soil conservation measures are aimed at preventing erosion, repairing damaged soil, and maintaining and improving soil productivity so that land can be used sustainably. Water conservation in principle is the utilization of rainwater that falls to the ground for agriculture to be used as best as possible, and regulate the flow time in order to avoid devastating floods and there is enough water in the dry season. The way that can be done is to give mulch to the cultivated plants whose area has high rainfall and make a rainwater reservoir in areas that have low rainfall. Every treatment given on a plot of land will affect the water system in that place and places - Downstream. So that all soil conservation measures are also water conservation measures [1].

Conditions of watersheds in Bali have decreased in quality, this is happening in some areas in Bali which experiencing floods, droughts, landslides and land degradation. Land degradation is caused by the erosion. Land will gradually reach critical level until very critical. The Gunggung watershed is an upstream of the Jinah Watershed and is included in the Oos Jinah river basin, which has a function and an important role in supporting the water needs of the Bangli and Karangasem Regency. Gunggung watershed which covers 727,278 ha passes 3 villages namely Menanga, Pempatan and Abang Batu Dinding village which located in Karangasem and Bangli Regency. Gunggung watershed has an area of 12.98% of the area of the Jinah basin. The area of the Gunggung watershed which is estimated to have a large erosion is the area that has a slope of 30-45% and greater than 45%.

Erosion research in Gunggung Watershed is focused on how to estimate the erosion rate and to cope with erosion by using best practice of landuse planning. Therefore this paper aims to predict erosion hazard at the Gunggung Watershed and to develop sustainable landuse planning to give contribution in soil conservation practice related to erosion hazard in Bali and Indonesia

2. Methodology

Erosion is the result of the interaction between function of climate, topography, vegetation and human to soil which expressed in the following equation: $E = f(i, r, v, t, m)$, while E: The magnitude of erosion; i: Climate; r: topography; v: plants; t: ground; m: man. These factors can be distinguished into two factors, factors that can be controlled by humans and factors that cannot be controlled by humans. Factors that can be controlled by humans are vegetation; while climate and topography are directly not controlled by humans. Meanwhile land can be controlled indirectly by certain processes [2].

Erosion measurements and forecasting are difficult to do, because the processes that affect erosion are complex. Based on several assumptions and simplifications, the measurement and forecasting of erosion can be done at a reasonable level of confidence [3]. The most widely adopted equation of erosion values in Indonesia is Universal Soil Loss Equation (USLE) proposed by Wischmeier and Smith [1].

The USLE method is a fairly practical method because the required parameters are easy to obtain and free to change depending on space and time. This method combines the main factors causing erosion and its quantitative relationship to predict the soil loss from sheet and flow erosion caused by rain and surface flow in a particular area [4]. The USLE method equation is:

$$\mathbf{E = R*K*LS*CP}$$

With A = Erosion or loss of land per unit of land area; R = rainfall Erosivity;
K = Erodibility of soil; L = Length of slope; S = slope;
C = Cropping/ farming method; P = Method of land management

The soil loss tolerance needs to be calculated as it is impossible to suppress erosion rates to zero on land cultivated for agriculture, especially steep soil. In principle, the soil loss tolerance is the rate of erosion that occurs does not exceed the rate of soil formation so there is still a top layer as a place to grow plants. the factors considered in determining the value of EDP is the depth of soil, physical characteristics and other soil properties that affect root development, prevention of trench erosion, depletion of organic matter, nutrient losses, and problems caused by sediment in the field [1].

If the erosion exceeds the rate of soil formation then conservation is needed. Soil conservation is the placement of each plot of land on the appropriate use of the soil's capability and treats it in accordance with the conditions required to avoid soil damage. Soil conservation methods are measures or treatments or facilities that can be used to prevent soil damage or to repair damaged lands. Soil conservation methods are basically divided into three groups, namely (1) vegetative methods, (2) mechanical methods, and (3) chemical methods. The vegetative method is the use of plants or residuals. Included in the vegetative method are tree planting, grass planting, crop rotation, mixed cropping, and use of plant or crop residues such as mulch and green manure. The mechanical method is the development of erosion prevention and manipulation of soil and earth surface structures. Including in the mechanical method is the processing of soil according to contour, planting and processing according to contour, planting in strips, making of bunds, terrace, channel switch, drainage channel, check dam and gully plug. Chemical methods in soil and water conservation are the use of synthetic or natural chemical preparations. Certain organic compounds can improve the aggregate stability of water effectively, but the materials used are still too expensive to be widely used [1].

2.1. Research Location

Gunggung watershed is located between UTM zone of 50 South and coordinates of 3024000 mE; 9079000 mS until 302600 mE; 9079000 mS and 3024000 mE; 9072000 mS until 3026000 mE; 9072000 mS. Gunggung watershed past 3 villages namely Abang Batu Dinding Village, Pempatan Village and Menanga Village in Banglidan Karangasem Regency, Province of Bali, Indonesia with total area of Gunggung Watershed is 727,278 ha.

2.2. Methodology

The method used in this study was field survey/ observation and data collection using field observation, then the result is analyzed at the laboratory of Soil Science and Environment, Faculty of Agriculture of Udayana University. The research stage is divide into7 stages: , (1) land unit determination, (2) Field observation and soil sampling, (3) Laboratory analysis and data collection, (4) Prediction of erosion using USLE (Universal Soil Loss Equation) method, (5) The tolerable erosion determination (EDP) then (6) determines the level of erosion hazard based on the depth of the soil, as well as the soil conservation plan if the erosion is greater than the allowed erosion, and (7) determining landuse management plan for sustainable agriculture. The flow diagram is shown in Figure 1.

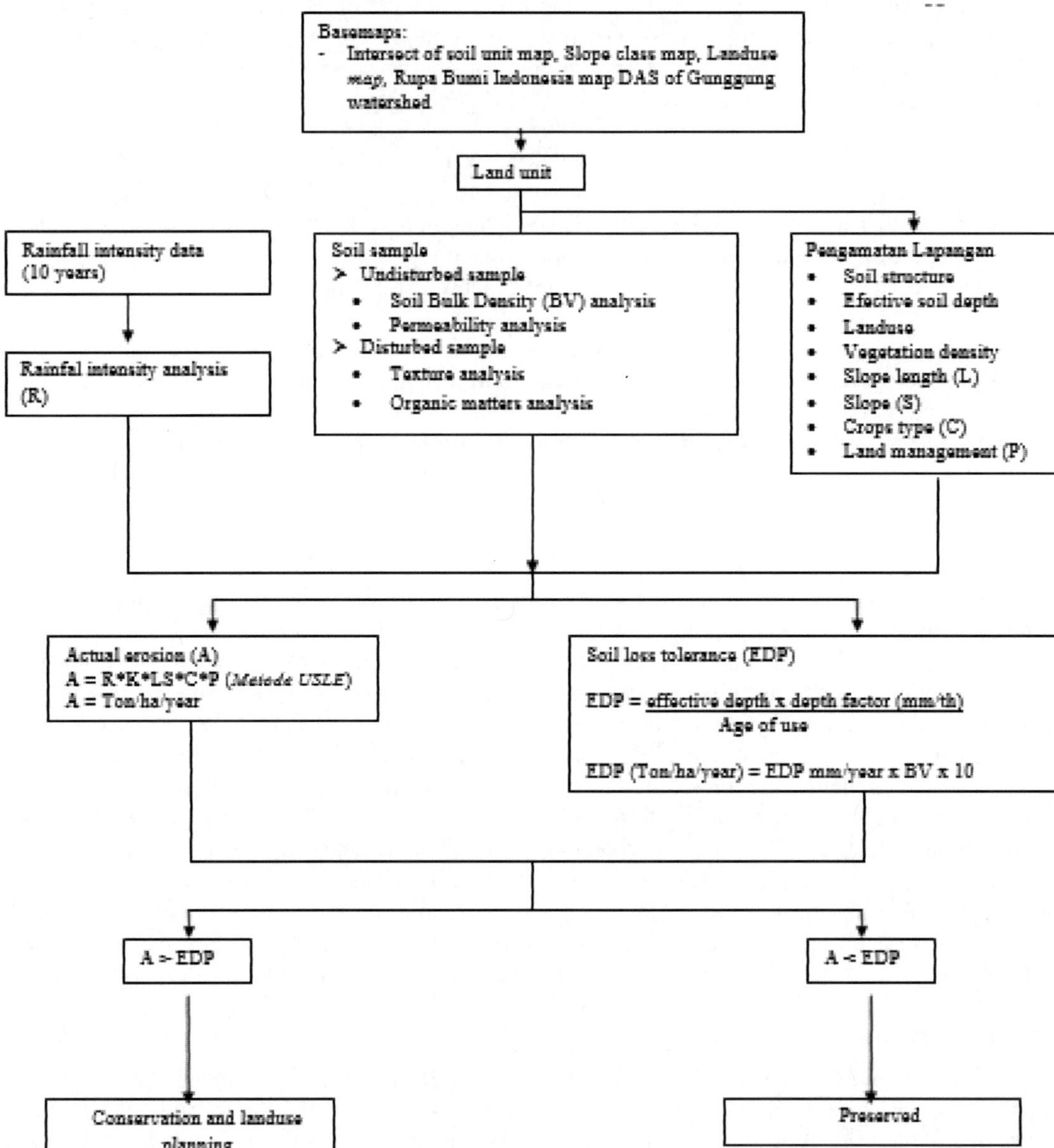

Figure 1. Research Diagram

The main aspect to be concerned is the tolerable erosion (EDP). If Actual erosion is more than EDP then conservation and landuse planning must adopted. If actual erosion is less than EDP then the recent soil condition and landuse practice can be preserved. In order to maintain the productivity of the soil, erosion that occurs should be lower than the tolerable soil loss [5]. The erosion process also brings about a loss of soil quality, and one of the ways of minimizing and even correcting the consequences of erosion is the adoption of conservationist practices [6].

2.2.1. Land units

The determination of land units is based on overlays on the soil type, slope classes and land use using QGIS software. From the overlay results obtained 17 units of land. Land unit that has the same class of slope, soil type and land use is grouped into 1 land unit. Each soil sample will represent one land unit. Full description presented in Table 1. Figure 2 shows the land unit map of Gungung Watershed

Table 1. Land Unit of Gunggung Watershed

Land Unit	Soil Type	Slope	Landuse
1	Regosol	>45%	Plantations
2	Regosol	30 – 45%	Plantations
3	Regosol	15 – 30%	Plantations
4	Regosol	8 – 15%	Moor
5	Regosol	0 – 8%	Moor
6	Regosol	30 – 45%	Plantations
7	Regosol	30 – 45%	Moor
8	Regosol	15 – 30%	Plantations
9	Regosol	15 – 30%	Moor
10	Regosol	8 – 15%	Plantations
11	Regosol	8 – 15%	Moor
12	Regosol	0 – 8%	Plantations
13	Regosol	>45%	Bush / scrub
14	Regosol	30 – 45%	Bush / scrub
15	Regosol	15 – 30%	Bush / scrub
16	Regosol	8 – 15 %	Bush / scrub
17	Regosol	0 – 8%	Bush / scrub

2.2.2. Field Observation and Sampling

The parameters observed in the field are

a. Length of Slope (L): measured from the tip of the slope to the change of slope shape.
b. Slope (S): measured using abney level
c. Soil Structure: observed by type of soil structure.
d. Effective Depth: searchable with a soil drill.
e. Dominant vegetation and vegetation density (C): observed visually
f. Land Management (P): observed field management of the field.
g. Ground sanding for texture analysis, permeability and soil organic matter.

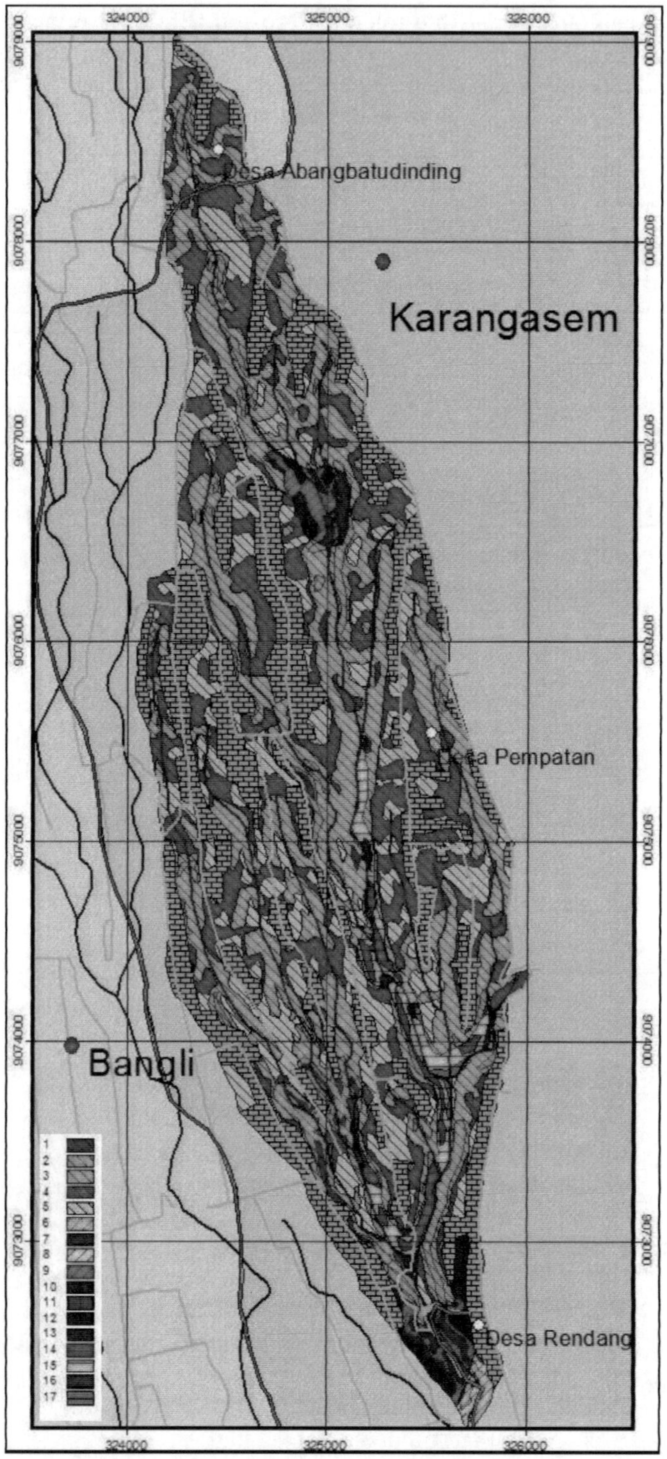

Figure 2. Land Unit Map

2.2.3. Soil Analysis

Soil analysis conducted in Soil and Environment laboratory of Faculty of Agriculture of Udayana University is land permeability, soil volume weight, soil texture analysis, and organic matter. Detailed soil analysis methods are presented in Table 2.

Table 2. Detailed Soil Analysis

No	Parameter	Methods	Unit	Equation
1	Permeability	De Booth method	cm/hour	$K = \frac{Q}{t} X \frac{L}{h} X \frac{1}{A}$ K = permeability Q = The amount of water flowing at each measurement t = Observation time (hour) L = soil thickness (cm) h = Surface water level from surface of soil sample (cm)
2	Soil texture	Pipet method	%	-
3	Organic matter	Walkey and Black method	%	$x = \frac{(B - A)N\ FeSO4\ x\ 3}{\frac{100}{100 - KL} XBerat\ tanah\ (g)} x\ 10\ x \frac{100}{77} x\ 100\%$ B = Blangko A = Sample KL = Soil moisture
4	Bulk volume	Ring Sampel method	gr/cm^3	Bulk volume(pb)= $\frac{Z}{A}$ Z = dry soil wight(gr) A = Volume (cm^3)

2.2.4. Erosion Prediction

The effects of processes is modified by biophysical environmental comprising soil, climate, terrain, ground cover and interactions between them. Susceptibility of soil to agents of erosion, soil erodibility, is determined by inherent soil properties, e.g, texture, structure, soil organic matter contents, clay minerals, exchangeable cations, water retentions and transmissions properties. Soil erodibility is a dynamic property and influenced by management [7]. The prediction of erosion (E) on this research will use the USLE (Universal Soil Loss Equation) as follows:

$$E = R*K*LS*CP$$

1) Rainfall factor (erosivity) (R)

As the primary means of water input to landscapes, rainfall can exert a significant influence on erosion. Precipitation is characterized by amount, duration, intensity (i.e., amount/ duration) and sequence (i.e., the order and timing of rainfalls). Rainfall intensity is usually the most important of the four factors affecting erosion. As intensity increases so does the kinetic energy of the raindrops, increasing detachment and transport of soil. The intensity and size of raindrops are important because the force of raindrop impact can compress or collapse soil pores and detach soil particles, which can further plug soil pores and increase runoff. Climate change models predict that the earth will experience more intense storm events, resulting in greater erosion rates [8]. Erosivity is influenced by environmental factor primarily climate including drop size distribution and intensity of rain, amount and frequency of rain, runoff amount and velocity, and wind velocity. Another non climate environmental factor affecting erosivity is chemical reaction leading to solusional weathering [7]. The R value represents the destructive force of rain or the erosivity of the yearly rain that can be calculated by the Bols equation [1].

$$R = 6{,}119\ (RAIN)^{1{,}21}\ x\ (DAYS)^{-0{,}47}\ x\ (MAXP)^{0{,}53}$$

IOP Conf. Series: Earth and Environmental Science **123** (2018) 012025 doi:10.1088/1755-1315/123/1/012025

R = Monthly rain intensity
RAIN = Average monthly rainfall (cm)
DAYS = Number of rainy days per month (days)
MAXP = maximum rainfall for 24 hours in the month (cm)

2) Soil Erodibility (K)

The soil erodibility value indicates soil sensitivity to erosion. The soil erodibility is affected by structure, texture, organic matter, and soil permeability. Soil intensity can be measured using the Wischmeier and Smith equations [1].

$$100K = 1{,}292\ [2{,}1M1{,}14\ (10\text{-}4)(12\text{-}a) + 3{,}25\ (b\text{-}2) + 2{,}5\ (c\text{-}3)]$$

K = Ground Erodibility
M = The percentage of very fine sand and dust (diameter (0,1 - 0,05) and (0,05 - 0,02) x (100 -% clay).
a = Percentage of organic matter
b = Soil structure
c = Soil permeability class

To use the above equations, soil analysis data is needed, sand percentage (2.0 - 0.10mm), percentage of very fine sand (0,10 - 0,05mm), percentage of dust (0,05 - 0,002mm), clay percentage Less than 0.002mm), percentage of soil organic matter, soil structure and soil permeability.

The soil structure in this research are divided into four types with different weight values, the types are very fine granular (<1mm), Fine granular (1 - 2 mm), medium until coarse granular (2 - 10mm) and block, blocky, plat, massif structure. The values are from 1 to 4. The soil structure value is presented in Table 3 [1].

Soil structure influences the ease with which it can be eroded. Soils with medium to fine texture, low organic matter content, and weak structural development are most easily eroded. Typically these soils have low water infiltration rates and, therefore, are subject to high rates of water erosion and the soil particles are easily displaced by wind energy [9].

Table 3. Soil Structure

Soil structure (diameter size)	Code
Very fine granular (<1mm)	1
Fine granular (1 - 2 mm)	2
Medium until coarse granular (2 - 10mm)	3
block, blocky, plat, massif	4

Meanwhile soil permeability is divided into six types with different values and wight. The types are very slow permeability, slow permeability, slow to medium permeability, medium permeability, medium to fast permeability and fast permeability. Soil permeability type is presented in Table 4.

Table 4. Soil Permeability

Permeability Class	Permeability (cm/hour)	Code
Very slow	< 0,5	6
Slow	0,5 -2,0	5
Slow - medium	2,0 - 6,3	4
Medium	6,3 -12,7	3
Medium - fast	12,7 - 25,4	2
Fast	>25,4	1

3) Slope Factor (LS)

As the slope increases, so does the probability that splashed soil will move downslope. On sloping land, there is usually net transport of soil downslope because displaced soil can travel further downhill than uphill due to gravity and slope angle [8]. LS is the ratio between the amount of erosion of a plot of land with the length of the slope and a certain steepness to the magnitude of soil erosion. Erosion rates are high especially on marginal and steep lands which have been converted from forests to agriculture to replace the already eroded, unproductive cropland [10]. The LS value of a soil can be calculated by the following equation:

$$LS = \sqrt{x} \times (0{,}0138 + 0{,}00965s + 0{,}00138s2)$$

Description: x is the length of the slope in meters and s is the steepness of the slope in percent.

4) Crops Management Factors (C) and Land Management (P)

Land areas covered by plant biomass, living or dead, are more protected and experience relatively little soil erosion because raindrop and wind energy are dissipated by the biomass layer and the topsoil is held by the biomass. For example, in Utah and Montana, as the amount of ground cover decreased from 100% to less than 1%, erosion rates increased approximately 200 times [10].

Land cover greatly affects the amount of erosion that occurs. The CP value determination is done by approach between the field condition and the CP value from the Bogor Research Center [1].

5) Erosion Rate

The erosion rates in the study area according to Finney and Morgan Erosion Method are divide into 5 classes, i.e, very low, low, medium, high and very high. The volume of erosion are varies between less than 15 ton/ha/yr for very low erosion rate until > 480 ton/ha/yr for very high erosion rate. Erosion rate based on Finney dan Morgan presented in Table 5 [1].

Table 5. Erosion Rate based on Finney dan Morgan

Erosion (Ton/ha/yr)	Erosion Rate
<15	Very low
15-60	Low
60-180	Medium
180-480	High
>480	Very High

6) Soil Loss Tolerance (EDP)

Soil loss tolerance is the tolerable rate of erosion in order to maintain a sufficient depth for the growth of a plant and not to exceed the rate of soil formation in order to enable high productivity in a sustainable manner. To find EDP can be calculated by Hummer equation as follows:

IOP Conf. Series: Earth and Environmental Science **123** (2018) 012025 doi:10.1088/1755-1315/123/1/012025

Edp (mm/ yr) = (effective soil depth x depth factor)/ soil use age

EDP (ton/ha/yr) = EDP (mm/yr) x BV x 10

2.2.5. Landuse Planning

Soil erosion, in particular, is regarded as one of the major and most widespread forms of land degradation, and, as such, poses severe limitations to sustainable agricultural landuse [11]. Landuse planning in this research was conducted by comparing the amount of erosion occurring and the amount of erosion that can be allowed on a particular land. If the permissible erosion is less than the predicted erosion, conservation measures need to be taken by improving the C and P values so that the erosion can be minimized.

To make it easier to choose the soil conservation method, calculated the maximum CP value to know what the maximum CP value in the soil conservation action planning is done so that the erosion is less than the permissible erosion.

3. Results and Discussions

3.1. Erosion Analysis

Rain erosivity result based on calculations using the Bols equation obtained that rain erosivity value ranged from 46.15 to 613.63. The lowest monthly rainfall erosivity occurred in August while the highest monthly rainfall erosivity occurred in January. The value of annual rain erosivity at Rendang station was 3045,40 ton / ha / cm of rain, annual rain erosivity in research location is high because of the high average of rainfall which reach 268,15cm / year. Based on the Oldeman's classification [9] monthly rainfall in the research area is included in the climate type D3. Climate type D3 shows that wet months occur 3 to 4 times consecutively in 1 year and dry months 4 to 6 times in a row. This type of climate suggests that the crops growth only able for once-time rice or one-time crops per year depend on water availability in the area.

The value of erodibility is the soil sensitivity to erosion, the higher the K value the higher the erosion occurs. Based on the calculation of soil erodibility value with the formula of Wischmeier and Smith [1] soil erodobility value in the study area ranged from 0.03 to 0.36. The value of this erodibility is categorized into very low values up to high erodibility. Very low soil erodibility values occur in land unit 2; 17, low erodibility occurs in unit 1; 3; 4; 5; 6; 8; 9; 13; 14, medium erodibility is occurring on land unit 10; 11; 12; 15; 16 and high value of erodibility occurred in land unit 7.

The value of erodibility is influenced by organic matter, composition of sand, dust, clay, structure and permeability value. The value of soil erodibity is rather high due to the high percentage of dust. Dust is easily transported by falling rainwater, so that the dust will close the soil pores and will increase surface flow. The increase in surface flow will increase the erosion that occurs. Land that has a dominant texture of sand, the possibility of erosion is small because sand allows a large infiltration and reduce overland flow. Soil that has a high organic element tends to improve soil structure so that the water holding capacity is high and able to reduce erosion. The granular soil structure allows the rain water to infiltrate the soil so that the surface flow is reduced. The permeability of the research area is very rapid, affecting the rapid flow of infiltrated water.

The length of slope factor and slope steepness (LS) affects the surface flow and the amount of erosion, among the most influential is the slope of the slope. The research area has a slope length of 12 m to 38 m, while the slope of the slope is around 5% to 57%. Wischmeier and Smith's equations indicates that the LS scores of the study areas ranged from 0.39 to 19.55. Lowest LS value on unit 12 and LS highest on land unit 13. The determination of plant management factor values and conservation measures is done by matching the situation in the field with the CP value of the soil research center. Land use in the research area is mostly mixed farming and some land is used as moor with poor conservation measures. Table 6 present the erosion rate in Gunggung Watershed.

Table 6. Erosion Analysis at Gungung Watershed

Land Unit	R	K	LS	CP	A (ton/ha/yr)	EDP (ton/ha/yr)	TE	Soil Depth (cm)	TBE
1	3045.4	0.13	18.28	0.04	289.48	19,903	B	70	SB
2	3045.4	0.03	8.68	0.12	95.16	22,300	S	100	S
3	3045.4	0.14	3.79	0.20	323.18	28,437	B	90	SB
4	3045.4	0.13	1.47	0.28	162.95	35,680	S	90	B
5	3045.4	0.14	0.69	0.28	82.37	33,345	S	100	B
6	3045.4	0.14	9.45	0.12	483.49	28,627	SB	95	SB
7	3045.4	0.36	5.98	0.28	1835.72	39,930	SB	110	SB
8	3045.4	0.15	2.31	0.12	126.63	27,467	S	80	B
9	3045.4	0.15	3.17	0.28	405.46	19,947	B	55	SB
10	3045.4	0.22	1.43	0.20	191.62	20,453	B	65	SB
11	3045.4	0.21	1.78	0.28	318.74	29,093	B	80	SB
12	3045.4	0.30	0.39	0.20	71.26	33,843	S	87	B
13	3045.4	0.20	13.62	0.12	995.48	24,320	SB	80	SB
14	3045.4	0.20	6.18	0.12	451.69	34,033	B	100	B
15	3045.4	0.22	3.56	0.12	286.22	24,547	B	70	SB
16	3045.4	0.24	2.04	0.12	178.92	25,982	B	85	SB
17	3045.4	0.06	0.92	0.12	20.17	41,873	R	110	R

TE: erosion rate based on Finney dan Morgan method, TBE :erosion hazard level based on the Erosion Hazard Classification method[1]; SR : Very low; R : Low; S : Medium; B: High; SB : very High;

The level of erosion occurs in the study area from low to very high. A low erosion rate occurs on land unit 17, the cause of the low rate of erosion is the low value of soil erodibility. The low soil erodibility value indicates that the soil is not easily eroded, also the area that tend to be flat causing greater infiltration. Medium erosion rate is happening on land unit 2; 4; 5; 8 and 12 because it is influenced by the value of vegetation and land management. Vegetation in the unit is a mixed vegetation with a very low density. Vegetation, which is less stratified and less dense, causing the erosion tend to occur because least canopy that able to block the rain drops to the ground. Rain that directly fall at soil surface will damage the soil aggregate and makes it more easily to be carried away by the overland flow. High erosion rate occurs in land unit 1; 3; 9; 10; 11; 14; 15 and 16 with the slope condition that affect the amount of erosion that occurs. Steep and long song slopes in these land units cause faster overland flow so aggregates of soil are quickly carried away by overland flow and become erosion. Very high erosion rate occurs in unit 6; 7; and 13 caused by the vegetation condition in the land unit. The vegetation mainly are is moors, mixed vegetation with very low density and shrubs. The less dense vegetation can cause the soil to be directly hitted by rain. The direct rainfall hit on the soil will faster damage the soil aggregate and will accelerate the erosion. Besides, the high erodibility value becomes the main factor of a very high erosion rate which happened in these land units. Map of erosion rate is presented in Figure 3.

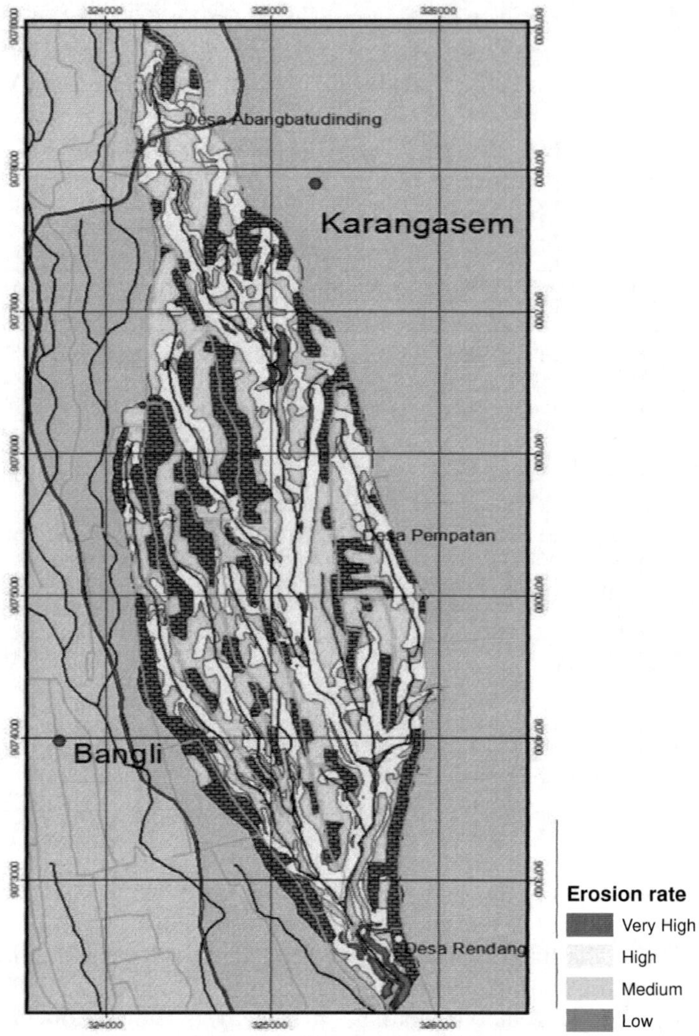

Figure 3. Erosion Rate Map

3.2. Landuse Planning

After the erosion rate calculated using the USLE equation, then it will be compared with the soil loss tolerance (EDP) in order to know the conservation measures. The land unit which erosion is smaller than the EDP is the land unit 17. Land unit 17 must be maintained properly, so that the present erosion conditions do not increase. Greater erosion value than EDP are land unit 1; 2; 3; 4; 5; 6; 7; 8; 9; 10; 11; 12; 13; 14; 15 and 16. Land units with erosion value greater than the soil loss tolerance should be given conservation measures.

Existing landuse on land unit 1 are mixed vegetation with high density, conservation measures in the form of traditional terraces. Landuse plan can be start with an improvement on the forest vegetation. Forest is recommended because in this land unit there are already has some plants with of economic value, then addition of some kind of woods with high litter will be very beneficial. The land unit 2 planning that will be improvement on the vegetation, ie the planting of a high density with mixed vegetation with a terrace of medium constructed bench. Land unit 3 existing land use is a low density mixed vegetation with a traditional terrace. Conservation planning that will be suggested is the improvement on vegetation and land management is a high density mixed vegetation with a terrace of medium constructed terrace. Land unit 4 existing landuse is moor with dominant plant of cassava with traditional terrace. Landuse planning will be intercropping cassava planting with cowpea with medium constructed terrace. Land unit 5 with land use of moor with dominant crop is cassava and land management of traditional terrace. Landuse planning is intercropping cassava with peanut with with medium constructed terrace.

Land unit 6 existing land use is a medium density mixed vegetation with a traditional terrace. Landuse plan to be suggested is a high-density mixed vegetation with a good constructed bench porch. Land unit 7 land use is moor with dominant crop is cassava and conservation measures in the form of traditional terrace. Land use planning that will be suggested is cassava planted with cassava peanuts and ground terrace planted with peanuts. Land use of land unit 8 are mixed density medium vegetation and conservation measures in the form of traditional terraces. It will be advised for landuse and conservation planning on land unit 8 is planting vegetation in the form of high density mixed vegetation with a terrace of medium constructed bench. Existing land use of land unit 9 are moor with traditional terraces. It will be better for conservation planning to intercropping cassava with peanuts with terraces of bunds on cassava plantations.

Table 7. Conditions of Existing Landuse, Landuse Planning and Conservation

Land Unit	Existing Landuse	Landuse Plan	Conservation
1	mixed vegetation with high density, traditional terraces	community forest	high litter
2	high density with mixed vegetation	high density mixed vegetation	medium construction bench
3	low density mixed vegetation with a traditional terrace	high density mixed vegetation	medium constructed bench terrace
4	moor with dominant plant of cassava with traditional terrace	intercropping cassava planting with cowpea	medium constructed bench terrace
5	moor with dominant crop is cassava with traditional terrace	intercropping cassava with peanut	medium constructed bench terrace
6	medium density mixed vegetation with a traditional terrace	high-density mixed vegetation	a good constructed bench porch
7	moor with dominant crop is cassava	high density mixed vegetation	ground terrace with cassava
8	mixed density medium vegetation with traditional terraces	high density mixed vegetation	medium constructed bench terrace
9	moor with traditional terraces	cassava with peanuts with terraces of bunds on cassava plantations.	terraces of bunds
10	mixed low density *vegetation* with traditional terraces	high density mixed *vegetation*	medium constructed bench terrace
11	*Moor with* a traditional terrace	high density mixed *vegetation*	medium constructed bench terrace
12	mixed medium density *vegetation* with traditional terraces	high density mixed *vegetation*	medium constructed bench terrace
13	shrubs with traditional terraces	community forest	high litter
14	shrubs with traditional terraces	high-density mixed *vegetation*	good constructed bench terrace
15	bushes with traditional terraces	high density mixed vegetation	medium constructed bench terrace
16	shrub with traditional terraces	high-density mixed *vegetation*	medium constructed bench terrace
17	shrub with traditional terraces	high-density mixed *vegetation*	medium constructed bench terrace

Existing landuse of land unit 10 are mixed low density vegetation with traditional terraces, for landuse planning is high density mixed vegetation and terrace of medium constructed bench. The landuse of land unit 11 is moor with a traditional terrace. For conservation planning is a high density mixed vegetation with a terrace of medium constructed bench. Landuse of land unit 12 are mixed medium density vegetation with conservation measures of traditional terraces, it will be planned for high density mixed vegetation with terrace of medium constructed bench. Landuse of land unit 13 are shrubs with conservation measures in the form of traditional terraces, and it will planned for a community forest with a high litter, because this unit is on a very steep slope. Landuse of land unit 14 is bush with a traditional terrace, then the conservation plan is a high-density mixed vegetation with a good constructed bench porch. Existing landuse of land unit 15 are bushes with traditional terraces, then it will be planned with high density mixed vegetation with good constructed bench terraces. The land unit 16 is a shrub with traditional terraces, it will be good for conservation planning is a high-density mixed vegetation with a terrace of medium constructed bench. Table 7 summarizes the conditions of existing landuse, landuse planning and conservation taken.

4. Conclusions

The level of erosion occurring in the study area from low to very high. A low erosion rate caused by the low value of soil erodibility. The low soil erodibility value indicates that the land is not easily eroded. Medium erosion rate is occurring because it is influenced by the value of vegetation and land management. Mixed vegetation with a very low density, which is less stratified and less dense, causing the erosion tend to occur because least canopy that able to block the rain drops to the ground. Rain that directly fall at soil surface will damage the soil aggregate and makes it more easily to be carried away by the overland flow. High erosion rate occurs with steep and long song slopes which cause faster overland flow so aggregates of soil are quickly carried away by overland flow. Very high erosion rate caused by the vegetation condition. The vegetation mainly are moors, mixed vegetation with very low density and shrubs. The less dense vegetation can cause the soil to be directly hitted by rain. The direct rainfall hit on the soil will damage the soil aggregate and will accelerate the erosion.

5. Acknowledgement

Writers would like to thank the Dean of Agriculture Faculty of Udayana University, Head of Center for Spatial Data Infrastructure Development (PPIDS) of Udayana University, fellow staf, researchers and professors at Udayana University for all the support, comments and suggestions.

6. References

[1] Arsyad S 2010 Konservasi Tanah & Air
[2] Hakim N M Y Nyakpa, AM Lubis, SG Nugroho, MR Saul, MA Diha *GB Hong, H. Bailey* 1986 Erosi
[3] Utomo W H 1994 Erosi dan Konservasi Tanah *Penerbit IKIP Malang* 194
[4] Sinukaban N 1989 Konservasi tanah dan air di daerah transmigrasi *PT. INDECO Duta Utama-BCEOM*
[5] Zulkarnain Z 2014 Soil Erosion Assessment of The Post-Coal Mining Site in Kutai Kartanagera District, East Kalimantan Province *Int. J. Sci. Eng.* **7** 130–136
[6] Telles T S, Dechen S C F, de Souza L G A and Guimarães M de F 2013 Valuation and assessment of soil erosion costs *Sci. Agric.* **70** 209–216
[7] Lal R 2001 Soil degradation by erosion *L. Degrad. Dev.* **12** 519–539
[8] Holz D J, Williard K W J, Edwards P J and Schoonover J E 2015 Soil erosion in humid regions: a review *J. Contemp. Water Res. Educ.* **154** 48–59
[9] Banuwa I S 2013 Erosi *Kencana Prenada Media Group. Jakarta*
[10] Pimentel D 2006 Soil erosion: a food and environmental threat *Environ. Dev. Sustain.* **8** 119–137
[11] Zuazo V H D and Pleguezuelo C R R 2008 Soil-erosion and runoff prevention by plant covers. A review *Agron. Sustain. Dev.* **28** 65–86

Spatial Analysis of Land Adjustment as a Rehabilitation Base of Mangrove in Indramayu Regency

Sodikin[1], S R P Sitorus[2], L B Prasetyo[3], and C Kusmana[4]

[1] Graduate Student of Natural Resources and Environmental Management Program (PSL) Bogor Agricultural University

[2] Department of Soil Science and Land Resources, Faculty of Agriculture Bogor Agricultural University

[3] Department of Forest Resources Conservation and Ecotourism, Faculty of Forestry Bogor Agricultural University

[4] Department of Silviculture, Faculty of Forestry Bogor Agricultural University

Email: sodikin.ips@uinjkt.ac.id

Abstract. Indramayu Regency is the area that has the largest mangrove in West Java. According to the environment and forestry ministry of Indramayu district will be targeted to be the central area of mangrove Indonesia. Mangroves in the regency from the 1990s have experienced a significant decline caused by the conversion of mangrove land into ponds and settlements. To stop the mangrove decline that continues to occur, it is necessary to rehabilitate mangroves in the area. The rehabilitation of mangrove should be in the area suitable for mangrove growth and what kind of vegetation analysis is appropriate to plant the area, so the purpose of this research is to analyze the suitability of land for mangrove in Indramayu Regency. This research uses geographic information system with overlay technique, while the data used in this research is tidal map of sea water, salintas map, land ph map, soil texture map, sea level rise map, land use map, community participation level map, and Map of organic soil. Then overlay and adjusted to matrix environmental parameters for mangrove growth. Based on the results of the analysis is known that in Indramayu District there are 5 types of mangroves that fit among others Bruguera, Soneratia, Nypah, Rhizophora, and Avicennia. The area of each area is Bruguera with an area of 6260 ha, 2958 ha, nypah 1756 ha, Rhizophora 936, and Avicennia 433 ha.

Keywords: Land, Mangrove, Rehabilitation

1. Introduction

Decline in the quality and quantity of mangrove forests can affect the economic life of coastal communities, such as decreased catches and reduced fishermen's income [1]. Indramayu district is one of the regencies located in West Java Province, this region has a coastal length of approximately 114.1 km and is very potential for the growth of mangrove and other coastal vegetation. Indramayu has a relatively wide forest potential of 43,027.41 ha. The types of commodities cultivated are for teak forests of 21,144.37 ha, brackish / mangrove forest (protected forest area) of 8,023.55 ha and eucalypt forests of

5,130.75 ha. The condition of mangrove forest in Indramayu Regency has been damaged due to changes in the surrounding environment and the direct and indirect pressure on the existence of the mangrove forest itself. Various rehabilitation efforts Have been conducted by the government and various NGOs to maintain the quality and quantity of mangrove forest in Indramayu district. Based on the research background, the formulation of the problem taken in this study is how the suitability of land for mangrove in Indramayu District, thus this study aims to see how the suitability of land for mangrove in Indramayu District [2,3].

2. Review of The Related Literature

2.1. Definition of Mangrove

Mangrove forests are typical forest types found along the coast or river estuaries, which are affected by tidal water, so often these forests are often called tidal forests, coastal forests, brackish forests, or mangrove forests. According to the Marine and Fishery Research Agency (2005), the characteristics of mangrove forest can be seen from various aspects such as floristic, climate, temperature, salinity, rainfall, geomorphology, hydrology, and drainage.

2.2. Evaluation of Land Suitability

Land is part of the landscape that includes physical understanding including climate, topography / relief, hydrology, even vegetation conditions that potentially will affect land use [4]. Land suitability is the depiction of the level of suitability of a plot of land for a particular use. The conformity class of an area may differ depending on the type of land use under consideration. Land suitability is at the unit level. Land suitability at the unit level is the subdivision of sub-classes. All units within a subclass have the same level of conformity in the class and have the same type of constraint at the sub-class level. Units differ from each other in the additional properties or aspects of management that are necessary and often a detailed distinction of their constraints. Knowing the limits in detail will facilitate the planning interpretation at the farm level. Land suitability symbols at the unit level are distinguished by Arab numbers placed after sub-class symbols, eg S3t-2, S3t-3. There are no restrictions on the number of units in a sub-class.

2.3. Method of Research

2.3.1. Setting of Research

The study was conducted in Indramayu District which is astronomically located at 107 ° 52 ° - 108 ° 36 ° East Longitude and 6 ° 15 °- 6 ° 40 ° South latitude. The method used is quantitative method. Determining the location of the research is done "purposive", that is the determination of the location chosen directly or intentionally on the grounds that at that location there are mangrove forests whose condition is damaged and at this time rehabilitation efforts have been done. The location of research in the form of maps is shown in Figure 1.

Data collection methods for mangrove land suitability analysis consist of salinity, water pH, sediment organic matter, flow velocity, substrate, land use, puddle frequency, maximum inundation, wave height, Land subsidence, Sea level rise, erosion level, land interaction to current And ups and downs.

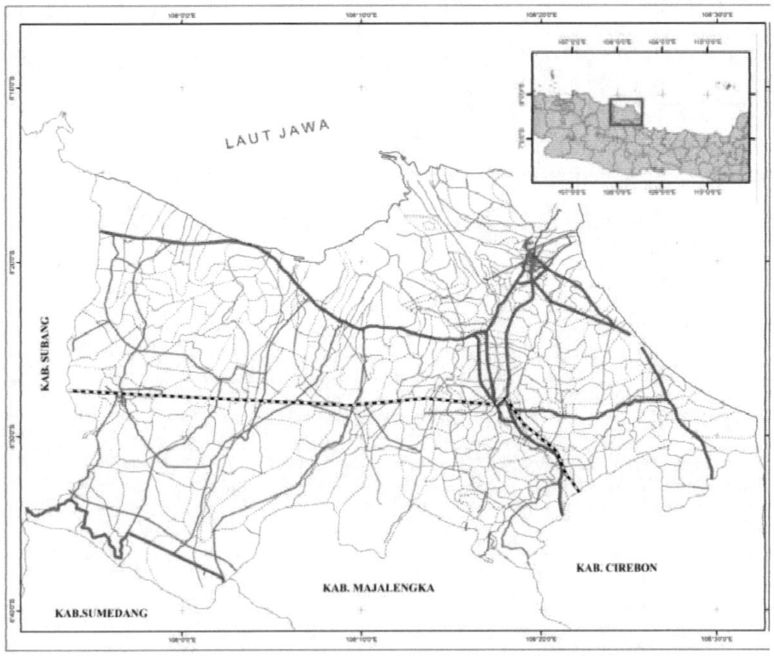

Figure 1. Map of District Administration Sukabumi

2.3.2. Analysis of the Availability and Suitability of Land for Mangrove Vegetation

The analysis of land availability for mangrove is done based on the matrix of mangrove land availability criteria as in Table 1.

Table 1. Matrix of Land Conformity Criteria for Mangroves

No	Parameter	Class	Value	Weight
1	Salinty ($^0/_{00}$)	29 – 33 25 - < 29 or > 33 – 37 < 25 or > 37	3 2 1	2
2	Water pH	7 – 8,5 6,5 - < 7 or > 8,5 –9,5 < 6,5 or > 9,5	3 2 1	1
3	Organic sediment material (%)	> 10,1 4,1 - 10 < 4	3 2 1	1
4	Frequency of Inundation (D/W)	20 10-19 <10 or >20	3 2 1	2
5	The maximum puddle (m)	< 0,5 0,5 – 1 > 1	3 2 1	2
6	Water flow (cm/dt)	< 1 1 – 10 > 10	3 2 1	1
7	Wave (m)	< 0,5 0,51 - 1 > 1	3 2 1	2
8	Substrate	Sand or silt	3	2

IOP Conf. Series: Earth and Environmental Science **123** (2018) 012026 doi:10.1088/1755-1315/123/1/012026

No	Parameter	Class	Value	Weight
		clay	2	
		Gravel	1	
9	Land use	Mangrove, swamp forest	3	1
		Aquaculture	2	
		Residential, industry	1	
10	*Land Subsidence* (cm/year)	< 1	3	2
		1 - 4	2	
		> 4	1	
11	*Sea level rise* (mm/year)	< 4,99	3	2
		5 – 9,99	2	
		> 9,99	1	
12	Erosion (m/year)	0	3	2
		-0,1 – (-2)	2	
		> -2	1	
13	Land interaction with the flow And waves	Protected	3	2
		Somewhat protected	2	
		Open	1	

Source: Ministerial Decree no. 51 / MENKLH / 2004); Landon, (1991); Khazali, (1999); DKP, (2008); Mazda, et al. (2003); IUCN, (2006); Dewanto, 2007); Gornitz et al. 1992); Dahuri, (2003); Yulianda modification, (2007)

Where:

Y = Final Value

Ai = Weight factor

Xn = Value of land suitability

The land suitability class is obtained based on the Equal Interval method [5,6] (Prahasta, 2002) to divide the range of attribute values into sub-ranges of the same size. The calculations are as follows:

$$I = \frac{(\Sigma\ ai.Xn) - (\Sigma\ ai.Xn)_{min}}{k}$$

Where:
I = Interval of land suitability class
K = Number of classes of land suitability desired
Based on the above formula and calculation obtained class interval and value (score) of land suitability as follows:
S1: Very Match, with value 55-66
S2: Quite Accordance, with a value of 44-54
S3: Appropriate Marginal with a value of 33-43
N1: Not Current, with a value of 22-32
N2: Not Permanent <22

After knowing the availability of land for mangroves, then the appropriate vegetation analysis in planting in the area begins with arranging the matrix of conformity criterion as shown in Table 2 as reference in analyzing the existing parameters so as to be able to make arrangements to the land suitability classes for growth Mangroves. Land suitability analysis for mangrove growth using weighted overlay method where each parameter is given the same weight then the result shows the suitability of area or area for mangrove growth.

Table 2. Matrix Selection of Mangrove Species on Suitable Land

Tidal Type/Inundation Class Watsin (1928)	Inundation Class (Salinity and Tidal Frequency) de Haan (1931)	Soil Texture Type	Dominantly *Mangrove* Type
All High tides	Brackish to salt, Salinity 10-20 ppt, Always flooded (1-2 times / day, minimum 20hari / month)	Coral, sandy, sandy clay	*Avicennia, spp* *Sonneratia, spp* *Rhizophora,spp*
	Brackish to salt, salinitas20-30ppt, always stagnant (1-2kali / hari, minimum 20 days / month)		
	Brackish to salt, salinity> 30ppt, always stagnant (1-2kali / hari, minimum 20 days / month)		
Medium high tides	10-19 days / month, salinity 10-20ppt	Dusty to dusty clay	*Bruguera gymnorrhiza*
	10-19 days / month, salinity 20-20ppt		
	10-19 days / month, salinity> 30ppt		
Normal high tides	9 days / month, salinity 10-20ppt	Dusty, clay dusty to clay	*Xylocarpus, spp* *Scyphiphora, spp* *Lumnitzera,spp*
	9 days / month, salinity 20-30ppt		
	9 days / month, salinity> 30ppt		
Spring tides only	A few days / month, salinity 0 ppt	Sandy until dusty clay	The marginal types in the environment such as *Xylocarpus moluccensis, Intsia bijuga, Nypa fruticans, Ficus retusa, Glochidion littorale.*
Storm highes tides only	A few days / month, salinity 0 ppt (Rarely stagnant pairs)	Sandy until dusty clay	*Oncosperma* spp, *Cerbera* spp.

Source: Kusmana et al. (2005)

3. Result and Description

3.1. General Condition of Research Area

Geographically, Indramayu Regency is located in 1070 52 '- 1080 36' East Longitude and 60 15 '- 60 40' South Latitude. Indramayu Regency has an area of 209,942 ha with 114.1 km long coastline stretching along the north coast between Cirebon and Subang regencies. Indramayu district has an average temperature between 22.9 ° - 30 ° Celsius, while the average rainfall during 2014 is 2,104 mm with the number of rainy days 103 days. The highest rainfall occurred in District Lohbener approximately 2,756 mm with the number of rainy days recorded 125 days, while the lowest rainfall occurred in District Terisi approximately 666 mm with the number of raindays recorded 52 days.

The soil type in the dominant study area is Inceptisol and Entisol, ie Typic Epiaquepts, Typic Endoaquepts, Aquic Dystrudepts and Typic Endoaquepts, Aquic Eutrudepts, Typic Udorthents. Ups and downs in Indramayu District are based on estimation data from two stations (Tanjung Priok and Cirebon), tidal types in the North Coast region of West Java, including in Indramayu Regency belonging to the mixed tidal type leading to semi-diurnal (multiple mix).

IOP Conf. Series: Earth and Environmental Science **123** (2018) 012026 doi:10.1088/1755-1315/123/1/012026

3.2. Analysis of Land Suitability for Mangrove in Indramayu District

Based on the result of overlay some parameters of land suitability for mangrove (salinity map, pH, sediment organic matter, puddle frequency, maximum inundation, velocity, wave, substrate, land use, land subsidence, sea level rise, coastal erosion / abrasion, Current) obtained several land suitability criteria such as land criteria according to 1, according to 2, suitable 3, not current, and not permanently appropriate. Land suitability for mangrove spatially as presented in Figure 2.

Based on Figure 2 it can be seen that the land suitability for mangroves in Indramayu Regency is very complex from very suitable to non-permanent in Indramayu District. The area of each class of land suitability as shown in Table 3.

Table 3. Land Suitability Area For mangroves in Indramayu District

District	**Area (ha)**				
	Very appropriate	**Simply Match**	**Appropriate Marginal**	**Not appropriate at this time**	**Not permanently suitable**
Sukra	0	103.06	132.26	4143.4	498.98
Patrol	0	25.00	92.98	3102.6	726.03
Kandanghaur	29.64	505.17	56.17	7179.5	808.39
Losarang	0	1298.46	0	9092.97	859.41
Cantigi	0	3958.6	348.91	3552.8	538.55
Pasekan	0	2040.93	2140.48	2449.67	233.275
Indramayu	0	1210.0	28.18	3151.7	837.65
Balongan	0	48.13	103.66	3359.9	330.152
Juntinyuat	0	7.08	22.50	4775.6	653.39
Karangampel	0	2.10	49.42	2598.2	456.39
Krangkeng	0	371.3	5.46	4740.82	518.44
Amount	29.64	9.569.83	2.980.02	48.147.16	6.460.65

Based on Table 3, it is seen that the area of land which has a very suitable category is 29.64 ha, it is quite appropriate that is 9,569.83 ha, according to the marginal is 2,980.02 ha, not according to the current of 48147.16 ha, and the non permanent suit 6,460,657 Ha. It can be concluded that the majority of Indramayu Regency owns the land with sufficient criteria suitable for mangrove. Based on the distribution of land that has the criteria according to one is only available in Kandanghaur District of 29.64 ha, whereas for other criteria almost spread in all subdistricts that become research area.

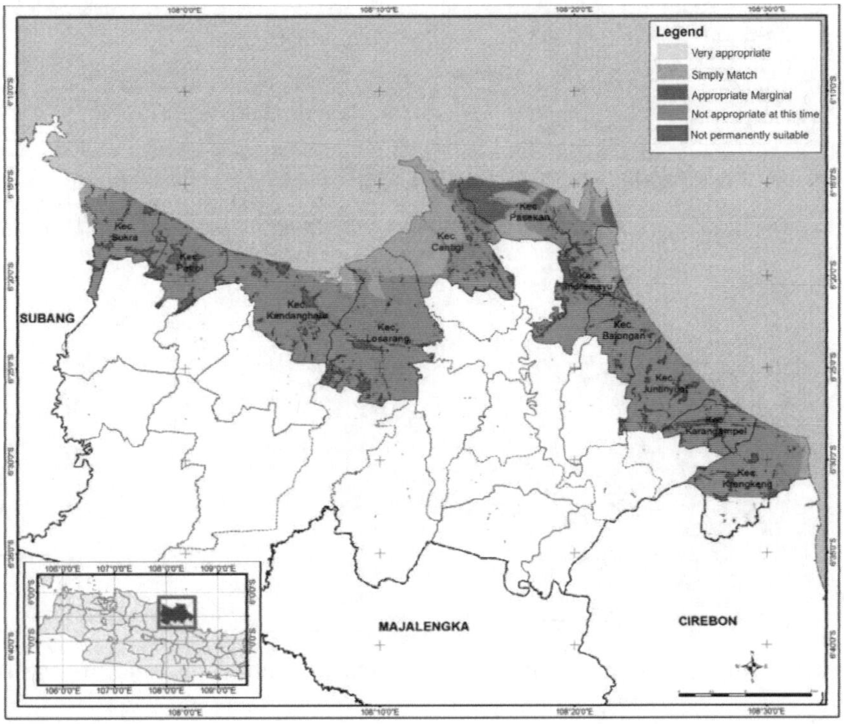

Figure 2. Map of Land Suitability for Mangroves in Indramayu District

3.3. Referral Type of Mangrove Vegetation suitable for Rehabilitation in Indramayu District

In conducting mangrove rehabilitation, feasibility analysis of land suitable for mangrove is needed, as this affects the growth of mangrove itself. In addition to the need for land suitability for mangroves, also the appropriate matching of mangrove vegetation species appropriate to plant in the area by taking into account and analyze the environmental factors that affect mangrove growth.

Based on the result of overlay analysis which has been done with criterion of environmental factors affecting mangrove growth sourced from matrix of land suitability for various types of mangrove resulted that in Regency of Indramayu there are five type which suitable to be made as type choice for rehabilitation, for example Avicennia, Bruguera Nypah, Rhizophora, and Soneratia. The spatial picture can be seen in Figure 3.

IOP Conf. Series: Earth and Environmental Science **123** (2018) 012026 doi:10.1088/1755-1315/123/1/012026

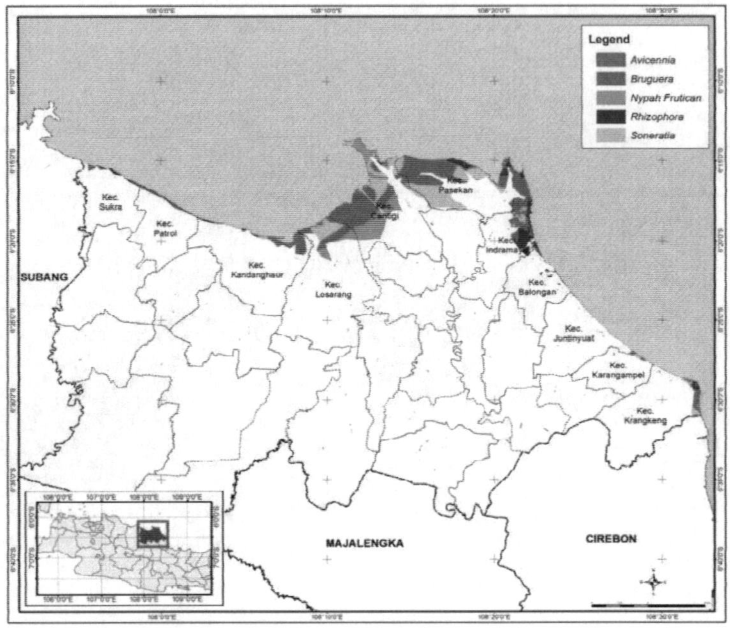

Figure 3. Map of mangrove type suitable for rehabilitation in Indramayu District

Table 4. Wide range of mangrove rehabilitation directions based on the type of vegetation

The appropriate type of mangrove	Area (ha)	%
Avicennia sp,	433	3.51
Bruguiera sp.	6,260	50.7
Nypa fruticans	1,756	14.2
Rhizophora sp.	936	7.58
Sonneratia sp.	2,958	24

Based on Table 4 it can be seen that the most suitable type of mangrove rehabilitation is the type of Bruguera which has an area of 6260 ha, then the second type is Soneratia area of 2958 ha, Nypah 1756 ha, Rhizophora 936, and Avicennia 433 ha. It shows that in Kabupaten Indramayu the type suitable for majority rehabilitation is suitable for Bruguera type.

4. Conclusion

4.1. Conclusion

The result of land suitability analysis for mangrove in Indramayu Regency obtained five categories of land suitability level that is land criteria according to 1, according to 2, according to 3, not suitable at present, and not permanent fit. The majority have land in Indramyu Regency has a suitable land criterion 2 for mangroves.

4.2. Recommendation

Local governments need to allocate programs related to increasing community participation in rehabilitation and conservation of mangrove forests. In addition, PERHUTANI needs to oblige people who have tenure in the forest area to apply silvofishery system to their ponds so that the mangrove area in Indramayu Regency will increase and have an effect on the environmental balance in the area.

5. References

[1] Mumby P J, Edwards A J, Arias-González J E, Lindeman K C, Blackwell P G, Gall A, Gorczynska M I, Harborne A R, Pescod C L, Renken H and Others 2004 Mangroves enhance the biomass of coral reef fish communities in the Caribbean *Nature* **427** 533–536
[2] Dahuri R 2003 *Keanekaragaman hayati laut: aset pembangunan berkelanjutan Indonesia* (Gramedia Pustaka Utama)
[3] Kusmana C, Wilarso S, Hilwan I, Pamoengkas P, Wibowo C, Tiryana T, Triswanto A and Yunasfi H 2003 Teknik Rehabilitasi Mangrove *Fak. Kehutan. Inst. Pertan. Bogor. Bogor*
[4] FAO 2005 *The Situation and Developments in The Forest Sector*
[5] Prahasta E 2008 Remote sensing: praktis penginderaan jauh dan pengolahan citra dijital dengan perangkat lunak ER Mapper *Inform. bandung*
[6] Kamal M, Hartono H, Wicaksono P, Adi N S and Arjasakusuma S 2016 Assessment of Mangrove Forest Degradation Through Canopy Fractional Cover in Karimunjawa Island, Central Java, Indonesia *Geoplanning J. Geomatics Plan.* **3** 107

IOP Conf. Series: Earth and Environmental Science **123** (2018) 012027 doi:10.1088/1755-1315/123/1/012027

Forest and Land Fire Prevention Through the Hotspot Movement Pattern Approach

T Turmudi[1], P Kardono[1], P Hartanto[1], and Y Ardhitasari[1]

[1]Centers for Research, Promotion And Cooperation, Geospatial Information Agency, Jl Raya Jakarta Bogor KM 46 Cibinong, Jawa Barat, 16911, Indonesia

Email: turmudi.pokja@gmail.com, turmudi@big.go.id

Abstract. Indonesia has experienced a great forest fire disaster in 2015. The losses incurred were enormous. But actually the incidence of forest and land fires occurs almost every year. Various efforts were made to cope with the fire disaster. The appearance of a hotspot becomes an early indication of the fire incident both location and time. By studying the location and time of the hotspot's appearance indicates that the hotspot has certain movement patterns from year to year. This study aims to show the pattern of movement of hotspots from year to year that can be used for the prevention of forest and land fires. The method used is time series analysis of land cover and hotspot distribution. The data used were land cover data from 2005 to 2016, hotspot data from 2005 to 2016. The location of this study is the territory of Meranti Kepulauan District. The results show that the highest hotspot is 425 hotspots occurs in the shrubs and bushes. From year to year, the pattern of hotspot movement occurs in the shrubs and bushes cover. The hotspot pattern follows the direction of unused land for cultivation and is dominated by shrubs. From these results, we need to pay more attentiont for the land with the cover of shrubs adjacent to the cultivated land.

Keywords: Hotspot, Landcover, Fire, Land and Forest, Time Series Analysis

1. Introduction

Indonesia has experienced a great forest fire disaster in 2015. The losses incurred were enormous. But actually the incidence of forest and land fires occur almost every year. Various efforts were made to cope with the fire disaster. The impact of these fires is smoke haze that blanketed the region known as the ASEAN Transboundary Haze Pollution. Losses caused by the huge smog covering the transport of water, sea, and land; health, education, economy, environment. As a result of peat fires in Sumatra accounts for the loss 2,8 billion dollar [1].

The identification of fires is obtained through hot spot data. A hot spot is indicated as the location of a fire [2]. Hot spot data is obtained through NOAA, MODIS, Hot spot data can also be downloaded from USGS and SIPONGI. In this activity the hotspot data used is hot spot data from MODIS image (Moderate Resolution Imaging Spectroradiometer) with spatial resolution 250 meters. Considering that at MODIS highest resolution, one pixel covers 250 m^2 [3].

Forest fires can occur because of the three elements of heat, oxygen, and fuel. Fire begins with combustion, which requires the mixture of heat, fuel, and oxygen in the right proportions or known as the fire triangle. In the broader dimension of the actual conditions in the field of fire incident is triggered by three factors: vegetation, landform, and climatology [4].

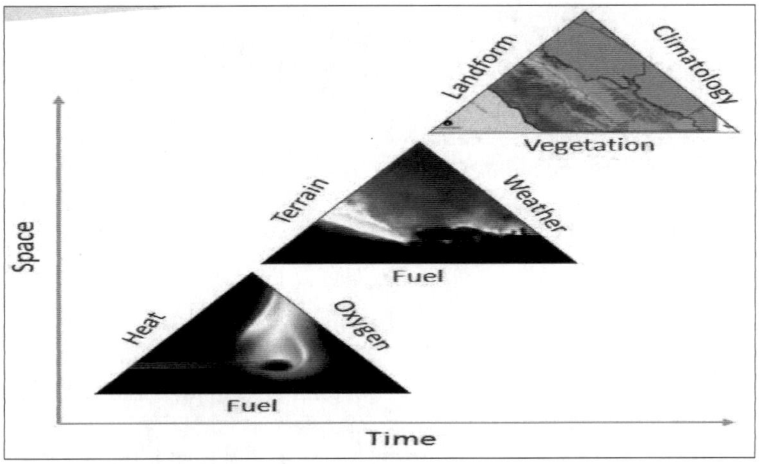

Source: Cochrane (2009) [3]

Figure 1. Triangel fire

According to Saharjo [5,6] the opportunities for forest and land fires are caused by factors: political will, law enforcement, coordination, human resources, knowledge about forest fire, early warning and early detection, human approach. In relation to this study, the focus of the study is more on early warning and early detection factors. Signs for fire occurrence based on the concept fire triangle, among them are the fuel in the broad scale is vegetation. The question is which vegetation is the most potential as fuel. This is important to understand, so we can be alert from the beginning of the possibility of forest and land fires.

Fire data were obtained through hotspot distribution data. Hot spot data is taken from year to year ie year 2005 until 2016. This is done so that pattern of movement of hotspot can be known. It is also important to know about the location of hot spots. Where hot spots happen and associated with what type of land cover.This study aims to show the pattern of movement of hotspots from year to year that can be used for the prevention of forest and land fires.

2. Methods

2.1 Study Area and Data

Regency of Kepulauan Meranti consists of 9 (nine) sub-districts are Riau Province (BPS.2017). Meranti Islands Regency is located face to face with Malaysia and Singapore State separated by Malacca Strait. Geographically Meranti islands Regency is located at coordinates between 0 ° 42 '30 "-1 ° 28' 0" N and 102 ° 12 '0 "- 103 ° 10' 0" E The total area of Meranti Islands Regency is 3,707.84 km2 or 4,26% of the area of Riau Province. Meranti Islands District, became interesting from the side of the fire disaster, because at the end of January 2014 re-occur forest and land fires in Riau starting from the region of Meranti Islands [5,7]. The data used are land cover data from 2005 to 2016, hotspot data from 2005 to 2016 as figure 3.

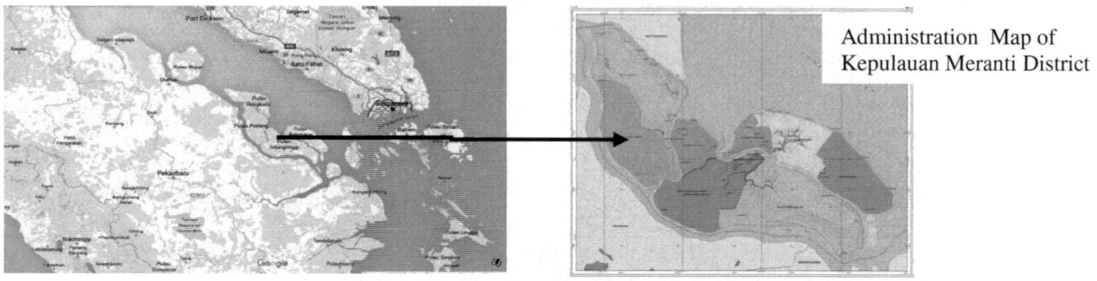

Figure 2. Location of Kepulauan Meranti District

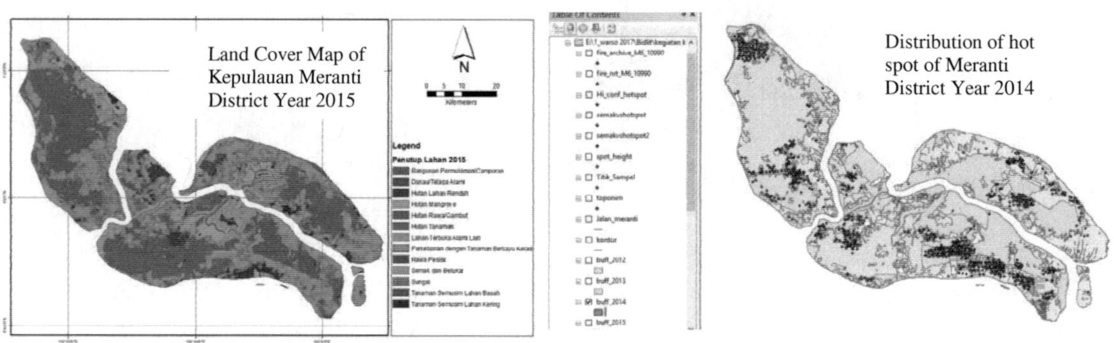

Figure 3. Some data that is used

The study was conducted with the stages of such activities on the flowchart (figure 4). Several stages are described in section below. Hotspots data for 2005, 2006, 2007, 2008, 2008, 2009, 2010, 2011, 2012, 2013, 2014. 2015 and 2016 were plotted on a 50 K RBI map, into a 50 K spatial hot spot data. The hot spot data were obtained from MODIS image with 250 m spatial resolution. Thus each pixel in the MODIS image represents an area of 250 m^2 which means each hotspot point represents a total area of 250 meters2. Confidence value of each hotspot selected is 85%.

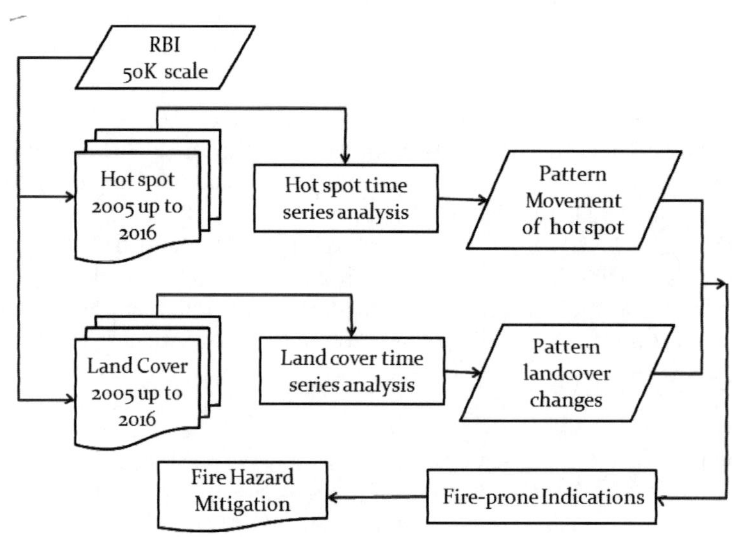

Figure 4. Flowchart Stage Research Activities

2.2 Hot spot data time series analysis

Analysis of hot spot time series data derived from several series (hot spot year: 2005, 2006, 2007, 2008, 2009, 2010, 2011, 2012, 2013; 2014; 2015; 2016. Hot spot distribution is from year to year changing position as well as extents. The fact is used to analyze the pattern of hot spot distribution movement [8]. The direction of hot spot movement leads to land with high fuel potential [5]. Based on triangle fire fuel for the occurrence of fire is vegetation. In this case, what is meant by vegetation is land cover.

To determine the area of effective fire, spatially, buffering of each hotspot [9]. Buffering was done at a distance of 250 meters from the hotspot. The distance between hotspots one with other hotspots is not the same, the fire area indicated by this buffering between hotspots is unified, but some are stand-alone.

Figure 5a. Distribution of hot spot

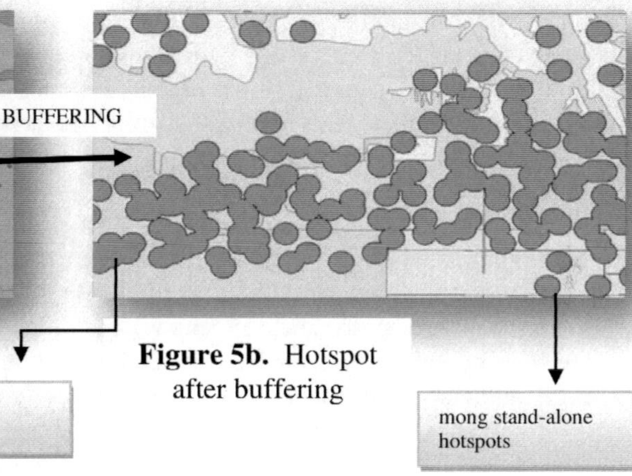

Figure 5b. Hotspot after buffering

2.3 Time series land cover analysis

Time series analysis on land cover was done to obtain information on hotspot occurrence in a land cover type. Based on triangle fire, vegetation is fuel. Through this analysis was obtained what type of land cover is most potential as a fuel for the incidence of forest and land fires.

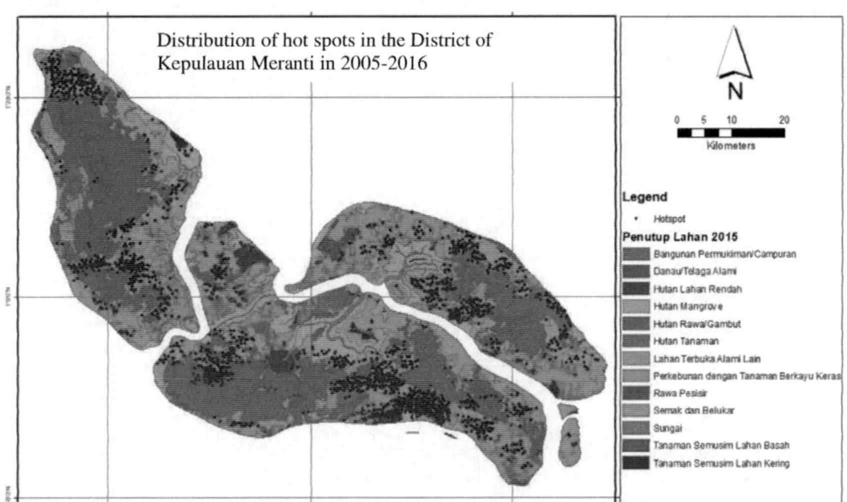

Figure 6. Map of distribution of hot spots in the District of Kepulauan Meranti in 2005-2016

2.4 Fire Prone Indication

Fire prone indication obtained through analysis of the pattern of movement of hotspots on land cover multi year ie from 2005 to 2016 Indication of this fire hazard can be: a. The first step to mitigate the fire disaster. b. Awareness of areas with potential fire; c. Preparation of an activity plan that includes the budget, equipment, and quantity and quality of human resources required [10]. Fire prone indication is indicated by map of the distribution of hotspots in the district of the Meranti Islands in 2005-2016. The spread of hotspots clumps a lot in shrubs and bushes and borders on plantations. Through fire prone indication can be known which administrative areas need to increase alertness to the possibility of fire that will come.

3. Results

The results show that the hotspot pattern follows the direction of unused land for cultivation and is dominated by shrubs and bushes. The highest hotspot is 425 hotspots occurs in the shrubs and bushes. From year to year, the pattern of hotspot movement occurs in the bush and bushes cover. In the 12-year time span (2005 up to 2016) the highest number of hotspots is always present in the type of land cover shrubs and bushes. This shows that trend the pattern of movement of forest fires always follows the lands of the shrubs and bushes. (See figure 7, and table 1).

In terms of location, it shows that the type of shrubs cover and bushes indicate adjacent or close to the plantation area. When referring to the fire practices caused by land clearing for new plantations / farms. The practice of clearing the land by burning , because of the consideration that the activity: a. easy to do; b. does not require expensive cost ; c. faster process; d. simple equipment; and it is assumed that the former land of fire is considered more fertile. (Saharjo. B H. 2016). Land clearing for plantation land also contributes to the occurrence of forest fires [11].

Table 1. Number of Hotspot in Kepulauan Meranti District

	Number of Hotspot per year											
Land Cover Type	2005	2006	2007	2008	2009	2010	2011	2012	2013	2014	2015	2016
Shrubs and bushes	8	18	4	28	158	16	8	18	29	425	33	71
Open land	3	1	0	0	31	1	0	1	0	277	13	3
Swamp forest	5	2	0	2	36	7	5	3	12	114	41	6

Source: Data MODIS YEAR 2005 UP TO 2016

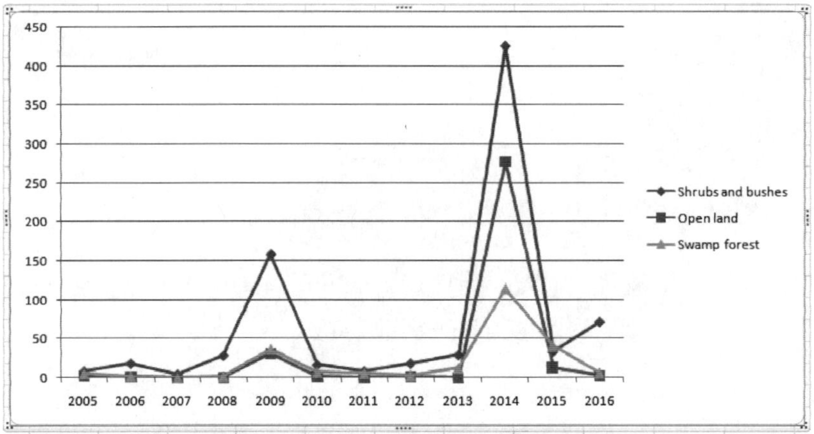

Figure 7. Chart Number of Hotspot events on a land cover 2005 up to 2016 in Kepulauan Meranti District

IOP Conf. Series: Earth and Environmental Science **123** (2018) 012027 doi:10.1088/1755-1315/123/1/012027

4. Conclusions

The existence of land cover in the form of shrubs and shrubs, especially those adjacent to plantation land, need to be wary of as a potentially high area of fire occurrence. From these results, we need to pay more attention for the land with the cover of shrubs and buses adjacent to the cultivated land. Based on this, it is necessary to always update the landcover data, especially shrubs and shrubs every year.

5. Acknowledgments

We are grateful to Centers for Research, Promotion And Cooperation, Geospasial Information Agency (BIG) for the data and financial support which allows us to write this scientific paper.

6. References

[1] Tacconi L 2003 *Kebakaran hutan di Indonesia: penyebab, biaya dan implikasi kebijakan*
[2] Fernando V 2014 Klasifikasi data spasial untuk kemunculan hotspot di Provinsi Riau menggunakan algoritme ID3
[3] NASA 2014 MODIS Data
[4] Cochrane M A and Ryan K C 2009 Fire and fire ecology: Concepts and principles *Tropical Fire Ecology* (Springer) pp 25–62
[5] Adinugroho W C, Suryadiputra I N N, Saharjo B H and others 2005 *Panduan pengendalian kebakaran hutan dan lahan gambut* (wahyu catur adinugroho)
[6] Saharjo B H 2016 *Pengendalian Kebakaran Hutan dan atau Lahan di Indonesia* (IPB press. Bogor)
[7] BPS 2017 *Kabupaten Kepulauan Meranti Dalam Angka 2016*
[8] Thariqa P and Sitanggang I S 2015 Spatial online analytical processing for hotspots distribution based on socio-economic factors in Riau Province Indonesia *Procedia Environ. Sci.* **24** 277–84
[9] Vetrita Y and Haryani N S 2012 Validasi Hotspot MODIS Indofire di Provinsi Riau *J. Ilm. Geomatika* **18**
[10] Heryalianto S C 2006 Studi tentang sebaran titik panas (HOTSPOT) sebagai penduga kebakaran hutan dan lahan di Propinsi Kalimantan Barat Tahun 2003 dan Tahun 2004
[11] Kayoman L 2010 Pemodelan Spasial Resiko Kebakaran Hutan dan Lahan Di Provinsi Kalimantan Barat

Analysis of Ricefield Land Damage in Denpasar City, Bali, Indonesia

R Suyarto[1], Wiyanti[1], and I N Dibia[1]

[1]Center for Spatial Data Infrastructure Development (PPIDS) Universitas Udayana, Bali, Indonesia

Email: r.suyarto@yahoo.co.id

Abstract. Soil as a natural resource, living area, environmental media, and factors of production including biomass production that supports human life and other living beings must be preserved, on the other hand, uncontrolled biomass production activities can cause soil damage, ultimately can threaten the survival of humans and other living things. Therefore, in order to control soil damage, first must inventories the soil condition data and its damage which then visualised in soil damage potential and soil damage status. The activities of the study are the preparation of a map of the initial soil conditions and the delineation of potentially land degradation distribution. Mapping results are used as work maps for verification on the field to take soil samples and create soil damage status. In general, Denpasar City have soil damage potential at very low, low until medium rate. Soil damage status in Denpasar City generally is low damage of bulk volume, total porosity, soil permeability and electrolyte conductivity which beyond limitation thresholds.

Keywords: Rice field land damage, Soil damage, GIS, Mapping

1. Introduction

Soil as a natural resource, living area, environmental media, and factors of production including biomass production that supports human life and other living beings must be preserved, on the other hand, uncontrolled biomass production activities can cause soil damage, ultimately can threaten the survival of humans and other living things. Government Regulation of the Republic of Indonesia No. 150 of 2000 on Control of Land Degradation for Biomass Production, Government Regulation No. 38 of 2007 on Governmental Administration of Government, Provincial Government and Regency / Municipality Government, clearly stipulates that Provinces and Regencies have the mandate, among others, to supervise the control ofsoil damage. This mandate is reinforced by the release of the No. 19 of 2008 on Minimum Service Standards (SPM) Environmental Area of Provinces and Regencies / Municipalities and Regulation of the Minister of Environment No. 20 of 2008 [1] on Technical Guidance SPM Environmental Area of Provinces and Regencies / Municipalities. It is therefore necessary to have a data containing soil conditions and soil damage status as stated in a minimum 1: 50,000 scale map for urban areas and a scale of 1: 100,000 for the district as a starting point for monitoring.

In the Government Regulation of the Republic of Indonesia No. 150 of 2000 standard criteria have been established for the destruction of land for the production of Biomass, including the parameters that must be established and measurement methodology. While the procedure of measuring the standard criteria of land damage for biomass production has been established through the Regulation of the State Minister of Environment No. 07 of 2006. The products of this law and other research will be as reference in the preparation of Soil Status Map [2–4]. The current problem is the lack of data on the condition and status of soil damage both extent and spatial distribution in various regions. Therefore, in order to control soil damage, first must inventories the soil condition data and its damage which then visualised in soil damage potential and soil damage status

2. Methodology

2.1. Area of Study

Denpasar City is the capital city of Bali Province, Indonesia. Denpasar City located between 08° 35" 31' -08° 44" 49' South Latitude and 115° 10' 23" -115° 16"27' East Longitude, which is bordered by: North: Badung regency, East: Gianyar Regency, South: Badung Regency and Lombok Strait, West: Badung Regency. The total area of Denpasar City is 12.78 Ha, including addition of reclamation of 380 Ha seashore beach [5].

2.2. Materials and Tools

2.2.1. Materials

Materials for this study are:

1) *Soil map*
As a basic map material can use the map of Rupa Bumi Indonesia (RBI) production of the National Survey and Mapping Coordination Agency (Bakosurtanal) [6].
2) *Spatial Plan Map (RTRW)*
RTRW map used is map RTRW lingkat Regency or City. The function of the RTRW map in the preparation of the initial soil condition map is to filter the effective working area to be surveyed and to see the soil conditions in the field based on the status of the land.
3) *Land map*
Land maps are needed as materials for assessing the potential for soil damage. The type (classification) of land used may vary, generally using the Soil Taxonomy (USDA) classification system and sometimes also the equivalent of Puslitan and FAO classifications.
4) *Slope Map*
Slope map is processed from topographic map. The slope of the land is closely related to the potential of erosion as the main factor causing soil damage to be used as material for assessing the potential for soil damage. Slope maps that are easily obtained are sourced from land unit maps.
5) *Rainfall Map*
The availability of this data is necessary in determining the potential for soil damage. The rain map is usually composed of isohyet maps. BMKG (Meteorology Climatology Geophysics Agency) at the provincial level sometimes also arrange rain maps. Another source is a rain map compiled by Bappeda in each regency, city or province.
6) *Landcover and Land Use Map*
In general, soil degradation in Indonesia occurs as an influence of human activities (land use), such as agriculture, forestry, mining, industry and so on. In estimating the potential for soil damage, the land use / cover map used is the most recent map that is still relevant to describe the condition
7) *Maps and Other Data*
Maps and other data such as critical land maps or direct reports from the public or relevant agencies on the existence of soil damage to certain areas, such information can be accommodated in the initial condition map if the position and distribution are known.

2.2.2. Methodology

The activities of the study were the preparation of a map of the initial soil conditions and the delineation of potentially land degradation distribution [7,8]. Mapping results were used as work maps for verification on the field. In principle, the initial conditions map (work map) presents alleged information of potential of soil degradation based on analysis of maps and secondary data. The potential for soil damage is assumed by two approaches, namely overlay of thematic maps and scoring of factors considered to have an effect on soil damage.

The process of preparing this work map consists of several steps, namely (1) screening of effective working area, (2) Storing potential land damage on thematic maps (3) overlay some of the thematic maps required to create the potential land damage, (4) field verification, (5) Preparation of Land Condition Map, (6) Creation of Land Damage Map. Figure 1 shows the methodology diagram.

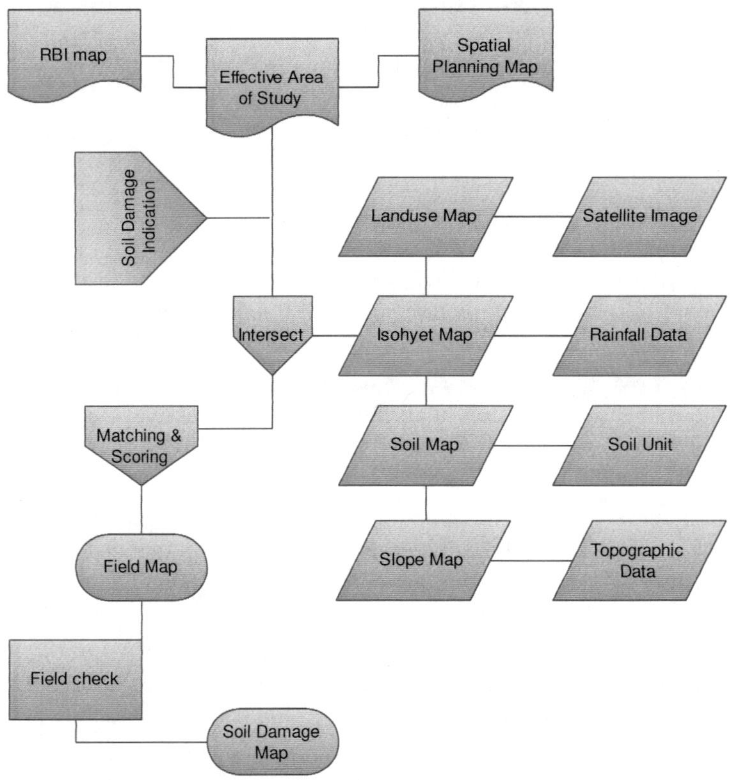

Figure 1. Methodology Diagram

1) *Intersect Process*
The intersect process will produce new polygons with attributes of land conditions in accordance with thematic constituent maps. In this polygon will be assessed potential land damage by considering the weighted score of potential damage to the land of its constituent maps.

2) *Matching and Scoring*
Scoring scores or weighting scores of potential damage to the soil obtained from the multiplication of the rating value with the weight value of each thematic map of soil maps, slope maps, rainfall maps and landuse maps. The rating value is set to range from 1 to 5. While the weight value for land use maps and soil maps are weighted by two (2) and the slope and rainfall maps are weighted by three (3).
The higher scoring scores obtained, the higher the potential the soil is damaged. The rating and scoring values of soil factor presented in Table 1, the rating and scoring values of slope factor

presented in Table 2, the rating and scoring values of rainfall presented in Table 3 and the rating and scoring values of landuse presented in Table 4.

Table 1. Assessment of Potential Land Damage Based on Soil Type

Soil Type	Class	Symbol	Rating	Score*Weight
Vertisol, with a quik humidity	Very low	T1	1	2
Oxisol	Low	T2	2	4
Alfisol, Mollisol, Ultisols	Moderate	T3	3	6
Inceptisols, Entisols, histosols	High	T4	4	8
Spodosol Andisol	Very high	T5	5	10

Description: * Aquents, Aquepts, Aquults, Aquoxs, etc. With the exception of Sulfaquept and Sulfaquent considered potentially high damage.

Table 2. Assessment of Potential Land Damage Based on Slope

Slope (%)	Soil Damage Class	Symbol	Rating	Score
0 – 8	Very low	L1	1	3
9 – 15	Low	L2	2	6
16 – 25	Moderate	L3	3	9
26 – 40	High	L4	4	12
> 40	Very high	L5	5	15

Table 3. Assessment of Potential Land Damage Based on Rainfall Intensity

Rainfall (mm)	Soil Damage Class	Symbol	Rating	Score
< 100	Very low	H1	1	3
1000 – 2000	Low	H2	2	6
2000 – 3000	Moderate	H3	3	9
3000 – 4000	High	H4	4	12
> 4000	Very high	H5	5	15

Table 4. Assessment of Potential Land Damage Based onlanduse (a)

Rainfall (mm)	Soil Damage Class	Symbol	Rating	Score
Forest Rice field Reed	Very low	T1	1	2
Mixed garden Shrubs/ bush Meadow	Low	T2	2	4
Production forest Cultivation	Moderate Table 4 (b)	T3	3	6
Moor (annual crop)	High	T4	4	8
Open ground/ clearing	Very high	T5	5	10

3) *Soil Damage Potential*

Potential damage calculated by grouping to the weighted score accumulation, the result of scoring value with the weight of each thematic map. This potential assessment is performed on the polygons produced by the intersect process. The value of the accumulated score ranges from 10 to 50. The maximum value occurs if all attribute values of each thematic map used are potentially very high against the soil damage. Based on the accumulation of the score, all the land to be assessed is classified into 5 classes of potential damage, ie very low, low, medium, high, and very high damage.

IOP Conf. Series: Earth and Environmental Science **123** (2018) 012028 doi:10.1088/1755-1315/123/1/012028

4) *Field Check*

Field verification is to prove whether or not the indication or potential damage to the land has been properly corrected. This activity is carried out in a priority order based on the potential for soil damage. The main priority is on land with the highest potential damage. In the field survey, there are several methods of land observation and sampling. Commonly used soil sampling methods are random sampling. This system is usually applied to fairly diverse land conditions where the land is divided into several relatively homogeneous land units through the overlay process of some thematic maps as previously mentioned and then in each random sampling unit.

Several standard parameters of ground damage criteria were measured directly in the field, while others were measured through laboratory analysis. For this purpose, soil sampling is required at each observation point. The verification and identification of soil damage parameters also allow for the change of polygons in the potential soil damage map. Location of soil sampling are focused on ricefield with assumption that ricefield will represent higher value to sustainable crops and agriculture development in general.

5) *Soil Damage Map*

The Soil Damage Map contains the values of ground damage criteria. The value included in the map legend is the range value of each parameter. The criteria for classifying the soil damage are presented in Table 5.

Table 5. Criteria for Potential Soil Damage

Symbol	Soil Damage Potential	Score
PR.I	Very low	< 15
PR.II	Low	15 – 24
PR.III	Moderate	25 – 34
PR.IV	High	35 – 44
PR.V	Very high	45 - 50

Data taken from the location of the activity to determine the criteria of the degree of soil damage consist of slopes, rainfall, erosion / sedimentation, thickness of solum, surface unity, soil fraction composition, degree of water grinding, fill weight, total porosity, pH (H_2O), electrolyte conductivity, redox and microbial quantities. These criteria are the minimum standard criteria set by the Environmental Ministry of Indonesia. The raw criteria are factors that may affect the quality of the soil for biomass production. Table 6 indicates the criteria for soil damage [1].

Table 6. Criteria for Soil Damage

No	Parameter	level of damage				
		B	BPAR	AR	ARPR	R
1	Slopes (%)	0 – 8	8 – 15	15 – 25	25 – 45	> 45
2	rainfall (mm/th)	< 1500	1500–2000	2000–1500	2500-2000	> 2500
3	Solum thickness (cm)	> 150	150 - 100	100 - 50	50 - 20	< 20
4	Surface Unity (%)	< 10	10 – 20	20 – 30	30 – 40	> 40
5	Composition of Soil Factionsand Koloid (%) sand (%)	 > 33 < 20	 < 33 – 28 > 20 – 40	 < 28 – 23 > 40 – 60	 < 23 – 18 > 60 – 80	 < 18 > 80
6	Water Smoothing Degree (cm/jam)	4,0 - 5,0	4,0 - 3,0 5,0 - 6,0	3,0 - 2,0 6,0 - 7,0	2,0 - 0,7 7,0 - 8,0	< 0,7 > 8,0
7	Content Weight (g/cm^3)	< 0,8	> 0,8 – 1,0	> 1,0- 1,2	> 1,2 – 1,4	> 1,4
8	pH (H2O) 1 : 2,5	6,0 – 5,5 6,0 – 6,5	5,5 – 5,0 6,5 – 7,0	5,0 – 4,5 7,0 – 7,5	4,5 – 4,0 7,5 – 8,0	< 4,0 > 8,0
9	Power Conductivity (mS/cm)	< 1,0	> 1,0 – 2,0	> 2,0 – 3,0	> 3,0 – 4,0	> 4,0
10	Number of Microbesa (cfu/g)	>10^8	< 10^8 - 10^6	< 10^6 - 10^4	< 10^4 - 10^2	< 10^2

3. Result and Discussion

3.1. Soil Damage Potential

The condition of agricultural land is indicated by the condition of existing land use and available facilities such as irrigation conditions. Ricefieldconditions in subdistrict of West and North Denpasar are ricefield / plantation land area of 256ha and 754 ha respectively from 24.13 km^2 and 31,12 km^2 area respectively. The area of ricefield in East Denpasar District is 694 ha and in South Denpasar is 896 ha, while the area of tegalan land in both subdistricts is 144 ha and 183 ha [9]. Mapping results are used as work maps for verification of the field. In principle, the initial conditions map (work map) presents information on potential soil damage based on analysis of maps and secondary data. Work maps were made by using overlay method between isohyet map, soil type map with slope map and landuse map. Spatial overlay process is done using QGIS device.

The analysis shows that the potential of soil damage in North Denpasar District and West Denpasar District has only two classes, which are low potency and medium potential. Soil that has the lowest damage potential has the highest area that reaches 1,769.55 ha or 96.69% from the research area. While the soil that has the potential for damage is only 60.64 ha or 3.31% of the area of North Denpasar District and South Denpasar District. Based on the location of the administration, North District has a potentially larger area of moderate damage compared to West Denpasar District. Potential land of moderate damage in North Denpasar District has an area of 57.12 ha while in West Denpasar District is only 3.53 ha. The common cause of the potential for low soil damage is due to the soil use and steep slope conditions. Types of land use of moor affected much of the potential for soil damage in the West District (Padang Sambian Kaja). While in sub district of North Denpasar, besides being caused by land use type (Ubung Kaja), it is also caused by steep slopes condition or has slopes exceeding 25% as happened in Peguyangan Kangin, Peguyangan Kaja and Tonja Villages.

South Denpasar Sub district and East Denpasar Sub district have only two classes that are low potency and medium potency. Soil that has the lowest damage potential has the highest area that reaches 1,769,55 ha or 96,69% from the research area. While the Soil that has the potential for damage is only 60.64 ha or 3.31% of the area of South Denpasar District. Based on the location of the administration, Southern Sub district has a land that has the potential of damage is more extensive when compared with the District of East Denpasar. Potential land of moderate damage in South Denpasar District has an area of 57.12 ha while in East Denpasar District is only 3.53 ha. Map of class potential distribution of soil damage in South Denpasar Sub district and East Denpasar Sub district shows that potentially moderate soil damage only reach 60,64 ha or 3,31% from study area. The common cause of the potential for low soil damage is due to the soil use and steep slope conditions. Types of land use moor affect much of the potential damage to land in Eastern District (Padang Sambian Kaja). While in sub district of South Denpasar, besides being caused by land use type of moor (Ubung Kaja), it is also caused by steep slopes condition or has slopes exceeding 25% as happened in Peguyangan Kangin, Peguyangan Kaja and Tonja Villages. Table 7 listed the area of soil damage potential for ricefield in Denpasar City. Meanwhile Figure 2 shows the soil damage potential in Denpasar City.

Table 7. Area of Soil Damage Potential for Ricefield in Denpasar City

Landuse	Sub-district	Village/ Kelurahan	Class	Area (Ha)
Ricefield	West Denpasar	Padang Sambian, Padang Sambian Kaja, Padang Sambian Klod, Dauh Puri Kauh, Pemecutan Klod,	Low	158.51
Ricefield	West Denpasar	Padang Sambian Klod	Medium	35.18
Ricefield	North Denpasar	Ubung Kaja, Ubung, Tonja, Peguyangan Kangin, PeguyanganK aja, Peguyangan	Low	503.41
Ricefield	North Denpasar	Peguyangan Kaja, Peguyangan Kangin, Tonja	Low	10.87
Ricefield	East Denpasar	Dangin Puri Klod, Sumerta, Kesiman Kertalangu, Sumerta Klod, Sumerta, Kesiman Kertalangu, Kesiman Petilan, Kesiman, Penatih	Medium	577.72

IOP Conf. Series: Earth and Environmental Science **123** (2018) 012028 doi:10.1088/1755-1315/123/1/012028

Landuse	Sub-district	Village/ Kelurahan	Class	Area (Ha)
		Dangin Puri, Penatih		
Ricefield	South Denpasar	Sesetan, Panjer, Renon, Sanur Kaja, Pemogan, Pedungan, Sidakarya, Sanur Kauh	Low	558.59
Ricefield	South Denpasar	Pedungan, Sidakarya, Pedungan	Very Low	0.94

4.2. Soil Damage Status

The result of analysis shows that in West Denpasar Sub-district there are 4 (four) parameters which have percentage value of relative frequency of soil over 10% ieBulk Volume, Soil pH, Electrolite Conductivity, and Total Soil Microbe. The situation shows that in West Denpasar Sub-district, the weight of the soil is relatively damaged. In addition, soil pH, soil DHL, and soil microbes also bring a relative frequency of soil damage of 20%, 40% and 60% respectively. Conditions in North Denpasar District have better condition than West Denpasar District. Bulk Volume is the only parameter that has a percentage value of the relative frequency of soil damage exceeding 10%. In North Denpasar District the relative frequency value of soil damage from soil contents weight is 100%.

Based on the criteria of soil degradation, relative frequency of soil degradation, and relative frequency score in North Denpasar Sub-district and West Denpasar Sub-district, the status of soil damage in both sub-districts can be calculated. The scores of soil damage for ricefield shows that the level of soil damage in North Denpasar District has a better total score of 4 with the status of low damage, whereas the total score of soil damage in West Denpasar District is 10 with the same status of low damage. The area of soil that has a score of 4 in North Denpasar District is 1,118, 47 ha, while the soil that has a score of 10in West Denpasar District is 711,72 ha. Relative frequency data of soil degradation and relative frequency scores for each standard of soil damage and total score for West Denpasar Sub-district and North Denpasar Sub-district are presented in Table 8.

Although the study area has a low degree of soil damage, the mitigation and recovery measures need to be considered, especially for the area of West Denpasar. Measures to address parameters that indicate damage or potential damage need to be considered based on the type of planted commodity. The parameter indicates potential damage to some locations but the range of values is still at a scale that allows the growth of plant roots. However, countermeasures are needed to improve the weight of the soil content so that plant growth becomes better. One way that can be done is to cultivate the land for moor land in accordance with the rules of soil conservation or when the paddyfield is not planted with rice paddy. Control measures that need to be done is the addition of organic materials. An increase in soil pH towards a slightly alkaline direction (> 6.5) needs to be controlled. Organic materials can be used because organic materials have a high buffer capacity to change soil pH.

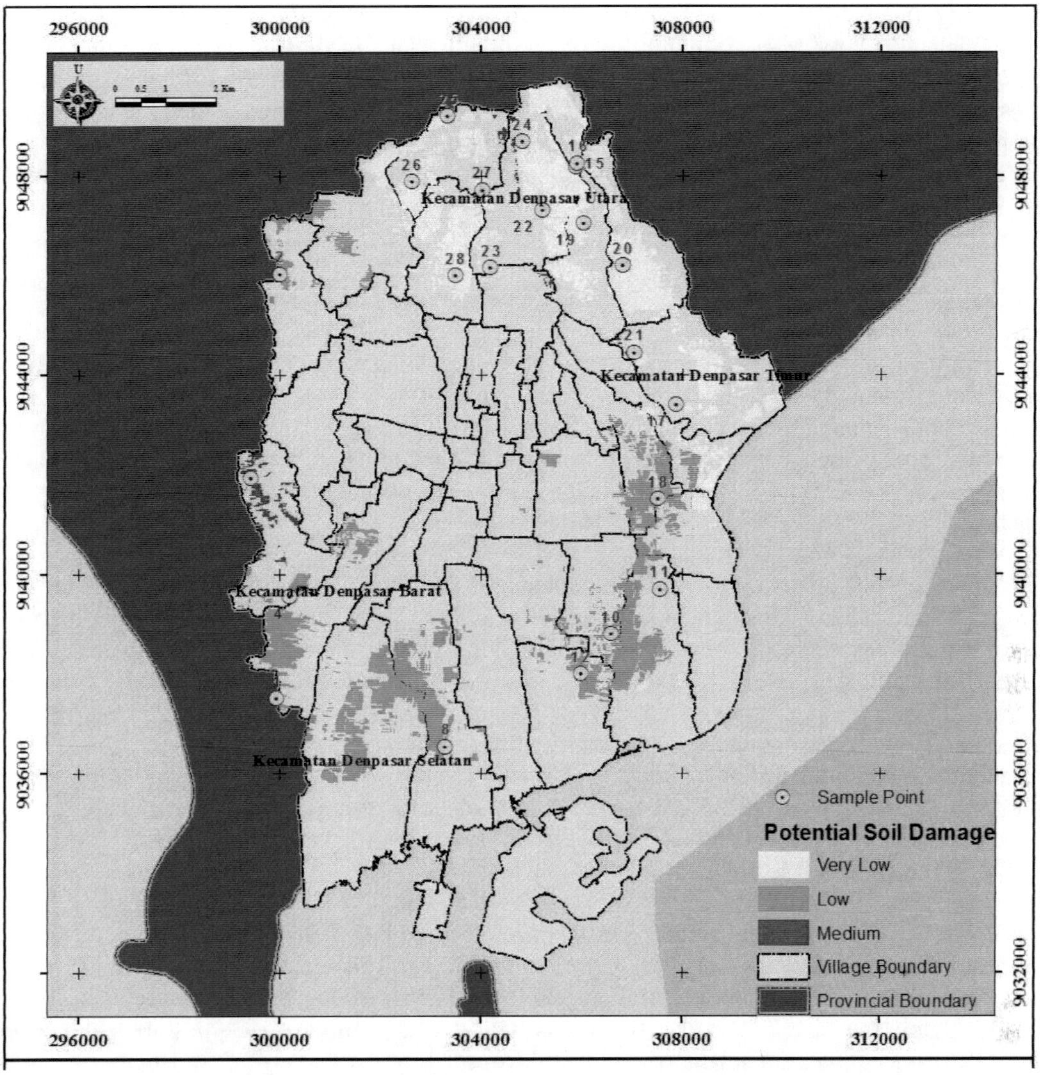

Figure 2. Map of Soil Damage Potential for Ricefield in Denpasar City

Table 8. Soil Damage Score for West Denpasar Sub-district and North Denpasar Sub-district

No	Criteria for Soil Damage	West Denpasar		North Denpasar	
		Relative Frequency of Soil Damage	Relative Frequency Score	Relative Frequency of Soil Damage	Relative Frequency Score
1	Slope (%)	0%	0	0%	0
2	Rainfall (mm/th)	0%	0	0%	0
3	Solum width (cm)	0%	0	0%	0
4	Surface rock (%)	0%	0	0%	0
5	Sand (%)	0%	0	0%	0
6	Permeability (cm/jam)	0%	0	0%	0
7	Bulk Volume (g/cm3)	100%	4	100%	4
8	pH (H_2O) 1 : 2,5	20%	1	0%	0
9	electrical conductivity (mS/cm)	40%	2	0%	0
10	Number of Microbes (cfu/g)	60%	3	0%	0
Total Score			**10**		**4**

IOP Conf. Series: Earth and Environmental Science **123** (2018) 012028 doi:10.1088/1755-1315/123/1/012028

Increasing the value of soil electrolyte conductivity indicates the occurrence of salinization process which at> 4 mS / cm can inhibit plant growth. The increasing value of soil electrolyte conductivity indicates an increase in the concentration of dissolved salt in the soil that may be caused by excessive water volume or increased solute salt concentration due to the fertilization or intrusion of seawater on coastal soils. When it is compared to the soil fertility status, soil in the study area still have a low DHL value so that specific control or recovery measures is not necessary unless there is a drastic increase in soil electrolyte conductivity value over a short period of time. However, organic materials can also be used because of the ability of organic materials to bind organic or inorganic ions in the soil environment thus preventing the process of salinization.

The improvement of the total microbial population of the soil is necessary to restore the soil biological fertility status because microbes are the key agents of biochemical processes in the soil. Microbial activity greatly determines the chemical, physical and biological properties of the soil. To restore the number of microbial populations the soil can be done by adding microbes among others with compost applications or regulating environmental conditions to be more suitable for microbial growth and development. One step that can be done is to use organic materials that have diverse qualities.

The condition of soil damage in South Denpasar Sub-district and East Denpasar sub-district indicates that the soil damage status in both Sub-districts can be calculated. The compilation result of the relative frequency score of soil damage levels indicates that the level of soil damage to in both Sub-districts has a total score of 2 to 8 with the status of low damaged soil.

Inceptisol is the largest land order in the city of Denpasar and also dominates the District of East Denpasar and South Denpasar. Inceptisol land in both sub-districts is used for mooring / gardens, ricefields, grasslands or associations between types of land use. In this study, Inceptisol soil is classified into land which is an association of ricefields. Inceptisol soil damage is low damaged by a score of 4 for inceptisol of ricefield association.

Alfisol, Entisol, and Mollisol are three types of minor land in Denpasar City. Alfisol and Molisolsoil are in the middle and north of Denpasar City. While the Entisolsoil located in the southern city of Denpasar. Usually these three types of soil are also associated with the dominant species of Inceptisol soil. In this study, the Inceptisol soil associations with the three minor soil types are grouped into 3 groups: Alfisol-Inceptisol, Entisol-Inceptisol, and Mollisol-Inceptisol.

The three associations of Inceptisol land with Alfisol, Entisol and land have a slightly damaged soil category with a score of 4 for the Alfisol-Inceptisol association, a score of 8 for the Entisol-Inceptisol association, and a score of 6 for the Mollisol-Inceptisol association. In general, the types of parameters that indicate the damage are total porosity, soil permeability and electrical conductivity beyond the critical threshold. The Entisol-Inceptisol Association has the highest level of damage with the highest score. The condition is due to the condition of Entisol soil that is still developing and its presence near the beach. Land entisol is located near the coast has a high electrical conductivity (DHL) and fast up to very fast soil permeability. The high electrical conductivity condition is a manifestation of high salt content and leads to a base pH of the soil. In total, extent of damaged soil with minor damage category is 2.586, 96 Ha from agricultural land area of 3,430.76 Ha or 75.404% of agricultural land suffered low damage. Table 9 listed the soil damage for Alfisol – Inceptisol soil. Table 10 listed the soil damage for Entisol – Inceptisol soil and Table 11 listed the soil damage for Mollisol – Inceptisol soil. Figure 3 shows the soil damage status in Denpasar City.

Table 9. Soil Damage for Alfisol – Inceptisol soil in South Denpasar and East Denpasar

No	Parameter	Limitation threshol	Samples/ Land unit					%	Score
			1	5	9	10	12		
1	Solum depth	<20 cm	>20	>20	>20	>20	>20	0	0
2	Surface rocks	>40 %	<40	<40	<40	<40	<40	0	0
3	Fraksi: Koloid sand Fraction	<18 % koloid; >80 % sand	14,97	3,52	3,47	5,64	6,91	0	0
4	Bulk Volume	>1,4 g/cm3	0,87	1,06	1,24	1,06	0,86	0	0
5	Total Porosity	<30 %; >70%	65,06	57,11	50,44	57,09	62,41	0	0
6	Permeability	<0,7 cm/jam; **Table 9** >8,0 cm/jam	54,89	5,41	0,48	0,37	29,43	100	4
7	pH (H_2O) 1 : 2,5	<4,5 ; >8,5	6,82	7,00	6,55	6,86	6,98	0	0
8	Electrolyte conducitvity	>4,0 mS/cm	1,18	1,05	0,16	0,95	0,97	0	0
9	Redoks	<200 mV	-	-	-	-	-	-	-
10	Microbe total	<10^2cfu/g soil	3 x 10^9	125x10^7	116x10^7	125x10^7	3x10^9	0	0
	Score total		4 (Low Damage)						
	Area		71,63 ha						

Table 10. Soil Damage for Entisol – Inceptisol soil in South Denpasar and East Denpasar

No	Parameter	Critical Threshold	Sample/ Land unit					%	Score
			3	7	8	11	15		
1	Solum depth	<20 cm	>20	>20	>20	>20	>20	0	0
2	Surface rocks	>40 %	<40	<40	<40	<40	<40	0	0
3	Fraksi: Koloid sand Fraction	<18 % koloid; >80 % pasir Kuarsitik	14,87	11,48	8,00	2,91	4,37	0	0
4	Bulk Volume	>1,4 g/cm3	1,21	0,27	0,99	0,72	0,66	0	0
5	Total Porosity	<30 %; >70%	56,05	87,53	61,55	69,96	69,12	20	1
6	Permeability	<0,7 cm/jam; >8,0 cm/jam	56,48	21,95	0,53	0,96	11,37	80	4
7	pH (H_2O) 1 : 2,5	<4,5 ; >8,5	6,90	7,65	7,54	7,35	7,35	0	0
8	Electrolyte conducitvity	>4,0 mS/cm	1,41	8,18	8,79	4,97	3,55	60	3
9	Redoks	<200 mV	-	-	-	-	-	-	-
10	Microbe total	<10^2cfu/g Tanah	117x10^7	21X10^7	44x10^7	135x10^7	73x10^7	0	0
	Score total		8 (Low damage)						
	Area		160,67 ha						

Table 11. Soil Damage for Mollisol – Inceptisol soil in South Denpasar and East Denpasar

No	Parameter	Critical Threshold	Sample/ Land unit				%	Score
			6	13	14	20		
1	Solum thickness	<20 cm	>20	>20	>20	>20	0	0
2	Surface Unity	>40 %	<40	<40	<40	<40	0	0
3	Composition of Fraction: Koloid Sand	<18 % koloid; >80 % Sand Kuarsitik	5,49	4,53	5,66	7,70	0	0
4	Content Weight	>1,4 g/cm3	0,83	0,83	0,77	1,01	0	0
5	Total Porosity	<30 %; >70%	67,06	67,62	69,57	60,08	0	0
6	Water Smooting Degree	<0,7 cm/jam; >8,0 cm/jam	5,57	0,53	0,48	2,86	50	2

	(permeabilitas)							
7	pH (H_2O) 1 : 2,5	<4,5 ; >8,5	6,70	6,76	7,11	6,90	0	0
8	electrical conductivity	>4,0 mS/cm	0,81	1,65	2,33	0,10	0	0
9	Redoks	<200 mV	-	-	-	-	-	-
10	Number of Microbes	<10^2cfu/g Tanah	$7x10^8$	$67x10^7$	$3x10^9$	$12x10^7$	0	0
	Score total				2 (Low damage)			
	Area				55,11			

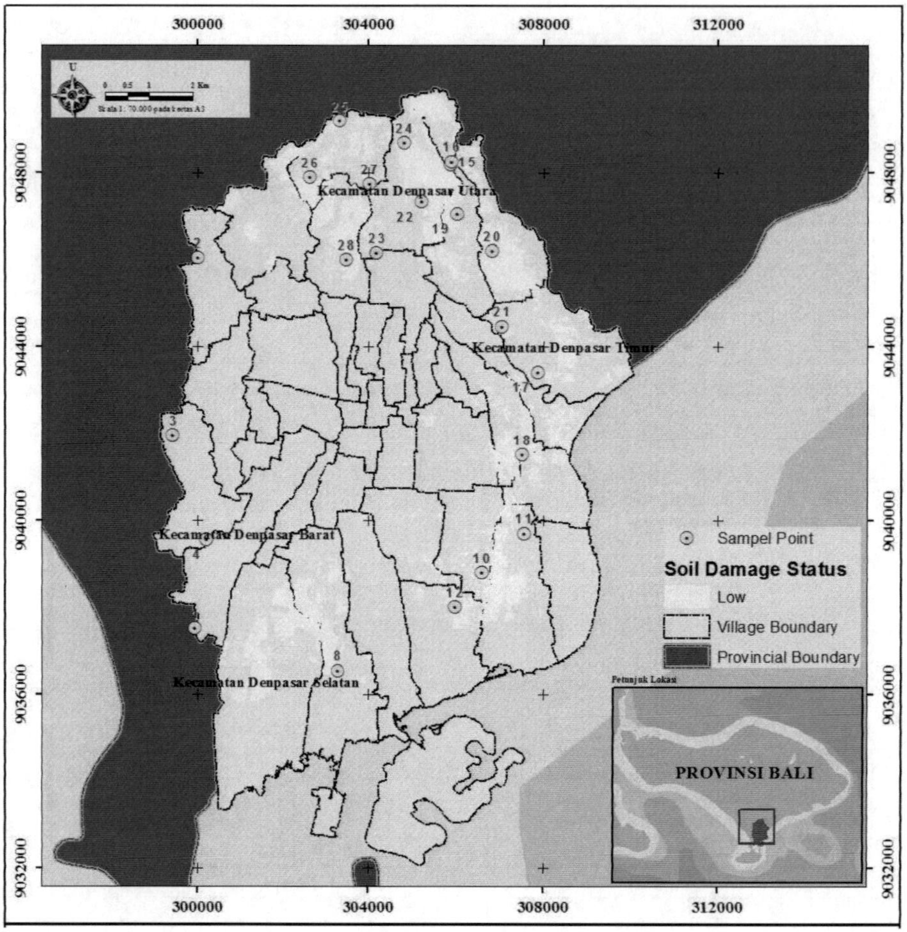

Figure 3. Soil Damage Status in Denpasar City

4. Conclusions

In general, Denpasar City have soil damage potential at very low, low until medium rate. Physical condition of ricefields in study area is relatively homogenous, ie soil condition, soil physical condition, soil chemical condition and soil biological condition. The result of analysis shows that in West Denpasar Sub-district there are 4 (four) parameters which have percentage value of relative frequency of soil over 10% ieBulk Volume, Soil pH, Electrolyte Conductivity, and Total Soil Microbe. Bulk Volume is the only parameter that has a percentage value of the relative frequency of soil damage exceeding 10%. In North Denpasar District the relative frequency value of soil damage from soil contents weight is 100%. The condition of soil damage in South Denpasar Sub-district and East

Denpasar Sub-district, indicates score of soil damage levels indicates that the level of soil damage to in both Sub-districts has a total score of 2 to 8 with the status of low damaged soil. Soil damage status in Denpasar City generally is low damage of bulk volume, total porosity, soil permeability and electrolyte conductivity which beyond limitation thresholds.

5. Acknowledgments

Writers would like to thank the Rector of UniversitasUdayana, Dean of Agriculture Faculty of UniversitasUdayana, Head of Center for Spatial Data Infrastructure Development (PPIDS) UniversitasUdayana, fellow staff, researchers and professors at UniversitasUdayana for all the support, comments and suggestions that makes this paper finished.

6. References

[1] Kementerian Lingkungan Hidup 2008 *Peraturan Menteri Negara Lingkungan Hidup Nomor 20 Tahun 2008*
[2] Wu J, Li Z, Gao Z, Wang B, Bai L, Sun B, Li C and Ding X 2015 Degraded land detection by soil particle composition derived from multispectral remote sensing data in the Otindag Sandy Lands of China *Geoderma* **241–242** 97–106
[3] Zanetti S S, Ceclio R A, Silva V H and Alves E G 2015 General calibration of TDR to assess the moisture of tropical soils using artificial neural networks *J. Hydrol.* **530** 657–666
[4] Medina J, Monreal C, Barea J M, Arriagada C, Borie F and Cornejo P 2015 Crop residue stabilization and application to agricultural and degraded soils: A review *Waste Manag.* **42** 41–54
[5] Dinas Kominfo Kota Denpasar 2015 *Sejarah Kota Denpasar*
[6] Suwahyuono, Hartono, Nurwadjedi, Sutrisno D, Suriadi A B, Radiarta N and Wibisono Y 2000 *Standard Data Tematik Sumber Daya Alam: Menuju Terbentuknya Basisdata Nasional* (Cibinong: PSSA Bakosurtanal)
[7] Lenka N K, Choudhury P R, Sudhishri S, Dass A and Patnaik U S 2012 Soil aggregation, carbon build up and root zone soil moisture in degraded sloping lands under selected agroforestry based rehabilitation systems in eastern India *Agric. Ecosyst. Environ.* **150** 54–62
[8] Zhang Y, Guo S, Liu Q, Jiang J, Wang R and Li N 2015 Responses of soil respiration to land use conversions in degraded ecosystem of the semi-arid Loess Plateau *Ecol. Eng.* **74** 196–205
[9] BPS Kota Denpasar 2012 *Kota Denpasar Dalam Angka 2012*

IOP Conf. Series: Earth and Environmental Science **123** (2018) 012029 doi:10.1088/1755-1315/123/1/012029

Spatial Modelling of Land Price in The Semarang City

W Widjonarko[1]

[1]Department of Urban and Regional Planning, Universitas Diponegoro, Indonesia

Email: widjonarko39@gmail.com

Abstract: Land has a very important role in supporting the population activity in both urban and rural areas. Demand for land tends to increase due to the increase in population, on the other hand the availability of land is limited. The increasing demand of land also occurred in the city of Semarang due to population growth and economic activity growth. The increasing demand for land in Semarang City has caused a shift in spatial demand patterns. The shift in land demand is due to limited supply of land in the area near to the city center, and the price become unaffordable for some residents of Semarang City. Due to the limitation of land supply in the city center has affected to the increasing demand of land in the suburbs. This phenomenon causes an increase in the price of land in the periphery of Semarang, and forms a land price pattern that resembles a circus tent, especially at a new center of activity on the periphery.

Keywords: Land, Land Price, Spatial Modelling

1. Introduction

The city of Semarang is developing very rapidly in the last decade, one of the indications is the growth of the population that grows significantly. BPS (Central Bureau of Statistics) [1] noted that population growth in Semarang increased by more than 350 thousand people, or about 35 thousand people per year. The population in Semarang City in 2015 was increased by 24.64% as compared to the population in 2005. This rapid population growth largely filled the spaces on the outskirts of Semarang City, creating a competition to gain space in the region - suburbs [2].

The population growth of Semarang City gives an influence to the demand of land on the periphery. Land demand on the periphery affects land price increase. This increase in land prices is a market phenomenon, where if demand increases while limited land supply will cause land price increase [3–5]. The three previous studies also see that the rise in urban land prices in addition to influenced by non-accumulated land supply factors is also influenced by access and availability of urban supporting facilities.

Demand for land on the outskirts of Semarang City causes changes in spatial land price patterns. The pattern of land prices that usually follow the distance function, which is farther away from downtown land prices will be lower starting to change, the location in the suburbs are dominated by activities that provide high economic opportunities tend to experience significant price increases [6–9]. This incident led to the emergence of new residential centers on the periphery with random patterns [10]. The development of residential areas on the outskirts of Kota Semarang has also provided a very significant implication for changes in land price patterns, especially land prices based on market prices. This change of spatial pattern will give an implication for efficient land management [11] and it is necessary to study the spatial pattern of land price (market price) in an effort to determine the instrument of control of spatial use as well as the instrument to optimize the source of regional income from the permit land.

Previous research in Semarang City and Magelang City used NJOP data of land as a foothold in developing land pricing model, then in this research will use market price as a foundation to form model of spatial land price in Semarang City. The use of market price as a foundation in the development of spatial model is based on the economic opportunities that can be obtained by the stakeholders and at the same time eliminate the potential losses from the Government of Semarang City against PAD revenue from various land related transactions.

2. Data and Methods

2.1. Data

This research used land price data according to market price in Semarang City. The process of collecting land price data was done by using the primary survey and survey method online by using the data of property selling price at several online trading sites. As a comparison of price data on the site online also conducted the excavation of land market information by using land buying and selling ads in local mass media. To find out the real price of the offer on the online site and mass media was done by phone verification process to get the exact price of the offered price. In general, the price of land price according to market price in Semarang City after going through the verification process is still in line with theories about urban land prices, where the price of land in the city center has the highest value and in the periphery has a lower value. A more complete description of land prices in Semarang City can be followed in the following tables and figures.

2.2. Methods

The process of data processing and analysis in making spatial pattern of land price in Semarang City used the aid of geographic information system software. Data processing in general as described in the previous description in the previous section is through georeferment process of land price data until the interpolation process. In detail the process of data processing and spatial pattern analysis of land prices can be followed in the following description.

2.2.1. Digitizing Land Price Data into Geographic Information System Using Google Earth

This process is done by finding the location geoposition of the land to be sold in the city of Semarang. After that the position is marked and coordinate system information is then saved in * kmz format.

2.2.2. Convert KMZ to Files in Geographic Information System Formats

After all land price data has georeferenced then the process of converting data from google earth into the geographic information system software environment can be started. At this stage, it is also done consolidation of data topology so ready to be done process of analysis to data.

2.2.3. Data Interpolation

The interpolation process uses four approaches, and then selected the most appropriate to the factual conditions.

IOP Conf. Series: Earth and Environmental Science **123** (2018) 012029 doi:10.1088/1755-1315/123/1/012029

2.2.4. Spatial Analysis of Land Price Pattern

At this stage, spatial statistical analysis was conducted to see the typology of land price and its form factor. This process is carried out by overlapping the land price map with the map of land price form factor such as the distribution map of the settlement area, the distribution map of the activity center, the regional accessibility map, the disaster condition map [12]. The result of this analysis was able to describe how big contribution of land price form factor to spatial pattern of land price (Tabel 1).

Table 1. Land Price by Market Price in Semarang City 2016

No	District	Location	Market Price (IDR)
1	Tugu	Desa Randu Garut (edge of Semarang highway Kendal)	3,243,243.24
2	Ngaliyan	Candi Kalasan	2,250,000.00
		Sekitar LP Kedungpane	725,000.00
		Candi Perak	1,650,000.00
		Bringin Putih	800,000.00
3	Mijen	Polaman	700,000.00
		Ngadirgo	175,000.00
4	Pedurungan	Jalan Wolter Monginsidi	3,500,000.00
		Jalan Fatmawati	1,900,000.00
		Klipang	1,165,413.53
5	Semarang Barat	Jalan Pamularsih, near simongan	8,650,519.03
		Jalan Abdurahman Saleh (near Balaikota)	4,972,375.69
		Jl. Indraprasta	1,900,000.00
6	Gayamsari	Jalan Elang Raya	2,250,000.00
7	Gunungpati	Jalan rejosari III, Desa Ngijo	453,091.68
		dusn Pingangan	75,000.00
		Ds Ngijo	300,000.00
8		Jalan Sultan Agung	25,000,000.00
9		Jalan Muara Mas Gang 8	1,750,000.00
10		Jl. Watulawang 6	1,436,950.15
11	Banyumanik	Jl. Permata Grafika	1,500,000.00
		Pudak Payung	875,000.00
12		Jl. Kedungmundu, Perum Sambiroto Baru	2,300,000.00
		Watugong, jalan alvita	1,700,000.00
		Ngesrep Barat	3,400,000.00
13	Tembalang	Jl. Mulawarman Selatan	1,900,000.00
		Jalan banjarsari barat	3,000,000.00
		jangli skip	1,650,000.00
		Perum Polri Mangunharjo	1,025,641.03
		Sendang Mulyo	1,174,089.07
14	Tlogosari	Perumnas, dekat apotik jamilah	833,333.33

Source: Market Survey, 2016

3. Results and Discussion

Based on the results of tracking the land market in Semarang City by using online buying and selling site and verification process with seller can be described spatial pattern of land price in Semarang City. The spatial pattern of land prices in Semarang City has not changed significantly in pattern and still follow the pattern of urban land price in the world, where the highest price is in the central area of main activity and decreases with distance and position to main activity center and road network. A more detailed description of the spatial pattern of land prices can be followed in Figure 1.

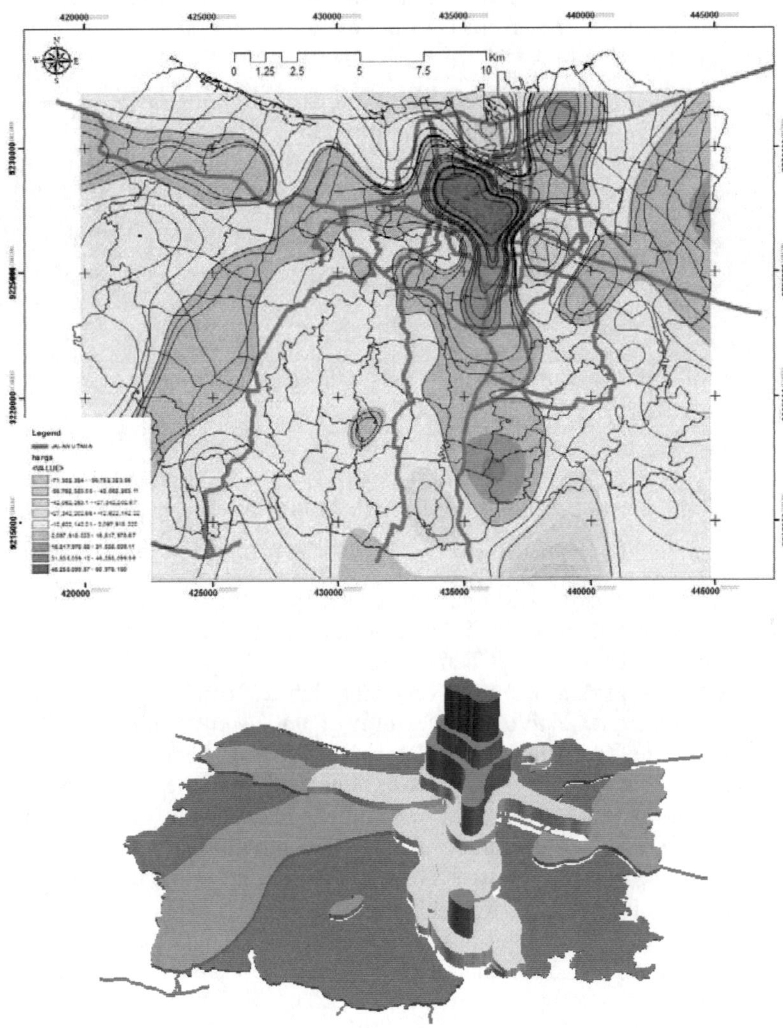

Source: olx.co.id, 2016, processed by the drafting team, 2016

Figure 1. Spatial Pattern Land Price (Market Price) Semarang City Year 2016

This spatial pattern still follows the economic patterns of the land, where the land with the opportunity to bring high economic benefits tends to have higher selling prices, if the supply of land in the region is increasingly limited. Simpang Lima area, Youth and Gajahmada area is the location with the highest land price value in Semarang City. This condition is inseparable from the highly prospective economic opportunities of trading businesses and services (malls, hotels, apartments) in these areas.

This economic opportunity is inseparable from the completeness of regional infrastructure and the accessibility of the area is very high when compared with accessibility in other areas in Semarang City. This condition will certainly provide a very high location advantage, because it can be a major magnet for economic activities in the city of Semarang. The proximity to the airport makes these areas currently filled with hotel business activities. Recorded in the area four- and five-star hotels are located. For example spatial pattern of land price can be seen from land price pattern on corridor intersection of five-pioneer independence, as can be followed in Figure 2.

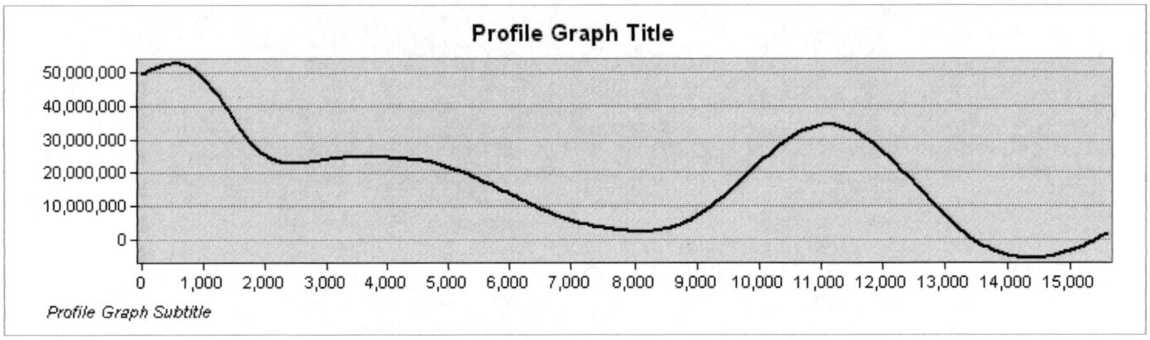

Source: olx.co.id, processed research team, 2016

Figure 2 Land Price Patterns on the Corridor of Simpang Lima-Perintis Kemerdekaan

In Simpang Lima area which is the main activity center the price of land reaches IDR 60 million per square meter, then decrease at a distance of 8 km from the intersection of five or around Jatingaleh to around IDR 8 million and rise back around Tembalang to around Rp 30 million per meter square. This pattern of increase cannot be separated from the very high demand of land in the area around Tembalang for the development of commercial activity in the form of trade, rental houses and apartments that are currently growing rapidly in the Tembalang area. Tembalang area has become a major magnet of property investment in Semarang City. This condition cannot be separated from the increasing demand of land that continues to increase in line with the growth of the number of students in Tembalang area which rose significantly from year to year. It is this student market that provides a derivative demand for land for commercial functions in the Tembalang Regions. Different conditions occur in Gunungpati area which is a region of higher education. In Gunungpati area although there is a spike in land prices in the area around Unnes Campus but the market value of land in the area is not as high as the price of land in Tembalang area. This condition is inseparable from the factor of land demand is much lower than the Undip Tembalang.

4. Conclusion

Based on the results of the analysis can be concluded that the spatial pattern of land prices in Semarang City has a common pattern that is identical with the pattern of urban land prices in general. This pattern also resembles the land price pattern in Semarang City in 1998 as the result of research conducted by Widjonarko [3] where the highest land price is in the central area of main activity and decreases as it away from the city center and the position on the network main road.

There is a tendency for land price increase in new centers in the outskirts of Semarang which is a new activity center such as in the Tembalang area, New Semarang Bukit Baru area, Pedurungan Area and Area of Sekaran. In these areas the pattern of land prices tends to be higher than the surrounding area. The patterns of land prices in areas that are not residential areas tend to not change the pattern, except in the un-built areas around the new activity center. This pattern is triggered by the demand for land in the area which tends to be higher than other areas in Semarang City, so that the market mechanism occurs. When there is a higher demand for land, on the one hand the supply of land is fixed then the consumer will be willing to pay a higher price.

5. Acknowledgments

Thank you to the Department of Urban and Regional Planning of the Faculty of Engineering Diponegoro University for the facilitation of funds for the study of Land Price Pattern of Semarang City.

6. References

[1] BPS 2015 *Kota Semarang dalam Angka 2015* (Semarang)
[2] Debolini M, Valette E, Francois M and Chéry J-P 2015 Mapping land use competition in the rural--urban fringe and future perspectives on land policies: A case study of Meknès (Morocco) *Land use policy* **47** 373–381
[3] Widjonarko W 1998 Model Harga Tanah Kota Semarang Sebagai Parameter Penentuan Tanah NJOP *Undergraduate Thesis, Diponegoro University*
[4] Raeka F and Sulistyarso H 2012 Model Perkembangan Nilai Lahan Perkotaan di Surabaya *J. Tek. ITS* **1** C48--C51
[5] Wijayanti P and Widjonarko W 2015 Model Harga Lahan Kota Magelang (Studi Kasus: Kota Magelang) *Tek. PWK (Perencanaan Wil. Kota)* **4** 727–736
[6] Elmanisa A, Kartiva A, Fernando A, Arianto R, Winarso H and Zulkaidi D 2017 Land Price Mapping of JABODETABEK, Indonesia *Geoplanning J. Geomatics Plan.* **4** 53–62
[7] Nichols J B, Oliner S, Mulhall M R and others 2010 *Commercial and residential land prices across the United States* (Divisions of Research & Statistics and Monetary Affairs, Federal Reserve Board)
[8] Albouy D and Ehrlich G 2013 The distribution of urban land values: Evidence from market transactions *Mimeogr. Univ. Illinois*
[9] Darin-Drabkin H 1977 *Land policy and urban growth* vol 16 (Pergamon)
[10] Irwin E G and Wrenn D H 2014 An assessment of empirical methods for modeling land use *Duke, JM Wu J. Oxford Handb. L. Econ.* 327–351
[11] Yu D and Yuefang S I 2015 Spatial Pattern and Bid Rent of Land Price in Beijing *Cent. Mod. Chinese City Stud. East China Norm. Univ.* 2012–2015
[12] Childs C 2004 Interpolating surfaces in ArcGIS spatial analyst *ArcUser, July-September* **3235** 569

IOP Conf. Series: Earth and Environmental Science **123** (2018) 012030 doi:10.1088/1755-1315/123/1/012030

Model of Numerical Spatial Classification for Sustainable Agriculture in Badung Regency and Denpasar City, Indonesia

N M Trigunasih[2], I Lanya[1], N N Subadiyasa[1], and J Hutauruk[2]

[1] Faculty of Agriculture, Universitas Udayana, Bali, Indonesia
[2]Center for Spatial Data Infrastructure Development (PPIDS) Universitas Udayana, Bali, Indonesia

Email: tri5963@yahoo.com

Abstract. Increasing number and activity of the population to meet the needs of their lives greatly affect the utilization of land resources. Land needs for activities of the population continue to grow, while the availability of land is limited. Therefore, there will be changes in land use. As a result, the problems faced by land degradation and conversion of agricultural land become non-agricultural. The objectives of this research are: (1) to determine parameter of spatial numerical classification of sustainable food agriculture in Badung Regency and Denpasar City (2) to know the projection of food balance in Badung Regency and Denpasar City in 2020, 2030, 2040, and 2050 (3) to specify of function of spatial numerical classification in the making of zonation model of sustainable agricultural land area in Badung regency and Denpasar city (4) to determine the appropriate model of the area to protect sustainable agricultural land in spatial and time scale in Badung and Denpasar regencies. The method used in this research was quantitative method include: survey, soil analysis, spatial data development, geoprocessing analysis (spatial analysis of overlay and proximity analysis), interpolation of raster digital elevation model data, and visualization (cartography). Qualitative methods consisted of literature studies, and interviews. The parameters observed for a total of 11 parameters Badung regency and Denpasar as much as 9 parameters. Numerical classification parameter analysis results used the standard deviation and the mean of the population data and projections relationship rice field in the food balance sheet by modelling. The result of the research showed that, the number of different numerical classification parameters in rural areas (Badung) and urban areas (Denpasar), in urban areas the number of parameters is less than the rural areas. The based on numerical classification weighting and scores generate population distribution parameter analysis results of a standard deviation and average value. Numerical classification produced 5 models, which was divided into three zones are sustainable neighbourhood, buffer and converted in Denpasar and Badung. The results of Population curve parameter analysis in Denpasar showed normal curve, in contrast to the Badung regency showed abnormal curve, therefore Denpasar modeling carried out throughout the region, while in the Badung regency modeling done in each district. Relationship modelling and projections lands role in food balance in Badung views of sustainable land area whereas in Denpasar seen from any connection to the green open spaces in the spatial plan Denpasar 2011-2031. Modelling in Badung (rural) is different in Denpasar (urban), as well as population curve parameter analysis results in Badung showed abnormal curve while in Denpasar showed normal curve. Relationship modelling and projections lands role in food balance in the Badung regency sustainable in terms of land area, while in Denpasar in terms of linkages with urban green space in Denpasar City's regional landuse plan of 2011-2031.

Keywords: Modelling; Numerical Classification; Sustainable Food Agriculture

1. Introduction

Increase in the number and activity of the population to meet the needs of their lives greatly affect the utilization of land resources [1–3]. Land needs for activities of the population continued to grow, while the availability of land is limited, so that there will be changes in land use [4,5]. Besides, the problems faced in the national food nowadays are the degradation of soil fertility [6], the import of rice, the competition of water use is increasing and the agricultural/rural infrastructure is still inadequate and the phenomenon of conversion of agricultural land nowadays.

The objectives of this research are (1) to determine spatial numeric parameter parameter of sustainable agriculture in Badung Regency and Denpasar City (2) to know the projection of food balance in Badung Regency and Denpasar City in 2020, 2030, 2040, and 2050 (3) to set the classification function Numerical spatial in making model of zonation of sustainable agriculture area in Badung Regency and Denpasar City (4) to determine the appropriate model of area to protect sustainable agricultural land in spatial and time scale in Badung Regency and Denpasar City.

2. Literature Review

Protection of sustainable food crops is a system and process in planning and establishing, developing, utilizing and nurturing, controlling and supervising the sustainable agriculture land of food and its area. Sustainable food crop is one of the government's policies in controlling the rate of conversion of agricultural land, especially rice fields in Indonesia. Sustainable food crops has the objective of (1) protecting sustainable agriculture and food lands, (2) ensuring the availability of sustainable food for agriculture, (3) realizing food self-sufficiency, sustainability and sovereignty, (4) protecting the ownership of agricultural land belonging to farmers (5) improving the welfare and prosperity of farmers and communities, (6) increasing the protection and empowerment of farmers, (7) increasing the provision of employment for decent living, (8) maintaining ecological balance and (9) realizing agricultural revitalization [7].

Within 14 years (1999-2013), there has been a change in the use of agricultural land to non-agricultural in Bali province of 4,906 ha. The average land use change per year occurs in Bali around 350 ha (0.41%). The highest rice field change during the fourteen years were in TabananRegency of 1,230 ha, followed by JembranaRegency of 1,078 ha, BulelengRegency of 677 ha, Badung Regency covering 672 ha, Denpasar with 659 ha, Regency Gianyar area of 497 ha and Klungkung Regency covering 173 ha. However, the addition of ricefield area occurred in 58 ha of Karangasem Regency and Bangli District (Mengani Village of KintamaniSubdistrict of 22 ha) [8]. The nine Regencies in Bali do not currently have a law on sustainable food crop protection mandated by Law 41 of 2009. This will have an impact on local food security, and the subak as a world cultural heritage will lose its existence as an irrigation system in Bali.

In terms of soil and water conservation, Subak has been in use since the seventh century with a semi-technical and simple watering system. Subakrice field also serves as flood control, rainfall catcher, especially located in the upper river. Rice crops as oxygen (O2) contributors on micro and macro scale [9]. The various requirements stipulated in Law no. 41 Year 2009, then should all Subakrice field in Bali need special attention about its existence. The existence of the need for non-agricultural development, such as housing, tourism, shops, industry and infrastructure facilities, limited land conversion is needed, especially in urban areas, government and tourism centers.

The act no. 41 of 2009 on the protection of sustainable agricultural land in Article 9 includes land suitability, availability of infrastructure, land use, technical potential of the land and the extent of the unity of land. The physical condition of the region consists of: the location position of the watershed, irrigation facilities, rainfall, the shape of the region /relief Slope, altitude, rainfall, suitability of agro-ecosystem lands, conformity to spatial planning and land use, are linked to areas that need to be protected in sustainable food agriculture [9].

IOP Conf. Series: Earth and Environmental Science **123** (2018) 012030 doi:10.1088/1755-1315/123/1/012030

3. Methodology

3.1. Area of Study

The research was conducted in Badung Regency and Denpasar City, Bali Province, Indonesia. Geographically, Badung Regency is located between 8^{0}14'20 "- 8^{0}50'52" South Latitude and 115^{0}05'03 "- 115^{0}26'51" East Longitude with an area of 418.52 km^2 (41,852 ha) or about 7.43% of Bali Island. BadungRegency is divided into 6 sub districts, namely: South KutaSubdistrict (this sub district is not included in the research because there is no rice field), Kuta, North Kuta, Mengwi, Abiansemal and Petang. From 6 sub-districts, Petang District has the largest area of 115 km^2, while Kuta District is the smallest area of 17.52 km^2. Badung Regency has 119 subak with 9,984 ha of rice field area [10]. The city of Denpasar is 127,78 km^2 wide and geographically located at 08°35'31 "- 08°44'49" South Latitude and 115°10'23 "- 115°16'27" East Longitude. Denpasar City is divided into 4 subdistricts namely North Denpasar, East Denpasar, West Denpasar and South Denpasar [11].

3.2. Research Stage

3.2.1. Preliminary Stage

This research activity began from literature study to get initial information about the condition of the research area from previous research results, either in the form of data from reports and maps that already exist. Literature studies was as a reference for field study planning. Preliminary studies include: analysis of data and information from satellite imagery and base maps (roads, rivers, irrigation canals, and administrative boundaries), land use maps, site planning observations. Interpretation of satellite imagery was for tentative landuse map making, area description, surveys and soil making sampling for agro-ecosystem land suitability requirements and for constructing thematic maps of land suitability of agro-ecosystems.

3.2.2. Data Analysis and Interpretation of Satellite Imagery

Interpretation of satellite imagery for the identification, description and classification of land use and landform analysis was done by using three analytical methods: (1) elemental analysis, (2) pattern analysis, and (3) physiography analysis. Element analysis is conducted with nine elements, namely: shape, and size, color and contrast, texture and pattern, shadow, location, and association. These nine elements are used for land use classification. Subsequently, a tentative map of landuse was made before field observation.

3.2.3. Land Resource Data Updates

Data updating focusd on map creation process on a scale of 1: 50,000, such as maps of conformity to regional landuse (strategic protected and conservation areas, agricultural cultivation areas, non-agricultural cultivation areas), land use (irrigated, mixed), watershed (upstream, middle and downstream), irrigation type (technical, semi technical, simple and rain-fed), shape / relief (flat, undulating, hilly and mountainous), slope of origin (> 40% 25-40%, and< 25%), rainfall / agroclimate zone, height (> 500, 100 - 500, and <100 m asl), land suitability agroecosystem (S1, S2, and S3), land productivity (5, 2.5 - 5 and <2.5 tons / ha / MT) and distance from the city center (> 5 km, 2.5 - 5 km, 2.5 km) and minimum area (> 5 ha, 2 - 5 Ha and <2 ha).

Analysis of spatial data and interpretation of Satellite Imagery 2015 was calculated by using Quantum GIS 2.10.1 application. Spatial data collection was in the form of satellite image in Denpasar City (Worldview Year 2015) and Badung Regency (Quick Bird Year 2013) and projecting into WGS 84 / UTM Zone 50 S coordinate system. Collecting and making vector data in Denpasar and suiting rice field with landuse plan by digitizing the regional plan of Denpasar in 2011-2031, landuse data creation by vectorisation of satellite image, watershed area by vectorisation of satellite image, irrigation field irrigation type using spatial data and attribute of subak Denpasar city, rainfall obtained from BMKG Region III Denpasar, agro ecosystem land suitability for paddy rice can be calculated from the results of the sampling point from subakin Denpasar City, land productivity obtained from subak database of Denpasar City and rice field sampling in Denpasar City. The distance from the city center was calculated by buffer method, the minimum area was calculated from the result of spatial distribution of ricefield in Denpasar city in 2015.

Collecting and building vector data in Badung regency started with suiting subakrice field with landuse plan by digitizing landuse of Badung Regency in 2013- 2033, building landuse data by digitizing satellite image, watershed location obtained from BPDAS UndaAnyar, irrigation system obtained from spatial data and attribute of subak of Badung regency, rainfall obtained from BMKG. The slope data is obtained from the topographical map, the height of the place was obtained from topographic maps. The suitability of agro-ecosystem land for ricefield was obtained from the result of sampling point of subak of Badung regency, land productivity was obtained from subak database of Badung regency and sampling of subak rice field. The distance from the city center is calculated by the buffer method (area) applied, the minimum area was calculated from the result of the width of subak area ofBadung regency of 2013. The data of each spatial per parameter was done by weighting and scoring, then all parameters were overlaid/intersected to combine spatial data values. After the overlay process then total scoring was calculated by calculating feature on the application. Population data in dbf format of overlay results was conducted to determine the class population. Then statistical analysis was done to get standard deviation, maximal, minimum, average and total value. The result of population statistical analysis obtained some modeling obtained from standard deviation [9], by formula:

$$s = \sqrt{\frac{n\sum_{i=1}^{n} x_i^2 - (\sum_{i=1}^{n} x_1)^2}{n(n-1)}}$$

s = standard deviation
xi = value of x number-i
$\bar{x}$ = average
n = number of data

Each modeling was analyzed by grouping the data of statistical analysis using select featured expression feature, so get some model that is model 1 to model 5. Each parameter and modeling was arranged with cartography rule with layout feature of composer manager then mappingper parameter in Denpasar and Badung regency as well as zonation map of subak land respectively in Badung regency and Denpasar city.

3.2.4. Establishment of Ricefield Area

The availability of image data in different time periods (2002 and 2015) makaes the land use balance was only calculated for Denpasar City area through overlay - intersect method in QGIS menu. The landuse map was generated from satellite image interpretation. Field check was conducted from the interpretation of satellite imagery for mapping of landuse in 2015.

Analysis of existing ricfield data of 2012 and landuse balance was calculated manually by comparing different year data with the same image data source. Land use classification refers to the National Land Agency classification system. This classification was based on a structured hierarchy, because the classification of cover / land use according to Bakosurtanal (2005) and the Ministry of Environment (2007) does not explain the data detail.

IOP Conf. Series: Earth and Environmental Science **123** (2018) 012030 doi:10.1088/1755-1315/123/1/012030

3.2.5. Criteria Parameter, Classification and Mapping of Sustainable Food Crops

The criteria and mapping of GIS-based PLPPB areas refers to the classification studies, by modifying the weights and scores according to the physical condition of the region [9,12]. The 11 parameters are:

a) The suitability of paddy field location with RTRW
b) Land use
c) The position of subak rice field in watershed (DAS)
d) Type of irrigated rice field
e) Rainfall
f) Region shape
g) Elevation of the place
h) Agro-ecosystem compatibility
i) Land productivity
j) Distance from city center
k) Minimal width of subak

Criteria for Weighting and Scoring of each Parameter forClassification of Agricultural Land in Badung Regency and Kota Denpasar listed in Appendix 1. From the results of the eleventh digit digital overlay parameters in Badung regency generate a total score in each data distribution. From the distribution of data, it can be seen the maximum value, minimum, average, and standard deviation. Overlay data can be grouped or made to ring population data based on total score value, with the following classes:

A. Lestari => x (> average + 1 standard deviation)
B. Buffer = x ((average-1 standard deviation) to (Average + 1 standard deviation))
C. Converted = <x (<average - 1 standard deviation)

Based on the population rings, there are classification of subak land zoning in several models:

1) *Model 1:*
A. Subak Lestari = (> average + 1st standard deviation)
B. Subak Buffer = ((average - 1 standard deviation) Up to (> average + 1st standard deviation))
C. Subak Convert = (<average - 1 standard deviation)

2) *Model 2:*
A. Subak Lestari = (> average + ½ standard deviation)
B. Subak Buffer = ((<average - ½ standard deviation) up to (> Averages + ½ Standard deviation))
C. Subak Convert = (<average - ½ standard deviation)

3) *Model 3:*
A. Subak Lestari = (> average)
B. Subak Buffer = ((<average - 1½ standard deviation) up to (> Average))
C. Subak Convert = (<average - 1½ dot deviation)

4) *Model 4:*
A. Subak Lestari = (> average - ½ standard deviation)
B. Subak Buffer = ((<average - 2 standard deviation) up to (> Average - ½ standard deviation)
C. Subak Convert = (<average - 2 standard deviation)

5) *Model 5:*
A. Subak Lestari = (> average - 1 standard deviation)
B. Subak Buffer = ((<average - 2 ½ standard deviation) up to (> Averages - 1 stdar deviation).
C. Subak Convert = (<average - 2 ½ standard deviation)

Based on the population ring can be made classification of subak land zonasi in some modeling that is:

1) *Model 1, consisting of sustainable subak, buffer and converted, is derived from the following formula:*
   A. Subak Lestari = (> average + 1/2 standard deviation)
   B. Subak Buffer = ((average - 1 / 2standard deviation) to (average + 1/2 standard deviation)
   C. Subak Convert = (average -1/2 standard deviation)

2) *(2) Model 2, consisting of sustainable subak, buffer and converted, is obtained from the following formula:*
   A. Subak Lestari = (> average + ½ standard deviation)
   B. Subak Buffer = ((average- ½ standard deviation) up to (Average + ½ standard deviation)
   C. Subak Convert = (average - ½ standard deviation)

3) *(3) Model 3, consisting of sustainable subak, buffer and converted, is obtained from the following formula:*
   A. Subak Lestari => x (average)
   B. Subak Buffer = ((average - 2 standard deviation) to (average)
   C. Subak Convert = (<average -2 standard deviation)

4) *(4) Model 4, consisting of sustainable subak, buffer and converted, is derived from the following formula:*
   A. Subak Lestari => x (> average + 1/2 standard deviation
   B. Subak Buffer = x ((average - 1 / 2standard deviation) to (Average + 1/2 standard deviation)
   C. Subak Convert = <x (average -1/2 standard deviation)

5) *(5) Model 5, consisting of sustainable subak, buffer and converted, is derived from the following formula:*
   A. Subak Lestari => x (> average + 1/2 standard deviation
   B. Subak Buffer = x ((average - 1 / 2standard deviation) to (Average + 1/2 standard deviation)
   C. Subak Convert = <x (average -1/2 standard deviation)

3.2.6. Food Balance Year 2020-2050 for Badung Regency and Denpasar City

The food balance was calculated based on projected population growth and land conversion. In Badung Regency the food balance is calculated for each Sub-district, while Denpasar city was calculated for whole city.

1) *Projected population growth*

Population projection is a scientific calculation based on certain assumptions of population growth variable. Population variables include: birth, death and migration, the three components of this variable that determine the size of the population and its characteristics in the future. Calculate the balance of food Year 2020-2050 Kota Denpasar and Badung regency. Calculate the projection of population growth by doing (1) collecting secondary data of Badung regency population per sub-district, (2) collecting secondary data of population of Denpasar City. Calculating the projection of Badung regency population was done by using Mathematical Method [13].

Formula: $Pn = Po (1 + rn)$

Namely: Pn = Population in year n
Po = Population in the early years
R = Population increase rate (%)
N = Period (time) between the initial year c

IOP Conf. Series: Earth and Environmental Science **123** (2018) 012030 doi:10.1088/1755-1315/123/1/012030

Data source of Petang sub-district using statistical data from Badung in 2006 - 2011, Abiansemal Sub-district using 2010-2015 data, Mengwi sub-district using 2010-2015 data, Kuta sub-district using data from 2012-2015, 2015. Calculating the projection of Denpasar residents by using the inter census formula [13].

Formula: Pm = Po + m / n (Pn-Po)

Namely: Pn = Number of population in year n
Po = Population in the early year
Pm = Population in estimated years
M = difference of year searched with initial year
N = year increment of 2 known censuses

For Denpasar City, data for 2000 and year 2010

2) *Calculating Projection of Land Function Distribution of Badung Regency and Denpasar City*

Calculating the landuse change of Badung Regency, calculated per sub-district by the formula: the area of land conversion per year = the area of land year n minus the n year land area minus 5 divided 5. The land area in the year was done by using the 2015 landuse data. The total landuse change model was calculated in 2020, 2030, 2040 and 2050. Calculation of the landuse function of Denpasar City was calculated thoroughly by using data of ricefield area of Denpasar City obtained from the digitization of Satellite Imagery of Ikonos Year 2002 and digitization of Satellite Image of Worldview Year 2015. Average land use change of Denpasar city per year was calculated by using the formula of ricefield area of 2015 minus area of rice field year 2002 divided by 13 year.

The land projection of 2020 will be calculated by subtracting the number of land use over 5 years since 2015. The land projection of 2030 will be calculated by subtracting the number of land use over 5 years, but if the land area is less than the urban green space area in the Denpasar City Spatial Plan Year 2011- 2031 (1563.62 ha), then the land area in 2030 is 1563.62 ha (equal with Denpasar City Spatial Plan Year 2011- 2031). The 2040 landuse projection is calculated by subtracting the area of urban green area with ten-year land conversion, but if the land area is smaller than the City of Denpasar (12,778 ha) multiplied by 30% or 833.4 ha then the land area in 2040 is 833.4 ha. In the law, open green space is grouped into two, namely private sector (garden / yard owned by public / private) and Public green space (managed and owned by local government). Private gren space by 10% of the total area, while Public green space is 20% of the total area. The projection for 2040 uses public green space by 30%. The land projection in 2050 is calculated by subtracting the projected 2040 results by less than ten years of landuse, but if the land area is smaller than the area of Denpasar (12,778 ha) multiplied by 20% (Public Revenue) or 555.6 ha, Land in 2050 is 555.6 ha. The description on the food balance table covers the year (2015, 2020, 2030, 2040 and 2050).

4. Results and Discussion

4.1. Result of Badung Regency

The results showed that the projection of food balance in Badung regency was in Petang District with productivity of 7 ton / ha, Abiansemal, Mengwi, Kuta Utara and Kuta respectively with productivity 8 ton / Ha, food balance condition in 2015 that is food deficit (-32843.44 ton of rice), year 2020 that is food surplus (25155.19 ton of rice), year 2030 that is food surplus (3401.79 ton of rice), year 2040 that is The food deficit (- 18434.78 tons of rice) and the year 2050 is the food deficit (-11824.82 tons of rice). The projection of food balance in Denpasar with productivity of 8 ton/ ha, food balance condition in 2015 is food deficit (-90931,35 ton rice), 2020 is food deficit (-112514.73 ton of rice) , 2030 is food deficit (-142065.17 ton of rice), 2040 is food deficit (-173229.00 ton of rice), and year 2050 is food deficit (-203716.97 ton of rice).

The area of sustainable food agriculture is the area of sustainable rice according to the projection of ricefield area with food balance. Badung regency includes:

a) Petang District by 2020, the area of sustainable rice is equal to the area of sustainable rice field in 2030, 2040 and 2050, which is 1,173 ha and entirely using model 5.

b) Abiansemal Subdistrict, area of 2020 sustainable rice field by 2,848,51 ha using model 5, year 2030 area of sawah sustainable for 2,808.80 ha using model 4, 2040 of sustainable area for 2,767,85 ha using model 1, And in 2050 the area of sawah sustainable for 2,727.52 ha using model 2.

c) Mengwi subdistrict in 2020 with area of rice field sustainable area of 4,348,05 ha using model 5, year 2030 area of sawah sustainable for 4,204.08 ha by using model 4, year 2040 area of sustainable land equal to 4,060.11 ha using model 3, And by 2050 the area of sawah is 3916.14 ha using model 1.

d) North Kuta sub-district by 2020 of sustainable area of 1,425.35 ha using model 5, 2030 of sustainable land area of 1,370.25 ha using model 4, 2040 of sustainable land of 1,315.15 ha using model 4, and In 2050 the area of sawah is 1,260.05 ha using model 4.

e) Kuta sub-districts 2020 sustainable land area of 18.5 ha using model 4, year 2030 of sustainable land area of 1.5 ha using model 3, 2040 of sustainable land for 0 ha using model 1 and 2050 Sustainable rice field of 0 ha using model 1.

Figure 1 shows the result of model 1, Figure 2 shows model 2, Figure 3 shows model 3, Figure 4 shows model 4, and Figure 5 shows model 5 for Badung Regency. Meanwhile data modeling in Badung Regency listed in Table 1, and Table 2 listed the data of suitability for rice field area in food balance by modeling in Badung Regency

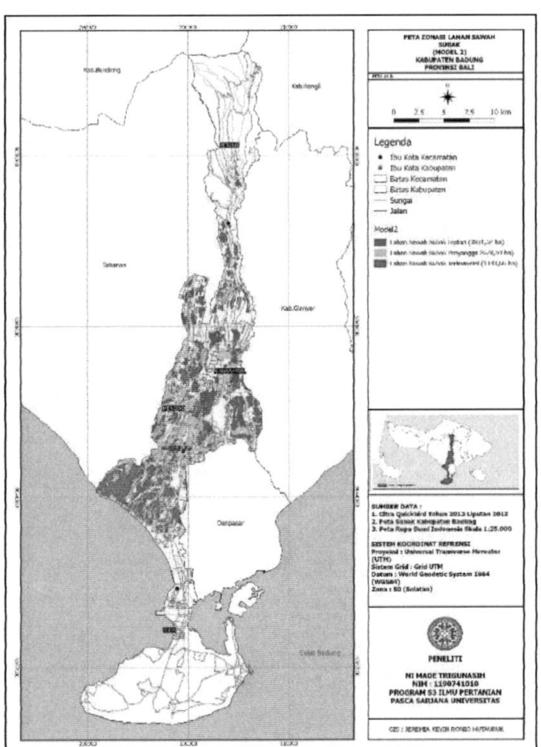

Figure 1. Model 1 Result for Badung Regency

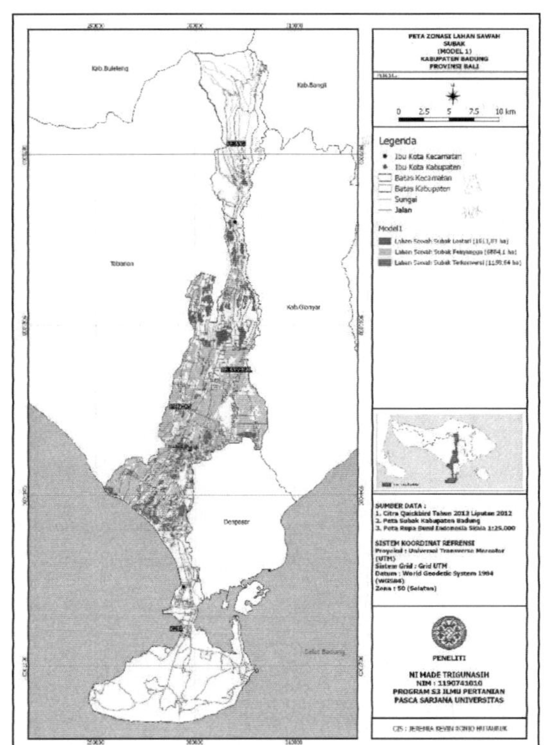

Figure 2. Model 2 Result for Badung Regency

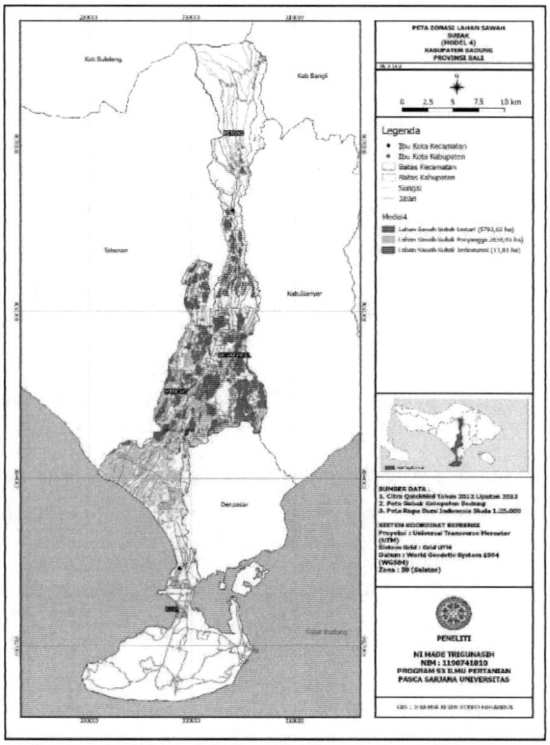

Figure 3. Model 3 Result for Badung Regency

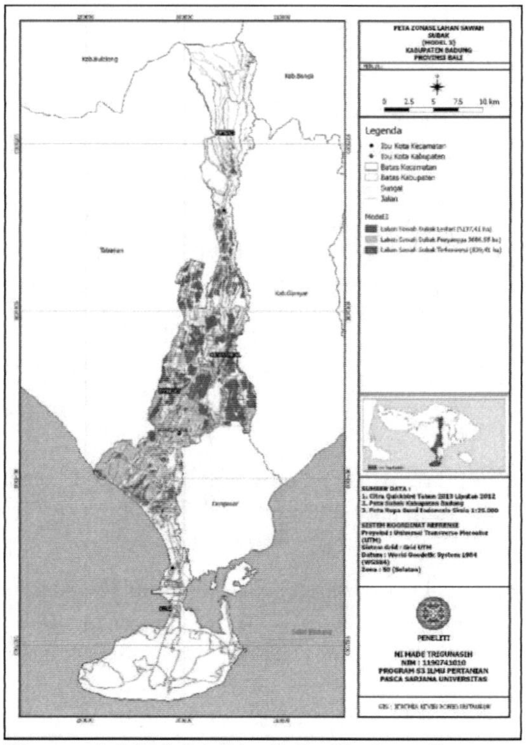

Figure 4. Model 4 Result for Badung Regency

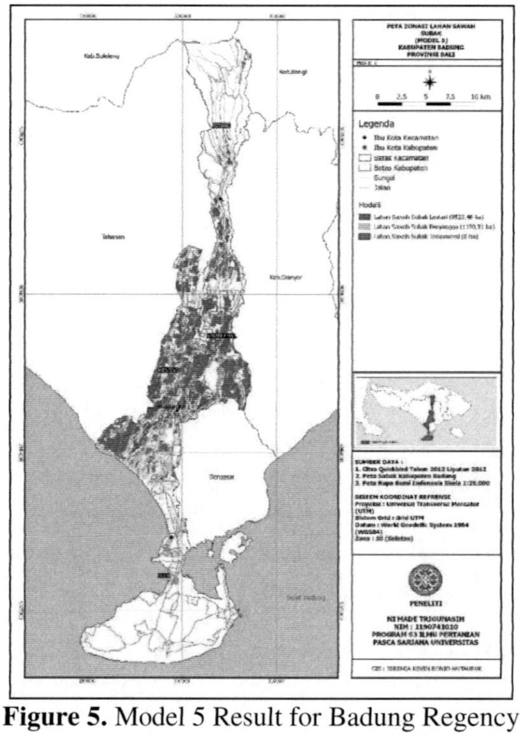

Figure 5. Model 5 Result for Badung Regency

Table 1. Area DataModeling in Badung Regency

Sub-district/ Model	Area (ha)		
	Lestari	Penyangga	Terkonversi
1	2	3	4
1. Petang			
a. Model 1	492.16	501.69	0
b. Model 2	675.83	244.07	19.19
c. Model 3	935.44	58.39	0
d. Model 4	973.75	20,09	0
e. Model 5	993.84	0	0
2. Abiansemal			
a. Model 1	705.43	2229.52	2.36
b. Model 2	1822.87	958.83	100.33
c. Model 3	2377.79	532.28	0
d. Model 4	2511.79	386.86	0
e. Model 5	2798.62	115.34	0
3. Mengwi			
a. Model 1	409.44	3530.75	359.89
b. Model 2	1332.82	1413.23	1651.76
c. Model 3	1804.36	2345.92	133.29
d. Model 4	2236.29	2061.99	0
e. Model 5	4141.34	155.31	0
4. Kuta Utara			
a. Model 1	0	637.26	728.68
b. Model 2	0	62.5	1342.91
c. Model 3	12,62	728.48	672,09
d. Model 4	64.5	1335.95	4.64
e. Model 5	647.6	758.09	0
5. Kuta			
a. Model 1	0	0	24.1
b. Model 2	0	0	24,1
c. Model 3	0	7,42	12,07
d. Model 4	0	16,93	7,17
e. Model 5	0,07	24,10	0

Table 2. Suitability for Ricefield Area in Food Balance by Modeling in Badung Regency

Sub-districts	Model - Year			
	2020	2030	2040	2050
1.Petang	Model 5	Model 5	Model 5	Model 5
2.Abiansemal	Model 5	Model 4	Model 1	Model 2
3.Mengwi	Model 5	Model 4	Model 3	Model 1
4.Kuta Utara	Model 5	Model 4	Model 4	Mode 4
5.Kuta	Model 4	Model 3	Model 1	Model 1

4.2. Denpasar City

City of Denpasar (urban), in 2020 the area of sawah sustainable is 1705,21 ha using model 5, year 2030 wide of sawah sustainable by 1563,62 ha using model 4, year 2040 sustainable area of 964,22 ha using model 3, And by 2050 the area of sawah is 556,6 ha using model 2. Factors affecting sustainable area, buffer and convert in Badung regency is suitability of paddy field with RTRW, watershed position, rainfall, place height, and shape of area (slope) whereas in Denpasar influence factor is suitability of paddy field with RTRW and paddy field position in the watershed.

Figure 6 shows model 1 result for Denpasar City, Figure 7 shows model 2 result for Denpasar City, Figure 8 shows model 3 result for Denpasar City, Figure 9 shows model 4 result for Denpasar City, and Figure 10 shows model 5 result for Denpasar City. Table 3 listed the data for area in modeling for Denpasar City. Meanwhile Table 4 listed the data of suitability of ricefield area in food balance by modeling in Denpasar City.

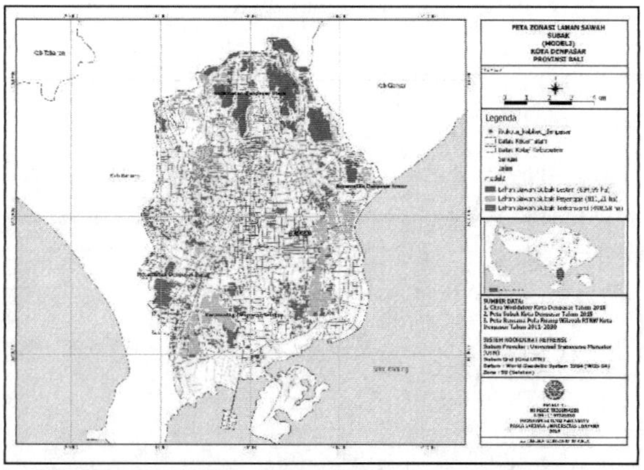

Figure 6. Model 1 Result for Denpasar City

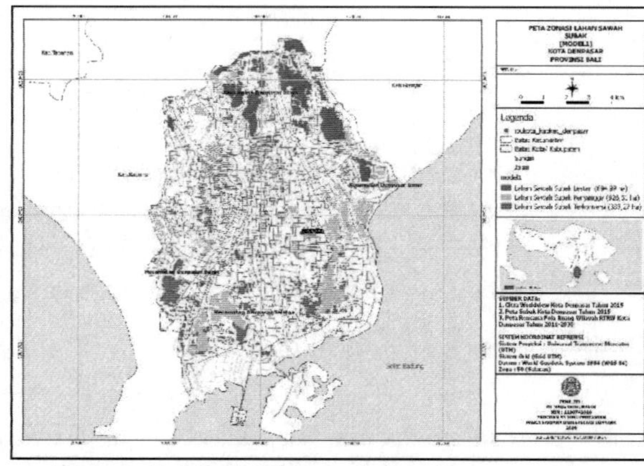

Figure 7. Model 2 Result for Denpasar City

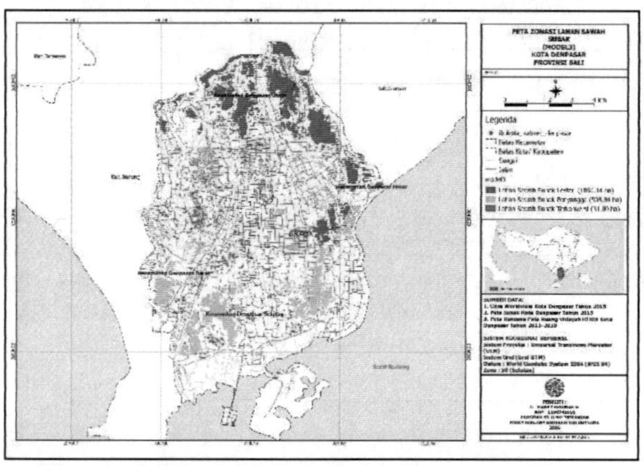

Figure 8. Model 3 Result for Denpasar City

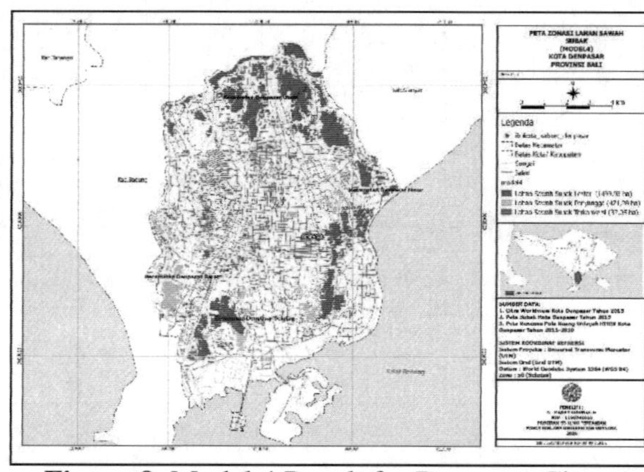

Figure 9. Model 4 Result for Denpasar City

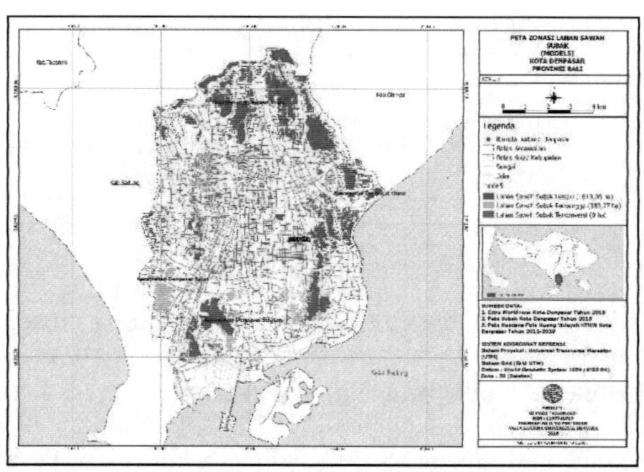

Figure 10. Model 5 Result for Denpasar City

IOP Conf. Series: Earth and Environmental Science **123** (2018) 012030 doi:10.1088/1755-1315/123/1/012030

Table 3. Data for Area in Modeling for Denpasar City

Model	Area (ha)		
	Sustainable	Buffer	Converted
Model 1	695,09	926,59	383,28
Model 2	695,05	811,30	498,61
Model 3	1064,28	908,88	31,80
Model 4	1500,80	472,11	32,05
Model 5	1621,58	383,38	0

Table 4. Data of Suitability of Rice field Area in Food Balance by Modeling

City	Model - Year			
	2020	2030	2040	2050
Denpasar	Model 5	Model 4	Model 3	Model 2

5. Conclusions

Based on the results and discussion can be concluded:

a) Classification parameter for rural area (Badung Regency) in the form of: suitability of paddy field with landuse plan, landuse, rice field position in watershed, type of irrigated ricefield, rainfall, area, height of place, suitability of agro-ecosystem land, ricefield production, Cities and minimal area of ricefields. Denpasar City Parameter same with Badung regency except parameter of area and height of place.

b) Projection of food balance at:
   1) Badung Regency has food deficit in 2015, 2040 and 2050, but in 2020 and 2030 experienced food surplus.
   2) Denpasar City 2015, 2020, 2030, 2040 and 2050 each experienced food deficit.

c) The numerical classification model for the KLP2B zonation
   Zoning area of Badung Regency in the form of:
   1) Subak sustainable in models 1, 2, 3, 4, and 5 each have an area of 1611.03 ha; 3831.52 ha; 5137.41 ha; 5782.03 ha and 8523.46 ha
   2) Subak buffer in models 1, 2, 3, 4 and 5, each has an area of 6884.1 ha; 2678.59 ha; 3686.95 ha; 3859.93 ha and 1130.31 ha.
   3) Subak converted on models 1, 2, 3, 4, and 5 each have an area of 1158.64 ha; 3143,66 ha; 829.41 ha; 11,81 ha and 0 ha.

   Zoning area for Denpasar City in the form of:
   1) Subak sustainable in models 1, 2, 3, 4, and 5 each have an area of 695.09 ha; 69505 ha; 1064.28 ha; 1500.8 ha and 1621.58 ha.
   2) Subak Buffers in models 1, 2, 3, 4 and 5 each have an area of 926.59 ha; 811.30 ha; 908.88 ha; 472.11 ha and 383.38 ha.
   3) Subak converted on models 1, 2, 3, 4, and 5 each have an area of 383.28 ha; 498.61 ha; 31.8 ha; 32.05 ha and 0 ha.

d) Sustainable agriculture model:
   The Regency of Badung includes:
   1) Petang sub-districts in 2020, 2030, 2040 and 2050 have a sustainable area of sawah equal to 1,173 ha and use model 5.
   2) Abiansemal sub-districts have sustainable paddy fields in 2020, 2030, 2040 and 2050 with area and model respectively of 2,848.51 ha (model 5), 2,808.80 ha (model 4), 2,767.85 ha (model 1), and 2,727.52 ha (model 2).

3) Mengwi sub-district has a sustainable rice field area in 2020, 2030, 2040 and 2050 with area and model respectively 4,348.05 ha (model 5), 4,204.08 ha (model 4), 4,060.11 ha (model 3), and 3916.14 ha (model 1).
4) North Kutasubdistrict has 2050 sustainable rice fields in 2020, 2030, 2040 and 2050 with area and model of 1,425.35 ha (model 5), 1,370.25 ha (model 4), 1,315.15 ha (Model 4), and 1,260.05 ha (model 4).
5) Kuta sub-districts have widespread sawah in 2020, 2030, 2040 and 2050 with the width and model of 18.5 ha (model 4), 1.5 ha (model 3), 0 (model 1), respectively, and 0 ha (model 1),
6) Denpasar city with sustainable rice field area of 2020, 2030, 2040 and 2050 respectively have wide and model of 1,705.21 ha (model 5), 1,563.62 ha (model 4), 964.22 ha (model 3) and 556.6 (model 2).

6. Acknowledgments

Authors would like to thank the Rector of Universitas Udayana, Dean of Agriculture Faculty of Universitas Udayana, Head of Center for Spatial Data Infrastructure Development (PPIDS) Universitas Udayana, fellow staf, researchers and professors at Universitas Udayana for all the support, comments and suggestions that makes this paper finished.

7. References

[1] Sarwuan T, Beka F T and Ogbole A 2016 Geographic Resource Monitoring and Assessment of Long-Term Land Cover Change *J. Sci. Eng. Res.* **3** 646–655
[2] Lanya I, Subadiyasa N and Hutauruk J 2017 Subak Land Information System Based On Remote Sensing And Geographic Information System In Denpasar City *Geoplanning J. Geomatics Plan.* **4**
[3] Semwal R L, Nautiyal S, Maikhuri R K, Rao K S and Saxena K G 2013 Growth and carbon stocks of multipurpose tree species plantations in degraded lands in Central Himalaya, India *For. Ecol. Manage.* **310** 450–459
[4] Liu Y, Yang R, Long H, Gao J and Wang J 2014 Implications of land-use change in rural China: A case study of Yucheng, Shandong province *Land use policy* **40** 111–118
[5] Cantarello E, Newton A C and Hill R a. 2011 Potential effects of future land-use change on regional carbon stocks in the UK *Environ. Sci. Policy* **14** 40–52
[6] Wu J, Li Z, Gao Z, Wang B, Bai L, Sun B, Li C and Ding X 2015 Degraded land detection by soil particle composition derived from multispectral remote sensing data in the Otindag Sandy Lands of China *Geoderma* **241–242** 97–106
[7] Sekretariat Negara Republik Indonesia 2009 *UU No. 41 Tahun 2009 Tentang Perlindungan Pertanian Pangan Berkelanjutan*
[8] BPS Provinsi Bali 2015 *Bali Dalam Angka 2015*
[9] Lanya I, Subadiyasa I N, Sardiana K and Adi I G R 2014 *Strategi Penentuan dan Pengendalian Alih Fungsi Lahan Pertanian Dalam Antisipasi dan Penanggulangan Dampak Negatif Pariwisata di Bali*
[10] BPS Kota Badung 2015 *Badung Dalam Angka*
[11] BPS Kota Denpasar 2015 *Kota Denpasar dalam Angka 2015*
[12] Subadiyasa I N N, Lanya I and Sardiana I K 2010 *Strategi Pengendalian Alih Fungsi Lahan Subak Berbasis Masyarakat dan Upaya Peningkatan Produktivitas Lahan di Kabupaten Tabanan, Bali*
[13] Muta'ali L 2015 Teknik Analisis Regional Untuk Perencanaan Wilayah, Tata Ruang dan Lingkungan *Yogyakarta Badan Penerbit Fak. Geogr.*

Appendix

Appendix 1

Criteria for Weighting and Scoring of each Parameter for Agricultural Land Classification in Badung Regency

No.	Parameter		Weight	Rating Score	Value
1	Ricefield suitability with landuse planning		11		
	a.	Strategic and protected areas		3	33
	b.	Agricultural Cultivation Area		2	22
	c.	Non-agricultural areas		1	11
					66
2	Penggunaan lahan :		10		
	a.	Irrigated ricefield		3	30
	b.	Moor		2	20
	c.	Mixed plantation		1	10
					60
3	Subak location in the Watershed area		9		
	a.	Upstream		3	27
	b.	Middle		2	18
	c.	Downstream		1	9
					54
4	Irrigation system		8		
	a.	Technical-semi technical irrigation		3	24
	b.	Simple irrigation/ non technical		2	16
	c.	Rain-fed		1	8
					48
5	Rainfall		7		
	a.	> 2500 mm/yr		3	21
	b.	2000 – 2500 mm/yr		2	14
	c.	< 2000 mm/yr		1	7
					42
6	Topography :		6		
	a.	Hilly to mountainous with slope> 40 %		3	18
	b.	Undulating to wavy with slope 25-40%		2	12
	c.	Flat to undulating with slope< 25 %		1	6
					36
7	Height:		5		
	a.	>500m asl		3	15
	b.	100 – 500 m asl		2	10
	c.	<100 m asl		1	5
					30
8	Agro ecosystem land suitability for ricefields		4		
	a.	High suitability (S1)		3	12
	b.	Medium suitability (S2)		2	8
	c.	Marginal suitability (S3)		1	4
					24
9	Productivity		3		
	a.	>5 ton/ha/MT		3	9
	b.	2,5 - 5 ton/ha/MT		2	6
	c.	< 2,5 ton/ha/MT		1	3
					18
10	Distance from city center :		2		
	a.	> 5km		3	6
	b.	2,5 – 5 km		2	4
	c.	< 2,5 km		1	2
					12
11	Minimum area		1		
	a.	> 10 ha		3	3
	b.	5-10 ha		2	2
	c.	<5 ha		1	1
					6

IOP Conf. Series: Earth and Environmental Science **123** (2018) 012030 doi:10.1088/1755-1315/123/1/012030

Appendix 2

Criteria for Weighting and Scoring of each Parameter
for Agricultural Land Classification in Denpasar City

No	Parameter		Rating		
			Weight	Score	Value
1	Land suitability with landuse plan		9		
	a.	Urban green space		3	27
	b.	Agricultural land		2	18
	c.	Non-agricultural land		1	9
					54
2	Land use		8		
	a.	Irrigated ricefield		3	24
	b.	Moor		2	16
	c.	Mixed plantation		1	8
					48
3	Subak location in the Watershed area		7		
	a.	Upstream		3	21
	b.	Middle		2	14
	c.	Downstream		1	7
					42
4	Irrigation system		6		
	a.	Technical-semi technical irrigation		3	18
	b.	Simple irrigation/ non technical		2	12
	c.	Rain-fed		1	6
					36
5	Rainfall		5		
	a.	> 2500 mm/yr		3	15
	b.	2000 – 2500 mm/yr		2	10
	c.	< 2000 mm/yr		1	5
					30
6	Agro ecosystem land suitability for ricefields		4		
	a.	High suitability (S1)		3	12
	b.	Medium suitability (S2)		2	8
	c.	Marginal suitability (S3)		1	4
					24
7	Productivity		3		
	a.	>5 ton/ha/MT		3	9
	b.	2,5 - 5 ton/ha/MT		2	6
	c.	< 2,5 ton/ha/MT		1	3
					18
8	Distance from city center :		2		
	d.	> 5km		3	6
	e.	2,5 – 5 km		2	4
	f.	< 2,5 km		1	2
					12
9	Minimum area		1		
	a.	> 10 ha		3	3
	b.	5-10 ha		2	2
	c.	<5 ha		1	1
					6

Drought disaster vulnerability mapping of agricultural sector in Bringin District, Semarang Regency

D R Lestari[1] and B Pigawati[2]

[1]Urban and Regional Planning Department, Diponegoro University, Semarang, Indonesia
[2]Geomatics and Computing Planning Laboratory, Urban and Regional Planning Department, Diponegoro University, Semarang, Indonesia

Email: dwitantri16@pwk.undip.ac.id

Abstract: Agriculture sector is a sector that is directly affected by drought. The phenomenon of drought disaster on agriculture sector has occurred in Semarang regency. One of districts in Semarang which is affected by drought is Bringin district. Bringin district is a productive agricultural area. However, the district experienced the most severe drought in 2015. The question research of this study is, "How is the spatial distribution of drought vulnerability on agriculture sector in Bringin district, Semarang regency?" The purpose of this study is to determine the spatial distribution of drought vulnerability on agriculture sector to village units in Bringin district. This study investigated drought vulnerability based on Intergovernmental Panel on Climate Change (IPCC) by analyzing exposure, sensitivity, and adaptive capacity through mapping process. This study used quantitative approach. There were formulation analysis, scoring analysis, and overlay analysis. Drought vulnerability on agriculture sector in Bringin district was divided into three categories: low, medium, and high.

Keywords: Agriculture, Drought, Semarang, Vulnerability

1. Introduction

The meaning of vulnerability is often equated with the meaning of poverty, when both are two things have different meanings. Vulnerability is potential difficulty which is faced by communities to cope with disasters, while poverty is related to income and consumption levels of society [1]. Policies to reduce vulnerability and to reduce poverty are two different things. Policies to reduce vulnerability focus on community security for disaster, while policies to reduce poverty focus on improving people's incomes.

Drought is a normal and rare phenomenon, but it causes amount of loss. Drought is a natural disaster that happens slowly and lasts long until the rainy season arrives, so it causes broad impact [2]. Drought is closely related to vulnerability on agricultural sector.

Vulnerability on agricultural sector has increased due to fluctuations of environmental conditions, so there is a need for increasing the resilience on agricultural sector [3]. Vulnerability can be formulated into mathematical formulations which comprise exposure, sensitivity, and adaptive capacity[4]. Exposure is the possibility of a disaster in an area [5]. Sensitivity is how a system will be affected when disaster comes [5] Adaptive capacity is how a system can survive from the impact of a disaster [5]. Efforts to manage drought are still ineffective while negative impacts on economic and social conditions continue to increase [6,7]. Effective efforts in response to drought hazard need to be done through enhanced adaptive capacity [8].

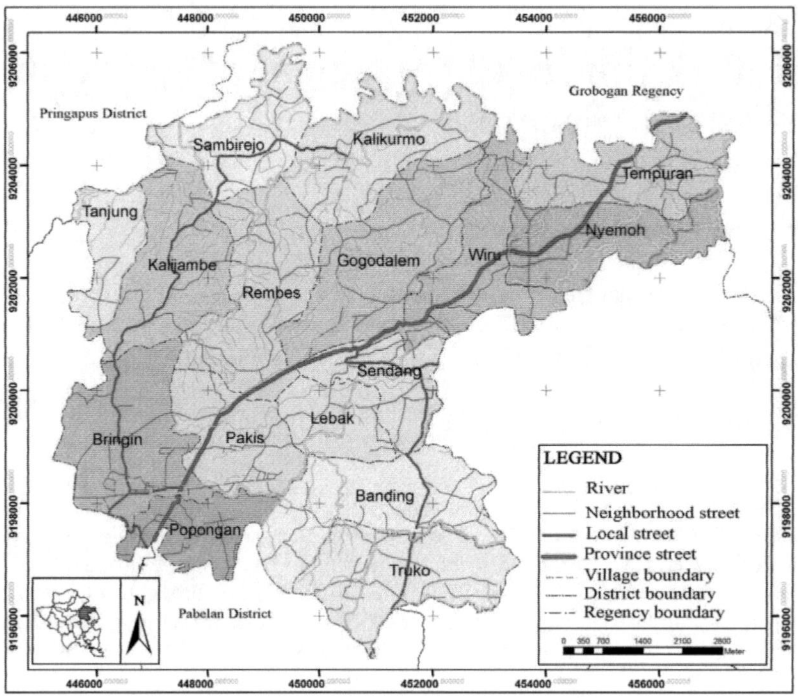

Figure 1. Administrative map of Bringin District.

Bringin district has the largest percentage of drought in Semarang regency. Bringin district is also a district which is dominated by people who work in agricultural sector. Drought in Bringin District causes adverse impacts on agricultural sector and also affects the welfare of farmers. This study can be a consideration to the preparation of regional action plan as drought mitigation in Bringin district. In the end, the establishment of Bringin District as a productive agricultural area can be realized.

2. Methods

2.1. Formulation Analysis

2.1.1. Average Annual Rainfall Calculation

Monthly rainfall data for 5 years was obtained from Meteorology Climatology and Geophysics Agency (BMKG). BMKG receives rainfall data from rain station which has function to identify rainfall. Rainfall on each month for 5 years is summed, then it is divided by the amount of dataas seen in equation (1).

$$\bar{X} = \frac{X_1 + X_2 + \cdots + X_n}{n} \tag{1}$$

Explanation

$\bar{X}$	=	Average
$X_1 + X_2 + \cdots + X_n$	=	Annual rainfall year-1, 2, …, n.
n	=	Amount of data

However, sometimes there is a loss of rain data. The loss of rain data can be calculated by using reciprocal methodas seen in equation (2).

$$.Px = \frac{\frac{PA}{(dxA)^2}+\frac{PB}{(dxB)^2}+\frac{PC}{(dxC)^2}}{\frac{1}{(dxA)^2}+\frac{1}{(dxB)^2}+\frac{1}{(dxC)^2}} \quad (2)$$

Explanation

Px	=	Average rainfall at rain station X (mm)
dxA	=	Distance between rain station X with rain station A (km)
dxB	=	Distance between rain station X with rain station B (km)
dxC	=	Distance between rain station X with rain station C (km)
PA, PB, PC	=	Rainfall at rain station A, B, C

[9]

2.1.2. Average Annual Dry Month Calculation

Dry month data were obtained from rainfall data. The selection of dry months was based on the criteria of Mohr. Dry month is a month that has average rainfall below 60 mm [10]. After knowing the number of months that have rainfall below 60 mm, then calculate the average dry months for five years.

2.1.3. Average Annual Actual Evapotranspiration Calculation

Actual evapotranspiration is calculated by using the Turc formula.

$$Ea = \frac{P}{\sqrt{0.9 + \frac{P^2}{L(t)^2}}} \quad (3)$$

Explanation

Ea	=	Average Annual Actual Evapotranspiration (mm)
P	=	Average Annual Rainfall (mm)
L(t)	=	$300 + 25t + 0.05t^3$
t	=	Average Annual Temperature (^{o}C)

[11]

Based on the formula above, temperature data is needed in order to know the evapotranspiration of water surface. Sometimes, temperature data is not available from the rain station. Therefore, the temperature in each rain station can be calculated by using the Mock formula.

$$dt = (Z_1 - Z_2)\, 0.006\ °C \quad (4)$$

Explanation

dt	=	The difference between temperature from the measuring station and reference station (°C)
Z_1	=	Elevation of the reference station (m)
Z_2	=	Elevation of the measuring station (m)

IOP Conf. Series: Earth and Environmental Science **123** (2018) 012031 doi:10.1088/1755-1315/123/1/012031

2.2. Scoring Analysis

Scoring is process of converting answers into numbers [13].

Table 1. Scoring. This table is based on other research. The variables are obtained from other combined studies [14–16] with modification.

Indicator/ variable	Criteria	Score	Category	Explanation
	Exposure			
Average annual rainfall (mm)	>2000	10	Low	-
	1501-2000	20	Medium low	
	1001-1500	30	Medium	
	500-1000	40	Medium high	
	<500	50	High	
Average annual dry month	0,0-2,0	10	Low	-
	2,01-4,0	20	Medium low	
	4,01-6,0	30	Medium	
	6,01-8,0	40	Medium high	
	>8,00	50	High	
Average annual actual evapotranspiration (mm)	<750	10	Low	Using Turc and Mock formula
	751-1000	20	Medium low	
	1001-1500	30	Medium	
	1501-2000	40	Medium high	
	>2000	50	High	
Geology	Breccia-andesit	10	Low	Geology map
	Clay sand, gravel, napal	20	Medium low	
	Clay, fine tuff, silt	30	Medium	
	Granite, solid	40	Medium high	
	Limestone	50	High	
Land cover	Forest	10	Low	Land cover map
	Agriculture	30	Medium	
	Settlements	50	High	
	Sensitivity			
Rainfed lowland	0-12,5%	10	Low	(rainfed low land area/ total area)*100
	12,6%-25%	20	Medium low	
	25,1%-37,5%	30	Medium	
	37,6%-50%	40	Medium high	
	>50%	50	High	
People working in agriculture	0%-20%	10	Low	(number of population who works in the agricultural sector/ total population who works)*100
	21%-40%	20	Medium low	
	41%-60%	30	Medium	
	61%-80%	40	Medium high	
	81%-100%	50	High	
	Adaptive Capacity			
Head of farmer families who have jobs outside the agricultural sector	0-20%	10	Low	(respondents who have jobs outside the agricultural sector/ total of respondents)*100
	21-40%	20	Medium low	
	41-60%	30	Medium	
	61-80%	40	Medium high	
	81-100%	50	High	
Farmer families who have family members working outside the agricultural sector	0-20%	10	Low	(respondents who have family members working outside the agricultural sector/ total of respondents)*100
	21-40%	20	Medium low	
	41-60%	30	Medium	
	61-80%	40	Medium high	
	81-100%	50	High	
Farmer families who have savings	0-20%	10	Low	(respondents who have savings/ total of respondents)*100
	21-40%	20	Medium low	
	41-60%	30	Medium	
	61-80%	40	Medium high	
	81-100%	50	High	
Farmer families who have livestock	0-20%	10	Low	(respondents who have livestock/ total of respondents)*100
	21-40%	20	Medium low	
	41-60%	30	Medium	
	61-80%	40	Medium high	
	81-100%	50	High	

IOP Conf. Series: Earth and Environmental Science **123** (2018) 012031 doi:10.1088/1755-1315/123/1/012031

2.3. Overlay Analysis

Overlay is an essential spatial analysis that combines two thematic layers [17]. Overlay analysis was performed after scoring process that resulted in exposure value (KT), sensitivity value (SN), and adaptive capacity value (KA). The scoring result of the exposure aspect was done in the unit of land. On the other hand, scoring on aspects of sensitivity and adaptive capacity was done in the administrative units of the village. To equate the two, the scoring aspect of the exposure results was made into the administrative units of the village by using the following formula.

$$KT = \left(TS_1 \times \left(\frac{LS_1}{LT}\right)\right) + \left(TS_2 \times \left(\frac{LS_2}{LT}\right)\right) + \cdots + \left(TS_n \times \left(\frac{LS_n}{LT}\right)\right) \quad (5)$$

Explanation

KT	=	Exposure
$TS_1, TS_2, \ldots, TS_n$	=	Score in the unit of land 1, 2, …, n
$LS_1, LS_2, \ldots, LS_n$	=	Area of score 1, 2, …, n (m^2)
LT	=	Total area of the village (m^2)

(Analysis, 2016)

After obtaining the value of exposure, sensitivity, and adaptive capacity, then calculate the value of vulnerability in each village.

$$Vulnerability = (Exposure + Sensitivity) - Adaptive\ Capacity$$

After obtaining the vulnerability value in each village, do the clashing to determine the low, medium, and high vulnerability level so that it has 3 classes.

$$X = \frac{(B - A)}{3} \quad (6)$$

Explanation

X	=	Interval
A	=	Lowest Data
B	=	Highest Data

Table 2. Vulnerability Classification Methods

Class	Formula
Low	A to A + X
Medium	(A + X) +1 to (A + X + 1) + X
High	[(A + X + 1) + X] + 1 to B

3. Discussion

3.1. Analysis of Exposure

The exposure analysis is the overlay of average annual rainfall (CH), average annual dry months (BK), average annual actual evapotranspiration (EA), geology (BP), and land cover (PL). Average annual rainfall, average annual dry months, and average annual actual evapotranspiration were obtained from five rain stations which are getas rain station, grenjeng rain station, tuntang rain station, bawen rain station, and kedungjati rain station. Spatial distribution of average annual rainfall, average annual dry months, and average annual actual evapotranspiration is distributed by Thiessen polygon method. The following section is a description of drought exposure in Bringin District.

IOP Conf. Series: Earth and Environmental Science **123** (2018) 012031 doi:10.1088/1755-1315/123/1/012031

3.1.1. Average Annual Rainfall

Villages that have low category of exposure are Kalikurmo, Gogodalem, Wiru, Nyemoh, and Tempuran. Those village has average annual rainfall >2000 mm. On the other hand, Bringin district is dominated by medium low category of exposure which has average annual rainfall 1501 mm to 2000 mm.

3.1.2. Average Annual Dry Months

Village that has medium low category exposure with average annual dry month 2,1-4,0 months are Kalikurmo, Gogodalem, Wiru, Nyemoh, and Tempuran. Bringin district is dominated by medium category exposure which has average annual dry months 4,1-6,0 months.

3.1.3. Average Annual Actual Evapotranspiration

Bringin District has two categories of average annual actual evapotranspiration which are medium (1001 mm to 1500 mm) and high (1501 mm to 2000 mm).

3.1.4. Geology

There are two types of constituent rocks in Bringin district which are volcanic rock and kerek formations. Kerek formation consists of clay and sandstone (score 20). Volcanic rock consists of volcanic breccia, lava breccia, lava (igneous rock), and tuff (rock containing volcanic dust). Kerek formation has score 20 and volcanic rock has score 10.

3.1.5. Land Cover

Bringin District is dominated by agricultural land cover (score 20) with percentage of 48.73%. Forest cover (score 10) has a percentage of 35.42% and settlements (score 30) has a percentage of 15.85%.

Table 3. Exposure Score

Village	Score					Total
	Rainfall	Dry month	Actual evapotranspiration	Geology	Land cover	
Bringin	20	30	30	12	22	114
Popongan	20	30	30	11	21	112
Pakis	20	30	30	13	28	121
Lebak	20	30	30	16	27	123
Banding	20	30	30	14	29	123
Truko	20	30	30	10	32	122
Nyemoh	13	23	37	17	25	115
Tempuran	10	20	40	20	26	116
Wiru	19	29	31	15	29	123
Sendang	20	30	30	14	31	125
Gogodalem	20	30	30	16	21	117
Rembes	20	30	30	18	25	123
Kalikurmo	19	29	31	20	25	124
Sambirejo	20	30	30	20	28	128
Kalijambe	20	30	30	19	26	125
Tanjung	20	30	30	18	25	123

Exposure score of each village is divided into three classes in ArcMap. The classification of this study is done manually, not based on certain standards. So, the result of this study is only valid for Bringin district, not valid for other area. The classification result can be seen in figure 2.

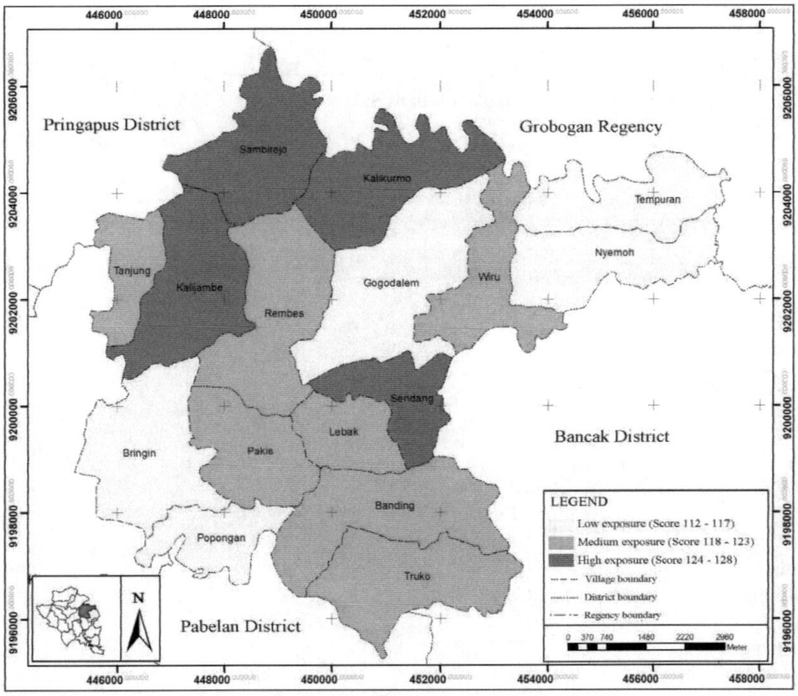

Figure 2. Exposure Map

The high exposure (score 124-128) is distributed in 4 villages, they are Sendang, Kalijambe, Sambirejo, and Kalikurmo. The high exposure in these four villages was caused by the constituent rocks which is dominated by kerek formations and land cover which is dominated by rice fields. Medium exposure (score 118-123) is distributed in 7 villages: Truko, Banding, Pakis, Lebak, Rembes, Tanjung, and Wiru. Medium exposure in these seven villages is caused by constituent rocks which are dominated by volcanic rocks. Low exposure (score 112-117) is distributed in 5 villages: Bringin, Popongan, Gogodalem, Nyemoh, and Tempuran. The low exposure in these five villages were caused by the volcanic rocks, the larger cover of forest land compared to other villages, and lower score of average annual rainfall and average annual dry month compared to other villages.

The village that has the highest exposure score is Sambirejo. The total score of exposure in Sambirejo is 128. Sambirejo is one of the villages that has the highest score in land cover aspect. Sambirejo is dominated by agricultural land. The percentage of agricultural land to the total land area in Sambirejo is 54%. That percentage is one of the highest percentage on agricultural land in Bringin district. The domination of agricultural land became one of the most influential factors to the exposure score of Sambirejo. Another factor that influences the high total score of Sambirejo exposure is the type of constituent rock. The whole area of Sambirejo has a type of kerek formation. This type of constituent rock has a higher exposure score than the type of volcanic rock. Kerek formation is good enough in absorbing and storing water, but not as good as volcanic rock. This causes the total score of exposure in Sambirejo is high.

Village with the lowest exposure score is Bringin with score 114. Bringin is dominated by volcanic rocks that have low category in exposure. The geology aspect score in Bringin is 12. This score is the second lowest score after Popongan. Bringin is dominated by forest (56%) that resulting score of 22 in land cover aspect. This score is the second lowest score after Popongan and Gogodalem.

3.2. Analysis of Sensitivity

Sensitivity analysis is the result of overlaying the rainfed lowland and population in the agricultural sector. Each condition in those variables has a score. Each village has total sensitivity score as listed in table.

Table 4. Sensitivity Score

Village	Score		Total
	Rainfed lowland	People working on agriculture	
Bringin	10	10	20
Popongan	10	20	30
Pakis	10	20	30
Lebak	10	30	40
Banding	30	40	70
Truko	20	50	70
Nyemoh	20	40	60
Tempuran	10	30	40
Wiru	10	40	50
Sendang	10	40	50
Gogodalem	10	40	50
Rembes	10	30	40
Kalikurmo	10	40	50
Sambirejo	10	40	50
Kalijambe	10	40	50
Tanjung	10	30	40

The sensitivity score of each village is calculated in ArcMap to know the spatial distribution. The distribution of low, medium, and high class is done manually, not based on certain standard, so the result of classification on this research is only valid for Bringin district, not valid for other study area.

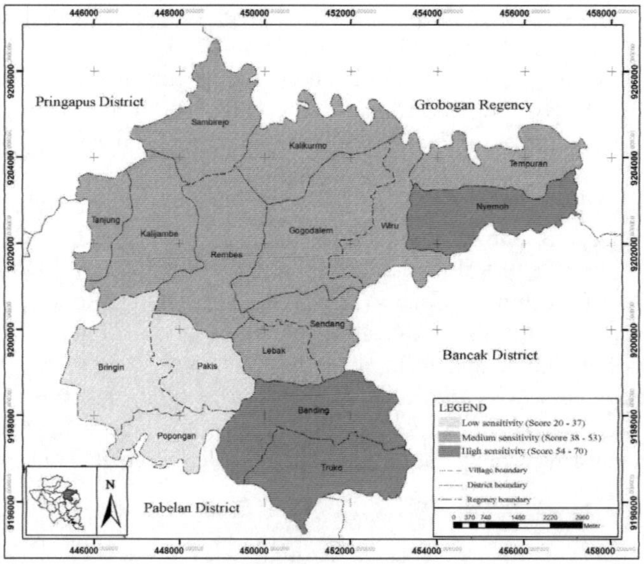

Figure 3. Sensitivity Map

High sensitivity (score 54-70) is distributed in 3 villages which are Truko, Banding, and Nyemoh. High sensitivity in those three villages caused by rainfed lowland and the number of people who work in agriculture sector which is higher than other villages. Medium sensitivity (score 38-53) is distributed in 10 villages: Sendang, Lebak, Rembes, Kalijambe, Tanjung, Sambirejo, Kalikurmo, Gogodalem, Wiru, and Tempuran. Medium sensitivity in those ten villages caused by the area of rainfed lowland which is not as extensive compared to other villages. Low sensitivity (score 20-37) is distributed in 3 villages which are Bringin, Popongan, and Pakis. The low sensitivity in these three villages is caused by the number of people working in the agricultural sector is lower than other villages.

The villages with the highest score on the sensitivity aspect were Banding and Truko with a total sensitivity score of 70. Banding has the highest score on the rainfed lowland, while Truko has the highest score on the population working in the agricultural sector. Meanwhile, Bringin has the lowest score of sensitivity with a total score of 20. Bringin has the lowest score of rainfed lowland and population working in the agricultural sector.

3.3. Analysis of Adaptive Capacity

Adaptive capacity analysis consists of analysis of head of farmer families who have jobs outside the agricultural sector, farmer families who have family members working outside the agricultural sector, farmer families who have savings, and farmer families who have livestock. Those four variables analyzed in this aspect of adaptive capacity are modifications of the prosperous family phasing indicator. This modification was done, so that the research results really describe the actual conditions in Bringin district. Each village has a total adaptive capacity score as listed in table 5.

Table 5. Score of Adaptive Capacity

Village	Score				Total
	Head of farmer families who have jobs outside the agricultural sector	Farmer families who have family members working outside the agricultural sector	Farmer families who have savings	Farmer families who have livestock	
Bringin	10	20	20	20	70
Popongan	20	20	20	10	70
Pakis	20	20	10	20	70
Lebak	20	10	10	10	50
Banding	20	20	10	20	70
Truko	20	10	10	10	50
Nyemoh	20	20	10	20	70
Tempuran	10	10	10	20	50
Wiru	10	20	10	10	50
Sendang	10	20	10	10	50
Gogodalem	20	20	10	10	60
Rembes	10	20	10	10	50
Kalikurmo	10	10	10	20	50
Sambirejo	10	20	10	10	50
Kalijambe	20	20	10	10	60
Tanjung	10	10	10	20	50

The total score of each village was divided into three classes which are low adaptive capacity, medium adaptive capacity, and high adaptive capacity. Low, medium, and high classification on adaptive capacity is done manually and not based on certain standards, so the results of classification in this study only valid to the study area. The classification is done by using ArcMap. The result of adaptive capacity classification can be seen in figure 4.

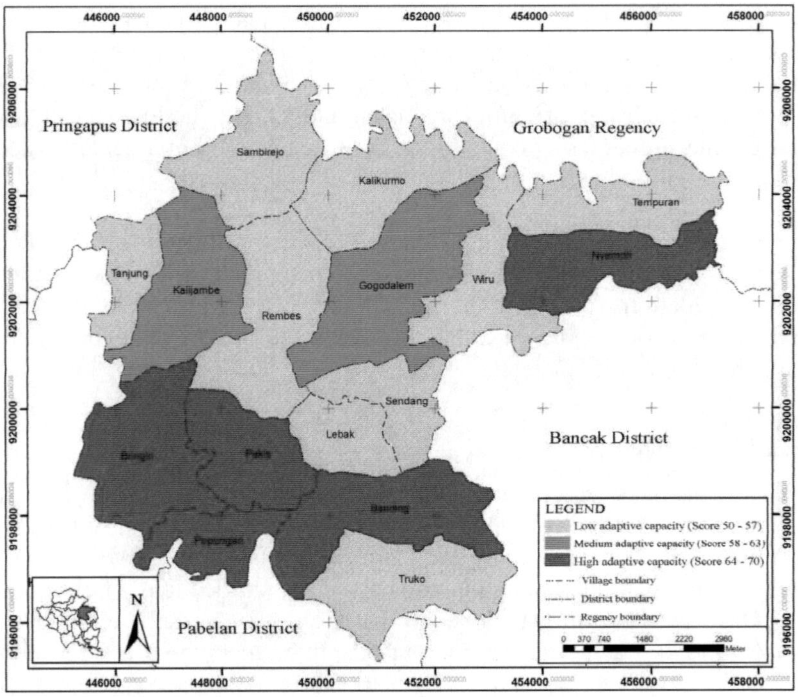

Figure 4. Adaptive Capacity Map

The villages with the highest adaptive capacity score (score 70) were five villages which are Bringin, Popongan, Pakis, Banding, and Nyemoh. The high adaptive capacity of these five villages was causes by the dominance of score 20 (low category) on the adaptive capacity variables. Villages that have medium category on adaptive capacity scores are two villages: Kalijambe and Gogodalem with a total adaptive capacity score of 60. Medium adaptive capacity in those village caused by score of 20 on job variable outside the agricultural sector and family members outside the agricultural sector, and score of 10 on the variable farmer families who have savings and farmer families who have livestock. The other nine villages: Truko, Sendang, Lebak, Rembes, Tanjung, Sambirejo, Kalikurmo, Wiru, and Tempuran have the lowest adaptive capacity score compared to the other seven villages. Those nine villages have total score of 50 on the adaptive capacity aspect. Adaptive capacity on those nine villages is low because of dominance of score of 10 (low) on adaptive capacity variables.

Although there are three classes of adaptive capacity, adaptive capacity for drought in Bringin district is in poor condition. This is evident from the scores on each variable in adaptive capacity that are in the low and medium low categories.

3.4. Analysis of Vulnerability

Analysis of vulnerability is the process of overlaying three maps which are exposure maps, sensitivity maps, and adaptive capacity maps. Each village has different scores as listed in table 6. The total score of each village is divided into three classes: high, medium, and low. The process of division is done manually, not based on certain standards, so the result of this study is only valid to the study area. The results of classification can be seen in figure 5.

Table 6. Vulnerability Score

Village	Exposure	Sensitivity	Adaptive capacity	Vulnerability
Bringin	114	20	70	64
Popongan	112	30	70	72
Pakis	121	30	70	81
Lebak	123	40	50	113
Banding	123	70	70	123
Truko	122	70	50	142
Nyemoh	115	60	70	105
Tempuran	116	40	50	106
Wiru	123	50	50	123
Sendang	125	50	50	125
Gogodalem	117	50	60	107
Rembes	123	40	50	113
Kalikurmo	124	50	50	124
Sambirejo	128	50	50	128
Kalijambe	125	50	60	115
Tanjung	123	40	50	113

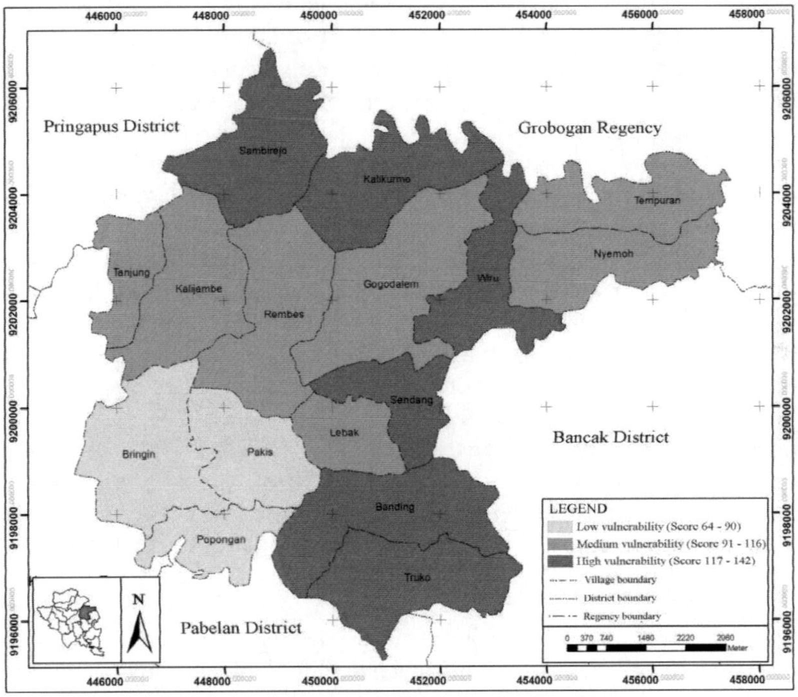

Figure 5. Vulnerability Map

High vulnerability (score 117-142) is distributed in 6 villages: Truko, Banding, Sendang, Wiru, Kalikurmo, and Sambirejo. High vulnerability in those six villages caused by high exposure and sensitivity scores and low adaptive capacity scores compared to other villages. The medium vulnerability (score 91-116) is distributed in 7 villages: Lebak, Tanjung, Kalijambe, Rembes, Gogodalem, Nyemoh, and Tempuran. Medium vulnerability in those seven villages caused by low exposure and sensitivity scores compared to other villages. Low vulnerability (score 64-90) distributed in 3 villages which are Bringin, Popongan, and Pakis. Low vulnerability in those three villages caused by low exposure and sensitivity scores and high adaptive capacity scores compared to other villages.

The village with the highest score is Truko. Truko has vulnerability score of 142. Truko has the highest sensitivity score of 70. In exposure aspect, Truko has medium exposure category with a score of exposure of 122. In adaptive capacity aspect, Truko has low category (Score 50). The village that has the lowest score of vulnerability is Bringin. Bringin has vulnerability score of 64. In sensitivity aspect, Bringin has the lowest sensitivity score in Bringin District. Bringin is also one of the villages with low category of exposure in Bringin District, with score of 114. In adaptive capacity aspect, Bringin has adaptive capacity score of 70. This score belongs to the category of high adaptive capacity in Bringin District.

Policymakers can refer to this vulnerability map within five years and review it in every 2 year. This is based on the Head of BNPB Regulation Number 4 of 2008 on Guidelines for the Preparation of Disaster Management Plan. The research must use data within the last five years (2011-2015). In addition, El-Nino phenomena is occurring within 3-7 years that also affects the planning period based on the results of this study.

4. Conclusion

The high exposure (score 124-128) is distributed in 4 villages, they are Sendang, Kalijambe, Sambirejo, and Kalikurmo. The high exposure in those four villages was caused by the type of constituent rock which is dominated by kerek formation and land cover that was dominated by rice fields (agriculture). The ability to pass water (save water) of kerek formation is lower than volcanic rock. That condition caused Kalijambe, Sambirejo, and Kalikurmo have high category of exposure aspect. Agriculture area has higher C factor (coefficient of running water) compared to forest, so that Sendang has high exposure category.

High sensitivity (score 54-70) is distributed in 3 villages which are Truko, Banding, and Nyemoh. Those three villages have high sensitivity because the percentage of rainfed lowland area and the population that works in the agricultural sector. Those variables have the highest score compared to other villages. Rice fields with rainfed irrigation have the highest risk of drought disaster and people working in agriculture are the most affected people from drought. These conditions cause Truko, Banding, and Nyemoh, which has a percentage of rainfed lowland area and the highest agricultural worker population in Bringin District, belongs to the high sensitivity category.

Low adaptive capacity (score 50-57) is distributed in 9 villages: Truko, Lebak, Sendang, Rembes, Tanjung, Sambirejo, Kalikurmo, Wiru, and Tempuran. Low adaptive capacity is caused by the dominance of score 10 (low category) in each variable. Scores on each variable in adapative capacity are in low and medium low categories. adaptive capacity for drought in Bringin district is in poor condition.

High vulnerability (score 117-142) distributed in 6 villages which are Truko, Banding, Sendang, Wiru, Kalikurmo, and Sambirejo. Those six villages have high vulnerability because of high exposure and sensitivity. In addition, low adaptive capacity also leads to high vulnerability in those six villages. High exposure is caused by the type of constituent rock dominated by the kerek formation and the type of land cover that is dominated by rice fields (agriculture). High sensitivity is caused by the percentage of rainfed lowland area and the population working in the agricultural sector. Low adaptive capacity is caused by the dominance of score 10 on each variable of adaptive capacity. This condition caused Truko, Banding, Sendang, Wiru, Kalikurmo, and Sambirejo have high vulnerability.

5. Acknowledgements

Bringin District Government, Climatology Station of Semarang, Department of Agriculture, Plantation, and Forestry of Semarang Regency, Statistic Agency of Semarang Regency, Development

Planning Agency of Semarang Regency, and Geology Agency of Ministry of Energy, Resources, and Mineral who gave data for this research..

6. References

[1] Chambers R 2006 Vulnerability, coping and policy (editorial introduction) *IDS Bull.* **37** 33–40
[2] Mujtahiddin M I 2014 Analisis Spasial Indeks Kekeringan Kabupaten Indramayu *J. Meteorol. dan Geofis.* **15**
[3] De Goede D M, Gremmen B and Blom-Zandstra M 2013 Robust agriculture: Balancing between vulnerability and stability *NJAS-Wageningen J. Life Sci.* **64** 1–7
[4] Metzger M J, Rounsevell M D A, Acosta-Michlik L, Leemans R and Schröter D 2006 The vulnerability of ecosystem services to land use change *Agric. Ecosyst. Environ.* **114** 69–85
[5] Brown J R, Kluck D, McNutt C and Hayes M 2016 Assessing drought vulnerability using a socioecological framework *Rangelands* **38** 162–168
[6] Cuevas S C, Peterson A, Robinson C and Morrison T H 2016 Institutional capacity for long-term climate change adaptation: evidence from land use planning in Albay, Philippines *Reg. Environ. Chang.* **16** 2045–2058
[7] Marhaento H 2016 Gis-Based Analysis for Assessing Landslide and Drought Hazard in The Corridor of Mt Merapi and Mt Merbabu National Park, Indonesia *Geoplanning J. Geomatics Plan.* **3** 15–22
[8] Lemos M C, Lo Y-J, Nelson D R, Eakin H and Bedran-Martins A M 2016 Linking development to climate adaptation: leveraging generic and specific capacities to reduce vulnerability to drought in NE Brazil *Glob. Environ. Chang.* **39** 170–179
[9] Fahmi I 2016 Analisis Pencarian Data Curah Hujan yang Hilang dengan Model Periodik Stokastik (Studi Kasus Wilayah Kabupaten Pringsewu) *J. Rekayasa* **19** 85–106
[10] Mohr E C J 1938 *De bodem der tropen in het algemeen, en die van Nederlandsch-Indië in het bijzonder* (Koninklijke vereeniging koloniaal instituut)
[11] Martin J A and Gray D M *Estimating Evapotranspiration on A Regional Scale* (Saskatoon)
[12] Mock F J 1973 Land Capability Appraisal Indonesia: Water Availability Appraisal *Basic study Prep. FAO/UNDP L. Capab. Apprais. Proj. Bogor, Indones.*
[13] Djaali H and Muljono P 2008 Pengukuran dalam bidang pendidikan *Jakarta, Grasindo*
[14] Paimin, Sukresno and Purwanto 2010 *Sidik Cepat Degradasi Sub Daerah Aliran Sungai* ed A N Gintings (Bogor: Pusat Penelitian dan Pengembangan Konservasi dan Rehabilitasi Badan Penelitian dan Pengembangan Kehutanan)
[15] Widyawati L 2016 Tingkat dan Sebaran Risiko Bencana Kekeringan di Kecamatan Kokap Kabupaten Kulonprogo *Geo Educ.* **1**
[16] Pujiono E and Setyowati R 2015 Penilaian Tingkat Kerentanan Sumber Daya Air terhadap Variabilitas Iklim di DAS Aesesa, Pulau Flores, Nusa Tenggara Timur *J. Penelit. Sos. dan Ekon. Kehutan.* **12** 177–195
[17] Eddy P 2009 Sistem Informasi Geografis: Konsep-Konsep Dasar (Perspektif Geodesi & Geomatika) *Inform. Bandung*

IOP Conf. Series: Earth and Environmental Science **123** (2018) 012032 doi:10.1088/1755-1315/123/1/012032

Dynamics Change of Vegetated Lands in A Highway Corridor during 37 Years (Case study of Jagorawi Toll Road, Jakarta-Bogor)

B P Perdana[1], Y Setiawan[1,2], L B Prasetyo[1]

[1]Department of Forest Conservation and Ecotourism, Faculty of Forestry,
Bogor Agricultural University, Kampus IPB Darmaga, Bogor 16680, Indonesia
[2]Center for Environmental Research, Bogor Agricultural University, Kampus IPB Darmaga, Bogor 16680, Indonesia

Email: bayuputraperdana07@gmail.com

Abstract. Recently, a highway development is required as a liaison between regions to support the economic development of the regions. Even the availability of highways give positive impacts, it also has negative impacts, especially related to the changes of vegetated lands. This study aims to determine the change of vegetation coverage in Jagorawi corridor Jakarta-Bogor during 37 years, and to analyze landscape patterns in the corridor based on distance factor from Jakarta to Bogor. In this study, we used a long-series of Landsat images taken by Landsat 2 MSS (1978), Landsat 5 TM (1988, 1995, and 2005) and Landsat 8 OLI/TIRS (2015). Analysis of landscape metrics was conducted through patch analysis approach to determine the change of landscape patterns in the Jagorawi corridor Jakarta-Bogor. Several parameters of landscape metrics used are Number of Patches (NumP), Mean Patch Size (MPS), Mean Shape Index (MSI), and Edge Density (ED). These parameters can be used to provide information of structural elements of landscape, composition and spatial distribution in the corridor. The results indicated that vegetation coverage in the Jagorawi corridor Jakarta-Bogor decreased about 48% for 35 years. Moreover, NumP value increased and decreasing of MPS value as a means of higher fragmentation level occurs with patch size become smaller. Meanwhile, The increase in ED parameters indicates that vegetated land is damaged annually. MSI parameter shows a decrease in every year which means land degradation on vegetated land. This indicates that the declining value of MSI will have an impact on land degradation.

Keyword: Highway Corridor, Vegetated Land, Patch Analysis, Landscape Metric

1. Introduction

Large scale road construction project executed by the government recently has affected the existing landscape condition. The planned toll road construction will connect the west end to the east end of Java Island. The construction requires large area hence the changes will greatly affect the exiting landscape condition. Development of transportation infrastructure is one of the factors that affect the level of landscape changes [1] in an area as the development of toll road and main road (motorway). Jagorawi toll road is one of the alternative toll roads that connect Bogor to Jakarta. The toll road becomes one of the factors that affect the development of developed land in the area.

Toll road is one of the required transportation infrastructures in an area to assist the accessibility of economy development and to connect regions. Development of toll road does not only provide positive impacts but also negative impacts to an area. The development positively induces the economy, trading, and governmental development, and improves the welfare of the community [2]. The negative impacts of the development are the increasing of land cover utilisations and changes, the decreasing of Green space, the increasing level of pollution, as well as conversion and change of undeveloped lands to developed lands.

The development of landscape changes that occurs can be ascertained through the use of technology of remote sensing and geographic information system (GIS) [3,4]. Remote sensing is a science or an art to obtain information about objects, areas or phenomena by analysing the data obtained through the use of devices, with indirect contact with the objects, areas or phenomena of the study. GIS technology uses optical images from satellite to measure and interpret the data on the field. Commonly used optical image is Land Satellite Image (Landsat). The image has sufficient frequency of shooting to monitor the land changes throughout the year. The objectives of this study are: 1) to examine land cover changes at Jagorawi Toll Road in 1978-2015, 2) to analyse landscape patterns around Jagorawi Toll Road and 3) to evaluate the changes on landscape pattern based on factor of distance from Jakarta to Bogor.

2. Methodology

2.1 Study Site

This study encompasses the Jagorawi highway (toll road) from Jakarta to Bogor. The distance is equal 2 to 42 km. In order to calculate a landscape pattern of that highway corridor, 2 km buffer right-left prepared (Figure 1).

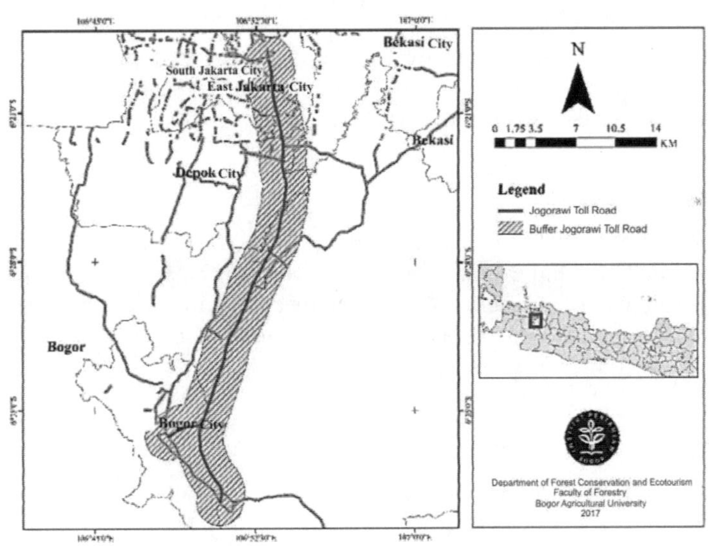

Figure 1. Study site

Jagorawi toll road is the oldest toll road in Indonesia that was built in 1972 and was first operated in 1978 as the main road that connects Jakarta to Bogor and Ciawi, West Java. Jagorawi toll road crosses over 28 villages, 9 villages are located in the administrative area of Jakarta and 19 other villages in the area of Bogor district. There are 11 gates along Jagorawi toll road such as Toll Gate (GT) TMII, Pasar Rebo, Cibubur, Cisalak Cimanggis, Gunung Putri, Citeureup, Sentul, South Sentul, Bogor and Ciawi.

IOP Conf. Series: Earth and Environmental Science **123** (2018) 012032 doi:10.1088/1755-1315/123/1/012032

Jagorawi Toll road has a flat topography (slope 0-2%), wavy (3-8%) and corrugated (9-15%). Jagorawi altitude area varies, approximately 20-300 meters above the sea level. Based on Schmidth-Ferguson climate classification, Jagorawi toll road area is classified with climate type A, which has an average rainfall greater than 2000 mm / year.

2.2 Materials and Tools

Tools used in this study are stationery and personal computer equipped with some software, such as: ArcGIS 10.3, Patch Analysis, ERDAS Imagine 9.1, Google Earth, and Ms. Office 2013. Materials used in this study are images from Landsat MSS 2 1978, landsat 5 TM 1988, 1995, and 2005, as well as landsat 8 OLI/TIRS 2015.

2.3 Data Collection

Data used in this study is the images of Landsat 2 MSS, Landsat 5 TM and Landsat 8 OLI / TIRS obtained by downloading from the internet while the field point obtained through ground check by recording the coordinates with a GPS.

2.4 Data Analysis

Data analysis was performed with classification of land cover and analysis of landscape pattern and gradient analysis through the use of patch analysis to describe the pattern of trend (trend) of degree of fragmentation in the region.

2.5 Land Cover Classification

Classification of land cover obtained from image processing through the use of Google Earth in 1978, 1988, 1995, 2005 and 2015 with supervised and non-supervised image classification. Land cover classified into developed land, water bodies, agricultural land, and vegetated land.

2.6 Landscape Pattern Analysis

This study used landscape metrics parameters such as Mean Shape Index (MSI), Mean Patch Size (MPS), Edge Density (ED), and Number of Patch (NumP). According to McGarigal et al. [5] these metrics landscape parameters are able to interpret the structure of spatial element, composition, and distribution in landscape. Landscape metrics include number of patches, edge density and average efficient shape in analysis of changes in land cover [6].

2.7 Gradient Analysis

Analysis of gradient is a method to compare patterns of trend (trend) from multiple locations by observing the slope of the line connecting between the two variables. The spatial patterns of land use reflect the human process that underlie and affect the ecology of urban environment [7]. The analysis performed by creating a 4x4 km swath along the observation area from Jakarta to Bogor. After that, each plot was analysed using the patch chart analysis to determine land cover changes throughout the observation area. Patch analysis parameters used are Mean Shape Index, Mean Patch Size, Edge Density, and Number of Patch.

IOP Conf. Series: Earth and Environmental Science **123** (2018) 012032 doi:10.1088/1755-1315/123/1/012032

Table 1. Landscape Metric Equation

Parameter	Formula
Edge Density (m/ha)	$\frac{\sum_{k=1}^{M} e\ ik}{A}(1000)$
Mean Shape Index	$\frac{\sum_{j=1}^{n}\left(\frac{p_{\ddot{U}}}{\sqrt[2]{\pi \times a_{\ddot{U}}}}\right)}{n_{\ddot{U}}}$
Mean Patch Size (ha)	$\frac{\sum_{j=1}^{n} a_U}{n_i}(\frac{1}{10000})$
Number of Patch (Unit)	N

Description: e ik: Total length of the patches in the landscape in m, n: total number of patches in the landscape, ni: number of patches in the landscape, m: the number of patches in the landscape beyond the limit unless the landscape, m ': the number of patches in the landscape and the landscape, j: number of patches, k: the same patch, A: total area.

3. Discussion and Result

3.1 Land use land cover changes

The term of land cover and land use is often used in the study of land cover changes. According to Lillesand and Kiefer [8], land cover related to the appearance of the Earth's surface, while the land use related to the human activity on the object. There are several classes of land cover around Jagorawi Toll Road such as water bodies including rivers, marshes, and reservoirs; developed land cover class, such as residential areas, economic sectors, and industrial area. Farm land class involves annual plants, and vegetated land class including grass, shrub, and tree areas.

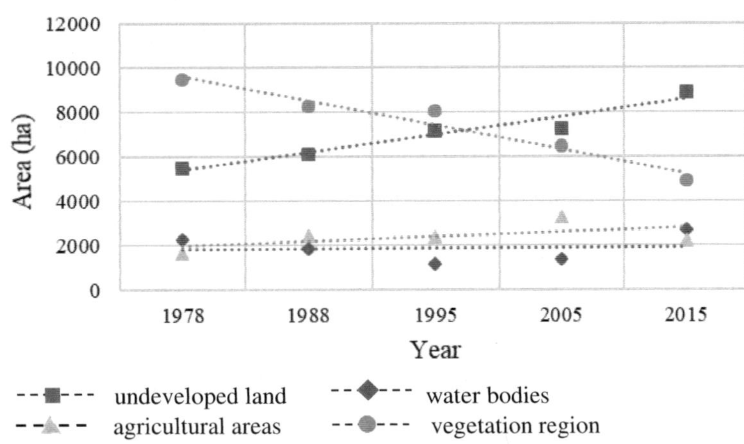

Figure 2. Dynamics of Changes in Land Cover Years 1978-2015

IOP Conf. Series: Earth and Environmental Science **123** (2018) 012032 doi:10.1088/1755-1315/123/1/012032

Land cover / land use changes that occurred in 1978-2015 had modified the land area. The undeveloped land in 1978 had an area of 5,465.16 ha or 29% in the year and continued to increase in 1988, 1995, 2005 and 2015 respectively as 6 114.15 ha (33%), 7,169.67 ha (38%), 7,275.96 ha (40%), 8,890.38 ha (48%). The vegetated land area in 1978-2015 decreased about 9 462.24 ha (50%), 8 279.46 ha (44%), 8,036.73 ha (43%), 6,441.56 ha (35%), and 4,901.85 ha (26%). The area of water bodies rose from 2,252.88 ha (12%) in 1978 to 2,654.64 ha (14%) in 2015. The total area of the farm land rose about 1,610.28 ha (9%) in 1978 and 2,207.16 ha (12%) in 2015 (Figure 2).

Figure 3a shows the changes of vegetated land around Jagorawi Toll Road. The vegetated land turned into developed land as much as 4,555.83 ha or 48% over a period of 37 years. The vegetated land turned into agricultural land area of 1,240.256 ha or 13%, the vegetated land turned into water bodies covering an area of 1,051.97 ha or 11%, and existing vegetated land area of 2,569.93 ha or 27%. This suggests that changes in the vegetated land most likely caused by existing development in the area such as the development of residential or industrial area. This is in accordance with statement that the factor of population density and development of motorway that connects the city can accelerate and expand changes of the lands around it.

Figure 3b shows the agricultural land cover changes that occurred around Jagorawi Toll Road. Visible changes could be seen in agricultural land around Jagorawi Toll Road. The agricultural land and undeveloped land transformed into an area of 448.99 ha or 20% over a period of 37 years. The agricultural land changed into land covering area of 142.46 ha of water bodies or 6%, agricultural land into vegetated land area of 1,263.61 ha or 58%, and the rest is agricultural land that does not change as much as 352.08 ha or 16%. According to Trisasongko et. al [9], the incidence of changes in land use from wetland to dry land was positively correlated with the growth of dry land in the region.

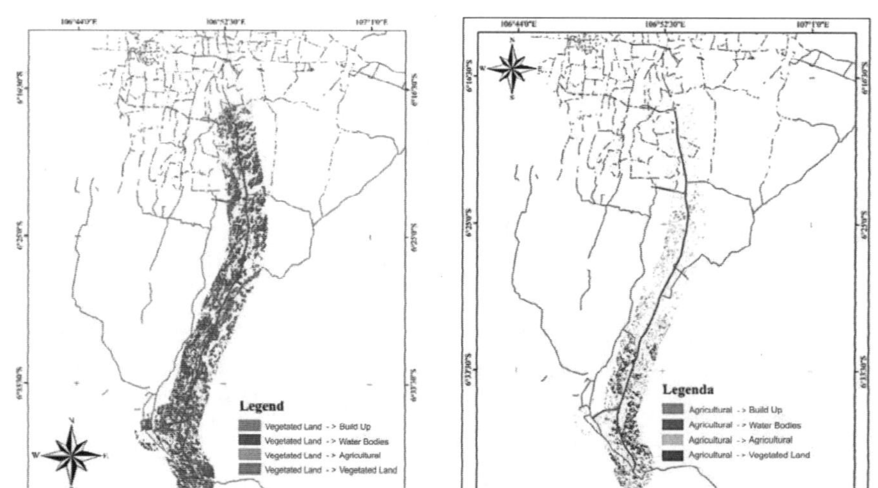

(a) (b)

Figure 3. Map of Land Cover Changes in Jagorawi in 1978-2015; a) Vegetated Land, b) Agricultural Land

IOP Conf. Series: Earth and Environmental Science **123** (2018) 012032 doi:10.1088/1755-1315/123/1/012032

3.2 Landscape pattern changes of vegetated lands

Landscape quantitative analysis based on fragmentation process is one of the necessary landscape processes that show the effect of human activity and human disturbance on the structure and function of landscape [10]. Landscape metrics provides quantitative method to explain spatial distribution and climate variation on urban temperature patterns in response to land covering pattern in urban landscape [11]. Quantitative metrics as indicators of environmental landscape can be used in a sustainable land use planning [11]. Land cover changes over a period of 37 years from 1978 to 2015 shows that the spatial characteristics of land use classes also changed over time.

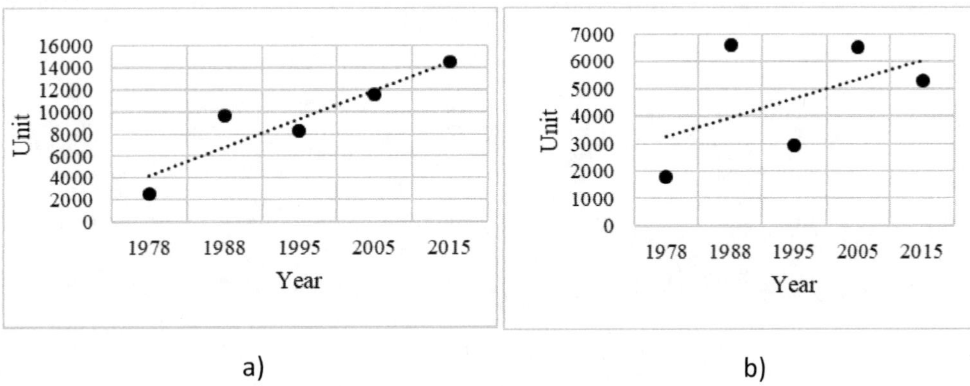

Figure 4. Number of Patch; a) Vegetated land, b) Agriculture land

Indicator of number of patches (NumP) is the total number of patches in the landscape if selected to analyse the 'landscape', or the number of patches for each individual class if selected to analyse the 'class' [5]. The results of NumP obtained in the vegetated land in 1978 was 2534 units, 9671 units in 1988, 8240 units in 1995, 11 576 units in 2005, and 14484 units in 2015. This shows the increasing trend in 1978-2015. Increasing trends of change also occur on the agricultural land for a period of 37 years. NumP obtained from the calculation results show the agricultural land in 1783 units in 1978, 6621 units in 1998, 2932 units in 1995, 6542 units in 2005, and in 5305 units in 2015 (Figure 4). Increased NumP indicates that the higher the value of NumP obtained the higher the level of fragmentation in the region. According Elkie et al. [12] says that the increase in NumP show increased fragmentation with a smaller patch size.

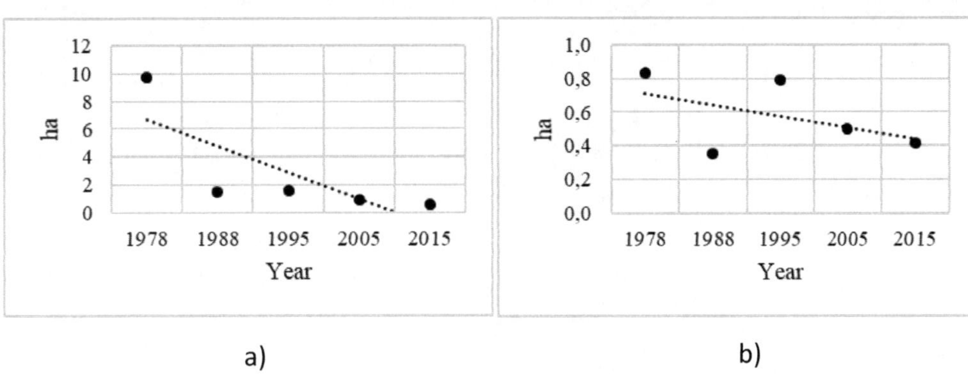

Figure 5. Mean Patch Size; a) Vegetated land, b) Agriculture land

Indicator of mean patch size is a measure of the average patch in particular class, or at the landscape level to the average size of all classes [13]. MPS results obtained in the vegetated land in 1978 was 9732 ha, 1.495 ha in 1988, 1,559 ha in 1995, 0.929 ha in 2005, and 0602 ha in 2015. This indicates a downward trend from the year 1978 to 2015. Impairment MPS in the vegetated land indicates that the vegetated land area within a period of 37 years continues to decline. According to Kamusoko et al. [14], MPS impairment shows the increase in the dynamics of land use patterns, so that the structure of the landscape becomes smaller as the result of land distribution system. MPS calculation results on the agricultural land also showed a decrease in area over a period of 37 years. The agricultural land in 1978 was 0829 ha, 0354 ha in 1988, 0791 ha in 1995, 0497 ha in 2005, and 0414 ha in 2015 (Figure 5).

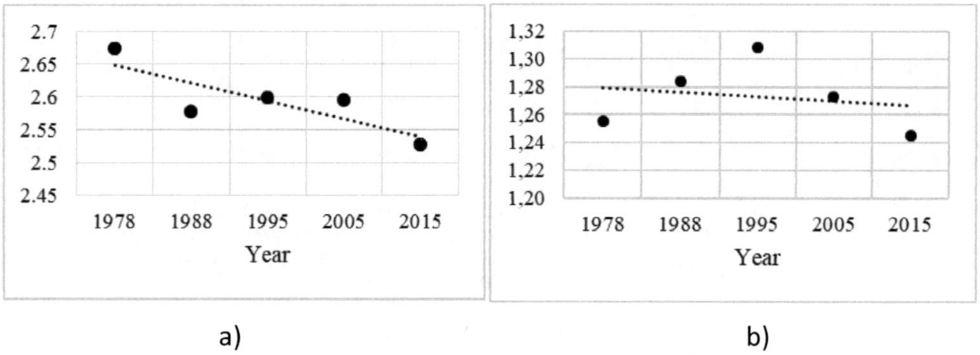

Figure 6. Mean Shape Index; a) Vegetated land, b) Agricultural land

Mean shape index is a parameter of the complexity of form in which the amount of the perimeter of each patch is divided by the square root of the area of the patch (ha) for each class (grade level) or all of the patches (landscape level) [5]. The comparison metric landscape obtained over the past 37 years shows that the average shape index (MSI) in the vegetated land was 2,674 in 1978, 2,577 in 1988, 2,600 in 1995, 2,595 in 2005 and 2,527 in 2015. This indicates a downward trend from 1978 to 2015 year. MSI calculation results in the agricultural land also shows declining trend change. MSI value in 1978 was 1,256, 1,284 in 1988, 1,308 in 1995, 1,273 in 2005, and 1,245 in 2015 (Figure 6). MSI parameters show a decline in each year, which means land degradation occur in the vegetated land and the agricultural land in 1978-2015. This indicates a decrease in the value of MSI have an impact on land degradation. According to Tazeh [11] who states that landform metric indicators can be an indicator that shows the processes of degraded lands.

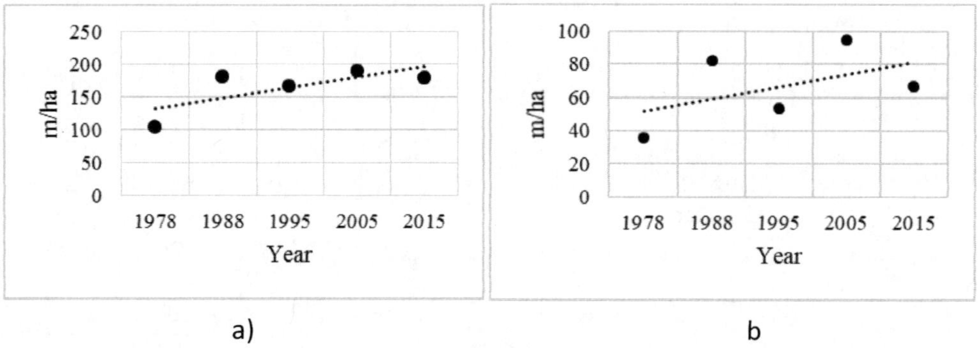

Figure 7. *Edge Density* a) Land vegetation, b) Agriculture

Indicator of edge density (ED) is the number of relative edges to landscape area [5]. ED is a standardised metric that is useful to compare the landscape with different sizes [12]. Edge density (ED) is the number of relative edges to landscape size . ED result was obtained from the calculation through the use of Patch Analysis on the vegetated in 1978 which was 104 078 m / ha, 181 108 m / ha in 1988, 167 550 m / ha in 1995, 190 518 m / ha in 2005, and 179 734 m / ha in 2015. This shows an increasing trend from 1978 to 2015. Increased ED also occurs on the agricultural land as indicated by the increasing trend in 1978-2015. ED value in 1978 was 35 665 m / ha, 82 260 m / ha in 1988, 53. 189 m / ha in 1995, 94 712 m / ha in 2005, 66 552 m / ha in 2015 (Figure 7). According to Tagil (2007), an increase in the perimeter of the ED will lead to an increase in the length and direction of damages to the land and to show that the landscape can be used in land use changes.

3.3 Gradient analysis on landscape patterns in the corridor Jakarta-Bogor

The graphic of NumP shows the increasing level of fragmentation in the 37-year period from 1978 to 2015. The vertical side (bottom to top) shows the rise of fragmented land each year. NumP charts in 1978 shows the smallest trend value compared to the other years. This is because in that year the degree of fragmentation around Jagorawi Toll Road was not too severe. Jagorawi Toll Road development impact that year has not been too significant, this can be proved from the presence of good vegetated land. In 1988-2015 on the graphic of NumP can be seen that the degree of fragmentation is increasing every year, which indicates vegetated land in 1988-2015 diminishing in number. In 2005-2015, NumP value shows a high increase, this is caused by the population growth due to urbanisation and continuous development in this year that lead to more fragmented vegetated land.

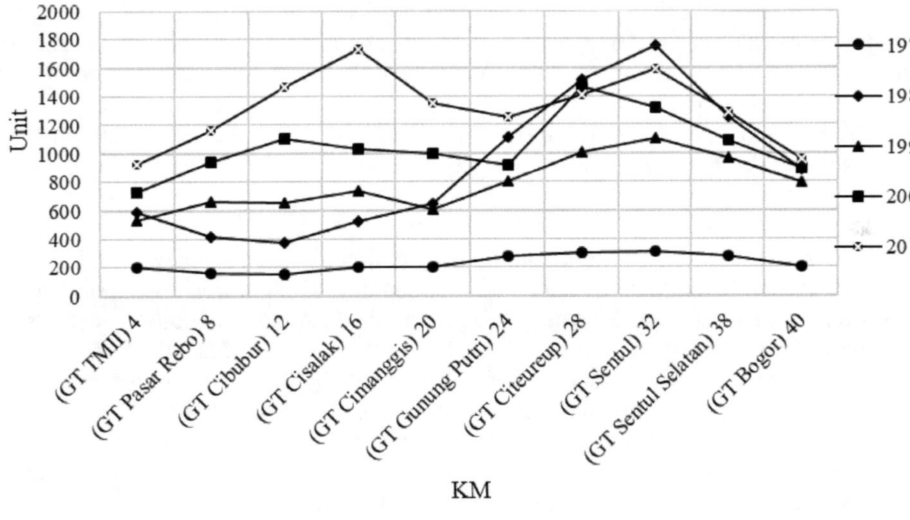

Figure 8. Number of Patch

On horizontal side (left to right) or a change in the landscape around the area of observation, the highest fragmentation of vegetated land at KM 28 and 32 in. Sentul area of Bogor District at km 32 is an area that has a high degree of fragmentation on vegetated land, this was caused by the developments on recreational area, economy area, gymnasium, and resident area. At KM 28, which is the Citeureup GT, the damage occurred on vegetated land in the area caused by the development of industrial area, economy area, and resident area.

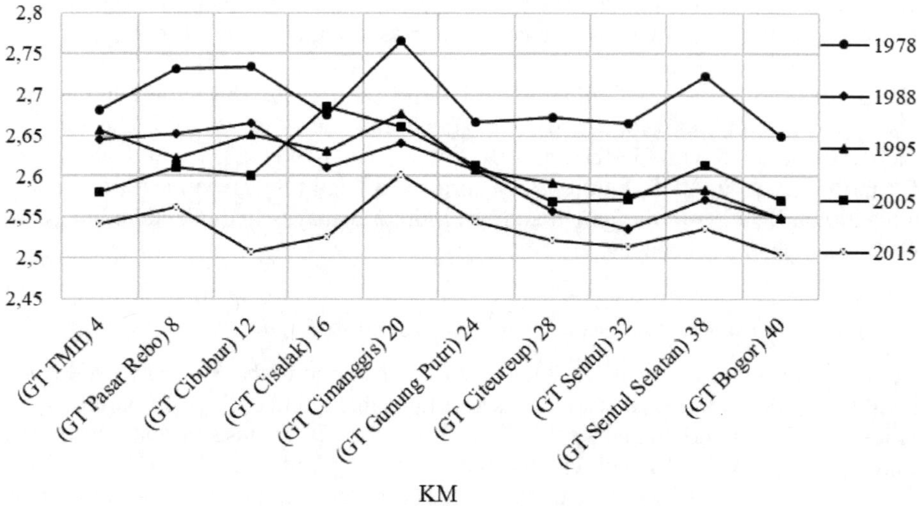

Figure 9. Mean Shape Index

MSI chart shows the pattern of land cover changes along Jagorawi Toll Road. The vertical side (top to bottom) shows a decrease in the value of the MSI over a period of 37 years. MSI chart in 1978 shows the highest value trend, because in that year Jagorawi toll road was still in the initial operation and the existing vegetated land had not yet suffered from severe degradation. The graphic results in 1988-2015 shows the value of MSI decreasing every year. This is shown on the graphic in 2015 where MSI value is the smallest among the other years due to the increasing development caused by the presence of the toll road that diminish the land use trend. Development of an area will generate traffic that affects transportation infrastructure, on the other hand, the presence of good transportation infrastructure will affect land use patterns [15].

The results on horizontal side (left to right) show the spatial patterns of vegetated land in the area of observation. Value trend on MSI parameter fluctuated as seen at KM 20 (Cimanggis Toll Gate) in each year, which means that in these areas, the existing land cover is still in good condition. The changes that occur before and after the KM 20 which shows a decrease on the graphic from MSI parameter indicates that vegetated land is degraded by the construction of this toll road. MSI chart increased at KM 38 (South Sentul) which means the vegetated land in the area is still in good condition. MSI chart in 1978 shows the highest value compared to year thereafter, because in that year Jagorawi toll road was still in the initial operation and the presence of vegetation on the land had not yet suffered severe degradation. Graphic results in 1988-2015 show the value of MSI decreasing every year. This is shown on the graphic in 2015 where MSI value is the smallest among the other years due to the increasing development caused by the presence of the toll road that diminish the land use trend.

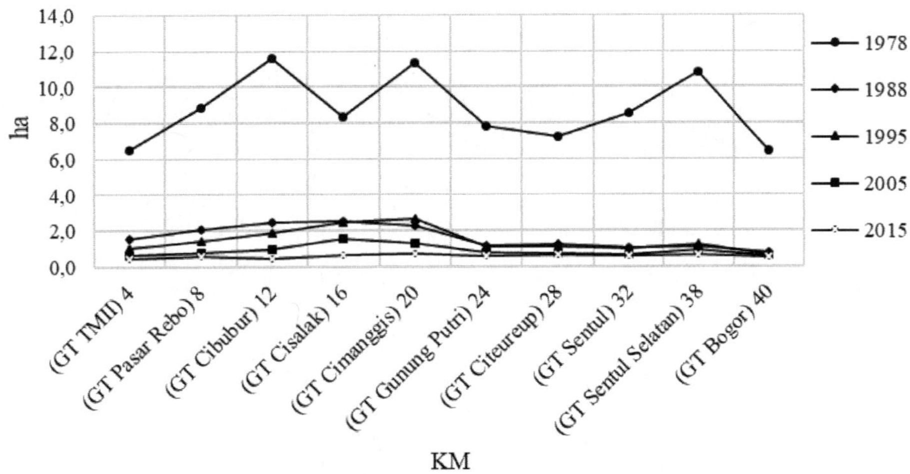

Figure 10. Mean Patch Size

The MPS graphic shows a decrease in the vegetated land area in 1978 to the following years. The vertical side (up and down) shows that the whole vegetated land area in 1978 still had a considerable area along the observation. This was because in that year the construction around the observations line was not too significant and did not affect the existing vegetated land. The 1988-2015 MPS graphic shows a severe deterioration in 1978, this can be seen from the graphic where the value decreases every year. It indicates the development around observation line was very rapid in those ten years. According to Akhirudin and Suharjo [16], the land conversion of land from undeveloped land to developed land is caused by residential, industrial, and other developments to support human life.

MPS graphics on the horizontal side (from right to left) shows the increase and decrease in the total area of vegetated land on observation line. During 1988-2015, visible pattern of land use change on the vegetated land along the observation. The vegetated land area that existed in the 1988-2015 did not experience significant changes each year, in contrast to the year 1978 that there was still a lot of vegetated land observation line. This was caused by during the time Jabodebatek was under rapid development in industrial, trading, and service sector that leaded to the decrease of vegetated land because it was considered less productive in term of value (land rent).

ED graphic shows the trend of the value of spatial patterns on the whole area of observation. These results show the pattern of landscape changes that occur over a period of 37 years. The vertical side (top to bottom) shows the increases and decreases in the value of ED in the area of observation. ED graphic in 1978 shows the smallest value compared to the other years. This suggests that in this year vegetated land was still in better condition compared to the years after, because in that year the Jagorawi Toll Road was still in its early stages of operation. After 1978, it is shown in the ED graphic that there us an increase in ED value throughout the years of 1988 to 2015 that indicates the damage to the vegetated land area due to the development undertaken around Jagorawi Toll Road. Figure 11 shows the change from 1978 to 2015 where the vegetated land continues to decline. The increase in ED parameter indicates that vegetated lands annually damaged. It is in accordance with the statement of Tagil (2007), the increase in the perimeter of the ED will cause an increase in the length and direction of the damage.

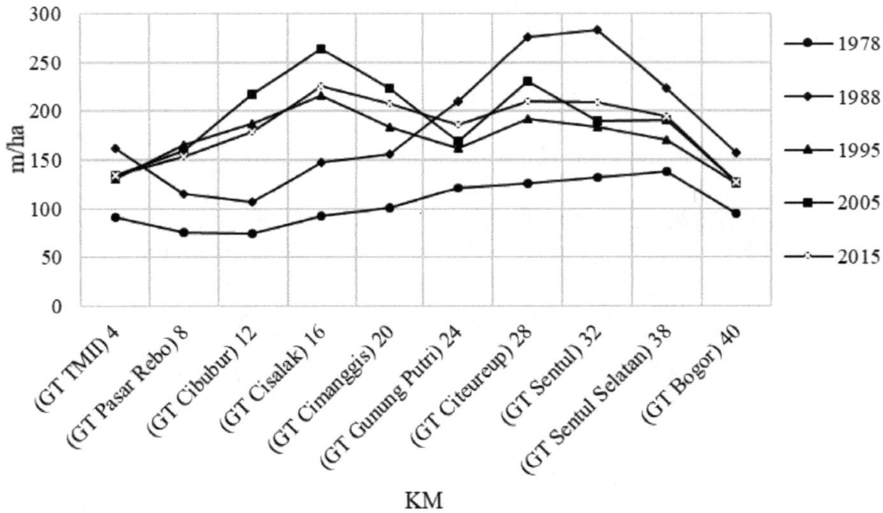

Figure 11. Edge Density

The horizontal side (left to right) shows the increased value of KM 16 and 28 where the vegetated land suffered severe damage compared to the other areas. This damage is caused by the influence of urbanisation such as the development of industrial buildings, the area of the economy, and the residential that leads to reduced vegetated land. KM 16 GT Cisalak, Depok, is an area that has a rapid development growth because it is in suburban area. The closer a developed area to the road then the wider the range will be as the result of the increasing of human activity. This is in accordance with the research of Masykuroh and Rudianto (2016) which states that the existence of toll gate has a significant impact on the land conversion of undeveloped land to developed land (economy and residential areas). The damage on the vegetated land also occurred at KM 28 GT Citereup which caused by the influence of development, industry, economy, and resident area.

4. Conclusion

Vegetation land turned into developed land during the last 37 years was 48% in area. Vegetated land turned into agricultural land was 13% in area, vegetated land turned into water bodies was 11% in area, and the existing vegetated land was 27% in area. Agricultural land turned into developed land was 20%. Agricultural land turned into water bodies was 6% in area, agricultural land turned into vegetated land was 58% in area, and the existing agricultural land was 16% in area.

Increased NumP indicates that the higher the value of the NumP obtained then the higher the level of fragmentation that exists in the area. MSI parameter indicates that the decrease in the value of the MSI will have an impact on land degradation. The decline in the value of the MPS shows increased land use pattern dynamics so that the structure of the landscape becomes smaller. Improvement on ED parameter indicates that the existing vegetated line along observation line is annually damaged.

The result of NumP graphic shows high fragmentation level on vegetated land in KM 28 and 32 in 1988-2015. MSI graphics on KM 20 and KM 36 increased in 1978-2015 which means the vegetated land is still in good condition. MPS graphics during the 1988-2015 did not experience significant changes each year, while in 1978 there was still a lot of vegetated land along the observation line. The ED graphic shows the increase value in KM 16 and KM 28 indicates the vegetated land in this area is more damaged compared to the other area.

Further research needs to be performed by identifying the factors that can affect land cover change around Jagorawi toll road. Based on the results of the calculation of landscape analysis, the upcoming development should take the principles of sustainable development into consideration in order to equalise ecological issue with economy issue. Furthermore, regulations and policies on maintenance of land use development should be released by the government.

5. References

[1] Hersperger A and Burgi M 2007 Driving Forces of landscape change in the urbanizing *Geoj. Libr.* **90** 45–60
[2] Arta F and Pigawati B 2015 the Patterns and Characteristics of Peri-Urban Settlement in East Ungaran District, Semarang Regency *Geoplanning J. Geomatics Plan.* **2** 103–115
[3] McRae B H, Hall S A, Beier P and Theobald D M 2012 Where to Restore Ecological Connectivity? Detecting Barriers and Quantifying Restoration Benefits *PLoS One* **7** 1–12
[4] Ohri A and Yadav P 2012 Urban sprawl mapping and land use change analysis using remote sensing and gis *Int. J. Remote Sens. GIS* **1** 12–25
[5] McGarigal K, Cushman S A, Neel M C and Ene E 2002 FRAGSTATS: spatial pattern analysis program for categorical maps
[6] Lausch A and Herzog F 2002 Applicability of landscape metrics for the monitoring of landscape change: issues of scale, resolution and interpretability *Ecol. Indic.* **2** 3–15
[7] Redman C L 1999 Human dimensions of ecosystem studies *Ecosystems* **2** 296–8
[8] Lillesand T, Kiefer R W and Chipman J 2004 *Remote sensing and image interpretation* (John Wiley & Sons)
[9] Trisasongko B H, Panuju D R, Iman L S, Harimurti A F, Ramly V A and Subroto H 2009 Analisis dinamika konversi lahan di sekitar jalur tol cikampek *Lap. Penelit. Kerjasama Pus. Penelit. Perenc. Pengembangan Wil. IPB dengan Kementeri. Lingkung. Hidup RI*
[10] Leitao A B and Ahern J 2002 Applying landscape ecological concepts and metrics in sustainable landscape planning *Landsc. Urban Plan.* **59** 65–93
[11] Tazeh M 2015 Investigation of Rangeland Changes Based on Landscape Metrics Analysis (Case Study: Kezab Rangelands, Yazd Province, Iran) *J. Rangel. Sci.* **6** 102–111
[12] Elkie P, Rempel R and Carr A 1999 Patch analyst user's manual. Ontario Ministry of Natural Resources *Northwest Sci. Technol. Thunder Bay, Ontario, Canada* 16
[13] McGarigal K and Marks B J 1995 Spatial pattern analysis program for quantifying landscape structure *Gen. Tech. Rep. PNW-GTR-351. US Dep. Agric. For. Serv. Pacific Northwest Res. Stn.*
[14] Kamusoko C and Aniya M 2007 Land use/cover change and landscape fragmentation analysis in the Bindura District, Zimbabwe *L. Degrad. Dev.* **18** 221–233
[15] Tamin O Z and Frazila R B 1997 Penerapan Konsep Interaksi Tata Guna Lahan--Sistem Transportasi Dalam Perencanaan Sistem Jaringan Transportasi *J. Perenc. Wil. dan Kota* **8** 11–18
[16] Akhirudin N H Suharjo. 2006. Identifikasi perubahan penggunaan lahan kota Surakarta tahun 1993--2004 dengan aplikasi sistem informasi geografis (SIG) *J. Penelit. Sains dan Teknol.* **7** 170–178

What is The Role of Land Value in The Urban Corridor?

A R Rakhmatulloh[1], I Buchori[2], W Pradoto[2], B Riyanto[3], and J Winarendri[2]

[1]Doktoral Program of Architecture and Urbanism, Diponegoro University, Jl. Prof. Soedarto, SH Street, Semarang, Indonesia 50275
[2]Department of Urban and Regional Planning, Diponegoro University, Prof. Soedarto, SH Street, Semarang, Indonesia 50275
[3]Department of Civil Engineering, Diponegoro University, Prof. Soedarto, SH, Street, Semarang, Indonesia 50275

Email: anita.ratnasari.r@gmail.com

Abstract. The high movement causes traffic congestion and indicates high movement intensity along the corridor. The higher attraction of the land use will encourage the higher attraction of movement and economic values in the location. This attraction is also affected by the high mobility in the corridor supported by available transport infrastructure. Thus this causes land values become increase significantly. Land use along the corridor can be seen as commercial function because this activity is able to survive in the premium location. The purpose of this research is to identify the effect of land use change toward land values in the commercial corridor. This research used positivistic method with descriptive analysis. The result shows that the land values change in commercial use in the corridor has different pattern of land use change pattern according to physical condition and land use which causes highly economic attraction. The new commercial land is influenced by the distance to city centre or CBD (Central Business District). Land use and public facilities that have local and city scope services do not give the significant impact to land values change.

Keywords: Urban Corridor, CBD, Location Analysis, GIS, Land Use

1. Introduction

The high number of movements in corridor has caused traffic congestion that indicates the high intensity of movements on the land along the corridor. The higher land use attraction on the land, the higher trip generation and economic attraction in those locations. This attractiveness is also supported by transportation infrastructure that can enhance the accessibility and mobility in the corridor.

The land use along the corridor can be seen as commercial function because this activity function is able to survive in the premium location. When it comes to urban areas, land values increase will be much related to strategic site (location factor) that associated with the ease of broad range or access to transportation system and its location to other land uses or urban configuration [1]. Thus, land value change becomes one of the factors of land use change and both have strong linkages. The closer or the easier access to corridor will certainly give impact on land value rise.

The corridor space that plays to connect city centre to surrounding small towns has apparently encouraged more problems. These problems are caused by feedback relationship (interrelationships) between both of them that result in human migration flows, goods and services among them. The major issue occurred is transportation problem and land use change rapidly happened in the corridor space. The purpose of this study is to identify the effect of land use change toward land value in the commercial corridor. Which in turn, this land value will hold land market mechanism toward functional change and physical land in the corridor. On the other hand, intensity of development in the corridor should be biased handled in order to avoid the bottle neck occurrence [2] that will cause many problems for the main city and surrounding areas which can eventually decrease the urban economic value.

2. Urban Commercial Corridor

The corridor space usually develop to be regions with high frequency of movements due to availability of main transportation network dan make the corridor as commercial area [3]. Corridor, according to Duany and Plater-Zyberk [3] in The New Urbanism is the liaison and also the separation between residential neighborhoods and districts that should not just be a remaining space but become an urban element characterized by the appearance of continuity. Duany and Plater–Zyberk [3] noticed the role of corridor in relation to desa-kota that plays as magnet in urban dynamic. The activity between these two elements enable the high movements in both. These high activities which encourage corridor to become a dynamically inclined space following the growth of both magnetic element [4]. Corridor is an element that accommodates the relationship between two elements with mutual support. Krier [5] argued that in the case of a corridor study linking several new towns and access to a main city will have an increased role beyond just circulatory connections.

Along its development, there are commercial corridors or commercial strips as urban and sub-urban connections that accommodate high-speed vehicles. In relation to this function, buildings on commercial corridors are built with sufficient "setback" for the vehicle parking lot. According to Manning [6] this phenomenon was also found in commercial corridors that connect the new city. The temporary building was seen to fill the space between the permanent functions that have been previously filled such as the industrial function.

The length of trip led to a high movement in the corridor as a hub of the city center with urban periphery. The problems of infrastructure constraints in this corridor area are called as the bottleneck of corridor issue [7]. At the same time, many people enter the corridor because of the time to go and get home from work is almost the same. On the other hand, the provision of public transport facilities and infrastructure services have no different from regular hours. These differences make the prominent disparity between transport demand and fully services during high traffics [2].In Minner and Shi's research [8] commercial corridors is a common element in a city known as linear commercial in which within the corridor consist of commercial property built on land close to and oriented to the main road, arterial road or other highways. Commercial corridor is also called as habitat for local businesses and the region targeted for redevelopment. With spatial analysis in research conducted by Minner and Shi [8] found that local businesses tend to thrive in regions along commercial corridors that are located close to the city center.

2.1. Land Use Change

Land use is every form of human intervention to the land in order to accommodate the various life needs of both material and spiritual [9]. Most experts agree that land change is a consequence or impact of economic growth and cities, high urbanization, population change and increased activities in urban areas.The main factors causing land use change [10] are population, land value, and transportation system. The change of these three main factors will cause the changes in the activity system and spatially embodied in the form of land use patterns. While Wu and Silva [11] explained

IOP Conf. Series: Earth and Environmental Science **123** (2018) 012033 doi:10.1088/1755-1315/123/1/012033

land change is driven by combined factors or spatial and non-spatial factors that spur the process of dynamics of land change with the interaction among factors occured. At the same time, the process of urbanization in cities and changes in physical, social and ecosystem aspects resulting from the impact of urban growth triggered land change.

In terms of land use change in the corridor, Arnott, Palma, and Lindsey [7] stated that the urban corridor grew out of the development of a decentralized metropolitan area. Residential growth to suburbs or urban periphery and daily commuting movements between sub-urban to city center (bottle neck corridor). Consequently, those movements caused bad congestions during rush hour which result in the in-efficiency of urban economy. Hence, it is necessary to adapt and expand the efforts to handle the congestion problem by providing persuasive solution such as telecommunication facilities support, computers, electricity network and public facilities. Eventually the corridor grew into a commercial area along the urban corridor that became the commuting line.

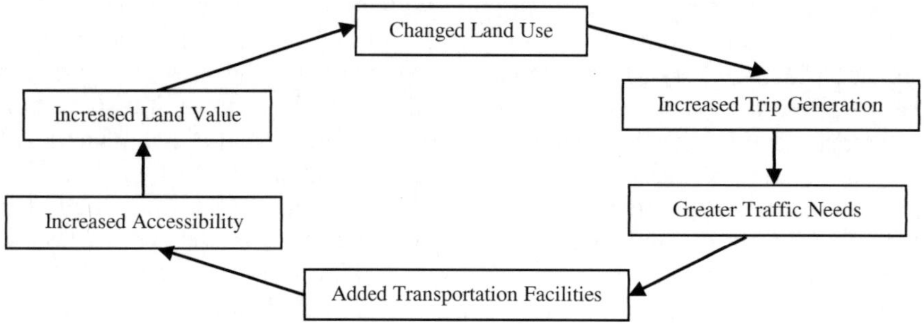

Source : Adapted from J. Paquette (1980) in [12]

Figure 1. Land Use Cycle and Transportation

A land use change will lead to an increase trip generation. This rise will lead to increased level of accessibility that will lead to an increased land value in the area. The increasing land value will eventually lead to the emerge of activities that compatible with the condition in region. Thus it can trigger the development of high rise building intensity on the land. When transportation access to the activity space (land parcel) is improved, then the activity space will be more interesting and usually become more developed. With the development of space activities, it will also increase the need for transportation. This increase then causes an overload transport that must be addressed.

2.2. Land Value

The advantage of having a good connection in the high accessible area will lead to increased land value [13]. Commercial properties or commercial buildings have the most expensive land value, followed by single-family residences, multi-family housing, and condominiums. For commercial purposes, properties close to the station reap the high premium and positive land value. This is consistent with the economic theory that commercial property generally raises demand due to accessibility benefits and its proximity to major transportation facility. This is supported by Bocarejo which shows that the introduction of the BRT (Transmilenio) transport system in Bogota, Colombia that has had positive influence to the value of commercial properties. The farther the commercial area with Transmilenio corridor or BRT system station, the lower the land value in the commercial area [14] Cervero [13] added that there was no evidence that rail investment caused land value change, but hedonic price results show a strong relationship between proximity to the transit system and land value. The strong relationships found due to reflection from substantial differences in land use, modes and corridor, revealed that government policies and decisions have an important role in creating land use and land value [13,15–17].

IOP Conf. Series: Earth and Environmental Science **123** (2018) 012033 doi:10.1088/1755-1315/123/1/012033

In contrast, Balchin & Pierre [18] explained that the demand for land is a reflection of the benefits or needs emerging from the use of some land by the communities as potential users. The greater the benefits derived from land use in these locations for various purposes, the higher the land price or the land rent. Thus, the bid-rent curve of the commercial area will have sharp bid-rent curve because it has the highest degree of accessibility. In contrast, the bid-rent curve of settlement areas shows the sloping curve. Hence, it can be concluded that changes in infrastructure provision have a significant effect on the land use change.

3. Methodology

This research used positivistic approach with quantitative technique and descriptive statistical analysis. Descriptive analysis was used to analyze the effect of land use change in spatial configuration that influence the land value along the corridor by examining the relationship land value changes as an indication of the urban corridor development. Land use data used Landsat Data in the period years of 1993, 2004, 2011 and 2015. To ilustrate our approach, two corridors were selected in Seemarang as the case study. The chosen places represent the typical urban and suburban environtmens: Corridor of Semarang – Ungaran dan Corridor of Semarang – Mranggen. The knowledge used include type of land use, transportation network provision and service in the corridor, land value, physical change (non-built area to built environment); functional changes (residential to non-residential); spatial changes (small to large area), economic social change (non-commercial to commercial sector such as industrial, trade and service sectors); demographic sector (low population density to high population density).

4. Results and Analysis

4.1. The Effect of Urban Land Configuration to Land Value

Spatial configuration in the Corridor of Semarang-Ungaran and Corridor of Semarang-Mranggen shows the existence of public facilities with regional and national scope. Thus, its existence gives important influence to land value in those corridors. In Corridor of Semarang Ungaran is influenced by Diponegoro University which is a national university and become an influence to development of residential and service centers in Tembalang and Banyumanik Regions. These service centers include education, health and transportation. While Corridor of Semarang Mranggen is influenced by urban settlement area in Pedurungan and Pucang Gading Region. In addition, the proximity to the CBD become the major influence for comprehensive public facilities provision in this corridor.

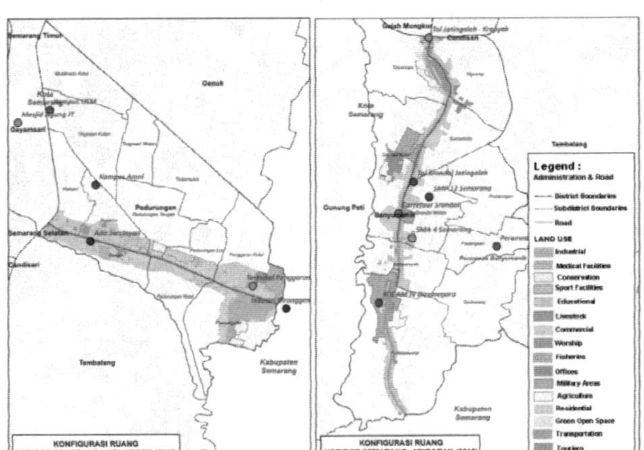

Figure 2. Spatial Configuration Corridor of Semarang-Ungaran and Corridor of Semarang Mranggen

IOP Conf. Series: Earth and Environmental Science **123** (2018) 012033 doi:10.1088/1755-1315/123/1/012033

4.1.1. Distance to City Centre

Simpang Lima is one of Central Bussiness Districts in Semarang. There are various activities in this areas such as trading, offices until entertainment activities. Simpang Lima become one of movement attractions for communities in and out of city. Distance to Semarang City Center will certainly affect the land value in the corridor. Based on this case, then the land value is selected as sample and analyzed its relationship with the distance to the city center. It can be seen from the graph that the land value on the property closest to the city center is more expensive than the property land value which is farthest from downtown.

Figure 3. Activity Points in Corridor of Semarang-Ungaran and Semarang-Mranggen

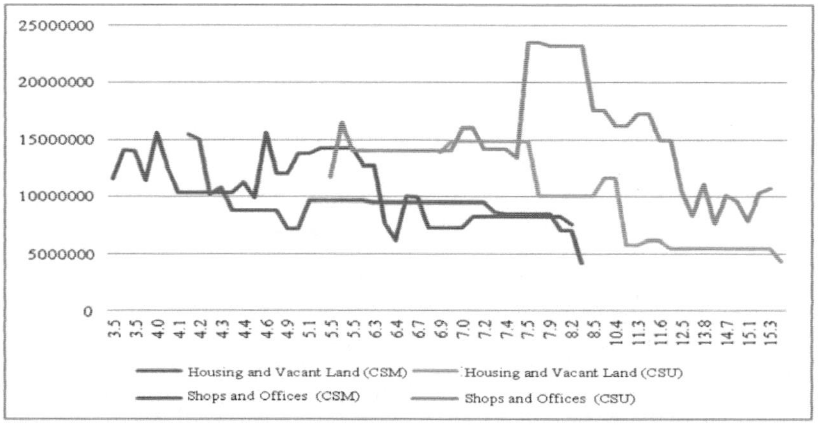

Figure 4. Fluctuation of Land Value toward Distance to City Centre

However, the graph also shows fluctuations of land value in the corridor where some properties with closer distance are cheaper (5.9 km) than the property further away from the city center (6.4 km). Corridor of Semarang-Mranggen has a land value pattern inversely proportional to the distance to CBD. The farther away from the city centre, the land value will decrease significantly. There is a difference of land value between shops and housing function. There is an increase of land value in the ADA Supermarket due to shophouses and Superindo market which became the trigger of crowd and human attraction. The land value rise in the area located in the intersection between Brigjen Sudiarto street and Soekarno Hatta artery road. Hence, there is no road median in this area and it enables people to access and reach the location in opposite direction without having to rotate. The increase of land value is about 30%, not as much as the land value happened in activity centers in Corridor of Semarang Ungaran.The housing function and vacant land do likewise, the pattern of land value shows declined considerably as further away from the city center.

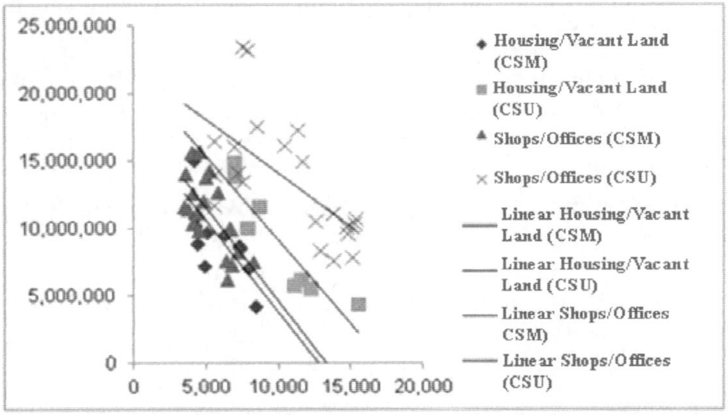

Figure 5. The Relationship between Land Value and Distance to CBD

There is prominent difference between land value for shops/offices use and housing/vacant land use in the corridor of Semarang Ungaran. The difference is almost 2 times higher between the land value for housing/vacant land compared to land value for the shop / offices. However, pattern of the two graphs still show almost the same. The land value climbed gradually at the Setiabudi road junction (corridor road) with Prof. Sudharto Road. Moreover, the land value rise caused by redevelopment program in the surrounding ADA Supermarket which is the entrance of toll road and T-junction connecting with Banyumanik residential area. The graph of housing and vacant land is steeper than the shop/offices due to economic value for housing/vacant land has small value. Thus the distance to city center has big influence on the land value.

4.1.2. Distance to Traditional Market

Beside CBD or city centre, activities in marketplace can influence land value. Market is divided into traditional and modern market. Gayamsari traditional market is located in the west of corridor of Semarang-Mranggen near the toll gate while Jatingaleh traditional market is located in the northern corridor of Semarang-Ungaran near the toll gate.

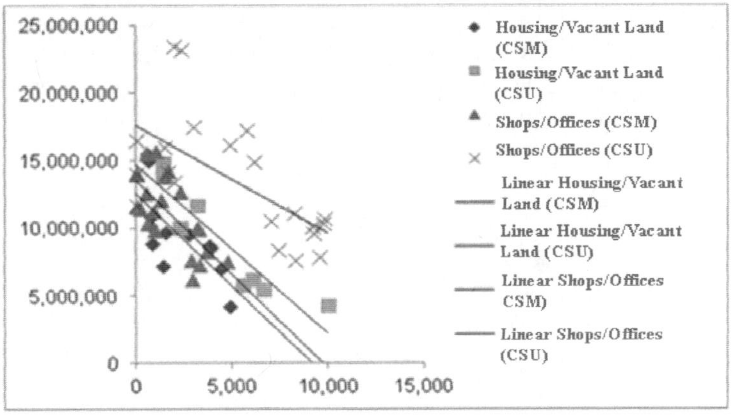

Figure 6. The Relationship between Land Value and Distance to Traditional Market

Figure above shows the dramatic decline of land value due to the far distance of properties from traditional market. The trendline of both shops/offices and housing/vacant land is predicted that it will

IOP Conf. Series: Earth and Environmental Science **123** (2018) 012033 doi:10.1088/1755-1315/123/1/012033

be same and remain stable in the corridor Semarang-Ungaran. However, when we viewed in detail, the land value goes down due to its location near traditional market particularly in Jatingaleh. The traffic congestion, slums and inconvenience condition surround the market make this area less attractive.

4.1.3. Distance to Modern Market

Modern market located in the corridor is Superindo. Superindo was built in the narrow street in the Semarang-Ungaran while another superindo was well-developed in main road of Semarang-Mranggen. Both corridors show the high land value for housing/vacant land if those properties located near modern market. The trendline tends to decrease continuously as far away as modern market and it indicates that land value of property is completely affected by the distance to modern market.As is presented in the chart shows the function of shops/offices is influenced by the distance to the modern market. Unlike the case in the Corridor of Semarang Mranggen with the graph shows a downward tendency. Sample of shops/offices buildings found at the surrounding modern market and there is slight difference of land value eventhough the distance is not equal.

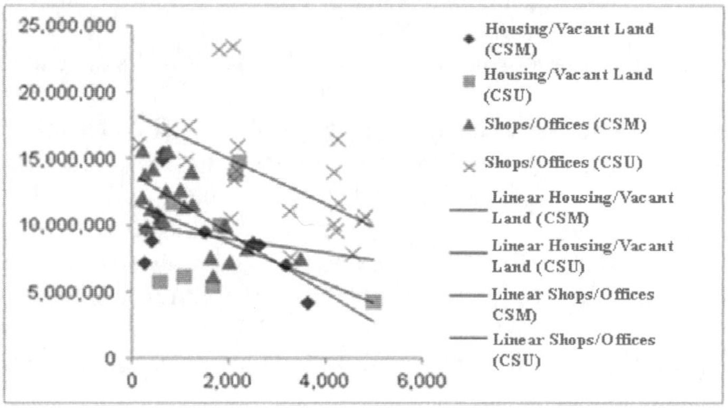

Figure 7. The Relationship between Land Value and Distance to Modern Market

4.1.4. Distance to Shopping Mall

There are many Shopping Malls are built in both corridors. ADA Supermarket Majapahit is located in the west of Semarang-Mranggen and Giant Central City while ADA Banyumanik and Carrefour Srondol are located in Semarang Ungaran. Each shopping center has different carrying capacity and services. There is an obviously difference from both the graphs. The graph represents inverse relationship between land value and distance to shopping mall in Semarang Ungaran. The land value slumped moderately as it closes to shopping malls (for housing.vacant land) while shops/offices has a positive relationship with the distance to shopping mall. When those located far from shopping mall, the land value will collapse sequentially.

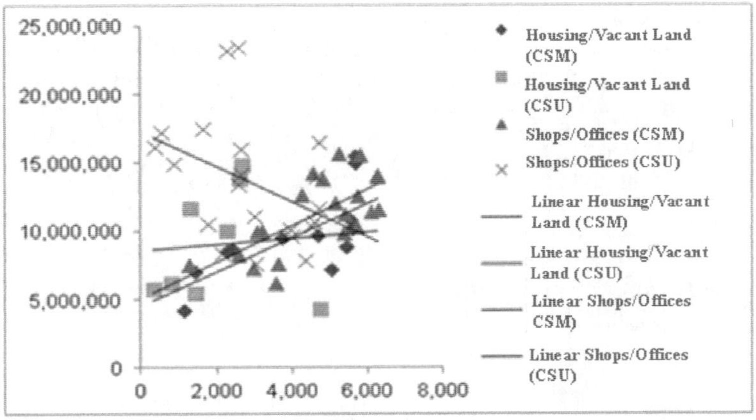

Figure 8. The Relationship between Land Value and Distance to Shopping Mall

4.1.5. Distance to University

Diponegoro University is the national college located 2,8 km away from Semarang-Ungaran while University of Semarang built 4,1 km away from Semarang Mranggen. Both universities are typically different. However, those colleges affect noticably toward land value change.

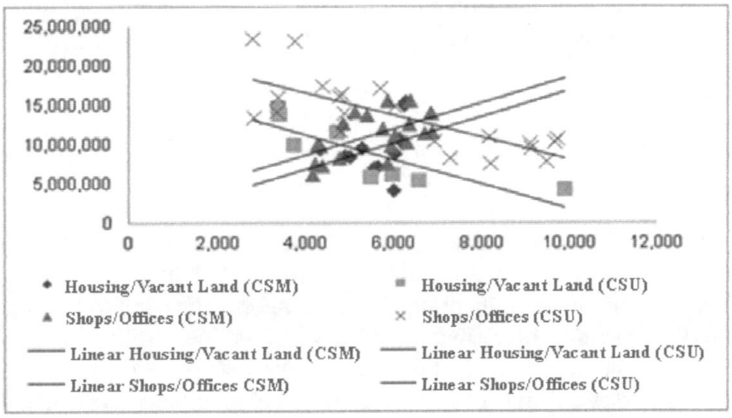

Figure 9. The Relationship between Land Value and Distance to University

Shops/offices are more commonly found than housing/vacant land in Semarang Ungaran corridor. Both lines indicate a downfall which means that the further away from the university has cheaper land value. University is as the main factor affecting the land value in Semarang Ungaran. Unlike in Corridor of Semarang Mranggen, University of Semarang plays as an attraction of human movement. Housing/vacant land is less commonly found around this corridors. In contrast, there are many shops/ retail buildings built arround the corridor. The graph ilustrates that land value is not influenced by University of Semarang because the value tends to drop sequentially when it is far away from the college.

4.1.6. Distance to Police Office

Police office is selected as one of factors affecting land value in both corridors. The nearest distance to police office is 1,8 km away from Semarang-Ungaran corridor while police office in Semarang-Mranggen is located in Pedurungan near artery road Soekarno-Hatta. Both corridors have the same trendline. The land values tend to decline when the properties located near police office. As the scatter plot suggest random distribution of each sample. The linear line fall significantly due to far distance to police office. Thus, police office has slight influence to land value change.

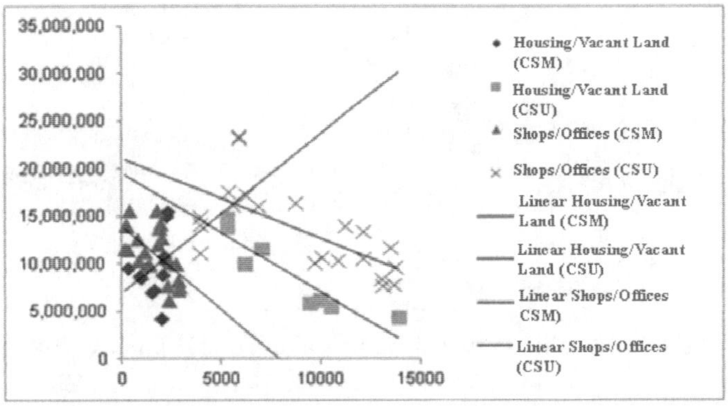

Figure 10. The Relationship between Land Value and Distance to Police Office

However, housing/vacant land in corridor of Semarang Mranggen tends to rise swiftly when it is located far away from police office. It can be concluded that police office has no influence on land value in both corridors. In conclusion, there is different pattern of land value for housing/vacant land and for shops/offices. The land value for housing/vacant land is tied to surrounding public facilities while shops/offices land value generated from its economic attraction. The land use of Diponegoro University in Semarang-Ungaran is the most predominant factor affecting land price. Furthermore, the distance to CBD or city center affecting land price generally eventhough shopping centers or modern markets also play important role in this case.

4.2. The Effect of Transportation Facility Toward Land Value

4.2.1 Distance to Toll Gate

The toll gate of corridor Semarang Mranggen is located near Gayamsari traditional market while corridor of Semarang Ungaran is located in ADA supermarket. Housing/vacant land has extremely low prices while shops/offices have high prices because shops/offices built in the commercial area. The entrance toll in the corridor of Semarang-Mranggen is located in front of Gayamsari. The existence of toll is accessible to connect and reach many places. The terminal, bus stops,station, bridges are less influential due to high use of motorized-vehicles and those transport facilities give nothing to land value change. The graph below illustrates that the high land value of properties in both corridors are mostly found at the proximity to toll gate. Both properties represent the same trendline.

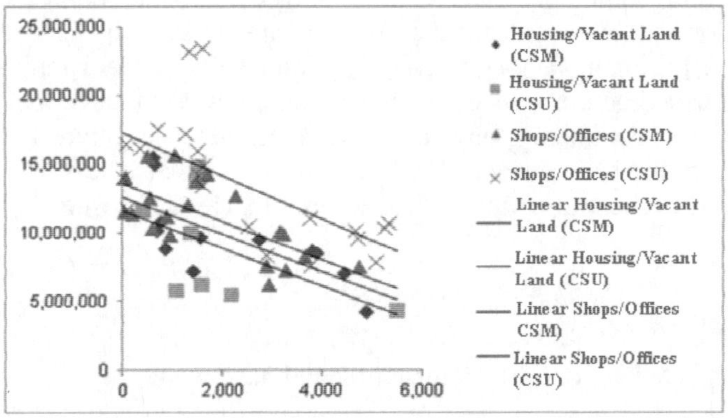

Figure 11. The Relationship between Land Value and Distance to Toll Gate

Although the toll gate gives impact on land value, there is a different pattern of land value in both corridors. For instance in corridor of Semarang-Ungaran where the land value goes down significantly due to traffic congestion and unaccessible condition around the toll gate. As a whole, the existence of toll gate gives enormous impact on land value change of all properties.

4.2.2 Distance to Bus Station

Public transport station in corridor of Semarang –Ungaran is called as Banyumanik Bus Station categorized as type C while bus station in corridor of Semarang Mranggen is Terminal Penggaron categorized as type B. The effect of bus stop on land value shows negative correlation that means the farther the property site from terminal the land value will be increasing except for shops/offices in corridor of Semarang-Ungaran because modern marketplace, shopping cemters such as ADA Supermarket and Carefoure are nearby.

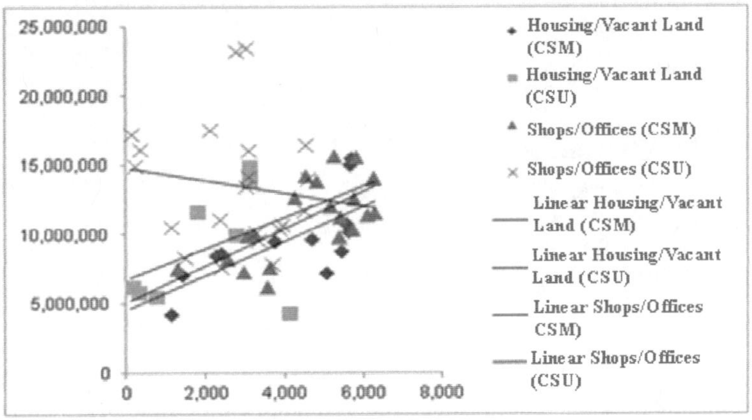

Figure 12. Land Value toward Travel Time to Terminal

4.2.3 Distance to Footbridge

The footbridge in the corridor of Semarang-Ungaran is located in Terminal of Banyumanik while another footbridge of Corridor Semarang-Mranggen is located in ADA Majappahit Supermarket. Footbridges are usually built near activity centers and this will therefore gives impact on land value due to shopping activities.

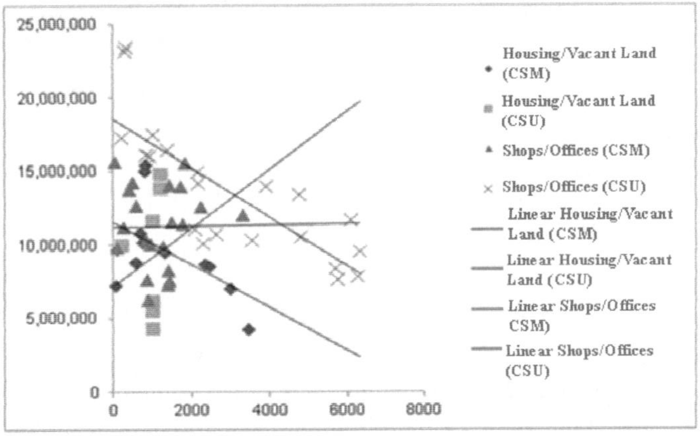

Figure 13. Land Value toward Travel Time to Footbridges

4.2.4 Distance to Bus Stop

Transportation system has a big role in daily human movements. Land value is also affected by accessibility aspect such as travel time and transportation cost. Shuttle bus in the selected corridors play role as supporting transportation infrastructure in those areas. The majority of travel time to bus stop from each property is 5 – 10 minutes. The longest time to reach bus stop is 16 minutes while the shortest time to reach bus stop is 1 minute from the whole samples. Thus, it can be concluded that the land value for the majority of properties located in the place with shortest time to reach bus stop around 2 millions – 25 millions per square meter. The scatter plot illustrates a decline linear graph along the increasing travel time. However, the downward trends for two type of properties are different. Shops/offices line tend to decline significantly compared to housing/vacant land. Furthermore, there no correlation between travel time to shuttle bus/bus stop and land value in corridor of Semarang – Mranggen.

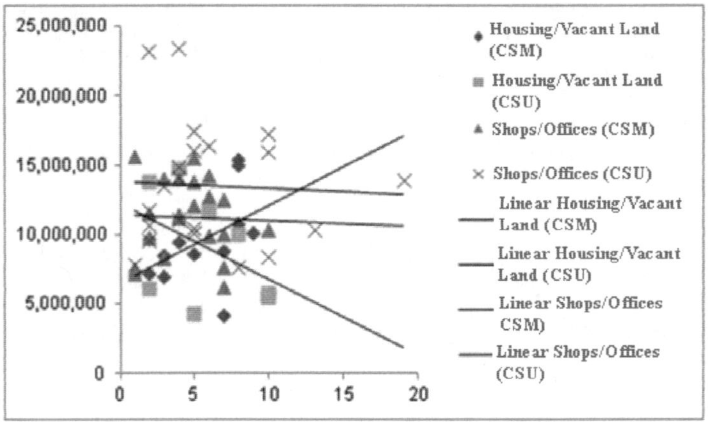

Figure 14. Land Value toward Travel Time to Bus Stops/Terminal

4.2.5 Parking Lot Area

The availability of parking lots has important impact on land value. Public parking lots affect the rise of land values in both corridors. However, the restricted parking lots in the corridor of Semarang-Mranggen recently causes unavailable parking area. The effect of restriction of parking lots causes shops and retail buildings are not managed well compared to building with parking lots.

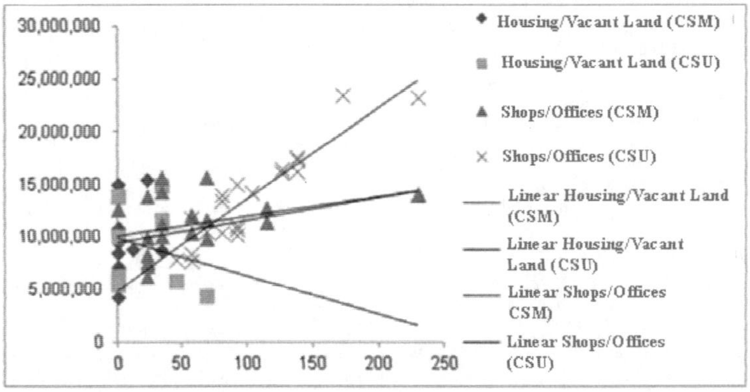

Figure 15. The Relationship between Land Value to Parking Lot

As can be seen in the scatter plot, there is a strong correlation between land value and parking lots located in shops/offices in both corridors. The number of parking lots which are able to accomodate more vehicles can raise the land value of shops/offices. In conclusion, the high dependency of motorized-vehicles, the existence of toll road give significant impact on land value. The provison of parking lots surround the properties used as attraction to encourage high land value in those areas. In contrast, public transportation infrastructure such as terminal, bus stops and footbridge give no impact on land value.

4.3. The Correlation between Land Value Change and Land Use

According to activity point above, it can be clearly seen that the activity points in corridor of Semarang-Mranggen located in traditional market of Gayamsari, ADA Supermarket Majapahit and Terminal Penggaron while in corridor of Semarang-Ungaran located between Prof Soedharto Street until the Toll Gate of Jatingaleh and the crossroads of Terminal Banyumanik. Land use changes between two corridors show the different pattern such as figure below:

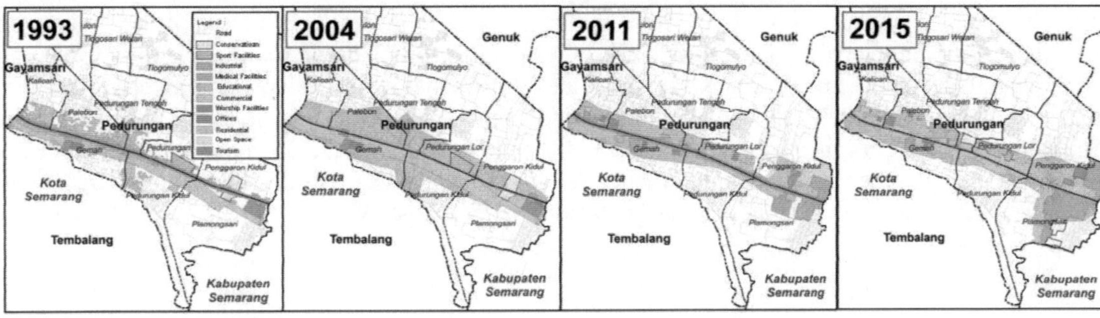

Figure 16. Land Use Change in Corridor Semarang-Mranggen in period years 1993-2004-2014-2015

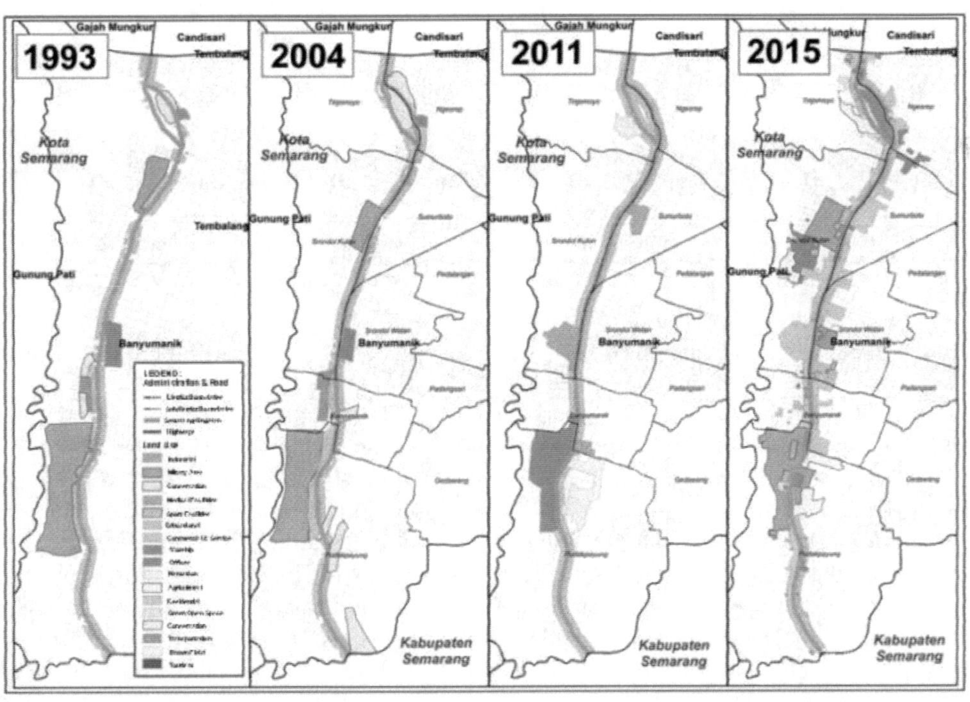

Source: Landsat Map 1993, 2004, 2011, 2015

Figure 17. Land Use Change in Corridor of Semarang Ungaran in the year 1993-2004-2014-2015

Corridor of Semarang Mranggen consists of many subdistricts such as Penggaron, Penggaron Kidul, Pedurungan Kidul, Pedurungan Tengah, Pedurungan Lor, Gemah, Palebon and Kalicari. The land use shows the prominent change from conservation and residential use to industrial use located in the eastern part of this corridor. The building density and the high intensity of public activities can be seen from traditional market of Gayamsari to Supriyadi junction. At this meeting point with Supriyadi lane connecting the corridor to Regional Trade Center of Pasar Johar and the Artery Street of Soekarno hatta is called as an outer ring road of Semarang. In addition, Supriyadi Street also connects to large-scale housing of perumnas Tlogosari with population density of 14,488 people/km2. Corridor Semarang – Ungaran consist of many subdistricts located in Kecamatan Banyumanik.The activity point sentralized in between Prof. Soedharto Junction untill toll gate of Banyumanik and Sukun Junction. The junction point located in Prof. Soedharto Street is aimed as connecting road to Diponegoro University causes the high land value at this point.

Furthermore, the toll gate of Banyumanik connects Terminal Kaligawe from the eastern part of Semarang and Terminal Mangkang from the west of City. The toll gate also connects the region to the southern part of Semarang such as Solo and Yogyakarta. Near to the toll booth is a Toll Road junction that connects to a large-scale settlement Perum Perumnas Banyumanik and also develops residential and residential development developed by other developers.We examined many properties such as warehouses, factories, homes, shops, shophouses and vacant lots. The land values for each sample is various start from 4,3 millions - 23 millions rupiah. The percentage of average land value around 69% per year while The percentage of average land value just about 30% per year in corridor of Semarang – Mranggen. The land value changes can be seen from the graph below:

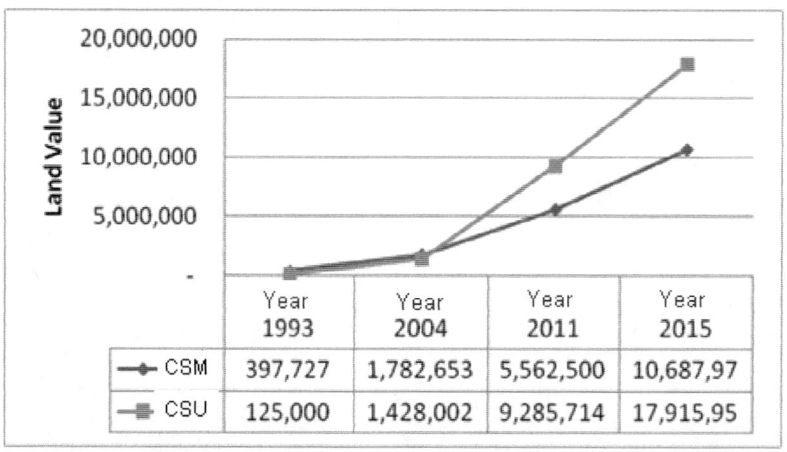

	Year 1993	Year 2004	Year 2011	Year 2015
CSM	397,727	1,782,653	5,562,500	10,687,97
CSU	125,000	1,428,002	9,285,714	17,915,95

Source: Suara Merdeka 1993, 2004 dan Primary Survei 2011-2015

Figure 18. Land Value in Corridor of Semarang Ungaran and Corridor of Semarang Mranggen in the year 1993-2004-2011-2015

The average land value in corridor of Semarang-Mranggen in the year 1993 was Rp 397.727 per square meter while at the same year in corridor of Semarang-Ungaran was Rp 125.000 per square meter. The land value in corridor of Semarang-Mranggen was higher than the land value in corridor of Semarang Ungaran because Semarang-Mranggen located near Semarang city centre (CBD). In 2004, both corridors had the same land value but in 2011, those move upward significantly with different percentage changes. The land value in corridor of Semarang Ungaran was higher than Semarang Mranggen. In 2015, the land value in Corridor Semarang Ungaran skyrocketed far above Corridor Semarang Mranggen. The difference between the values around 80%.The land value changes in both corridors can be seen from the table below:

Table 1. Land Value and Land Use in Corridor of Semarang Ungaran and Corridor of Semarang Mranggen in the year 1993-2004-2011-2015

Land Use	Year 1993		Year 2004		Year 2011		Year 2015	
	KSM	KSU	KSM	KSU	KSM	KSU	KSM	KSU
Land Price	397,727	125,000	1,782,653	1,428,002	7,500,000	9,285,714	9,687,976	17,915,958
Residential	1,137,918	677,576	965,048	1,708,482	738,399	1,635,657	203,251	1,214,071
Trading & Services	554,097	-	680,856	524,313	869,387	597,138	1,125,474	876,856
Military Areas		474,899		474,899		474,899		474,899
Industrial	37,679		533,123		533,123		812,184	
Public Facilities	-	-	243,086	51,420	281,204	51,420	281,204	193,289
Conservation	734,644	1,606,639	42,225	-	42,225	-	42,225	-

Source: Suara Merdeka 1993, 2004 and Primary Survey

The residential land use in 1993 - 2015 in both corridors decreased gradually. The drastic decline was located in the Corridor Semarang Mranggen. Residential land use changed into industrial and services use. The large changes in the main corridor was followed by the land use behind the corridor. The land use change that encourage several construction caused the lack of green space. Hence, the land value drop considerably due to the bad impact environmentally. The lack of green space in in corridor of Semarang Mranggen located in along the river (Watershed), while in the corridor of Semarang Ungaran located in Gombel Hills. It can be seen that along the connecting road will develop into premium area with high values and it will apply land use mechanisms in along the corridor. The land will be transformed into the land use with certain economic value that commonly known as commercial use.

5. Conclusion

According to the analysis above that the land value is divided into two categories such as commercial property and non commercial property. Commercial property consist of high economic activities that can pay a high land value. While non-commercial property is used for personal interests that give no economic value for both landowner and property manager. Both commercial and non-commercial properties, both have different responses and characteristics toward urban space configuration and the existing transport infrastructure.

Both commercial and non-commercial properties are influenced by distance to city centre and human activities out of corridor (commuting activities). The result is in accordance with Land Rent [19,20] and Bid Rent Teories [18,21] that the land value increase the road junction to human activity centers (such as University) and toll gate to Perumnas residential areas in both corridors. Non-commercial properties have highly sensitive characteristic toward space configuration changes compared to commercial properties because non-commercial properties have no economic attraction and it is necessary to have sufficient access to public facilities or other high-attractive land uses. Thus, non-commercial land value is influenced by surround public facilities and accessible road. On the other hand, the toll gate, bus stops, terminal and footbridges have no effect on land values.

This condition is very unique when compared with some theories or studies in other countries that most land values are affected by the proximity to public transportation facilities. However, those theories seem impossible to apply in Indonesia due to high dependence on private cars or motorized-vehicles and only very few people who use public transportation. Public transport in Indonesia is only used by captive movement or urgently certain conditions. The bad services and condition of public transport is usually dirty, unmanaged and unconfortable for disabled people need to be improved.

6. References

[1] Xiao Y, Orford S and Webster C J 2016 Urban configuration, accessibility, and property prices: a case study of Cardiff, Wales *Environ. Plan. B Plan. Des.* **43** 108–129
[2] Arnott R and DePalma E 2011 The corridor problem: preliminary results on the no-toll equilibrium *Transp. Res. Part B Methodol.* **45** 743–768
[3] Duany A and Plater-Zyberk E 1994 The neighborhood, the district and the corridor *New Urban. Towar. an Archit. Community, McGraw-Hill, New York, xvii-xx*
[4] Bentley I 1985 *Responsive environments: A manual for designers* (Routledge)
[5] Krier R and Rowe C 1979 *Urban space* (Academy editions London)
[6] Manning R 2009 *Using indicators to encourgae development: Lessons from the millennium development goals* (DIIS Reports/Danish Institute for International Studies)
[7] Arnott R, De Palma A and Lindsey R 1990 Economics of a bottleneck *J. Urban Econ.* **27** 111–130
[8] Minner J S and Shi X 2017 Churn and change along commercial strips: Spatial analysis of patterns in remodelling activity and landscapes of local business *Urban Stud.* 42098016684274
[9] Vink A Land-use in advancing agriculture
[10] Meyer M D and Miller E J 1984 Urban transportation planning: a decision-oriented approach
[11] Wu N and Silva E A 2010 Artificial intelligence solutions for urban land dynamics: a review *CPL Bibliogr.* **24** 246–265
[12] Yang J and Gakenheimer R 2007 Assessing the transportation consequences of land use transformation in urban China *Habitat Int.* **31** 345–353
[13] Cervero R 2006 Effects of light and commuter rail transit on land prices: Experiences in San Diego County *Univ. Calif. Transp. Cent.*
[14] Bocarejo J P, Portilla I and Pérez M A 2013 Impact of Transmilenio on density, land use, and land value in Bogotá *Res. Transp. Econ.* **40** 78–86
[15] Nelson A C 1993 Disamenity influences of edge cities on exurban land values: a theory with empirical evidence and policy implications *Urban Stud.* **30** 1683–1690
[16] Gibbons S and Machin S 2005 Valuing rail access using transport innovations *J. Urban Econ.* **57** 148–69
[17] Iacono M and Levinson D 2017 Accessibility dynamics and location premia: Do land values follow accessibility changes? *Urban Stud.* **54** 364–381
[18] Balchin P N, Kieve J L and Bull G H 1988 *Urban land economics and public policy* (Springer)
[19] Barlowe R 1978 Land resource economics: the economics of real estate
[20] Weber C, Petropoulou C and Hirsch J 2005 Urban development in the Athens metropolitan area using remote sensing data with supervised analysis and GIS *Int. J. Remote Sens.* **26** 785–796
[21] El-Barmelgy M M, Shalaby A M, Nassar U A and Ali S M 2014 Economic land use theory and land value in value model *Int. J. Econ. Stat.* **2** 91–98

The Settlements Growth in Mijen District, Suburb of Semarang

B Pigawati[1], N Yuliastuti[1], and F H Mardiansjah[1]

[1]Departmen of Urban and Regional Planning Engineering, Faculty of Engineering, Diponegoro University, Indonesia

Email: bitta.pigawati@gmail.com

Abstract: Semarang is one of metropolitan cities in Indonesia. As common in metropolitan cities, Semarang has problems regarding the availability of urban space, especially for settlements. This is related to the increase of population in Semarang. The selection of settlements should consider the suitability of space usage. This study aimed to analyze the growth of Semarang settlements in 2006-2015, distribution patterns, characteristics, directions and factors affecting growth. The location of the research is Mijen District located in Suburb of Semarang. This research used a quantitative descriptive spatial approach by using remote sensing technique and Geographic Information System (GIS). The results showed that some of the growth sites of settlements in Mijen District, the suburb of Semarang are located in areas which not suitable for settlements. There are several types of settlement patterns in Mijen District. Accessibility is the major factor driving the growth of settlements. An integrated development policy is needed to maintain a sustainable balance of urban settlement development.

Keywords: Settlement Growth, Suburban Area, Semarang City

1. Introduction

City is the center of creativity. It is also human containers and an object which is perceived and enjoyed by millions of people of widely diverse class and character [1]. City is as a place of livelihood activity [2].It is also reflects the vitality and human opportunities to achieve what they are dreaming about as well as symbolizes of socio-economic progress. Furthermore, this phenomenon causes population growth in urban areas quite rapidly. Limitations of land for settlements as well as the high level of land use competition in the city center results in shifting land use to settlements to the peripheral area. As a result, in the peripheral area develops settlements scattered area, irregular houses, and disintegrated each other.In addition, urban suburbs have a type of development—the low income groups—they look for living opportunities near the city [3].

Initially, humans choose the area for settlements in areas that fit with their needs and ensure their lives. However, due to the increasing population growth and the economic limitation for people to choose places to live, it results unsuitable areas that becomes their residence, and also. Unstable land, too oblique, unhealthy areas, dirty areas are built into settlements [4]. The suburbs are composed of low population categories but due to the migration of the high cities gives rise to fairly complex dynamics in the area [5,6].Moreover, this is also supported by the existence of competition for development activities for settlements and non-settlements, these unplanned and unorganized activities are done by the public, private and the government. This development forms are sometimes lacks of appropriate direction with irregular spatial structuring of the region, with the occurrence of ineffective use of space and complicate the subsequent planning pattern.

The changes of physical development and land use can cause various problems, one of them is environmental problem of settlement. Further, it has an effect on environmental quality of settlement [7]. Moreover, city's physical development can be identified based on the widespread area of the built area which is mostly used as a residential area [8].

Residence is generally called as settlement and it is specifically referred to as a house [9,10].Settlement is an area dominated by residential environments with its primary functions as residential areas equipped with infrastructure, environmental facilities and workplaces. Further, it provides limited services and employment opportunities to support livelihoods, so that the functioning of these settlements can be efficient and effective [11]. Hence, the high demand for residential and limited settlement land can lead to decreasethe quality of settlements. Quality of settlement is the ability of a settlement to interact with its supporting elements. The elements are residents, facilities and environmental infrastructure. In the context of residential environment, the quality of settlements is decreased or can not be seen from the development of settlement areas that are more oriented towards achieving the quantitytarget only [12].

The topographic condition of Semarang City varies from flat to hilly affects the pattern and distribution of residential areas. Initially, residents chose the location of the settlement by considering a safe place, not indisasters prone area and have good access to the city center or workplace. Accessibility means to demonstrate the ease of movement or scope from one place to another within an area and it has to do with distance [13]. Furthermore, accessibility has been accepted as part of the spatial aspect for various activities. Hence, a better accessibility means a greater land use change. The high land use change is also due to the ease of accessibility in the research area, thus encouraging land conversion from agriculture to non-agricultural.

However, in subsequent developments, the distribution of settlements tends to be unfocused. The housing developers compete to build housing in a location that is expected to give big profits. This condition seems to have a good impact for the city as it looks more advanced, but this phenomenon actually leads to the backwardness of urban conditionsFactors affecting the development of settlements are the physical factors of nature, social factors and economic factors [4].

Implementation of spatial arrangement is the effort of achieving the objectives of spatial planning through the implementation of spatial planning, space utilization, and control of space utilization [14]. The policy of urban settlement utilization of Semarang City has been explicitly stated in the spatial plan of Semarang City by setting location and its distribution based on the carrying capacity of the environment and should not be converted to other function. The suitability between the spatial use and the development of the field is the main key to see how far the spatial policy can accommodate development of each region.

Remote sensing refers to the activities of recording, observing, and perceiving (sensing) objects or events in faraway (remote) places. In remote sensing, the sensors are not in the direct contact with the objects or events being observed [15]. Public can utilize this technology to conduct various studies or plans related to the environment and natural resources. The settlement environment can be identified by the image generated by the remote sensing technology. This is a technological advances support in obtaining data that has a high degree of detail. Quickbird image is a high resolution satellite image generated by Remote Sensing Techniques that can be used to clasify regular and irregular settlements [16].

Research Sites were located in Mijen District. Mijen District is located in the suburb of Semarang City. It is a district with the widest area in Semarang City that is 57.55 km^2 consisting of 14 sub disctricts. Administratively, Mijen District is adjacent to Kendal Regency in the west. This research aims to examine the growth of settlement areas in Mijen district, suburb of Semarang.

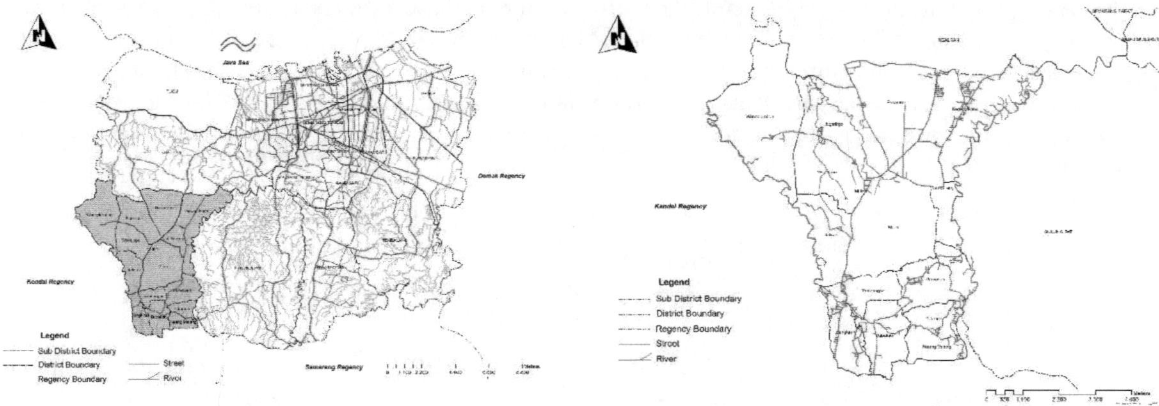

Figure 1. Administrative map of Mijen District and Semarang City

The land use in Mijen District consist of settlement of 1157 ha (20,10%), Industrial of 122 ha (2.12%), commercial of 37 ha (0.64%), non built area 4444 ha, and others 2 ha (0.03%).

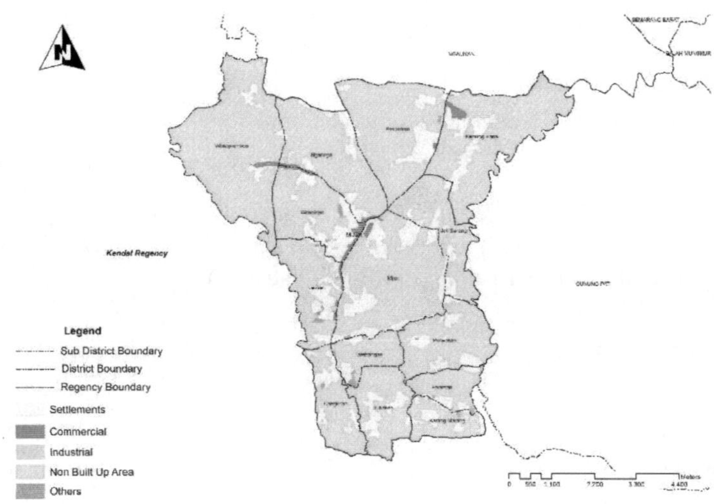

Figure 2. Landuse map of Mijen District

2. Methods

Research on Development of Settlement in Mijen District of Semarang City used quantitative descriptive method. All data and information were analyzed quantitatively with spatial approach by using overlay analysis technique with utilizing remote sensing imagery and Geographic Information System. Stages of analysis performed are as follows:

a. Analysis of residential population characteristics based on socio-economic and population condition was aimed to describe the characteristics of residents by using education level indicator,

IOP Conf. Series: Earth and Environmental Science **123** (2018) 012034 doi:10.1088/1755-1315/123/1/012034

type of livelihood, population density and population growth rate. Population Growth Rate is the rate of population growth per year within a certain period used to know the change of population between two time periods.This numbers was expressed as a percentage of the base population.To calculate population growth rate, geometric method was used.

b. Analysis of environmental characteristics of settlements based on facilities and Infrastructure Facility by using indicators of trading facilities, educational facilities, health facilities, street infrastructure, drainage infrastructure. This analysis aims to describe the characteristics of settlements in Mijen District. Using the Indonesian National Standards (SNI) which contains the standard of meeting the needs of facilities and infrastructure of settlements, determined by scoring techniques.

c. Analysis Pattern of settlement distribution nearest neighborhood analysis [17]was determined by calculating magnitude of adjacent neighbor parameter (T) using formula:

$$T = \frac{Ju}{Jh} \qquad Jh = \frac{1}{2\sqrt{p}}$$

Explanation

T:	Index spread of nearest neighbour
N:	Number of settlements
∑J:	Distance between settlements
Ju:	∑J/N
A:	Area
p:	∑N/A (Density)
Jh:	1/2√p

T = 0–0.7Clustered pattern
T = 0.71 –1,4Random pattern
T = 1.41 –2.15 Uniform pattern

d. Analysis of the Development of Settlements based on indicators of changes in the settlement area Year 2006 and 2015 was done with overlay techniques. This analysis will result in extensive growth of settlements and their distribution in Mijen District.

e. Analysis of land suitability based on indicators of the existence of disaster prone areas and land use directives as stated in the Spatial Planning Document done by overlay techniques. This analysis aimed to find out whether the settlement developments are still in the suitable zone for settlements.

3. Results and Discussion

3.1 Population and Social Economy

Population is the object of a plan. To plan the settlement area it is necessary to know the condition of the people who are the residents of the settlement area. The total population in Mijen District in 2006 was 45248 inhabitants. Residents in large numbers are located in Wonolopo Sub disctrict (5493 inhabitants). The total population in Mijen District in 2015 was 61405 inhabitants. Residents in large numbers are located in Jatisari Sub disctrict.

The population density of Mijen District in 2006 was 786 inhabitants/km^2 and in 2015 it became 1307 inhabitants/km^2 with growth rate 3.45%. The area that has the greatest population density wasJatisari Sub disctrict.

IOP Conf. Series: Earth and Environmental Science **123** (2018) 012034 doi:10.1088/1755-1315/123/1/012034

Table 1. Population Density and Growth Rate of Mijen District 2006-2015

No	Sub Districts	Population Density (person/km^2)		Population Growth Rate 2006-2015	
		2006	2015	(r)	(%)
1	Cangkiran	911	1,307	0.041	4.09
2	Bubakan	902	1,061	0.018	1.81
3	Karangmalang	1,037	1,122	0.009	0.88
4	Polaman	976	1,115	0.015	1.48
5	Purwosari	776	963	0.024	2.43
6	Tambangan	955	1,151	0.021	2.09
7	Jatisari	2,078	4,476	0.089	8.9
8	Mijen	939	1,320	0.039	3.85
9	Jati Barang	1,019	1,268	0.025	2.46
10	Kedungpani	745	911	0.023	2.25
11	Pesantren	131	197	0.047	4.67
12	Ngadirgo	884	1,133	0.028	2.79
13	Wonolopo	1,360	1,857	0.035	3.53
14	Wonoplumbon	370	411	0.013	1.27
Mijen District		786	1,307	0.035	3.45

Source: Analysis, 2017

Population in Mijen consisted of 38046 educatedpeople ≤ Junior high school, 11761 people are Senior High School graduated, and 4899 people are College graduated. The inhabitants of Mijen District are mostly educated ≤ Senior High School (81%). This condition can occur because there is educational facilities ≤ Junior high school that well spread in each sub disctrict, Mijen District. Residents in Mijen District whose college education 9% is located in Jatisari Sub disctrict. Residents in Jatisarisub disctrict who mostly work as civil servants and Indonesian Armed Forces have an average income of Rp 3,000,000. The average income of the population in Jatisati is the highest income compared to the other sub disctricts. Therefore, they are able to send their children to college. Jatisari sub disctrict is also located in a strategic place supported by good road infrastructure, so it facilitates people to reach the location of higher education facilities which are not available in Mijen District.

Mostly, people in Mijen District work as farm workerof 9947 people (43%), with monthly income aroundRp 1,500,000. This condition is related to the available of large agricultural land/non-built land in Mijen District, which is 77% of the total area of Mijen District. Population working as farmers is spread in WonoplumbonSub disctrict 1602 people (16%), Wonolopo 1563 people (16%), and Purwosari 1445 people (15%). Another occupation ofMijen population is as industrial workers of 6255 people (27%), with total monthly income is Rp 1,900,000.

This industrial worker lives in Mijen Sub disctrict (21%) and Ngadirgo Sub disctrict (16%). In these two areas there is a lot of land used as an industrial area. The average income of the population in Mijen District is Rp 2,000,000 per month. The condition of Mijen District which works as farmers and industrial workers indicates that Mijen District has the characteristic of mixing area of sub disctrict and city. Therefore, Mijen Districtbecome the initial transition from sub disctrict to the city.

IOP Conf. Series: Earth and Environmental Science **123** (2018) 012034 doi:10.1088/1755-1315/123/1/012034

3.2 Settlement Environment

The completeness of facilities and infrastructure in an area will be an attraction for residents to choose the location as an ideal residence. Facilities are elements in a residential environment that serve to support the implementation and development of social, cultural and economic life (Ministry of Law and Human Rights, 2011). The attractiveness of a region which has comprehensive facilities and infrastructure, as well as close to the workplace is another factor causing the growth of residential areas. To describe the completeness of residential area facilities in Mijen District, see the following criteria:

Category low	: contain ≤ 1 kind of facility score 1
Category moderate	: contain 2 kind of facility score 2
Category high	: contain ≥ 3kind of facility score 3

a. Educational facilities

Educational facilities in Mijen District consist of Kindergarten of 34, Primary school of 36, Junior high school of 9 and Senior High School of 8. Distribution of educational facilities includedas high category is in Cangkiran, Bubakan, Polaman, Purwosari, Jatisari, Kedungpani, Ngadirgo, Wonolopo, and Wonoplumbonsub disctrict. The distribution of educational facilities included as moderate category is located in Karangmalang, Mijen, Jatibarang and Pesantren Sub disctrict. There was no low category in Mijen Districts.

b. Trading Facilities

The trading facilities in Mijen District consist of Market of 1, Store/shop of 597, Cooperative (Economic Enterprise) of 9. Trading facilities including the high category are only located in Ngadirgo Sub disctrict. The other sub disctricts include in moderate category. There was no sub disctrict in Mijen District in low trading facilities.

c. Health Facilities

Health facilities in Mijen District consist of Polyclinic of 16, Community Health centers of 5, Clinic/Doctor of 16. Health facilities including high category are located in Pesantren and WonolopoSub disctrict. Health facilities including moderate category are located in Jatisari and Ngadirgo Sub disctricts. The distribution of low category health facilities are located in Cangkiran, Bubakan, Karangmalang, Polaman, Purwosari, Tambang, Mijen, Jatibarang, Kedungpani, and Wonoplumbon sub disctricts.

d. Settlement facilities in Mijen District

The determination of categorization of settlement facilities was done by summing the score of educational, trade and healthfacilities. The characteristics of settlements in Mijen District based on residential facilities are as follows:

1. Settlements with high facilities are located inJatisari, Pesantren, Ngadirgo, Wonolopo Sub disctrict
2. Settlements with medium facilities are located in Cangkiran, Bubakan, Karangmalang, Polaman, Purwosari, Tambang, Mijen, Jatibarang, Kedungpani, Wonoplumbon sub disctrict
3. Settlements with Low-facilities are not found in Mijen District.

3.3 Settlement Infrastructure in Mijen District

There are three types of roadin Mijen District; collector roads, local roads, and environmental roads. Generally, the road infrastructure in Mijen District is in good condition. Inadequate road infrastructure is located in Jatibarang and Wonoplumbon sub disctrict. The condition of drainage in Mijen District is mostly good. Inadequate drainage infrastructure is located in Karangmalang, Purwosari sub disctrict and Wonoplumbon sub disctrict.

Source: Observation, 2017

Figure 3. Road Infrastructure and drainage

Based on the condition of road and drainage infrastructure in Mijen District, it can be concluded that the areas that have good and optimal working infrastructure condition are located in Cangkiran, Bubakan, Polaman, Tambang, Jatisari, Kedungpani and Ngadirgo. In other sub disctricts, the infrastructure is not functioning optimally.

3.4 Characteristics of Settlements in Mijen District

Settlement clustering is generally caused by social class and the ability of land access to build residential dwellings. People prefer settlements based on environmental characteristics such as location, education, crime rate, environmental quality, socio-demographic composition facilities, settlement characteristics and land prices.

There are 3 (three) types of Settlement Characteristics in Mijen District determined based on the completeness of facilities and infrastructure condition, see follows:

1. Settlement area which has very complete facilities with good infrastructure condition and optimal function can be found in Tambangan, Jatisari, Kedungpani, Pesantren, and Ngadirgo sub disctrict.
2. Settlement area that has adequate facilities with good infrastructure condition and optimal function is located in the Cangkiran, Bubakan, Polaman, and Wonolopo sub disctrict.
3. Settlement Area that has adequate facilities and good infrastructure condition but not function optimally are in Karangmalang, Purwosari, Mijen, Jatibarang and Wonoplumbon sub disctrict.

3.5 Settlement Patterns

There are several factors that influence the formation of settlement patterns, such as geographical and natural, man-made and site factors.Selection of settlement sites is influenced by accessibility, availability of space and land, availability of facilities and infrastructure of settlements, physical and environmental conditions [18].

The distribution patterns of settlements are distinguished into compact and not compact. Not compact distribution pattern of settlement that spreads randomly and systematicly provides opportunities for settlement development due to physiographic factors and the accessibility of settlers. On the other hand, compact distribution pattern of settlement that is uniform and clustered is supported by the availability of land for settlements. The factors that influence distribution pattern of settlement are geographical, socio economic, and population factors [19]. This study also utilized the same variables, but tried to apply to different areas ie in the suburban area.

Settlement is a source of information about humans and their activities. The settlement patterns give the impression of the physical dispersal of settlements along with the density of its inhabitants [20]. Social status is one of the factors that drive the clustering of settlements. High-income societies and those with high social status have a better chance of choosing a settlement environment according to their interests and tastes whereas low-income communities have limited access to their settlement environment [21].

IOP Conf. Series: Earth and Environmental Science **123** (2018) 012034 doi:10.1088/1755-1315/123/1/012034

One of the methods to measure settlement patterns is nearest neighborhood analysis model [17], by calculating the nearest neighborhood parameters (T).

Table 2. Settlement Patterns of Mijen Sub District 2015

No	Sub Districts	∑N	∑J	Ju	A	p	2√p	Jh	T	Patterns
1	Cangkiran	5	2.53	0.506	2.76	1.81	2.69	0.371	1.36	Random
2	Bubakan	2	1.34	0.670	2.09	0.96	1.96	0.511	1.31	Random
3	Karangmalang	6	2.35	0.392	2.03	2.96	3.44	0.291	1.35	Random
4	Polaman	3	2.04	0.680	1.61	1.86	2.73	0.366	1.86	Uniform
5	Purwosari	5	2.96	0.592	4.67	1.07	2.07	0.483	1.23	Random
6	Tambangan	2	1.34	0.670	3.58	0.56	1.49	0.669	1.00	Random
7	Jatisari	6	3.09	0.515	2.21	2.71	3.30	0.303	1.70	Uniform
8	Mijen	7	5.84	0.834	4.74	1.48	2.43	0.411	2.03	Uniform
9	Jati Barang	6	4.18	0.697	2.27	2.64	3.25	0.308	2.27	Uniform
10	Kedungpani	2	0.98	0.490	5.83	0.34	1.17	0.854	0.57	Clustered
11	Pesantren	3	4.06	1.353	6.80	0.44	1.33	0.753	1.80	Uniform
12	Ngadirgo	4	5.36	1.340	4.91	0.81	1.81	0.554	2.42	Uniform
13	Wonolopo	3	2.91	0.970	4.04	0.74	1.72	0.580	1.67	Uniform
14	Wonoplumbon	3	2.98	0.993	10.01	0.30	1.09	0.913	1.09	Random

Source: Analysis, 2017

Based on nearest neighborhood analysis, settlement pattern in Mijen District can be classified into 3 (three) pattern types as follows:

- Settlements with random patterns are located in Cangkiran, Bubakan, Karangmalang, Purwosari, Tambangan and Wonoplumbonsub disctrict.
- Settlements with uniform patterns are found in Polaman, Jatisari, Mijen, Jatibarang, Pesantren, Ngadirgo and Wonolopo sub disctricts.
- Clustered settlements are found only in Kedungpanisub disctrict.

Figure 4. Settlement Pattern of Mijen District 2015

3.6 Settlement Growth in Mijen District

Settlement area in Mijen District was 909 ha in 2006, while in 2015 it was 1157 ha. During the Period 2006 to 2015 there was a broad increase about 248 ha (27%). The increasing of settlement area was mostly found in Pesantren village (56%) and Mijen village (48%)

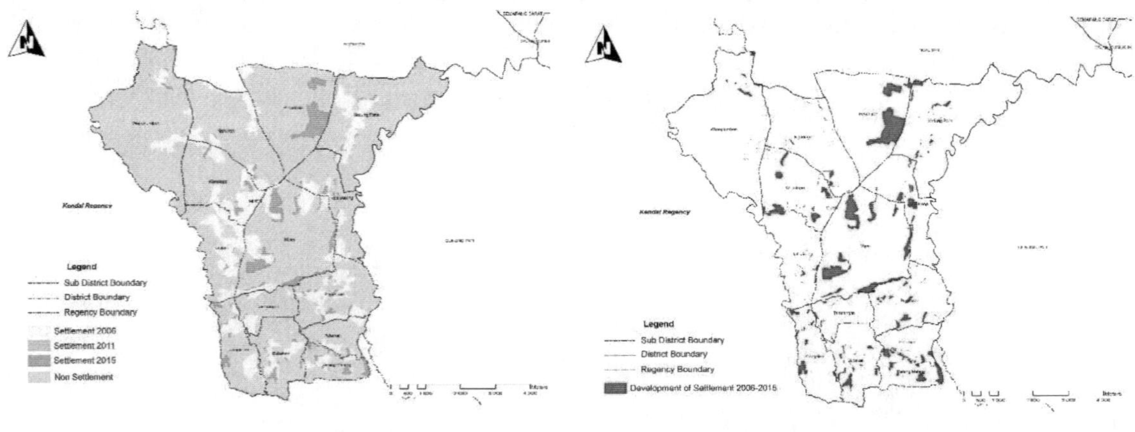

Figure 5. Settlement Growth in Mijen District 2006-2015

Table 3. Settlement Growth in Mijen District 2006-2015

No	Sub District	Settllement(ha)		Δ SettllementYear 2006-2015	
		2006	2015	ha	%
1	Cangkiran	60	63	3	5
2	Bubakan	33	47	14	42
3	Karangmalang	37	56	19	51
4	Polaman	34	44	10	29
5	Purwosari	114	133	19	17
6	Tambangan	73	96	23	32
7	Jatisari	81	96	15	19
8	Mijen	83	123	40	48
9	Jatibarang	43	51	8	19
10	Kedungpani	81	107	26	32
11	Pesantren	27	42	15	56
12	Ngadirgo	73	87	14	19
13	Wonolopo	117	135	18	15
14	Wonoplumbon	53	77	24	45
Mijen District		909	1157	248	27

Source: Analysis, 2017

Regular settlements are settlements which plans are well laid out, ranging from home units as well as physical environmental conditions [22]. Regular settlements in Mijen District in 2006 were 59 ha and 2015 to 110 ha in total. During the period of 2006 until an increase occurred, the regular settlements area was 86%. Distribution Regions of regular settlements in 2006 were mostly found inJatisari, Kedungpani and Mijen Sub disctrict. Distribution Regions of regular settlements in 2015 were located in the same location as in 2006. This condition occurred because these 3 sub disctricts had a good road/access infrastructure, so that it had the potential for expansion of residential areas.

The irregular settlement area of Mijen District was 850 ha in 2006, while in 2015 it was 1047 ha. During the period 2006 to 2015 there was an increase of irregular settlement area around 23%. The distribution of irregular settlements in 2006 was predominantly located in Wonolopo and Purwosari village. Distribution of the irregular settlement area in 2015 was found in the same location with 2006. This condition occurred because these two villages are not available for vacant land suitable for settlement, so there is no expansion for residential area.

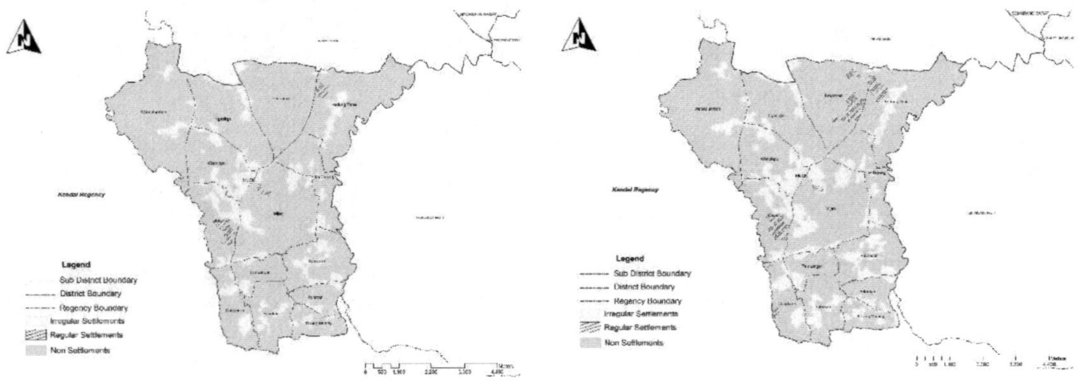

Figure 6. Settlement Area of Mijen District Year 2006 and 2015

3.7 Suitabilityof Settlement Development with Spatial Planning Document

Spatial Planning Document is a government guide document to determine spatial arrangement. Rapid and unfocused settlement growth causes unsuitable of settlement location construction.

Table 4. Settlement Suitability of Mijen Sub District 2006-2015

No	Sub District	Total Growth	Suitable	Unsuitable
		ha	ha	ha
1	Cangkiran	9.34	2.38	11.72
2	Bubakan	3.40	8.19	11.59
3	Karangmalang	0.85	18.40	19.25
4	Polaman	4.24	1.73	5.98
5	Purwosari	6.79	4.08	10.88
6	Tambangan	1.70	0.98	2.67
7	Jatisari	6.79	0.19	6.98
8	Mijen	14.43	53.85	68.28
9	Jatibarang	10.19	1.13	11.32
10	Kedungpani	9.34	0.90	10.24
11	Pesantren	47.54	11.73	59.27
12	Ngadirgo	1.70	2.82	4.52
13	Wonolopo	14.43	5.17	19.60
14	Wonoplumbon	0.85	4.86	5.71
Mijen District		131.59	116.41	248.00

Source: Analysis, 2017

Width of settlement's development area in Mijen District according to Spatial Planning Document is 131,59 ha (53.6%), it is mostly located in Pesantren sub disctrict 36.13%, while in Wonolopo and Mijen sub disctricts were respectely 10.97%. The extents of settlements' development which were not in accordance with the Spatial Planning Document was 116.41 (46.94%).

Most of them were located in Mijen Sub district (53.85%) and 18.40% in Karangmalang sub disctrict. The location of residential areas in accordance with the Spatial Planning Document has accommodated the hazard-prone constraints of flood and land movement. The development of settlements located in locations that are inconsistent with the Spatial Planning Document need to receive serious attention. Since this location is located in the area designated as a catchment area. If this location is ignored, it will cause a flood disaster.

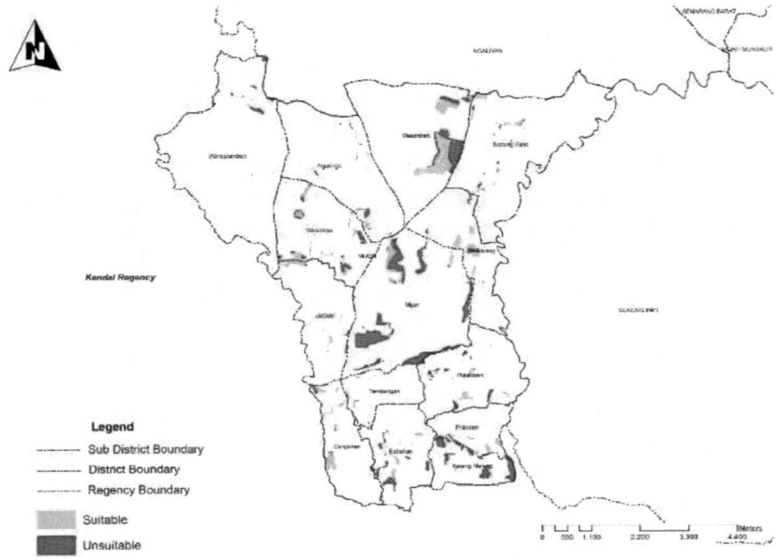

Figure 7. Suitability of Settlement Growth in Mijen District Year 2006-2015

4. Conclusion

In this study used applying population, socio-economic, geographical, and settlement infrastructure factors as variables affecting the development of settlements. Remote sensing and GIS were used to determine the development of settlements in Mijen sub district in 2006 - 2015. The results show that there were 3 settlement patterns, the most dominant is the uniform pattern in 7 sub districts. The types of environmental characteristics of settlements generally indicate the conditions of location. It has adequate facilities with good infrastructure conditions and optimal function.

Settlement area in Mijen District was 909 ha in 2006, while in 2015 it was 1157 ha. During the Period 2006 to 2015 there was a broad increase about 248 ha (27%). The increasing of settlement area was mostly found in Pesantren village (56%) and Mijen village (48%). Width of settlement's development area in Mijen District according to Spatial Planning Document is 131.59 ha (53.6%), it is mostly found in Pesantren sub disctrict (36.13%), while in Wonolopo and Mijen sub disctricts were respectelly 10.97%. The extents of settlements' development which is not in accordance with the Spatial Planning Document are 116.41 (46.94%). Most of them are located in Mijen 53.85% and 18.40% in Karangmalang sub disctrict. The location of residential areas in accordance with the Spatial Planning Document has accommodated the hazard-prone constraints of flood and land movement. The non-conforming location should be returned to its original function and controlled for its use in accordance with the relevant rules. It is certainly done to keep urban development sustainable.

5. Acknowledgements

Thanks to all staff inMijen District Office, BPS, Bappeda (Regional Planning Agency) as the institution of data providers and Department of Urban and Regional Planning of Diponegoro University for providingus opportunity to conduct research.

6. References

[1] Lynch K 1960 *The image of the city* vol 11(MIT press)
[2] Andreasen M H and Møller-Jensen L 2017 Access to the city: Mobility patterns, transport and accessibility in peripheral settlements of Dar es Salaam *J. Transp. Geogr.* **62** 20–29
[3] Daskalova D and Slaev A D 2015 Diversity in the suburbs: Socio-spatial segregation and mix in post-socialist Sofia *Habitat Int.* **50** 42–50
[4] Sumaatmadja N 1988 *Studi Geografi, suatu Pendekatan Analisa Keruangan* (Bandung: Alumni)
[5] Tritsch I and Le Tourneau F-M 2016 Population densities and deforestation in the Brazilian Amazon: New insights on the current human settlement patterns *Appl. Geogr.* **76** 163–172
[6] Krisjane Z, Berzins M, Ivlevs A and Bauls A 2012 Who are the typical commuters in the post-socialist metropolis? The case of Riga, Latvia *Cities* **29** 334–340
[7] Wesnawa I G A 2010 Perubahan Lingkungan Permukiman Mikro Daerah Perkotaan Berbasis Konsep Tri Hita Karana di Kabupaten Buleleng Bali *Forum Geogr.* **24** 111–118
[8] Chapin F stuart and Keiser E 1979 *Urban Landuse Planning* (United States America: The Board of Trustees of The University.)
[9] Hudson F S 1974 *Geography of Settllements* (London: Mc. Donald and Evan)
[10] Hammond C W 1979 *Elemen of Human Geography* (London: Goarge Allen and Unwin)
[11] Anon 1997 Kamus Tata Ruang
[12] Soemarwoto O 1983 *Ekologi Lingkungan Hidup dan Pembangunan* (Jakarta: Djambangan)
[13] Bintarto and Hadisumarno S 1989 *Metode Analisa Geografi* (Jakarta: LP3ES)
[14] Ditjen Penataan Ruang 2007 Pedoman Kriteria Teknis Kawasan Budi Daya: Peraturan Menteri Pekerjaan Umum No. 41/PRT/M/2007 1–60
[15] Weng Q 2010 *Remote Sensing and GIS Integration :Theories, Methods, and Applications* (New York: McGraw-Hill Company)
[16] Pigawati B 2011 Use of Satellite Imagery for Study of Settlement Area in Semarang City **25** 140–151
[17] Hagget P 1975 *Geography: A Modern Synthesis* (New York: Harper and Row Publisher)
[18] Budihardjo E 1986 *Arsitektur dan kota di Indonesia* (Alumni)
[19] Yang R, Xu Q and Long H 2016 Spatial distribution characteristics and optimized reconstruction analysis of China's rural settlements during the process of rapid urbanization *J. Rural Stud.* **47** 413–424
[20] Zee D Van Der 1979 *Human Geography of Rural Settlement and Population* (Enchesde: ITC)
[21] Pacione M 2005 *Urban Geography: a Global Perspective (2nd Edition)* (London: Routledge)
[22] Arta F and Pigawati B 2015 The Patterns and Characteristics of Peri-Urban Settlement in East Ungaran District, Semarang Regency *Geoplanning J. Geomatics Plan.* **2** 103–115

Urbanization and Land Use Changes in Peri-Urban Area using Spatial Analysis Methods (Case Study: Ciawi Urban Areas, Bogor Regency)

D L Cahya[1], E Martini[1], and K M Kasikoen[1]

[1]Esa Unggul University, Indonesia

Email: darmawan@esaunggul.ac.id

Abstract. Urbanization is shown by the increasing percentage of the population in urban areas. In Indonesia, the percentage of urban population increased dramatically form 17.42% (1971) to 42.15% (2010). This resulted in increased demand for housing. Limited land in the city area push residents looking for an alternative location of his residence to the peri-urban areas. It is accompanied by a process of land conversion from green area into built-up area. Continuous land conversion in peri-urban area is becoming increasingly widespread. Bogor Regency as part of the Jakarta Metropolitan Area is experiencing rapid development. This regency has been experienced land-use change very rapidly from agricultural areas into urban built up areas. Aim of this research is to analyze the effect of urbanization on land use changes in peri-urban areas using spatial analysis methods. This research used case study of Ciawi Urban Area that experiencing rapid development. Method of this research is using descriptive quantitative approach. Data used in this research is primary data (field survey) and secondary data (maps). To analyze land use change is using Geographic Information System (GIS) as spatial analysis methods. The effect of urbanization on land use changes in Ciawi Urban Area from year 2013 to 2015 is significant. The reduction of farm land is around -4.00% and wetland is around -2.51%. The increasing area for hotel/villa/resort is around 3.10%. Based on this research, local government (Bogor Regency) should be alert to the land use changes that does not comply with the land use plan and also consistently apply the spatial planning.

Keywords: Land Use Change, Peri-Urban Area, Spatial Analysis Methods, Urbanization

1. Introduction

Rustiadi [1] said that the sub-urbanization in the peri-urban happening more and more around the major cities in Indonesia. The expansion of the urban area to the suburb is often considered as a process that is contradictory, since it is always accompanied by a process of conversion of agricultural land which is very productive. Peri-urban as the area around the big cities is a dynamic zone, due to the conversion of agricultural land is continuously creating mixed land use, as stated by Rakodi in Adell [2].

The urban population in Indonesia has increased significantly since 1971 until 2010. In 1971 the urban population in Indonesia reached 20.7 million people, or 17.42% of the total population of Indonesia, but in 2010 reached 92 million people or 42, 15% [3], proved to have occurred on a large scale urbanization in Indonesia [4]. Considering the limited urban land for urban residents as place for living, then as an alternative to choose peri-urban as a place for living, because it is still possible to

IOP Conf. Series: Earth and Environmental Science **123** (2018) 012035 doi:10.1088/1755-1315/123/1/012035

carry commuters daily. As a result, the conversion of agricultural land that occurs in the area around the major cities in Indonesia are very apprehensive. On the other hand, the city grows large and uncontrollably.

In the peri-urban area, which is an alternative place for living of the urban population, the conversion of agricultural land changes into residential land. This condition will reduce agricultural product such as rice for the Indonesia country, in turn, it can be turned into a net importer of rice, which should be self-sufficient rice needs.

In the core cities, the limited supply of amenities of life such as housing, water, electricity and others, makes the working population in the city choose to live outside the city, and the best choice is peri-urban area [5], because while still enabling the commuter daily, dualism atmosphere of cities and villages is still very strong, so the peri-urban area is become a comfortable place to live. The impact will occur daily commuter movements that requires the availability of transport that is fast, easy, cheap and safe, which can accommodate the movement of population in large numbers.

Limitations of the Government to provide land for housing and its various amenities, then the choice of the population residing in peri-urban area needs to be supported by the availability of transport facilities. The dynamics of peri-urban area can also be caused by other things such as an alternative tourist location, comfortable apartment, vacation home, although in the core city population had homes. Conversely, for the working population in the peri-urban area can also perform commuting to the core city, as do various activities, such as trade, schools and others.

Bogor Regency as a part of Jakarta Metropolitan Area is experiencing rapid development [6]. Most of Bogor Regency is an agricultural region. Along with the development of Jakarta Metropolitan Area, most of the population looking for an alternative place to stay in this regency. As a result of land use changes occur very rapidly, from agricultural areas into urban areas.

This situation needs to watch out for in turn agricultural land will be reduced so as to reduce agricultural product and may eventually disrupt food security. This study aims to analyze the dynamics of land use changes in peri-urban areas of Bogor Regency using case study of Ciawi Urban Area that are experiencing land use change from agriculture area into build up area quite rapidly in the last few years [7,8].

2. Data and Methods

To achieve the objectives, the stages of research that will be done is the preparation, data collection, and analysis. Preparation was done by compiling survey instruments, data requirements, and other preparation.

The data used in the analysis of the dynamics of land use changes were primary data and secondary data [9–14]. Sources of primary data consist of field observations, interviews with stakeholders, while secondary data in both softcopy and hardcopy documents and maps. Secondary data sources consisted of relevant institutions, libraries and electronic media.

Analysis of the dynamics of land use change was done by super-impose method using ArcGIS software [15]. The research was conducted in May through November 2016 with a case study of Ciawi Urban Area, Bogor Regency.

IOP Conf. Series: Earth and Environmental Science **123** (2018) 012035 doi:10.1088/1755-1315/123/1/012035

3. Results and Discussion

3.1. Characteristics of Ciawi Urban Area

3.1.1. Administrative

Ciawi Urban Area consists of 11 villages in two districts, namely the Distric of Ciawi (Village Banjarwangi, Banjarwaru, Bantarsari, Bendungan, Bitungsari, Ciawi, Pandansari, and Telukpinang), and the District of Megamendung (Village Gadog, Sukakarya, and Sukamahi). This area has boundaries as follows: (a) North: District of Sukaraja, (b) East: District of Cisarua, (c) South: District of Caringin, (d) West: City of Bogor, For more details about the area can be seen in Table 1, Figure 1 and Figure 2.

Table 1. Administrative of Ciawi Urban Area (GIS Analysis, 2016)

District	Village	Area (Ha)
Ciawi	Banjarwangi	108.08
	Banjarwaru	148.54
	Bantarsari	122.03
	Bendungan	155.87
	Bitungsari	156.05
	Ciawi	145.65
	Pandansari	225.40
	Telukpinang	124.46
	Total 1	**1,186.07**
Megamendung	Gadog	191.97
	Sukakarya	3.35
	Sukamahi	257.89
	Total 2	**453.22**
Total		**1,639.29**

3.1.2. Position of Ciawi Urban Area

The accessibility of Ciawi Urban Area is high enough, it will attract the development of urban activity. The road fromBogor City - Ciawi - Puncak - Cianjur is the primary road that acts as a regional road. By looking at natural resources that aboundan in Ciawi Urban Area, it is a potential as a main attraction for developing tourism activities, as well as the attraction of Puncak area makes Ciawi is crossings for tourists headed to the summit. The problem posed is congestion peaked on holidays such as Saturday and Sunday. With the implementation of the system through the Puncak does not have a significant impact to break the bottleneck.

Neighborhood of Ciawi Urban Area was considerable influence on the development of Urban Ciawi for their system of movement that occurs. Cibinong city which is the capital city of Bogor and Bogor influence on activities in the Ciawi Urban Area. Judging from its functions, Cibinong City is the center of government, trade centers and regional scale services and settlements, giving rise to the movement patterns in the surrounding area. While the Bogor City is the center of trade settlements and regional scale. The orientation of the marketing activities of agricultural production and plantation activities in addition to meeting the needs of local scale also meets the needs of the regional scale.

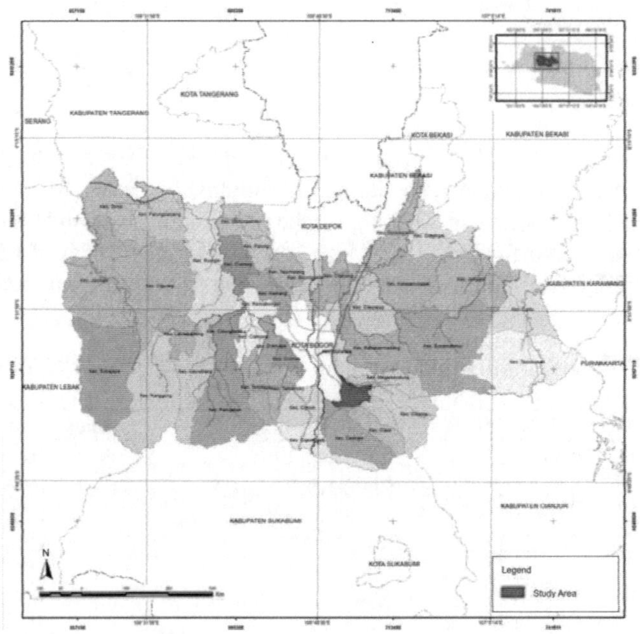

Figure 1. Orientation Map of Ciawi Urban Area

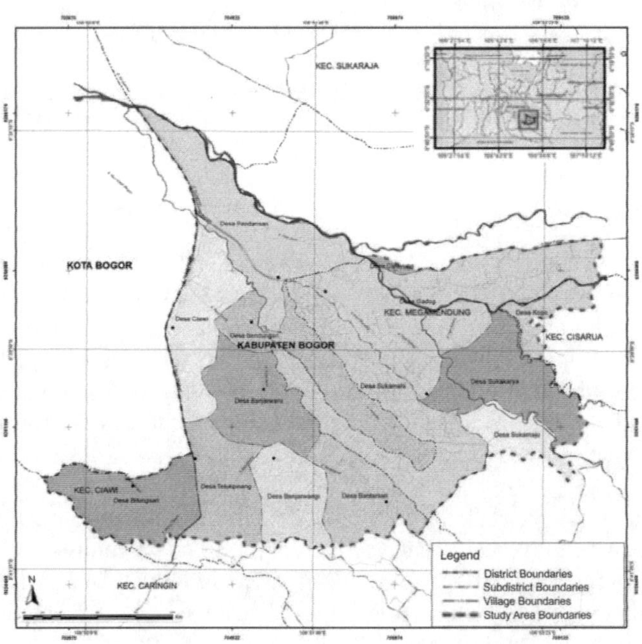

Figure 2. Map of Ciawi Urban Area

3.1.3. Population

Residents of Urban Ciawi progress was being decreased from the year 2009 had a population of 85 436 inhabitants, in 2010 amounted to 85 453 inhabitants, in 2011 amounted to 85 036 people, the year 2012 amounted to 86 615 inhabitants and in Year 2013 amounted to 87 804 inhabitants. Total population in Ciawi Urban Area in Year 2013 was the most numerous in the Village Dam and the number of people are in the village at least Banjarwangi.

IOP Conf. Series: Earth and Environmental Science **123** (2018) 012035 doi:10.1088/1755-1315/123/1/012035

Total migration of population Ciawi Urban Area in Year 2013 were more likely Ciawi Urban Area residents who moved out of Urban Ciawi compared to newcomers. Judging from the number of people coming was as many as 575 people, while the total population was 730 people moved.

3.1.4. Housing and Settlement

The development of some housing in the Ciawi Urban Area, such as Housing Vimala Hills, a residential development in the Ciawi Urban Area attracts investors to invest, mainly by class developer PT Agung Podomoro Land, Tbk through its subsidiary PT. Adhi Putra Prima. With the concept of the development of villas and resorts which provide 80% of the land as a green open space, Vimala Hills will have 690 residential units.

In Vimala Hills will also be built by Pullman, it is adapted the resort concept. This hotel will be the first five star hotel in Ciawi Urban Area, Bogor. Besides villas and hotels, Vimala Hills will also be equipped with commercial areas, flower gardens, fishing village, deer park, tea house and strawberry fields, children playground, and others. Vimala Hills will also have a traditional market selling vegetables and fruits are organic crop farmers around the site.

Trade and services in the Ciawi Urban Area is still within their jurisdiction. Meeting the needs of people concentrated in Ciawi Urban Area Ciawi market. Ciawi Urban Area community needs can already be met by the availability of local market this. However, problems arising from market activities are centralized bottleneck at one point that is the intersection of Jalan Raya Ciawi and Sukabumi, Jalan Raya Bogor Ciawi and City. With the lack of land market made the proliferation of street vendors (PKL) along the streets around the market.

Ciawi Urban Area trading activities are not only limited in Ciawi Market Center, but also linear along the street. Since their natural tourism area Puncak revive growth in trade and services along the main road. With affordable location make the activities of trade in services is growing series. The dominance of trade and services along this road are the sellers of souvenirs and harvest local area.

The development of trade and services activities makes a special attraction for tourists, because tourists can easily get souvenirs typical of the region.

3.1.5. Infrastructure

Based on its status, Urban Ciawi consists of Environmental Road, County Road, National Roads and Highways. On the other hand, based on its function consists of a Local Road, primary arterial roads, collector roads primer. In general, the road network in Ciawi Urban Area form a grid pattern is radial. This is because the road network Ciawi Urban Area forming concentric radial field with the intersection as a center for further growth of unit concentrically connecting with the surrounding area.

Condition of roads in Ciawi Urban Area good enough and have asphalt pavements. Road conditions consist of two (2) lanes for two (2) directions undivided (2/2 UD), but there are also some points consisting of two lanes in one direction. With the peak tourist appeal makes Ciawi Urban Area be crossing travelers heading to the summit so that the impact of congestion is high enough, by looking at such conditions it is applied peak one-way lanes. But the one-way system that was unable to break the bottleneck that ada. Hence, to look for a solution to be able to parse the existing congestion.

3.2. Dynamics of Land Use Changes

3.2.1. Land Use In Year 2013

Land use in Ciawi Urban Area in Year 2013 (Figure 3) can be categorized into 21 types of land use. With the dominance of the largest land use is a garden area of 919.12 ha (56.07%) and the housing area of 335.90 ha (21.71%). From the data of land use in the Ciawi Urban Area in Year 2013,

IOP Conf. Series: Earth and Environmental Science **123** (2018) 012035 doi:10.1088/1755-1315/123/1/012035

it can be said that the land contained in the Ciawi Urban Area still dominated by build-up area of 1,069.19 ha (65.22%). While non build-up area of 570.10 ha (34.78%).

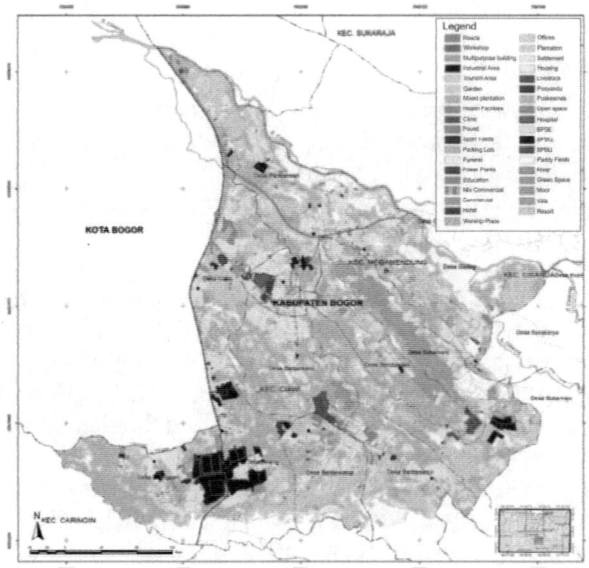

Figure 3. Map of Land Use in Ciawi Urban Area Year 2013

3.2.2. Land Use in Year 2015

Land use in Ciawi Urban Area in Year 2016 (Figure 4) could be categorized into 21 types of land use. With the dominance of the largest land use is a garden area of 919.12 ha (56.07%) and the residential area of 335.90 ha (21.71%). From the data of land use in the Ciawi Urban Area 2015 can be said that the land contained in the Ciawi Urban Area still dominated by build-up area of 966.19 ha (58.94%). While non build-up area of 673.09 ha (41.06%).

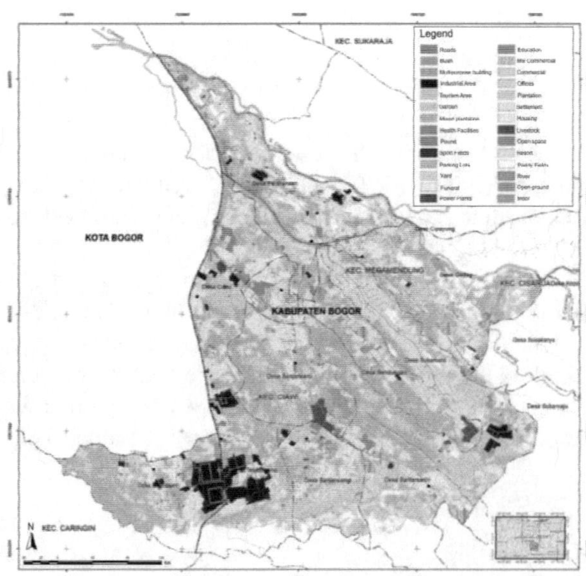

Figure 4. Map of Land Use in Ciawi Urban Area Year 2015

3.2.3. Dynamics of Change of Use of Land

Based on the data of land use Ciawi Urban Area in Year 2013 and 2015, it can be seen that within 2 (two) years there has been a significance change in land use is quite significant, namely the reduced land area of 65.62 Ha garden (-4%) and wetland at 41, 23 Ha (-2.51%) and an increase in land area hotel/villa/resort is quite large, amounting to 50.90 ha (3.1%). For more details about changes in land use in the Ciawi Urban Area Year 2013 to the Year 2016 can be seen in Table 2 and Figure 5.

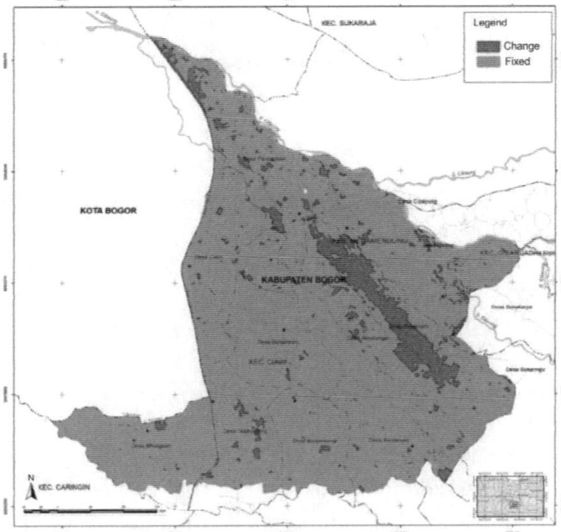

Figure 5. Map of Land Use Change in Ciawi Urban Area Year 2013-2015

Table 2. Land Use Change in Ciawi Urban Area Year 2013-2015

No	Land Use Types	Area (m^2)	Area (m^2)	Land Use Change (m^2)	Percentage (%)
1	Roads	489,506.05	702,393.25	212,887.20	1.30
2	Hotel/Villa/Resort	131,198.76	640,159.17	508,960.41	3.10
3	Industrial Area	518,171.47	569,018.83	50,847.36	0.31
4	Tourism Area	4,834.56	6,545.01	1,710.45	0.01
5	**Plantation**	**9,191,169.47**	**8,534,963.54**	**(656,205.93)**	**(4.00)**
6	Health Facilities	31,319.02	31,625.37	306.35	0.00
7	Water Body	64,140.21	76,455.53	12,315.32	0.08
8	Commercial	268,545.60	311,578.11	43,032.51	0.26
9	Sport Fields	9,499.54	10,216.67	717.13	0.00
10	Parking Lots	6,714.15	17,996.88	11,282.73	0.07
11	Funeral	10,539.84	24,751.59	14,211.75	0.09
12	Power Plants	13,027.48	13,027.48	0.00	0.00
13	Education Facilities	139,259.08	152,338.22	13,079.14	0.08
14	Warehouse	18,705.05	18,705.05	0.00	0.00
15	Worship Place	37,284.11	43,258.50	5,974.39	0.04
16	Office	221,717.22	249,469.93	27,752.71	0.17
17	Settlement	3,559,048.74	3,717,519.74	158,471.00	0.97
18	Housing	158,907.91	169,890.01	10,982.10	0.07
19	**Livestock**	**109,460.45**	**105,392.09**	**(4,068.36)**	**(0.02)**
20	**Paddy Fields**	**1,349,901.35**	**937,645.09**	**(412,256.26)**	**(2.51)**
21	River	59,910.89	59,910.89	0.00	0.00
	Total	**16,392,860.95**	**16,392,860.95**		

Souces: GIS Analysis, 2015

4. Conclusion

It can be concluded that the dynamics of land use changes in Ciawi Urban Area of the Year 2013 to the Year 2015 is significant, namely the reduction of farm land area of 65.62 hectares (-4.00%) and wetlands of 41.23 ha (-2.51%) and increase in land area hotel/villa/resort is quite large, amounting to 50.90 ha (3.10%). Land use changes from green area into build-up area in Ciawi Urban Area mostly occur along the main road leading to Puncak.

5. Acknowledgments

Thanks to Kemenristekdikti which has provided funding this research for two consecutive years with a competitive grants scheme in Year 2015 and 2016.

6. References

[1] Rustiadi E, Saefulhakim S and Panuju D R 2009 *Perencanaan dan pengembangan wilayah* (Yayasan Pustaka Obor Indonesia)
[2] Adell G 1999 Theories and models of the peri-urban interface: a changing conceptual landscape
[3] BPS 2010 *Data Kependudukan Indonesia* (Jakarta)
[4] Ginsburg N S, Koppel B, McGee T G and Others 1991 *The extended metropolis: Settlement transition is Asia* (University of Hawaii Press)
[5] Tacoli C and Others 2006 *The Earthscan reader in rural-urban linkages.* (Earthscan)
[6] BPS 2015 *Kabupaten Bogor Dalam Angka 2013-2015*
[7] BPS 2014 *Kecamatan Ciawi Dalam Angka 2013-2014*
[8] BPS 2014 *Kecamatan Megamendung dalam Angka 2013-2014*
[9] Cahya D L 2014 Model Sistem Dinamik Perubahan Guna Lahan Pertanian Perkotaan di Kabupaten Bogor *Forum Ilmiah* vol 11
[10] Doyle J K and Ford D N 1998 Mental models concepts for system dynamics research *Syst. Dyn. Rev.* **14** 3–29
[11] Forrester J W 1994 System dynamics, systems thinking, and soft OR *Syst. Dyn. Rev.* **10** 245–256
[12] Forrester J W 1993 System dynamics and the lessons of 35 years *A systems-based approach to policymaking* (Springer) pp 199–240
[13] Kasikoen K M 2011 Tipologi Perkotaan di Kabupaten Cilacap *Planesa* **2**
[14] Kasikoen K M 2010 Pemanfaatan Model Sistem Dinamis Dalam Memperkirakan Jumlah Penduduk Di Masa Datang Studi Kasus: Kabupaten Cilacap *Planesa* **1**
[15] Tian L 2015 Land use dynamics driven by rural industrialization and land finance in the peri-urban areas of China: "The examples of Jiangyin and Shunde" *Land use policy* **45** 117–127

Spatial Collaboration Model among Local Governments in Ratubangnegoro Region in the Boundary Area of Central Java and East Java Provinces, Indonesia

H Wahyono[1] and L Wahdah[1]

[1]Department of Urban and Regional Planning, Universitas Diponegoro, Indonesia

Email: hadi.wahyono@pwk.undip.ac.id

Abstract: In Indonesia, according to Law No. 23/2014 on Local Government, a local government can conduct cooperation with other local governments that are based on considerations of efficiency and effectiveness of public services and mutual benefit, in order to improve people's welfare. Such cooperation can be categorized into mandatory and voluntary cooperation. Cooperation shall be developed jointly between the adjacent areas for the implementation of government affairs which have cross-local government externalities; and the provision of public services more efficient if managed together. One of the parts of the area that is directly related to the implementation of the policy liabilities of inter-local government cooperation which is mandated is the he province boundary areas. The public management of the provincial boundary areas is different from the central province area. While the central province area considers only their own interests, the management of boundary development must consider the neighboring regions. On one hand, the area is influenced only by its own province policy, while on the other influenced by neighboring regions. Meanwhile, a local government tends to resist the influence and intervention of neighboring regions. Likewise, neighboring local governments also tend to resist the influence and intervention of other local governments. Therefore, when interacting on the boundary, inter-local government interaction is not only the potential for cooperation, but also conflict-prone regions. One of the boundary area provinces attempt to implement the collaborative planning approach is the boundary area of Central Java Province and East Java Province, which is known as Ratubangnegoro Region. Ratubangnegoro region is one of the strategic areas of both provinces. In order to the interaction between the region could take place, there are regencies in the region have formed and joined the Inter-Local Government Cooperation Agency (BKAD-Badan Kerjasama Antar Daerah) Ratubangnegoro. Based on the explanation, this article explains the institutional model of collaborative planning contained in BKAD Ratubangnegoro. The model is much more complex, because it involves two different levels of government, provincial and regency government hierarchies. Institutional model of cooperation in the regency boundary area should be different from that of between provinces. The results of this are expected to be input from related parties of the inter-regional cooperation institution, particularly cooperation in the area of the provincial boundary with the province, which is implemented by the regencies contained therein.

Keywords: Spatial Collaboration, Local Government Cooperation, Local Government Boundary, Ratubangnegoro.

IOP Conf. Series: Earth and Environmental Science **123** (2018) 012036 doi:10.1088/1755-1315/123/1/012036

1. Introduction

The boundary area of local government is not only an area having a boundary line of local government, but also a place of mutual interaction between neighboring local governments. The local governments utilize such interactions to accommodate their own mutual interests with their neighbors [1,2]. If they use the interactions for the common good, in the form of cooperation, it can be mutually beneficial. However, as every local government tends to hold their own autonomy, they tend to resist any intervention coming from its neighbors [3]. Therefore, if in the interaction, the local government rejects the influence of its neighbors and uses it only for its own sake, then the interaction is conflict-prone. In short, the interaction is not only potential for cooperation, but also vulnerable to local conflicts, which can be institutional, social, economic, spatial and land use conflicts [4]. In other words, the interaction is ambiguous, as it can be profitable, but also unprofitable.

Newman [5] defines border or boundary as a marker indicating the spatial restrictions, which can be functionally or administratively. Functionally, the boundary is the termination of function space as a place or container. While, administratively, the space should be limited so it do not disturb neighbors' space as the space of administrative power. Furthermore, Newman [5] states that the boundary area is the area that has boundary line. Based on such understanding, it can be formulated that boundary area can be defined as an area which has boundary line that ends the influence of neighboring areas. Where the boundary line is imaginary, the area became a place which influences between the neighboring areas interact. Based on this, there are two main characteristics which has a borderline, and the interaction between the influence of a region and its neighboring region.

A local government boundary area is a consequence of the limitation of the jurisdiction of local government in the care of the local region and the interests of society, as well as regional development [6]. In order to a local government authority does not conflict with the neighboring regions, the government sets an administrative boundary line [7]. Collaboration between local governments is a consequence of the nature of the line is a translucent effect, which become a place of interaction influence a local government and neighbors. Since it is influenced passable, the area around the line of the region of an interaction of mutual influence between neighboring regions [7,8].

Inter-local government cooperation is a manifestation of a local government effort to implement collaborative planning in the utilization of different regional externalities. The collaborative plan aims to conduct among relevant parties to develop a solution to a problem faced together [9,10]. Innes [11] explained that the parties use collaborative planning to accommodate their aspirations and reconcile the differences between them to build solutions together. Through the process, participants can share the planning and building of empowerment to express and realize their aspirations. The application of the collaborative planning approach in inter-local government interaction aims to determine the right strategy to build cooperation between regions. Given the boundary area is ambiguous. The approach is used not only to promote the regional potentials for cooperation, but also to negate potential regional conflicts.

In Indonesia, according to Law No. 23/2014 on Local Government, the local governments can conduct cooperation with other local governments based on considerations of efficiency and effectiveness of public services and mutual benefit, in order to improve people's welfare. Such cooperation can be categorized into cooperation mandatory and voluntary cooperation. Cooperation shall be developed jointly between the adjacent areas for the implementation of government affairs which have cross-regional externalities; and the provision of public services more efficient if managed together.

One part of a local government that is directly related to the implementation of the policy liabilities of inter-local government cooperation which is mandated to the local government boundary area. The area is located in the area around the local governmental boundary having cross-regional implementation of government affairs. The public management of the area is different from the central local governmental area. While the central area only considers their own interests, the management of border development

IOP Conf. Series: Earth and Environmental Science **123** (2018) 012036 doi:10.1088/1755-1315/123/1/012036

must consider the neighboring regions. On one hand, the area is influenced only by its own regional policy, on the other hand it is influenced by neighboring regions.

One of provincial boundary areas that has a local government cooperation, is the boundary area of Central Java (Jawa Tengah, shortened as Jateng) Province and East Java (Jawa Timur, shortened as Jatim) Province, which is known as the Ratubangnegoro Region. The region is one of the strategic areas of both the provinces. In order that the inter-local government cooperation to work well, all four regions agreed to establish the Inter-local Government Cooperation Agency (Badan Kerjasama Antar Daerah, shortened as BKAD) Ratubangnegoro. BKAD Ratubangnegoro is a cooperation institution among regions of Tuban and Bojonegoro regencies, the entrance of Jatim Province; as well as Blora and Rembang regencies, the entrance of Jateng Province. The name of Ratubangnegoro is an acronym derived from the initials of the four regencies.

The four regions are located in the boundary areas of Jateng and Jatim northern provinces, where Tuban and Rembang Regencies are in northern coastal region and passed traffic lines between provinces in the Java Northern Region (Pantura Jawa). While Bojonegoro and Blora regencies are in position south of the regency before, that partly has the potential of forests and oil and gas. With the potential and geographical conditions such that the mobility and community activities among the four regencies are often past the boundaries of the regency and even provincial administrative regions. This of course led to consequences in various aspects, in which each pulse of public life, government and the business community in the region bordering regency will affect each other and have socioeconomic and environmental impacts.

If viewed from the direction of development of the region, Bojonegoro Regency has been seen as a better developing since it is supported by the activities of the economic potential, which allows the development of synergies between the regions. Bojonegoro Regency has an urban area geared to the development of agricultural production centers to increase agricultural yields and the development of the petroleum processing industry. As for Tuban Regency, directed the development of the region as a center of activity and commerce services that need to be developed industrial activities that include small industries, agriculture, agro-industry and mining. Bojonegoro and Tuban regencies equally included in the development of regions.

The local governmental cooperation institutionalized in the BKAD Ratubangnegoro has tasks to coordinate, plan and implement the interaction of local governments in the Ratubangnegoro region. In performing its duties, the institution up a joint secretariat to facilitate their activities in implementing inter-local government cooperation, as well as provide feedback and suggestions to the institution. What's interesting is that the existence of such institution has been able to cut the route of administration should be done by all four members of the regency. In normal conditions, when they wanted to cooperate with the area that is outside the province, the governor should involve. However, the role of the governor can be done by BKAD Ratubangnegoro. The institution has been able to make the interaction which was more flexible, efficient and effective.

This article aims to analyze the institutional model of collaborative planning contained in the BKAD Ratubangengoro. Such institutional model of cooperation in the local boundary area is different from the model between regions within a province. The model is much more complex, because it involves two different levels of government, both the provincial and regency government hierarchies.

2. Data and Methods

The study discussed in this article used a qualitative research approach. The approach was used to examine in detail and in-depth a collaboration model between local governments involved in BKAD Ratubangnegoro (Figure 1). Characteristics of these objectives in accordance with the opinion of Creswell [12] which states that qualitative research is very suitable for research that aims to build knowledge based on buitt perspective a constructive depth of experience of the individual, social values and history; as well as a participatory perspective narrow source.

Specifically, the study used a case study explanation as a research strategy, as described by Yin [13]. According to Yin [13], the explanatory case study attempts to answer research questions that are explanative. Case study is a study of the object to be positioned as a 'case', which is done completely, thoroughly and deeply by using various sources of data [13]. Yin [13] looked at a case study as one of the qualitative research strategy that aims to disclose detailed and thorough review of the object studied. Because the position is an important case, Yin [13] looked at the case study is not a research method which should be selected, but the most important is choosing the right research object to be positioned as a case.

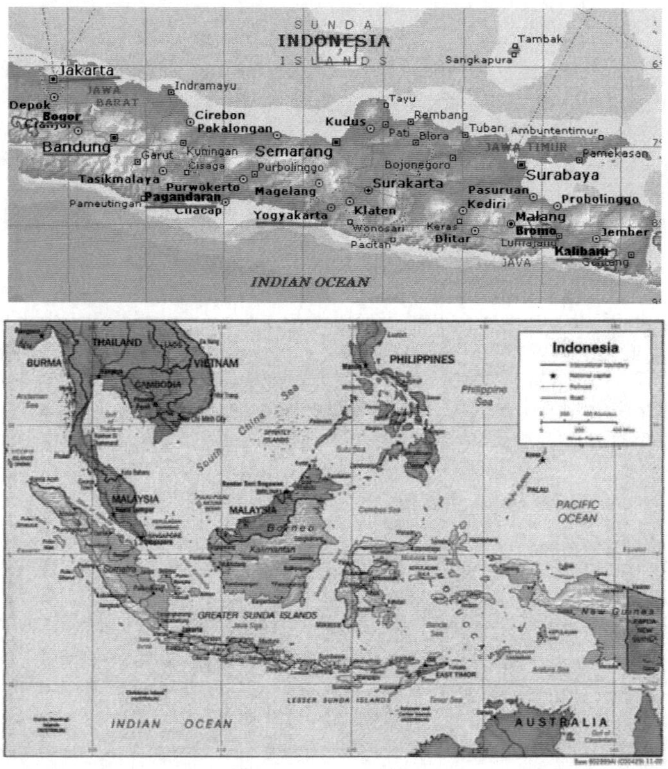

Figure 1. Map of the Ratubangnegoro Location in Indonesia (http://tragom.me/map-world-java.html)

Yin [13] reported a case study aims to examine in detail and depth of the research object. Case study method in this study used a thematic assessment method it is a method that examines the formation of certain patterns of data obtained. These patterns are used to build the theme and sub-themes, and eventually the researchers were able to determine the main theme of which is supported by the facts of the results of data collection in the field. In general, the process is as follows: Identification stage: the data that have been collected through the study of literature, observation, and in-depth interviews, were identified. Data that is not in accordance with the needs of neglected or discarded; Interpretation stage: the data already categorized links with each other to be interpreted. The stage will be done by the interpretation of the data and linking goals and objectives of the research; Concluding stage: the stage will be used when data have been interpreted, then do, concluding the results that's then used to construct a model; The object of this research is located in Ratubangnegoro area, which is an area consisting of four

IOP Conf. Series: Earth and Environmental Science **123** (2018) 012036 doi:10.1088/1755-1315/123/1/012036

regencies, which are Rembang, Blora, Tuban and Bojonegoro Regencies, located in the boundary area of Central Java and East Java Province, in Java Island of Indonesia.

3. Results and Discussion

BKAD Ratubangnegoro is a cooperative institution of four government regencies of two provinces, namely Blora and Rembang regencies in Central Java Province; and Tuban and Bojonegoro regencies of East Java Province. All four regencies are located in the boundary area of both provinces and the four mutually benefit one another. The initiation of institution is motivated by the fourth regencies that feel they have similar potential for development, as well as mutually supportive and complementary advantages and disadvantages of each. In addition, the economic movement of peoples' lives in the four regencies is very dynamic, and often crosses the boundaries of administrative regions so that they feel that the inter-local government cooperation they should do. To that end, on July 6, 2006, the four Regents of the four counties made a mutual agreement to synergize their programs. Through inter-local government cooperation, they can build and develop the synergies of the potential they have. The purpose is to synergize BKAD Ratubangnegoro establishment of governance, development programs and community service in order to provide the greatest possible benefits for the welfare of the people and maintain the unity and integrity of the Unitary Republic of Indonesia.

The establishment of BKAD Ratubangnegoro was marked by signing a Letter of Agreement between the Regents of Bojonegoro, Tuban, Blora and Rembang with No. 188/ 05/ 412.12/ 2006, No. 180/ 2530/ 4.4.012/ 2006, No. 640/ 3109/ 2006, No. 130/ 1988/ 2006 on Regional Development Cooperation. Furthermore, the agreement is followed up with the signing of the Joint Decree of the Regents of Blora, Rembang Regent, Regent of Tuban and Bojonegoro Regent No. 130.1/4884/2010, No. 188.45/ 147/ KPTS/ 414.012/ 2010, No. 01/ 2010 and No. 253/ 05/ 412.11/ 2010 on the Establishment of BKAD Ratubangnegoro, on 21 September 2010.

To follow up the establishment of this institution, the four regencies governments agreed to form the joint secretariat Ratubangnegoro, through the signing of the Joint Decree Regent of Blora, regent of Rembang, Regent of Tuban and Bojonegoro Regent No. 130.1/ 4885/ 2010, No. 188.45/ 148/ KPPS/ 414 012/ 2010, No. 02/2010 and No. 254/ 05/ 412.11/ 2010 on the Establishment of a joint secretariat Ratubangnegoro on September 21, 2010. The area become a joint secretariat is designated in turns between regions, for 2 (two) years, with a corresponding sequence Ratubangnegoro acronym. Until now, such joint secretariat position has been held on the year Blora Regency 2011-2012, Tuban in the year 2013-2014, the Government of Rembang Regency Year 2015-2016, and Bojonegoro in the year 2017-2018.

Based on mutual agreement between the regions, BKAD Ratubangnegoro has scope for cooperation in some fields, such as health; social; mining and energy; spatial planning; forestry and plantations; public works; trade and industry; transportation; capital investment; living environment; employment; tourism; and education and culture. Nonetheless, to date, the agreement made in BKAD Ratubangnegoro areas of cooperation includes health, mining, energy, forestry and plantations, investment and employment, as well as cooperation in tourism and culture.

Various agendas have been carried out by BKAD Ratubangnegoro, all in Blora, Tuban, Bojonegoro and Rembang. Some cooperative activities that have been done, include the development of an investment in 2010; the development of education and culture in 2012; social development and employment in 2012; tourism development in 2012; the synchronization of development/ maintenance of public works of roads and bridges in 2012; enforcement of regulations as well as the disruption of public order and peace of the community in 2013; the handling of health care in 2013; the handling of service of artificial insemination in 2014; traffic handling animals and animal products in 2014; veterinary services in 2014; disaster prevention in 2014, synchronization of spatial area in 2014, motor vehicle inspection program in 2014, and exchange of tourism and creative economy in 2014.

IOP Conf. Series: Earth and Environmental Science **123** (2018) 012036 doi:10.1088/1755-1315/123/1/012036

Although it has been established ten years ago, BKAD Ratubangnegoro still faces many challenges and critical issues concerning sustainability. The main issues is incompatible of the regulatory foundation for the establishment of the institution, the Government Regulation No. 50/2007 on Procedures For Local government cooperation, with Law No. 23/2014 on Local Government. The new law states that cooperation between regions in the local boundary area is mandatory, The need for adjustment of cooperative activities between the different areas in the region the boundary. On the other hand, the old law does not regulate explicitly on it.

The Letter of Agreement of BKAD Ratubangnegoro has been signed for ten years ago, but the follow-up implementation does not work because of constraints on financial reasons. The existence of the Joint Secretariat should be strengthened in establishing inter-local government cooperation on the boundary. Urgency MOU between the parties, among others, the weak point in the exercise in the Cooperation Agreement, cooperation based on the needs, an MOU was made in the context of supervision within a certain period.

In carrying out this authority, the province did not just play the role of implementing and regulating the field directly and inter-local government cooperation, but also provide assistance and support, cooperation among regions in specific fields such as the development of infrastructure and facilities area, investment, industry and trade, and agriculture. Thereby formally, cooperation between regions should be arranged or facilitated by the province.

According to Yunus [4], there are two main issues in the discussion of institutional interaction between regions, which are related to regional autonomy; And related to development financing. The first issue relates to how to manage two or more regional autonomy meetings well, so as to avoid conflicts between regions. Inter-regional autonomy meetings can create regional development synergies, but are otherwise vulnerable to conflicts of inter-regional interests that result in conflict between the regions.

The second issue relates to the coordination and integration of financial utilization of each region in financing inter-regional interaction activities. In Indonesia, the main obstacle faced is the absence of clear legislation governing it. According to existing regulations, the financing of joint development activities and responsibilities must be undertaken by each region. Meanwhile, often the financing coming from these areas is merged into one to create greater synergies, so the responsibility is difficult to specify one by one based on their origin and source.

Adopting the Warsono's explanation [2] about the variety of cooperation between regions, the range of inter-regional interaction institutions can be grouped into two models. The first model is the institutional model of inter-regional networks. This institutional model of inter-regional interaction is based on this network, it considers that each region should be free and independent to interact with each other. This model places inter-regional interaction as an embodiment of collective action that must be managed by the regions involved. Therefore, there is no central authority structure and purpose of such cooperation which is not the result of agreement from those regions. While the second model is the institutional model of relations between local governments. This model has a formal and hierarchical interaction character, which views, inter-regional interaction as a coherent activity with clear objectives. The process is structured from above, directed to a specific purpose, decision-making is dominated by centralized authority and has clear goals and values. In other words, this model is built on formal rules and regulations.

In practice, in Indonesia, institutional interaction between regions consists of five different models. If they are grouped into two different models developed based on Warsono's opinion [2], then there are two kinds of models that include institutional, inter-regional networks, namely the model of networking institutions and coordination forums. The other is inter-regional relationship that includes three models, namely the model of cooperation forum forums, cooperation institutions and business entities. The following table is an explanation of the five kinds of institutional interactions between regions can be seen on the Table 1.

As mentioned before, interregional institutional interaction in BKAD Ratubangnegoro consists of two types, namely the interaction between provincial governments, namely between the Provincial Government of Central Java and East Java Provinces; and inter-regency government interactions, consisting of 4 regencies, which are Blora and Rembang are in Central Java Province, and two regencies, Tuban and Bojonegoro in East Java Province.

The model of interaction in the provincial level interaction is networking forum. The function of inter-regional institutional networking model is very limited, that is only in an effort to facilitate information sharing activities between the provinces that benefit to the regions. The objective of the activity is to align the cross-regional development plan. A harmonized development plan can reduce conflict between the provinces and even increase profits for the regions that interact. While the scope of the information shared is usually limited to technical implementation and budgeting. Nevertheless, this model can be utilized to promote strong tolerance among regions that can be inter-regional binding forces.

Because it lacks a clear institutional form, this institutional model has weak regulatory authority, and its inter-regional interaction is fluid and flexible. In other words, this model is informal because it is not contained in a clear organized form. As a result, often information-sharing activities do not result in binding agreements, as regions have only an obligation to provide information on their own areas and use information from other regions for their own local interests, without any other obligations or expenses, as long as the information shared provides mutual benefits. As a result, violations of the consequences of sharing information between regions that occur usually do not result in sanctions.

The institutional model of the network is generally the beginning of the development of a more sophisticated model of cooperation, particularly those embodied in clearer and binding institutions. Without being preceded by information sharing, inter-regional interactions will not begin and develop. In the subsequent institutional model development, the interaction is not only about sharing information, but also sharing other things, such as ideas, ideas, resources, benefits and even risks.

Meanwhile, at the regency level cooperation model interaction cooperation is the cooperation joint secretariat model. The model is an institutional model that provides a forum in the form of inter-regional forums for sharing information, coordinating and working together to utilize shared resources to achieve mutually agreed goals. This institutional model is the development of a coordinated institutional model developed by adding other activities, such as the implementation of joint programs to facilitate in creating joint activities that are integrated in local internal policies.

However, this institutional model has the character of liquid and flexible member-to-person relationships. Although it does not provide and administer sanctions for violating areas, it is possible to set up an incentive structure and a set of rules that can influence its members to implement what has been mutually agreed. The strength of this model lies in the magnitude of the long-term contribution provided by the cooperation forum on the internal policy of the region as it can have a significant effect. So the solidity and sustainability of the cooperation forum with this model can be more guaranteed.

In Indonesia, the cooperation forum is a forum of cooperation institutions established in areas called the Inter-regional Cooperation Agency (BKAD). Basically, BKAD was formed to accommodate the cooperation of the Coordination Team for Regional Cooperation (Tim Koordinasi Kerjasama Antar Daerah, shortened as TKKSD) which is an institution established in a region headed by a regional secretary who has a member consisting of the Local Government Working Units (Satuan Kerja Perangkat Daerah, shortened as SKPD) involved in inter-regional cooperation. The team is in charge of preparing the implementation of inter-regional cooperation undertaken by the relevant SKPD. The interaction between the teams was conducted in a cooperation forum managed by the provincial TKKSD.

Cooperation is an interaction act between parties to exchange information, carry out joint activities, and share resources, for mutual benefit and achieve common goals. Cooperation is a more advanced measure of coordination with the addition of resource sharing characteristics. Through sharing of resources, cooperative action can alter and bind the parties involved. Successful use of resource sharing

IOP Conf. Series: Earth and Environmental Science **123** (2018) 012036 doi:10.1088/1755-1315/123/1/012036

can provide mutual benefits, but on the contrary, failure to utilize shared resources can be at risk of shared losses.

The development of institutional model of this cooperation forum has a higher level of difficulty than the institutional model of the network and the coordination forum. In Indonesia, the difficulties faced by these interacting areas were solved through assistance by provincial TKKSD. Nevertheless, the development of this model requires internal support and readiness to open up to interacting interactions from other regions.

Table 1. Local Government Interaction Institution Models

Type of Interaction	Definition	Relationship	Characteristics	Resources	Model of Institutional
Networking	Exchanging information for mutual benefit	Informal institution	Minimal time commitments, limited levels of trust, and no necessity to share turf; information exchange is the primary focus	No mutual sharing of resources necessary	Networking forum
Coordinating	Exchanging information for mutual benefit, and altering activities to achieve a common purpose	Semi-formal institution	Moderate time commitments, moderate levels of trust, and no necessity to share turf; making access to services or resources more user-friendly is the primary focus	No or minimal mutual sharing of resources necessary	Coordination forum
Cooperating	Exchanging information for mutual benefit, and altering activities and sharing resources to achieve a common purpose	Formal institution	Substantial time commitments, high levels of trust, and significant access to each other's turf; sharing of resources to achieve a common purpose is the primary focus	Moderate to extensive mutual sharing of resources and some sharing of risks, responsibilities, and rewards	Cooperation joint secretariat
Collaborating	Exchanging information for mutual benefit, and altering activities, sharing resources, and enhancing the capacity of another to achieve a common purpose	Formal institution	Extensive time commitments, very high levels of trust and extensive areas of common turf; enhancing each other's capacity to achieve a common purpose is the primary focus	Full sharing of resources, and some sharing of risks, responsibilities, and rewards	Cooperating agency
Integrating	Exchanging information for mutual benefit, and altering activities, sharing resources, and enhancing the capacity of the institution to achieve a common purpose	Formal institution	Long time commitments, very high levels of trust and extensive areas of common turf; enhancing each other's capacity to achieve a common purpose is the primary focus	Full sharing of resources, and full sharing of risks, responsibilities, and rewards	Joint venture agency

Source: Developed from Roberts, 2004

4. Conclusion

As mentioned before, interregional institutional interaction in BKAD Ratubangnegoro consists of two types, namely the interaction between provincial governments, namely between the Provincial Government of Central Java and East Java Provinces; and inter-regency government interactions, consisting of 4 regencies, which are Blora and Rembang are in Central Java Province, and two regencies, Tuban and Bojonegoro in East Java Province.

The model of interaction in the provincial level interaction is networking forum. The function of inter-regional institutional networking model is very limited, that is only in an effort to facilitate information sharing activities between the provinces that benefit to the regions. The objective of the activity is to align the cross-regional development plan. A harmonized development plan can reduce conflict between the provinces and even increase profits for the regions that interact. While the scope of the information shared is usually limited to technical implementation and budgeting. Nevertheless, this model can be utilized to promote strong tolerance among regions that can be inter-regional binding forces.

Meanwhile, at the regency level cooperation model interaction cooperation is the cooperation joint secretariat model. The model is an institutional model that provides a forum in the form of inter-regional forums for sharing information, coordinating and working together to utilize shared resources to achieve mutually agreed goals. This institutional model is the development of a coordinated institutional model developed by adding other activities, such as the implementation of joint programs to facilitate in creating joint activities that are integrated in local internal policies.

5. Acknowledgments

The author would like to thank the Faculty of Engineering, Universitas Diponegoro which has funded the research whose results are written in this article.

6. References

[1] Visser J A 2002 Understanding local government cooperation in urban regions: toward a cultural model of interlocal relations *Am. Rev. Public Adm.* **32** 40–65
[2] Warsono H 2009 Regionalisasi Dan Manajemen Kerjasama Antar Daerah (Studi Kasus Dinamika Kerjasama Antar Daerah Yang Berdekatan di Jawa Tengah) *Yogyakarta Progr. Dr. Ilmu Adm. Negara, Univ. Gajah Mada*
[3] Clark G L 1984 A theory of local autonomy *Ann. Assoc. Am. Geogr.* **74** 195–208
[4] Yunus H S 2006 *Megapolitan: konsep, problematika, dan prospek* (Pustaka Pelajar)
[5] Agnew J A, Mitchell K and Toal G 2008 *A companion to political geography* (John Wiley & Sons)
[6] Haris S 2005 Desentralisasi dan Otonomi Daerah: Desentralisasi, Demokratisasi & Akuntabilitas Pemerintah Daerah
[7] Rumford C 2006 Theorizing borders
[8] Seyler W C 1974 Interlocal relations: cooperation *Ann. Am. Acad. Pol. Soc. Sci.* **416** 158–169
[9] Healey P 1997 *Collaborative planning: Shaping places in fragmented societies* (UBc Press)
[10] Margerum R D 2002 Collaborative planning: Building consensus and building a distinct model for practice *J. Plan. Educ. Res.* **21** 237–53
[11] Innes J E 2004 Consensus building: Clarifications for the critics *Plan. theory* **3** 5–20
[12] Creswell J W 2013 *Research design: Qualitative, quantitative, and mixed methods approaches* (Sage publications)
[13] Yin R K 2009 Case study research: Design and methods fourth edition *Los Angeles London SAGE*

Industrialization Impact on Worker Mobility and Land Use in Peri Urban Area (Case study of Semarang District, Indonesia)

H B Wijaya[1], H Kurniawati[1], and S T E W Hutama[1]

[1]P5 UNDIP, Department of Urban and Regional Planning, Diponegoro University, Kampus UNDIP Tembalang, Semarang 50275, Indonesia

Email: holibinawijaya@yahoo.com

Abstract. In many cases, industrialization has stimulated the urbanization process massively. It tends to attract substantial number of labor migrants from nearby region to fulfill the demand of workers. The paper reports the research result of industrialization phenomena in Semarang district, Indonesia. It carried out the survey by taking 250 samples of migrant workers. The result shows that the presence of labor-intensive industries becomes the most influence factor for many migrants, rather than the driving factors from the place of origin. The attraction factor could cover regional and across the province force, as indicated that all migrant respondents come from both inside and outside of Central Java Province. Furthermore, based on the land-use distribution of the migrant settlement area, it indicates a growing land-use change, both of land cover and land functions.

Keywords: Industrialization, Peri urban area, Worker Mobility

1. Introduction

Industry sectors play significant role in enhancing the economic development in most of developing countries, since the primary sectors have low contribution to accelerate the national economic. The share of this sector in commodity value added has a highly significant positive correlation of income per capita [1,2]. Manufacture and labor-intensive industries are so powerful to build up the economic development by its contribution in non-mining export, absorbing the surplus manpower, influence the emergence of new employments [3].

The presence of this manufacturing in Semarang District also contributes to the improvement of economic income. Though, this manufacture performance replaced the farming sector as a mainstay priority sector in 1990s [4]. Some studies stated that the manufactures were growing faster in Semarang District as a result of the expansion of industrial development from Semarang City [5–9]. The higher land-price in Semarang City gives more pressure to the industrial development. Based on the business principal, the manufacture expands to the Peri Urban area that has the lower land-price, just like in Semarang District. It applies to the new manufacture and the expansion of the old ones.

Industrialization in peri urban plays an important role in the process of urbanization, which the most relevant trigger for the urbanization is the population growth, especially in migration cases [10]. The characteristics of manufactures in Semarang District are labour-intensive industries, which are mostly of garment industries. The substantial number demand of workers could not be fulfilled by the

IOP Conf. Series: Earth and Environmental Science **123** (2018) 012037 doi:10.1088/1755-1315/123/1/012037

local supplies. The industries in Semarang district provide many job creations for local and non-local people from inside and outside districts, whether urban or rural areas. This demand and supply on job opportunities create the worker mobility and migration flow inside Semarang District.

These immigrants have the highly contribution to the urbanization. They support the population growth which is followed by the growing needs and activities. The urban transformation begins when these ascending needs are met. It has driven the transformation of socio economic, the population density, and land use conversion for the most[11]. Study about the expansion of urbanization from the downtown to the periphery of Semarang City stated that the transformation specifically can be seen in (1) the majority of livelihood from the farming sectors to non-farming sectors; (2) the cultural blending between the local people and the migrants; (3) land conversion from agricultural into residential, industry, and business[12].

This study aims to identify the industrialization impact in Semarang District, specifically on migrant and worker mobility and the land use conversion. This paper addresses the following research question: (1) Does the industrialization become the pull factor for migrants?; (2) Where is the location scope of the origin of migrants? (3) How is the land conversion in industrial land surrounding?
The three main questions identify how industrialization process attracts the migrants and has further influence on the urbanization, with the main indicator in land use conversion.

2. Methods

The research used primarily data that was collected from sample workers in Semarang District. Data collection was done through field survey that was conducted in May to early June 2017. The field survey was based on the questionnaire that has been made to get their profile about their life and working experience. There were several criteria applied to choose and balance respondents' i.e. Male and female; Work in Semarang District; Age 15-29 year; Migrant (born or live outside Semarang District).

By using the criteria, there were 249 respondents get for this research that were divided into 2 agreed migrant characteristics: migrant and commuter. The migrant itself is respondents who used to live outside Semarang District and has been living in Semarang District by May 2017. Based on these criteria, we found 198 respondents. While the rest of the respondents, as many as 49 respondents, were commuters whose characteristics live outside Semarang District and need to commute between Semarang District and their hometown every day.

After conducted data collection, the data were analysed with 3 kinds of analysis. The first step was answering Research question no. 1 which was using quantitative analysis. This analysis was using SPSS software with descriptive statistics analysis. The second one was spatial analysis with GIS software. This analysis aims to analyse the location of migrant origin distribution before they moved to Semarang District. The last analysis was spatial analysis using Landsat Map. This analysis was used to analyse the land use change from open space to build environment.

3. Results

3.1. Migration Factor

Survey of migrants in Semarang District focuses on identification the main reason why the migrants move and stay or commute in Semarang Districts. All respondents are migrants in the age 17 to 29 years old, where the majority is the 19 (17,7%) and 29 (14,1%) years old migrants. Its result is that 69,1% respondents move and commute to Semarang District because of the chances to get a better job (**Figure 1**). Most of them work as employee (93,2%), which is mostly in industrial sector (66,7%), then the rest are entrepreneurs (6,8%). This result indicates that the presence of industrial sector in Semarang District become the pull factor for migrants.

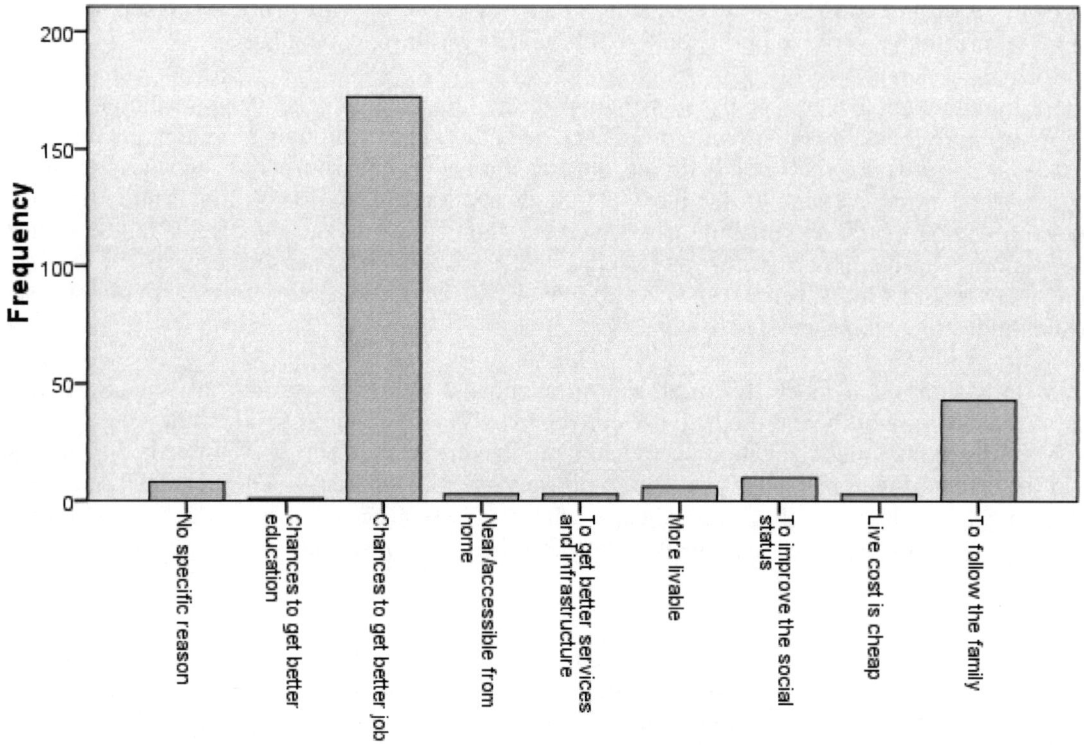

Figure 1. The main reason to move and commuting to Semarang District (Pull Factors)

The data shows the pull factor is associated with the push factor from many migrants. Most of respondents (80,3%) stated that they migrate from their origin to work or seek a job (**Figure 2**). Based on survey, many respondents get the information about the presence of manufactures in Semarang District that offers the abundant job creation that suited with their background qualification. These job opportunities and the job needs are met. Mostly respondents take the migration preference in Semarang District because of the job oriented, which is get a job or new better job.

The migrant's respondents are relatively low educated, so that they have less bargaining in job position. There are only 3 respondents (1%) that passed the diploma/bachelor degree, then the rest take the education in the level elementary school (2%); junior high (27,7%); senior high (30,5%); and vocational high school (38%). There is almost 60% of respondents work in manufacturers as machine operator, and some others are marketing/sales service; administration; cleaning/home service; public service; experts/skilled worker; and unskilled worker. Some respondents that passed the vocational high or senior high school have the skill qualification on garment; automotive; information, technology, and communication; administration; and fashion design. Yet, they have already had the skill and competence in some sector, but it is almost useless in the job. Most of respondent stated that the employers have the specific rules and qualification, so that the skill and experience are mostly useless because there will be specific training at first.

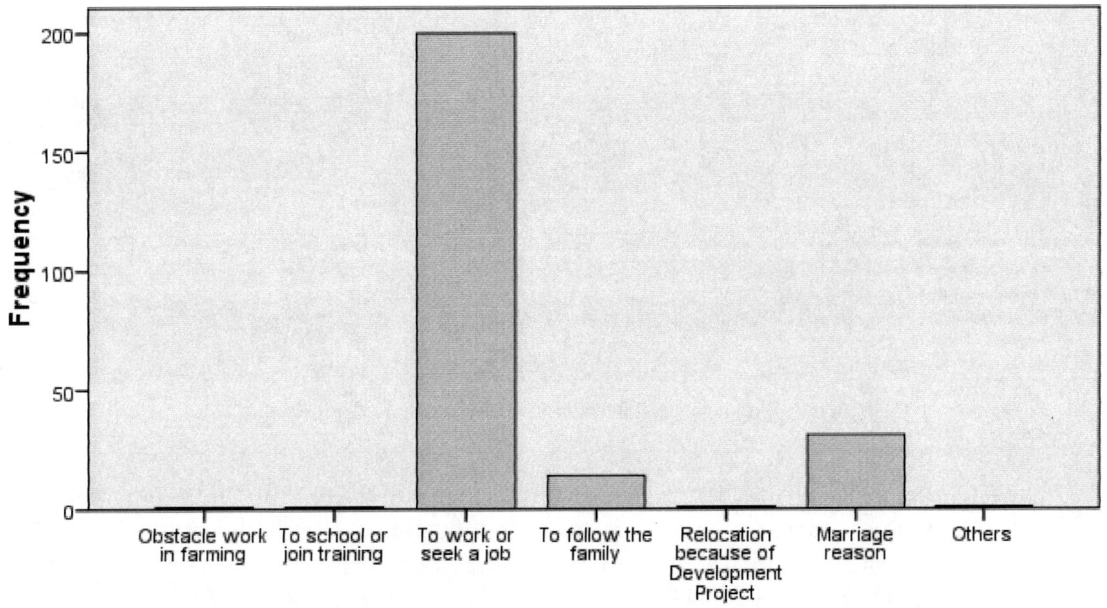

Figure 2. The main reason for migrating (Push Factors)

The presence of industry becomes the most influence factor to pull migration inside Semarang District. The industrial sector has developed since 1980s and was growing faster because of expansion from urban area in Semarang City. Many urban areas, just like Semarang City, Jakarta, or Surabaya, loss their competitiveness to this industrial expansion, in terms of highly wages, higher lands price, and the supporting infrastructures. The industrial expand to the peri urban area to get lower in operational costs, specifically in Central Java Province. There are some advantages, which is the affordable land supply, the lower wages for workers, the easy investment program, and some local government supporting program in infrastructure and services.

3.2. The Origin Distribution of Migrants

The attraction of industrial sector of Semarang District reaches multi provincial level on national scope. It can be seen by the origin of migrants in Semarang District. There are two types of migrants for the respondent in this study, which is migrants and commuters. Some migrants come from Eastern Java (4%), Yogyakarta Province (0,8%), West Java (0,4%), and Jakarta Province (0,4%), while the most come from nearby Semarang District in Central Java Province (**Figure 3**). It means that these migrants live in there before moving and live in Semarang District. Most respondents (61,8%) migrate in the age 15-20 years old, when they passed the junior or senior high school, rather than the rest are migrants that find new better job in the above 20 years old. Based on in-depth interview with some migrants, they get the job vacancy information from their friends, families, senior in high school, or colleagues or just follow them to seek the job vacancy in Semarang District.

The existence of migrants in Semarang District commonly creates less mobility. The reason is most migrants live in a boarding house (56%) and they reach the workplace on foot. They get motorcycle or public transport to spend some holiday. For this reason, the migrants also have less contribution to the traffic jam and fuel consumption.

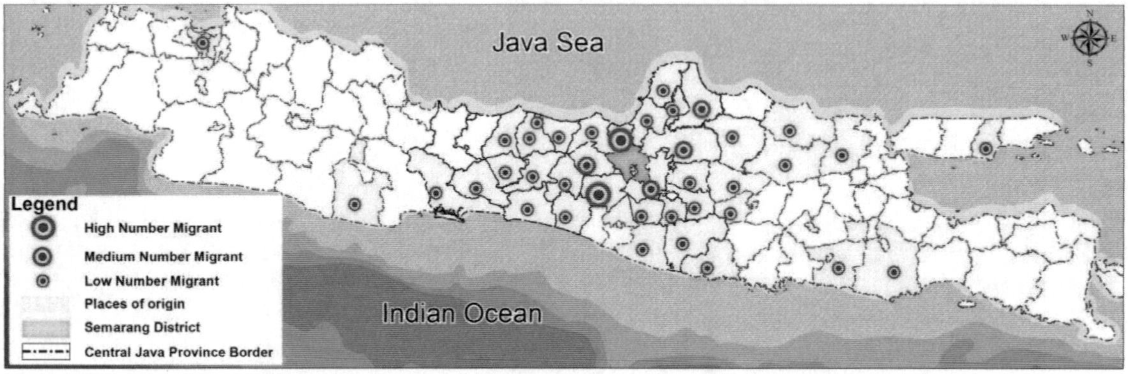

Figure 3. The origin distribution of Migrant Respondents

Besides migrants, there are also 49 commuter's respondents that work in Semarang District and travel from home to the destination every work time. These commuters live in nearby of Semarang District, that is the most live in Semarang City (80,4%); Salatiga (11,8%); Demak (3,9%); Kendal (2%) and Magelang District (2%) (**Figure 4**). Most of commuters comes from Semarang City, especially people in periphery of Semarang City and Semarang District, such as in Banyumanik and Gunung Pati Sub District. It is an anomaly case that these people prefer to work in District, rather than in the urban area of Semarang City. It occurs mostly because of working matters, such as the supply of job that suit with the qualification, and the closer location from home (we found that the Semarang District is closer to the periphery area of Semarang City).

The preference for commuting is because the easily accessible for travelling, that is in term of good road infrastructure, the readiness of public transportation, and the distance between home and work destination is relatively close. In fact, all commuters' respondents take the motorcycle to reach the work destination with the most travel distance is about 10-29 km and the travel time duration is 30 minutes till 1 hour. This situation describes that the commuters have a highly contribution to the traffic jam and fuel consumption.

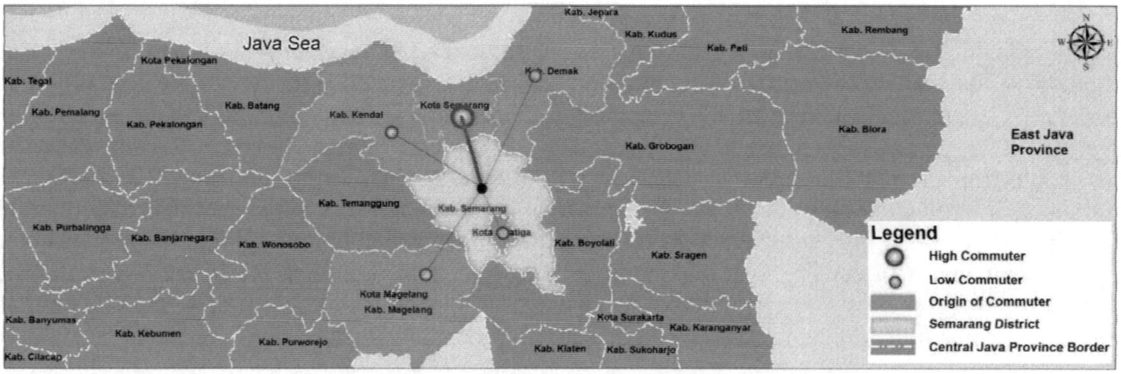

Figure 4. The origin distribution of Commuter Respondents

The existence of manufacturer in Semarang District attracts the workers till the national scope areas. The fundamental reason of this is the higher job creation in Semarang District. This pull factor is met with the push factor, where there are lack opportunities to get a job in the origin area. Inevitable, this migration reason is demand-supply interaction. So, what about the impact of these migrants phenomenon to the origin area and also to the destination that is Semarang District. Undeniable the presence of migrants promotes the growing needs to be fulfilled, such as housing, food and consumption, clothing, as well the life styles. A study stated that it is not economic growth, but the population growth that highly contribution to the urbanization process and also causes the development problem[10]. It sounds like disadvantage for the Semarang District, but in terms of local economic development, it contributes to the positive improvement. A study results that the job creation (similar case of industrialization in Semarang District) becomes the main driver to the recoveries of economic development in United States, than the out migrations ones [13]. The wisely policy should be aware to optimize this positive impact to the local economic growths, and well-prepared for the disadvantage that may threaten the sustainable development.

3.3. Indicator for Urbanization: Industrial Expansion and Land Conversion

In line with the increasing number of manufacturers, the land conversion in Semarang District raises up as well. In 20 years, the number of manufacturer increases significantly. They agglomerate near the urban area (that is Semarang City) at first, but they expand to the south part of Semarang District, follow the heavy business track, that is Semarang City to Solo. The expansion of built up area follows the concentration of industry location. The built-up area covers the residential, amenities, trade and business area, etc. The most developed areas are in Ungaran Barat, UngaranTimur, Bergas, Bawen, and Ambarawa Sub District, where it lays the agglomeration of industries.

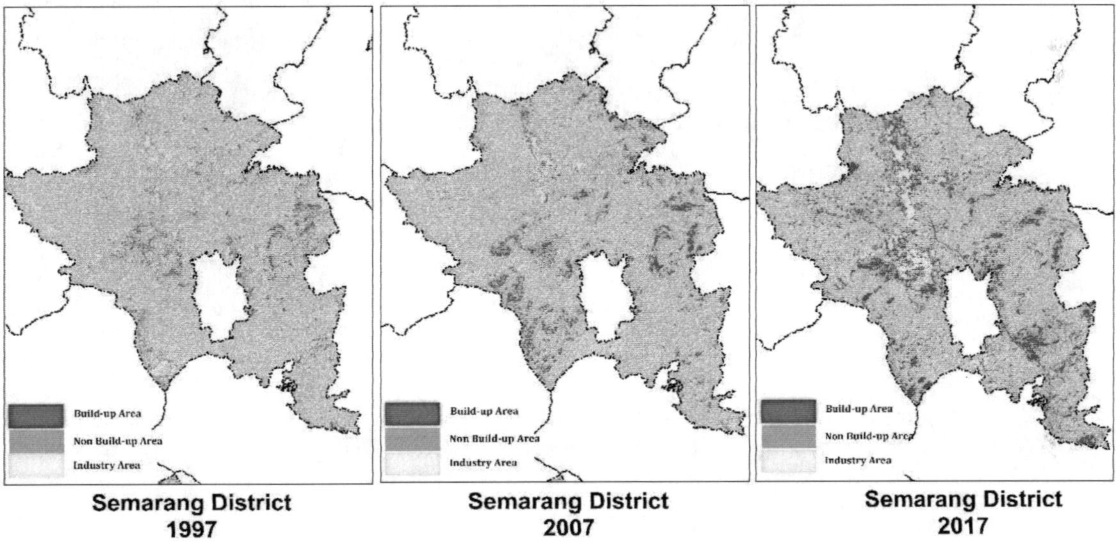

Figure 5. Land conversion in Semarang District in 20-year time series

There is not only the growing built up area, but also the land use transformation. Based on observation and interview with the dwellers in surrounding of PT. Semarang Garment manufacturer, there are many changes in the land use. The land used to be field transformed to the manufacturer building, beside that many houses extend to boarding house, bike and motorcycle custody, food stalls, shop groceries, mobile phone services, etc. There are many migrants live in the boarding house near the workplace. Most of them go to work on foot and they often go home to their boarding house to take a lunch at the working break time. After its development in 2002, the running activities of this garment manufacturer absorbs many workers, which is local people and migrants. There are many changes in the surroundings of the manufacturer location, to meet the daily needs of workers. It is not only about the land conversion and land use transformation, but also the social and economic impact in local area. For now, local people are more aware to the mobile technology and information, beside the opportunities to increase the income from the new business and service activities.

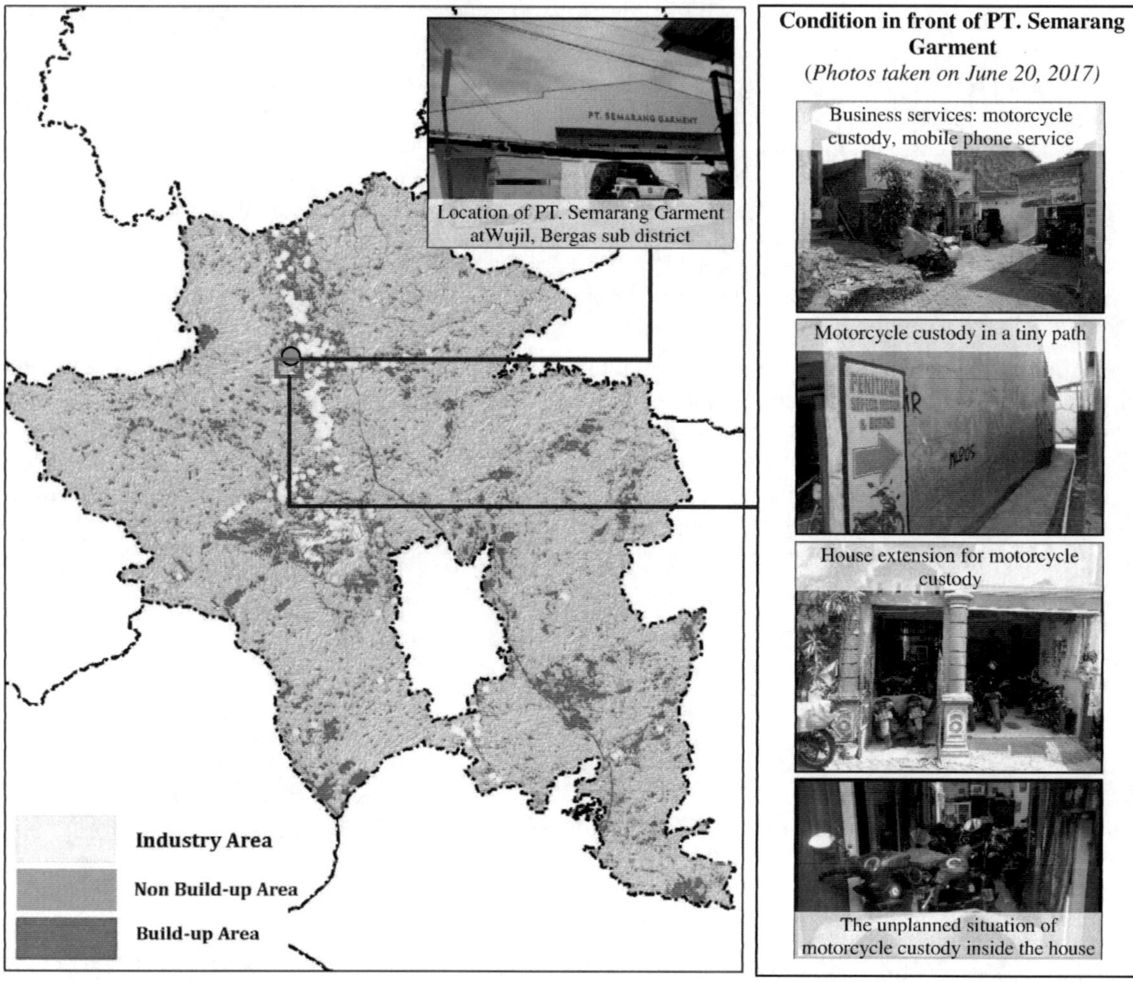

Figure 6. Land use transformation in the surrounding of PT. Semarang Garment Manufacturer in Wujil, Bergas sub district

4. Results and Discussion

Land use conversion and population growth often become the visible indicator for the urbanization process, besides the other invisible indicators, i.e. social and economic indicators [11,14,15]. The research result shows that the industrial development drives the population growing, the improvement of built-up area, the land use transformation, as well the social and economic change of local people. It begins with the manufactures development in some agricultural or non-agricultural land that promotes the growing land conversion into built-up areas. For further effect, it also absorbs many workers from local people or migrants that promote the growing population. In the process of increasing daily needs, some local people extend their residential land use into business and services for workers. Besides that, the presence of migrants has influence the social transformation by the diffusion of mobile technology and information. In the economy context, there are several opportunities to increase the income from the new business and service activities. For the case in Semarang District, industrialization has driven the urbanization process.

The phenomenon of commuters from Semarang City exposes the happenings of Desakota context. The emergence of Desakota was driven by the economic impetus and it occurs as interlock between bottom-up rural industrialization and top down urban expansion[16]. We found there is preference of peri urban people for accessing the job opportunities in District area, rather than in Centre Business or

Service area of Urban. The mega urban context frame it as an urban expansion, when the city cannot supply the needs (in terms of job opportunities), the dwellers will stick to their more appropriate area, especially in the district areas. There are mostly people in periphery of Semarang City that work in Semarang District, due to it is closer distance to their home. In the other side, the Desa kota area exposes that the industrialization raises the job opportunities that becomes the attraction point for peri urban dwellers. The evidence supports the argument that economic impetus has become the most attraction point for the Desakota area, in case of these commuters from urban area.

The rural urbanization process in developing countries is growing faster, and will be greater in the future. It is driven by the strong beliefs of the contemporary economic wisdom that industry and services sectors promote the higher returns than agriculture sector. The industrialization growths follow the economic principal, where the land prices become the most stimulant factor. There are many cases in developing countries that this rural-urban transition are growing uncontrollable/unplanned, so that it creates the more sensitive to issues of sustainable development, especially in terms of environment, energy demand, food security, regional inequality, and some social economic problem, just like urban poverty [17]. For this reason, there is a need the wisely policy to support the urbanization process to promote the important component of sustainable development. Some scholars suggested the importance of studies to follow up the rapid industrialization in rural area to understand the rural growth and to smoothly promote the rural urban development in balance and sustain [18,19]. The clear policy to redefining the spatial patterns is needed to straight the position and role to support the sustainable development.

5. Acknowledgement

The paper reports part of collaboration research result of Small Cities, Migration and Youth Employment Outcomes in India and Indonesia between JustJobs Network– JJNand P5 Undipin 2017.

6. References

[1] Szirmai A 2012 Industrialisation as an engine of growth in developing countries, 1950-2005 *Struct. Chang. Econ. Dyn.* **23** 406–420

[2] Haraguchi N, Cheng C F C and Smeets E 2017 The importance of manufacturing in economic development: Has this changed? *World Dev.* **93** 293–315

[3] Damayanthi V R 2008 No Title *J. Indones. Appl. Econ.* **2** 68–89

[4] Dewi N and Others 2006 *Analisis Pertumbuhan Ekonomi di Kabupaten Semarang Tahun 1996--2003.* (Universitas Negeri Semarang)

[5] Abdullah 2010 *Pengaruh Perkembangan Industri Terhadap Pola Pemanfaatan Lahan di Wilayah Kecamatan Bergas Kabupaten Semarang* (Universitas Diponegoro)

[6] Irianta F . G 2008 *No Title* (Universitas Diponegoro)

[7] Buchori I, Sugiri A, Maryono M, Pramitasari A and Pamungkas I T D 2017 Theorizing spatial dynamics of metropolitan regions: A preliminary study in Java and Madura Islands, Indonesia *Sustain. Cities Soc.*

[8] Buchori I and Sugiri A 2016 An empirical examination of sustainable metropolitan development in Semarang City, Indonesia *Aust. Plan.* **53** 163–177

[9] Sugiri A, Buchori I and Soetomo S 2011 Sustainable Metropolitan Development: Towards an Operational Model for Semarang Metropolitan Region *Int. J. Environ. Cult. Econ. Soc. Sustain.* **7** 301–324

[10] Jedwab R, Christiaensen L and Gindelsky M 2015 Demography , urbanization and development : Rural push , urban pull *J. Urban Econ.*

[11] Deng X, Huang J, Rozelle S, Zhang J and Li Z 2015 Impact of urbanization on cultivated land changes in China *Land use policy* **45** 1–7

[12] Prihanto T 2010 Perubahan Spasial Dan Sosial-Budaya Sebagai Dampak Megaurban Di Daerah Pinggiran Kota Semarang *J. Tek. Sipil dan Perenc.* **12**

[13] Greenaway-McGrevy R and Hood K K 2016 Worker migration or job creation? Persistent shocks and regional recoveries *J. Urban Econ.* **96** 1–16

IOP Conf. Series: Earth and Environmental Science **123** (2018) 012037 doi:10.1088/1755-1315/123/1/012037

[14] Surya B 2016 The Processes Analysis of Urbanization, Spatial Articulation, Social Change and Social Capital Difference in the Dynamics of New Town Development in the Fringe Area of Makassar City (Case Study: In Metro Tanjung Bunga Area, Makassar City) *Procedia - Soc. Behav. Sci.* **227** 216–231
[15] Wisdiawan R A 2014 *Dampak Pembangunan Kawasan Industri di Kabupaten Semarang Terhadap Pekerja Wanita*
[16] WU B-S 2009 Modeling the Dynamics of Desakota Regions: Global – Local Nexus in the Taipei Metropolitan Area *Geography*
[17] Mcgee T G 2009 The Spatiality of Urbanization: The Policy Challenges of Mega-Urban and Desakota Regions of Southeast Asia
[18] Long H, Zou J and Liu Y 2009 Differentiation of rural development driven by industrialization and urbanization in eastern coastal China *Habitat Int.* **33** 454–462
[19] Tian L 2015 Land use dynamics driven by rural industrialization and land finance in the peri-urban areas of China: "The examples of Jiangyin and Shunde" *Land use policy* **45** 117–127

IOP Conf. Series: Earth and Environmental Science **123** (2018) 012038 doi:10.1088/1755-1315/123/1/012038

Planning of Agro-Tourism Development, Specific Location in Green Open Space Sarbagita Area, Bali Province

I Lanya [1], N Subadiyasa[1], K Sardiana[1], and G P Ratna Adi[1]

[1]Study Program Agroecotechnology, Faculty of Agicultural, Udayana University, Denpasar, Bali

E-mail: indahnet@yahoo.co.id

Abstract. Tourism development has a negative impact on agricultural land in Bali, resulted in the transfer of rice field of 800 ha/year. Subak rice field area as a world cultural heritage, requires conservation strategy, increasing economic and environmental value, through integrated agriculture development with tourism. Tourism destination planning in the form of tourist destination (TD) and tourism object (TO) by raising local genius, at specific location, is expected to preserve nature and culture, as well as the economic value of the region. Research Methods: (1) identification of agrarian cultures, (2) field survey, (3) mapping of site specific TD/TO plans, and (4) compile documents of agro-tourism road map based on local genius. Seven subak areas in the green open space area have the potential to develop new TD/TO, namely: (1) Gedon2Subak in Tanah Lot area, is developed for the preservation of agriculture, the implementation of the zoning plan of the sacred, madya and nista areas, (2) the Kerdung and Penatih Subak areas, developed for urban farming in Denpasar City, (3) Cangi south Subak area, built for agro-tourism plasmanutfah banana and Cemagi Let Subak area developed agro-tourism food crops and horticulture, (4) Erjeruk Subak area, developed tourism plasmanutfah coconut.

Keywords: Tourist Destination, Tourism Object, Local Excellence, Agro-Tourism Zoning Plan

1. Introduction

Sarbagita area, covering the City of Denpasar and three districts, namely: Badung Gianyar, and Tabanan, designated as a national strategic area of tourism in Bali. These four areas are a very rapid area of tourism development and infrastructure of accommodation facilities. As a result, there was a change in the function of rice field in Bali Province 1000 ha year -1.The main cause of land supply for non-agricultural development in Bali Spatial Plans (SP), generally irrigated rice fields of Subak system [1].

The positive impacts of agro tourism are: conserving plasmanutfah, nature and agrarian culture, as well as increasing the value of regional ekonowi. Determination the strategy of determining and controlling the function of land to anticipate and mitigate the negative impact of tourism in Bali 2014-2015, produces Subak sustainable land zoning, buffer and converted, and draft of Local Regulation on Sustainable Land Agricultural Protection (SLAP) [1]. The results need to be continued with research that can improve the economic value of SLAP area through integration of agriculture with tourism and able to control land conversion, such as building agro-based tourism based on specificlocations.

There is a continuing conflict of interest in land use between agriculture and tourism. In Sarbagita area, there are center of natural and cultural tourisms, such as: agrowisata in Soka, Bedugul area, and Jatiluwih Tabanan regency. Badung Utara area. Kopi Luak and Sekar Bumi in Gianyar Regency. Agro tourism that developed only in Jatiluwih which is designated as world cultural heritage area (UNESCO, 2012) [2], Ceking Village in Gianyar and in Kertalangu Tourism Village in Denpasar.Agro tourism has been widely developed in other areas, such as in Malang, Bogor, North Sumatra and other areas of specific location with certain types of plants. The development is very rapid, considering the needs of people in downtown who need fresh oxygen. The results of field observations in Bali is indicating that the growth of integrated tourism activities with agriculture.Ceking tourist attraction that enjoys the natural beauty of terracing rice fields, coffee mongoose, sekar bumi in Gianyar regency. Kertalangu Tourism Village is in Denpasar City, Agro tourism in North Badung, and in Tabanan Agrowisata in Soka and Jati Luwih and Soka areas. The rice fields in Jatiluwih is unique withterrace form, it combines with fresh air, are the DTW attraction set by UNESCO in 2011 as a cultural heritage. Nevertheless, the cultural attractions of agrarian and cultural attractions make the supporting material of traditional ceremony (*banten*) not yet developed in all agro tourism in Bali.

2. Material and Methods

Research location is in Sarbagita area (Denpasar City, Badung Regency, Gianyar Regency, and Tabanan Regency) Bali Province. Especially subak areas that has been designated as subak sustainable and potentially developed as a tourist attraction or tourist destination. Each municipal district was selected a subak region with high accessibility to be planned as agro-tourism (Figure 1). Materials and tools used include WorldView satellite imagery in 2015, a set of computers and software Arc-GIS 10.2.

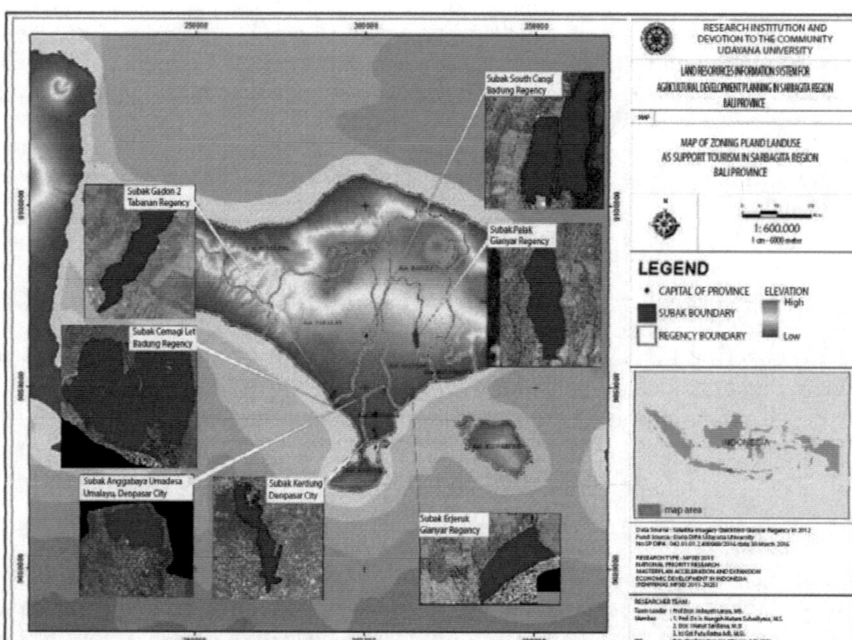

Source: Data Analysis, 2016

Figure 1. Map of Agro-tourism Development Site Location in Sarbagiata area

IOP Conf. Series: Earth and Environmental Science **123** (2018) 012038 doi:10.1088/1755-1315/123/1/012038

3. Results and Discussion

3.1. Inventory of Agrarian Culture in Subak Rice Field as Tourism Cultural Attraction of Agriculture

Subak Institution in Bali has an organizational structure headed by a Pekaseh. Pekaseh's task is to regulate and supervise the distribution of irrigation water and to lead the religious ceremony (ritual) of subak. A pekaseh is assisted by a Petengen who acts to help Pekaseh perform daily duties related to farming activities.

Subak generally manage irrigation and subak land as the main producer of paddy rice. Subak in Bali also has an additional function as the executor of ritual activities (upakara). This activity makes the subak in Bali very distinctive and different compared to other agricultural organizations in other areas. The subak system in Bali is based on the *Tri Hita Karana* concept, which means harmony between human and God, with fellow human beings and the environment. The rituals performed in Subak are held in Pura Subak (Pura Bedugul), although some are carried out in the upper reaches of the canal and in each rice field. The existence of Pura Subak will always be related to one of the duties and functions of subak in Bali that is ritual (Upakara) which must be continuously done in Pura subak. Rice Subak in Bali generally grows rice with two harvest time, while for three times planting done with rice-rice-palawija pattern. Rice fields that have twice the harvest time of the year generally grow local rice or commonly called Balinese rice.

The result of inventory and description of subak in Sarbagita area is based on accessibility, integration with tourism object, and potential of possible location that can be developed. Subak Pekaseh response obtained seven Subak [2,3], namely: (1) Subak Kedung, Pedungan Village, (3) Subak Cangi Selatan, Sembung Village, Abiansemal Sub-district, Badun Regency, (4) Sebak Cemagi in Mengwi, Badung, (5) Subak Gadon 2, Beraban Village, Kediri Sub-district, Tabanan District, (6) Subak Erjeruk Pantai Lebih, More Villages, Sukawati Sub-district, and (7) Subak Abian Petak in the village of Payak, Payangan Gianyar Regency. The location and the seventh position of the subak are presented in Figure 1.

The process of conducting the farming business of rice field subak in Bali performs a series of activities by using ceremonial means (Upakara) which goes synergistically with the process of land preparation and planting rice / palawija activities. Pekaseh Interview conducted onJuly 2016, Documented,on July 2016, the procession of adat / religious (upakara) conducted within the Subak region there are 13 ceremonial processes, namely: (1) *Mapag Toya* (ritual petition for grace from irrigation water), (2) *Matekap (plowing),*(3)*Ngewiwit* and*nandur*,(4) *Ngewiwit* (ritual begins to plant rice), (5) *Nandur* (planting) , (6) *Nyaab Toya* (ritual begging grace from irrigation water, (7) *Biukukung* (ritual when rice starts Contained / bunting), (8) *Mesaba Gede* (ritual at the time of harvest), (9) *Mesaba Alit* (ritual at the time of rice harvest in each field), (10) *Manyi* (harvest), (11) *Nini rice* (Ritual after harvest), (12) *Mebanten Rice* (ritual when inserting rice into barns, (13) Storing to barns.

Traditional post-harvest activities include: After harvest in paddy fields, rice is brought to their respective homes, The tool used is sanan, where the tool is made of coconut wood (seseh) and at the ends are made taper. The way to carry it is to be carried (negen) from rice fields to their homes. Further in the dry to dry milled, cleansed from straw and stored in the barn. Rice will be taken from the granary to taste for food for several days, followed by pounding rice (nebuk) to get rice. The traditional process of pounding rice by using "dimple" as the base of the collision made of stone or wood perforated and "lu" as a tool pounder made of wood, with the end duberi bracelets of iron. When it becomes rice will be cooked traditionally, using firewood in the traditional kitchen.

IOP Conf. Series: Earth and Environmental Science **123** (2018) 012038 doi:10.1088/1755-1315/123/1/012038

3.2. Subak Rice Field Zoning

Subak in Bali other than is as regulator of irrigation system and as preserver of agriculture culture in Bali, it also saves potency as agotourism agri-tourism area) with its natural and cultural potency. The purpose of the existence of agrotourism in addition to supporting economic subak area, as well as a supporter of nature conservation and rice fields Subak.

It is said to be a supporter because nature is the power of tourism subot agrotourism. As the purpose of this research is to accelerate other than that. This research took four subak locations in Sarbagita area (Denpasar, Badung, Gianyar and Tabanan), namely:
1. Kerdung Subak, in Pedungan Village, District South Denpasar, Denpasar City.
2. Cangi Selatan Subak, in Sembung Village, Abiansemal Sub-district, Badung Regency.
3. Gadon 2 Subak, in Beraban Village, Kediri Sub-district, Tabanan District.
4. Erjeruk Subak, in District Sukawati, Gianyar Regency.

Subak alternatives considered for agro tourism are:
1. In Denpasar there are four subak, three in Penatih area (Anggabaya, Umalayu and Umadesa) East Denpasar subdistrict and one subak (Kerdung) in Pemogan Village, Denpasar Selatan District.
2. In Badung regency there are two Subak (Cemagi Let, in Munggu Village and Subak Cangi Selatan in Sembung Village of Mengwi Subdistrict.
3. In Tabanan district one subak (Subak Gedon 2) in Beraban village Kediri district.
4. In Gianyar Regency there are two subak (Petak, Petak Village, Gianyar Subdistrict, and Subak Erjeruk in Sukawati Village) Sukawati Subdistrict

The agro-tourism zoning plan is presented in Figure 2 for urban agriculture agribusiness in Kerdung Subak, Denpasar City. Figure 3 for banana agro-tourism in Cangi Selatan Subak, Badung regency. Picture 4 for Figure 5 for agro-tourism of the sacred area in Gedon Subak2, in Tabanan District. These four subak have excellent accessibility to visit, because they are located in the tourist route: Kuta-Sanur-Gianyar, Kuta-Tanah Lot, Denpasar-Sangeh-Bedugul. Similarly, these locations are designated as green open spaces or areas of agricultural land. The development of agro ecotourism in subak will be able to preserve the nature and agrarian culture. It also triggers the development of an orange economy, which is a culture-based economy and local wisdom. The flowers in the main / sacred zone are indispensable for religious ceremonies, and coconut and banana are always needed for banten (traditional ceremonies and angama).

Figure 2. Map Plan zonasi agro urban tourism agriculture agribusiness Kerdung Subak, Denpasar City

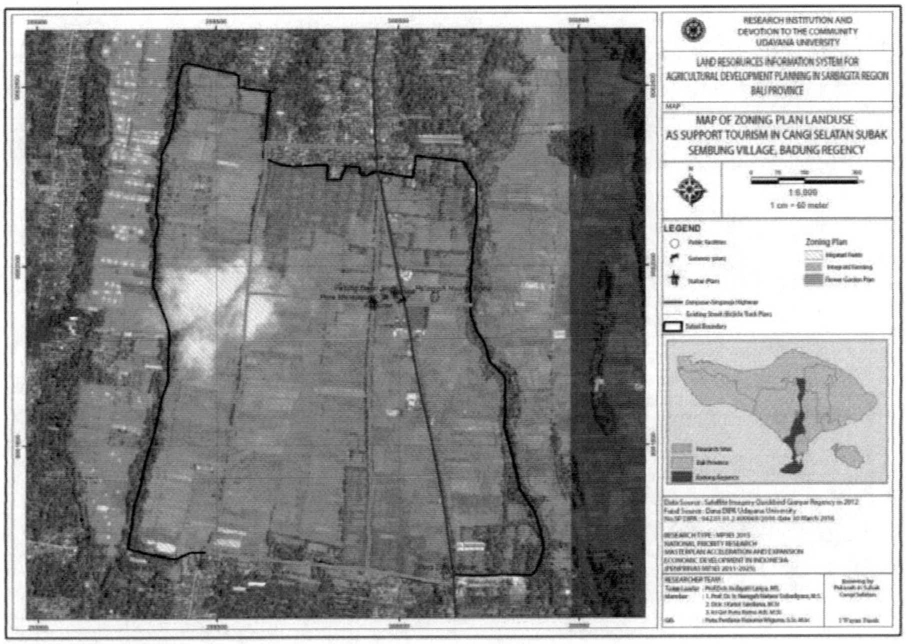

Source: Data Analysis, 2016

Figure 3. Map of zonation plan of banana agro-tourism in Cangi Selatan Subak, Badung Regency

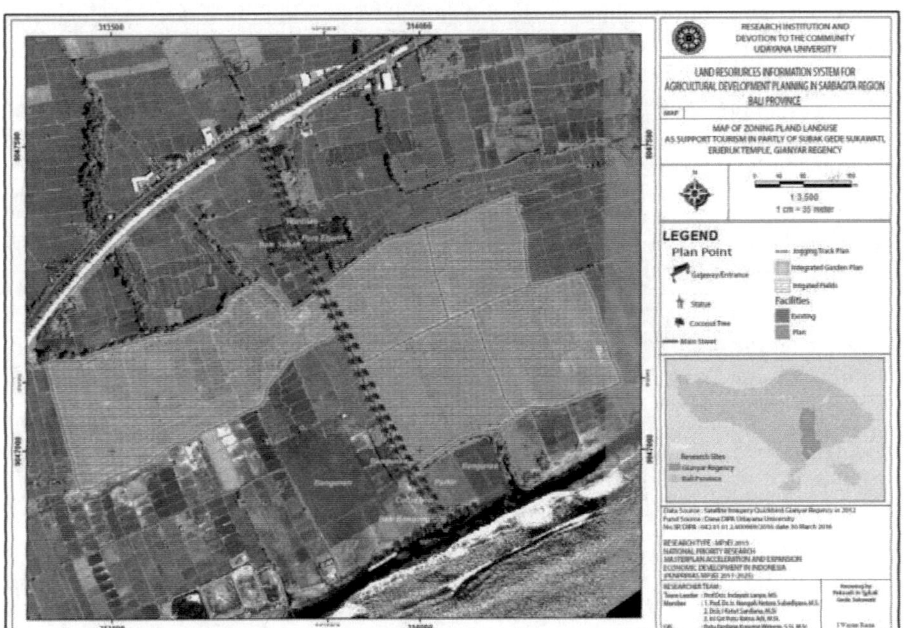

Source: Data Analysis, 2016

Figure 4. Map of Zoning Plan of Coconut Agro Tourism, Subak Erjeruk, Pantai Purnama, Sukawati, Gianyar regency

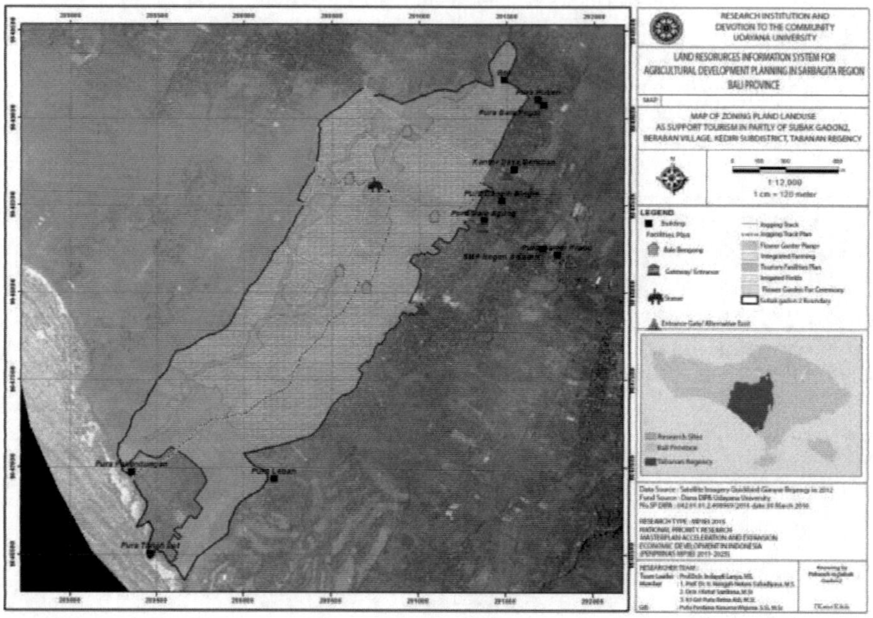

Source: Data Analysis, 2016

Figure 5. Map of Subak Gedon 2 zonation plan for the implementation of Tanah Lot Temple area: main zone, middle zone, and nista zone in Kecamatan Kediri, Tabanan District

Agro tourism for coconut and banana do not yet exist in Bali. Both of these plants are multi-purpose and multi-function tropical plants from all parts of the plant. The coconut ecosystem is growing very well by the beach. Therefore it is very suitable to be developed in Subak Erjeruk (Purnama Beach). Bananas also grow well at <500 m altitude from sea level, very suitable to be developed around Sangen tourism object. Unlike agro ecotourism in Denpasar City, it is necessary to diversify agriculture, in the form of urban agriculture that combines agriculture with hobby. Such as combination of rice field with jogging tract, fishing pond, rice mina, fish pond, flowers and horticultural crops supporting tourism.Vision, Mission, Goals and General Goals Agro tourism in the Sarbagita area in accordance with the results of the study [4,5] are presented below.

3.3. *Vision, Mission, Goals and General Goals Agro tourism in Sarbagita area*

Vision:

"The creation of subak bali tourism object based on farming communities and WebGIS by raising local wisdom, superior, competitive, and increase the economic value of the region"

Mission:

1. Realizing land resource information system (SIDL) and supporting road supporting map of agro tourism in specific location.
2. Increasing the economic value of subak area through increasing production and quality of agro-tourism supporting products based on tri mandala (holy zone, madya, and nista).
3. Preservation of subak as a cultural heritage by increasing the value of environment, agrarian culture and orange economy.
4. Realizing new agro tourism object in Sarbagita area and information based on Web_GIS (www.wisatasubakbali.com) with mengangangkat local wisdom.

5. Interwoven helmet cooperation (subak community, government, academics, and tourism practitioners).

Aim:
1. Establish SIDL and agro-tourism support roadmap on GIS-based location-specific.
2. Increasing the economic value of subak area through increasing production and quality of agro-tourism supporting products based on *tri mandala*.
3. Preserving Subak as cultural heritage duania, environmental value, agrarian social culture, and orange economy.
4. Building a new agro tourism object in Sarbagita area and information based Web_GIS (*www.wisatasubakbali.com)*with raise local wisdom.
5. Interwoven helmet cooperation (subak community, government, academics, and tourism practitioners

Target:
1. SDL documents and land use zoning plans in accordance with agriculture-based tourism potentials of GIS information.
2. Increased economic value of subak region through increased production, and quality of plant flowers (sacred zone), plasmanutpah paddy rice (middle zone) and integrated agriculture: mixture of food crops with livestock (zone nista).
3. Lestariinya subak as a natural conservation area, agrarian social culture, preservation of rice field plasmanutfah, banana and coconut, environmentally sustainable agriculture.
4. The realization of new tourism objects Dewi Sri, banana and coconut plantation, and urban agriculture based on Web_GIS, by raising the local wisdom of Bali (agrarian cultural attraction) with *www. Pariwisatasubakbali.com.*
5. Interwoven helmet cooperation (subak community, government, academia, and tourism practitioners).

4. Conclusion

Potential tourist destination (DTW) in four subak that can be developed in Sarbagita area (Denpasar, Badung, Gianyar and Tabanan) seven subak areas, namely:

a. City Dempasar two subak, namely: Subak Kerdung planned to be developed for Agroekowisatawisata Tanaman Pangan dan Hortikultura located in the route Kuta / Nusadua-Sanur / Denpasar).
b. Badung regency two subak, namely: Subak Cangi Selatan planned: Agrowisata Plasmanutfah Tropical Banana, Sangeh –Bedugul.
c. Tabanan District Subak, Subak Gadon 2, Tanah Lot Area Tabanan regency is planned to develop agriculture culture agriculture Subak system for Tourism Object Various Types of Local and Excellence of Rice to Add Tanah Lot attractions plus).
d. Gianyar Regency two subak, namely: Subak Erjeruk, Purnama beach planned to be developed for coconut plasmanutfah attractions.

5. Acknowledments

Thanks To the Rector of Udayana University, which has provided the opportunity in the conduct of research and support in this publication. Similarly to Mr. Chairman Institute for Research and Community Services and to Mr. Dean of the Faculty of Agriculture, University of Udayana we say thank you, which provide the opportunityfor the research. We thank the Research Executive Team and those who influence helped this research.

6. References

[1] Lanya I, Subadiyasa N N, Sardiana K and Adi G P R 2015 Numerical classification, Subak zoning and land transfer function rice field in the Province of Bali based on Remote Sensing and GIS *Procedia Environ. Sci.* **24** 47–55

[2] Lanya I, Subadiyasa N and Hutauruk J 2017 subak land information system based on remote sensing and geographic information system in denpasar city *Geoplanning J. Geomatics Plan.* **4**

[3] Roth D 2014 Environmental sustainability and legal plurality in irrigation: the Balinese subak *Curr. Opin. Environ. Sustain.* **11** 1–9

[4] Lanya I, Subadiyasa N N, Sardiana K and Adi G P R 2017 Transfer function control strategy of Subak rice field land and agricultural development in Denpasar city *IOP Conference Series: Earth and Environmental Science* vol 54p 12004

[5] UNESCO 2012 *Subak System as Manifestation of the Tri Hita Karana Philosophy*

Riverine Settlement Adaptation Characteristic in Mentaya River, East Kotawaringin Regency, Kalimantan Province

L Esariti[1], N Yuliastuti[1], and N K Ratih[1]

[1]Department of Urban and Regional Planning, Diponegoro University, Semarang

Email: adoeng@gmail.com

Abstract. The research looks for the importance of riverine settlement preservation as one of the efforts to carry out sustainable management of a traditional settlement. East Kalimantan, more specifically riverine settlement in Mentaya River is known as one of the traditional settlements that put river as the central of their livelihood activities. The theory of Rapopport [1] was used to investigate the importance of cultural aspect influence to the settlement process, and to seek for the behavioural and environment relationship in determining the pattern of adaptation process. Mix method approach was conducted by utilizing in depth interviews among 40 respondents within three districts, namely in Baamang, Mentaya Seberang and Mentawa Baru Ketapang subdistrict. The result shows that culture dominantly affect the process of settlement adaptation, especially the aspect of family structure, social network, and kinship. The adaptation pattern is influenced significantly by physical environment, type of physical condition of the houses, economic condition and the degree of heritage preservation motive. It sums up that adaptation process could be effective if the integration of culture, activities and government regulations is performed.

Keywords: Adaptation, Riverine Settlements, Behavioural and Environmental Interactions

1. Introduction

The word of sustainability is nowadays labelled in a planning and design cycle to reflect the effectiveness of the program. In Indonesian context, sustainability has become one of three pillars of national development, together with the gender and the good government pillars. Budihardjo and Sujarto [2] noted that sustainable development as a process of change where resource utilization, investment direction, development orientation and institutional change are always in balance and in they were reinforced to meet now and then human needs and aspirations. In many development applications,the most important things of setting the sustainability is by designing indicators to measure in which stages these process is conducted. Mc Larren [3] defined the concept of sustainable development as a process of achieving a set of desired conditions. It could be the condition of equity, natural environment protection, economic and social welfare, and also community well-being.

Mentaya riverine settlement located in the riverbank of Mentaya River of East Kotawaringin Regency in Kalimantan. According to its historical background, the beginning of growth of Sampit City has been started since Banjar Sultanate era in 1500 AD and progressed to become an urban area after the emergence of two largest wood and rubber processing factories in Kalimantan, namely PT NVBDH (Brengsel Factory) and PT Sampit. In the early days of the kingdom, people opened up land for gardening and extracted natural products as well as built houses to keep the land. At that time the

IOP Conf. Series: Earth and Environmental Science **123** (2018) 012039 doi:10.1088/1755-1315/123/1/012039

Sampitcity is still a village by the river with a population of about 500 people. After occupying the area for a long time and feeling the life is getting better, they began to invite families and relatives from their home region (Banjarmasin) to settle in the new city. This leads to the development of riverine settlement.

At the time of Dutch colonizationin the mid-19th century, the urban growth embryo began to form in accordance with the development of seaports and road frames. The Dutch designed the roads and pathways by linear pattern and surrounded by settlement which is grouped according to ethnicity (Chinese, Javanese, Dayaknese, Dutch and other immigrant ethnicity) In the post-independence period, the Sampit Urban Area became bigger because of the construction of the Brengsel and PT Sampit timber plantations as well as several other timber companies in areas that revive economic activity in Sampit City. The settlement development is also influenced by cultural and religious values, both in social relations and in the design of houses and residential areas. These historical remains can be found in the form of buildings, dwellings, and infrastructure such as the *RumahBahari*, the City Park, the Evangelical Church, the Brengsel Plant, the harbours, markets, and the main road network on the banks of the Mentaya River

Adaptation is the process and the way in which it adapts to its environmental conditions [4]. The purpose of adaptation is for people to live harmoniously and well-integrated with their environment. The river as the philosophy of life and the orientation of the Dayaknese (indigenous Kalimantan) became one of the characteristics of settlements on the banks river. In other words, the adaptation that occurs can be seen from two aspects. First, adaptation to environmental conditions; how the characteristics of river edge activities can be accommodated. Furthermore, adaptation to the condition of their house; how is the settlement process form a spatial organization pattern that makes residents feels at home. Finally, efforts were made to achieve ideal conditions of the house and the desired environment to reach their settlement sustainability. Brocklesby and Fisher [5], Widodo [6], Diem [7] stated that the relationship between behaviour and environment are in the settlement process. These literatures identify the importance of meaning to the cultural aspect and see this relationship as a unity of the environmental quality system. Such interactions are understood as adaptations.

In a traditional settlement and riverine especially, the most important aspect of the sustainability is the process of adaptation and focuses on the harmonious relationship between man and nature. Previous studies did not focus on the human environment relationship in regards to traditional riverine settlement. Rather than, seeing riverine settlement generally related to climate change and natural resilience context. This study aims to identify and to analyse the characteristics of adaptation of settlement of Mentaya River is as effort to support the sustainability of traditional settlement of Kalimantan river bank. Adaptation becomes the main analysis to be discussed, since adaptation will show the processes, motives and ways to settle for the riverine settlements in order to maintain its livelihood and its efforts to improve the quality of life[8][9][10]. The research focuses on the mixture of indicator of natural environment protection and individual wellbeing in the adaptation of traditional settlement. It is undoubtedly that culture plays an important role in the settlement, as culture contributes to the meaning of identity and the characteristics of a settlement [1][11][12]. We argue that sustainability of a riverine settlement will be effective if in the adaptation process involves the inetgration of culture, activities and government regulations. This paper therefore explores the livelihood adaptation process in Mentaya riverine settlement has revealed the influences of behaviour and environment relationship. It contributes to current debates on riverine settlement livelihoods adaptation by emphasizing on the process, motive and adaptation technique conducted by the population.

2. Data and Method

The mixed method approach was conducted in this research, which qualitative approach was used more dominant than the quantitative one. The exploratory comparative analysis identifies the process, motive and adaptation technique of Mentaya River settlement to the condition of house and environment. Another analysis included a superimpose map analysis to determine the characteristics of riverine settlement pattern.

The data presented in this paper were derived from semi-structured interviews among 40 respondents located in three sub districts, namely Baamang, Mentaya Seberang and Mentawa Baru Ketapang corridors. The in-depth interview technique is utilized by purposive sampling, which is determining criteria of respondent based on:

- Living in a house distance of 0-20 meters from the Mentaya River.
- Residents of riverbank settlements aged 15-60 years.
- Already settled for at least 10 years

Interview guideline was divided into 3 major sections that fit the research objectives, answering (a) Mentaya river traditional settlement characteristic, (b) process, type and motive of their settlement adaptation and (c) occupants expectation to their settlement sustainability.

3. Results and Discussion

3.1 Respondent Profile

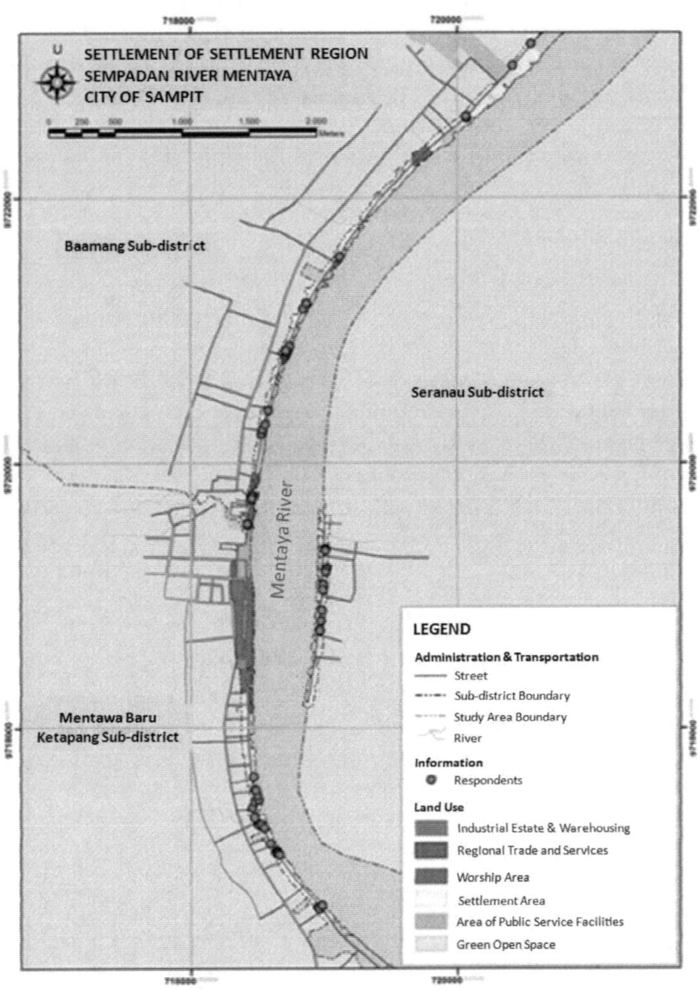

Figure 1. Map of Respondent Distribution Location

The profile of the respondents describes the economic condition, family structure and home occupation process. Based on the interview, it is depicted that in general, the family structure is traditionally formulated. For example the head of the household is still dominated by man (58%). He is also the only breadwinner in the family (73%). Most of the respondents mentioned that, they inherited the houses from their parents and occupied by nucleus family and the extended family. The ownership then transferred into self-titled (95%) which guarantee the ownership of the house belongs to the nucleus family.

The educational background shows that the level of education is in the medium level, the majority are graduated from senior high school (73%) and followed by graduated from universities (18%). From employment conditions, many of the income earner works at informal private sector (40%) and government employee (20%). Almost all the households being interviewed are originally native (88%) inhabitants and have long settled in the houses for more than 30 years.

3.2 *Characteristic of Mentaya Riverine Settlement*

3.2.1 Natural Conditions

Discussion of these natural conditions includes hydrological, geological, and topographic conditions in Sampit City. Hydrological conditions or surface water in the city of Sampit is flooded by a large river of Mentaya River which empties into the Java Sea. Mentaya River flows from north to south of Sampit City, has an average width of 400 meters with a river length of ± 400 km and an average depth of 6 meters. The settlement along the Mentaya River is located in lowland with altitude of land at Corridor I Baamang of 32 mdpl, Corridor II Ketapang 34 mdpl, and Mentaya Seberang Corridor of 3 mdpl. The soil types found in the suburbs of the Mentaya River are podsolic and organosol (peat) which reflect the suitability for built environment.

3.2.2 Infrastructure Condition

The basic needs of infrastructures in the Mentaya Riverine settlement are complete because of the availability of electricity, clean water, road network, drainage, telecommunication network, and other public facilities such as shops, houses of worship, educational facilities and transportation facilities. However, the presence of public open space and green belt area is still less prevalent in some settlement areas so that residents take advantage of the house front or the edge of the road for a place to socialize. Other basic needs that are still lacking are waste management services. Based on the observation, the residents process their waste conventionally, by burning the waste piles. It is due to the unavailability of communal trash in their environment. This is quite disruptive because if the garbage is not properly managed it can adversely affect the river environment. The emergence of waste piles could trigger floods, create unpleasant odors, leads to slum, and disrupt the smoothness of water transport.

In accordance with Sampit Urban Spatial Plan Year 2016-2036, the research location is designated as a Regional Activity Centre[13]. This means that the location of the settlement is strategic because it is close to the municipal service centre. The proximity and the completeness of the infrastructure provided, also the quality of the infrastructure building condition is at a sufficient level to perform its services as the activity centre (Pusat Kegiatan Wilayah).

3.2.3 Settlement Pattern and Conditions

Mentaya riverine is an unplanned settlement with patterns following the circulation of roads and rivers. It is a linear pattern filled with a regular grid settlement. The regular shape is due to the flat topographical conditions, so that the road growth becomes parallel with a uniform shape. The design of the house is still traditional, and constructed from ironwood. Most of the houses directly adjacent to the river and it is located on the river body (in this case located on the east side of Baamang and Ketapang Corridor) are pillar house. On the other side of Mentaya River, the majority of the construction is landed houses. The house is above the river oriented back to the river, while the house opposite the land oriented facing the river. Observation reveals that generally the houses in Mentaya river settlements are pillar houses (*Rumah Panggung*). Only few of them are raft houses (*Rumah Lanting*), which is almost all of them was abandoned.

The difference between pillar houses and raft houses is on the nature and development [14]. The nature of the raft houses is to be used for economic reasons. Initially as the river is for major accessibility of the population, they use the raft houses for boat moorings. Later on, it has been used for houses, stall/ shop and for fish cages. On the other hand, the pillar houses are originally used for a landmark for Kalimantan Province. It is popular to be named *Rumah Betang*, where the major construction is made from ironwood, which has special feature as an anti-termite material. According to Dayaknese belief, it has to be built facing east (the direction of sunrise) with the stern face to the west (the direction of the sunset).It is considered a symbol of hard work to survive from sunrise to sunset (Source: Interview, 2017).

Settlement in Mentaya Seberang

Raft houses (*rumah lanting*) in Mentaya Seberang

Settlement in Baamang Hulu

Pillar houses (*Rumah Panggung*) in Ketapang

Traditional house called *Baanjung*, architectural design is combination from Dayak, Banjar and Dutch culture

Figure 2. Type of houses in Mentaya River Settlements

IOP Conf. Series: Earth and Environmental Science **123** (2018) 012039 doi:10.1088/1755-1315/123/1/012039

3.3 Process, Type snd Motive Of Mentaya Riverine Settlement Adaptation

The process of the adaptation is accelerated by heterogeneity of characteristics of economic activities from the population[12], in order to enhance individual and collective abilities in improving the quality of life. Initially the land was bought and the house was built by the great grandfather. The division of space is based on the number of family members and economic ability. Usually the number of rooms within a house is designed based on extended family context. The minimum number of rooms are 7, namely 1 living room, 1 family room, 3 bedrooms, 1 room for kitchen and dining room, and 1 bathroom / WC is allocated outside the house (they called it *Batang*).

The housing design is to divide the public and private area by putting the public area in the front of their house, as a welcoming area. And the private area is connected to service area in the back of the house. Between the public and private area is separated by a blind curtain shows the sense of trust between family members. Following the development of the family member to be more independent, several of the children moved out to the houses near to their workplace. There is also belief that a family should survive and sustain by themselves. Together, they can solve and ease all the problems arose.

For more than 30 years living at home, no renovation has ever been done. Most of the households mentioned the lack of funding as one of the factors not doing renovation. However, the most common renovation is done on the roof, which replaces the roof material from shingles (*sirap*) to tin roof.

This kind of renovation is sometimes followed by strengthening their house platform by checking the durability of the wood against the water erosion. Another type of adaptation is by preserving their first houses as the house preserved and is enriched with their family history. Although many of them moved out from the first houses, but they keep on maintained the originally of the housing design, which was dominantly influenced by Dutch and Dayaknese architectural design. The motive is to preserve cultural architecture value as their settlement identity.

To enhance livelihood, many of the households decided to stay at the riverine along with their nucleus family. Their motive is to stay close with their ring of support [7,10,11]. Several adjustments to their houses arrangement has been made to adapt to the level of the development of the family needs. For example, extending the room space to welcome a new born baby, changing the room usage to improve the economic performance by having a home based enterprises such as tailor, food stall and barbershop. These activities are as adaptation motive of putting their home as an investment material. In this context, home is seen as a commodity to improve occupants' economic status.

On the other hand, adjustments also have been made to their environment. The concept is to raise their comfortability and having a harmonious relationship of human and environment [4]. The interview result reveals several of adaptation types conducted by population such as:

- Moving out trends to the lowlands. As family getting bigger, each nucleus family wants to have their own independent household authority. They move out and directed to the lowland, avoiding the riverine activities agglomeration. They understood that household activities could emerge physical burden for the river preservation.
- Learning, conducting and preserving positive behaviour towards environment management, such as no littering habit, collaborative unpaid work (*gotong royong*) for maintaining environmental hygiene, planting shade tree to improve home comfort and so forth.
- Maintaining good mutual relationship with the neighbour. For example: forming a collection watch among residents towards their security of the property. As it is evidenced by statistical record of criminal activities happened due to the increasing number of immigrants occupying the Mentaya River.

- Inhibiting the further occupation of raft houses (*Rumah Lanting*) for safety reason. This is also intended for slowly reducing the slum characteristic caused by unmanaged physical appearance of riverine settlements.

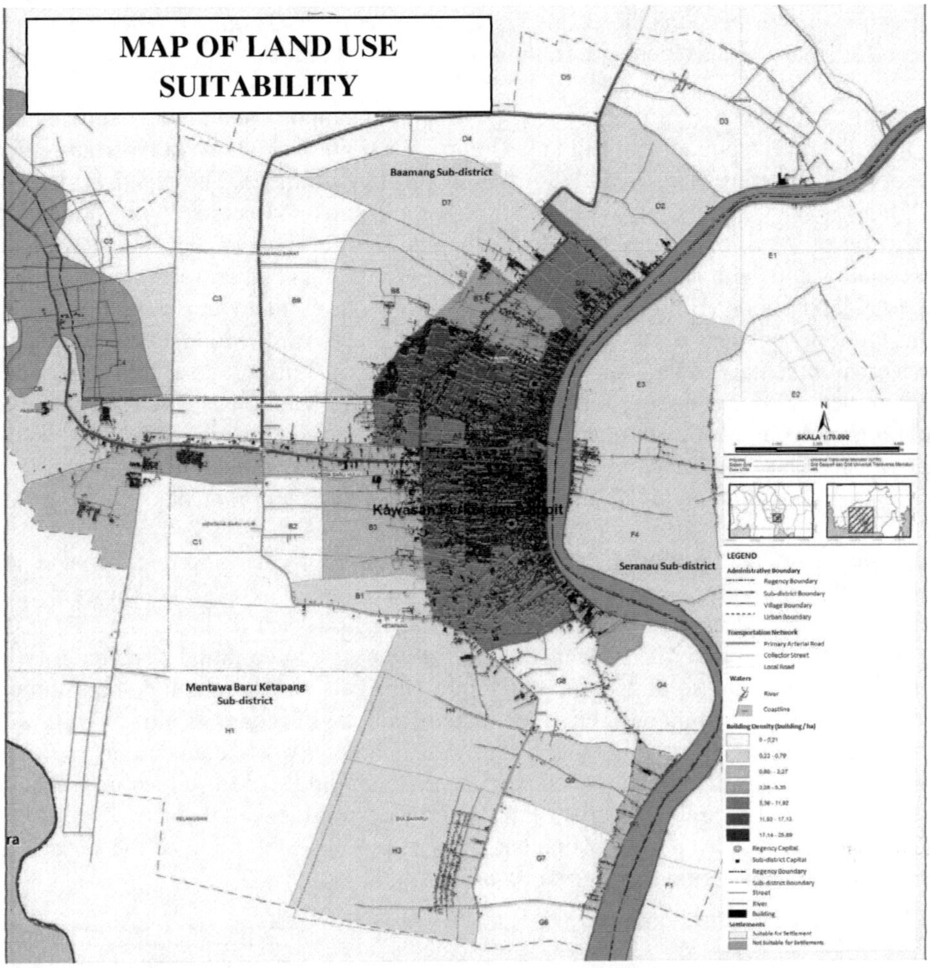

Figure 3. Map of Overlay of Building Density Map and Area Development Trend

3.4 Mentaya Riverine Settlement Sustainability

The settlement pattern is formed by at least six elements [11], namely (a) ownership security, (b) accessibility, (c) privacy, (d) social interaction, (e) infrastructure and (f) harmonious relationship between human and environment[11]. Evidence shows that security of tenure develops the relationship between residents and settlement. The security tenure protects the improvement of the property value, including also culture and traditional value of the home design. The knowledge and local wisdom of using their native building material such as ironwood proved to be sustained and extend the durability of building construction period. The privacy of able to decide based on family interest put them to have more social bonding between nucleus and extended family. This improves their quality of live in performing collective capability.

Physically, as it can be seen from the overlay map analysis result, the development of the riverine communities gained governmental support. Based on the Spatial Plan Regulation found in SIPP Year 2015 document, the spatial structure and detail development is directed to follow its organic formation alongside the Mentaya River[13]. The map also shows that river is still dominantly used as the main orientation of economic activities in the city of Sampit. SIPP also as law abided regulation that suggests the local government to establish infrastructure provision suitable with the characteristic of the Mentaya riverine settlement. For example, policy needed for several traditional preserved houses such as the ownership protection, the detail regulation (*Petunjuk Teknis*) for heritage conservation actions, integrating tourism activities with traditional attractions features.

The connection between main activities, government support and culture proves to be key ingredients to maintain livelihood[15][16][17]. The main activities of Mentaya riverine settlement is a traditional oriented activities as it depends on the natural environment. The physical development is regulated within the carrying capacity of Mentaya River and surrounded area, while social activities is encouraged to be preserving of cultural value into way of life. The local wisdom to always have harmonious relationship with natural environment proven to make them always feels comfortable living in the neighbourhood. Bounded with the location requires them implementing good habit to maintain the environment and always integrate economic activities with the characteristics of the region as an environmentally friendly riverine settlement area. Furthermore, the community expect that the government will has to be consistent in performing settlement liveability. For example: various programs of economic development in tourism should be integrated and be a focus of Mentaya river development. This can be done by promoting river tour (*Wisata Susur Sungai*), religious attractions (*Safar / Ramadhan Festival*) and garden tour (*Wisata Kebun Rakyat*).

It is clear, that in order to perform livelihood; that is the feeling of being comfort, safe and secure in the riverine settlement, human capital, natural capital and social capital should be performed as key factors in the settlement adaptation process [18]. Human capital such as local knowledge, skills and capacity to adapt has been proven to influence the effectiveness of livelihood sustainability in term of capability to create practical solutions for economic survival. Natural capital that is the physical characteristic as a river has maintained settlement activities to be in accordance to its natural limitation and advantages. Finally, social capital namely family structure, network, connections (neighbourhood, kinship), relation of trust and mutual understanding and support, shared values and behaviours are dominant features that enhance livelihood sustainability. It also evidenced that the adaptation pattern is influenced significantly by physical environment, type of physical condition of the houses, economic condition and the degree of heritage preservation motive.

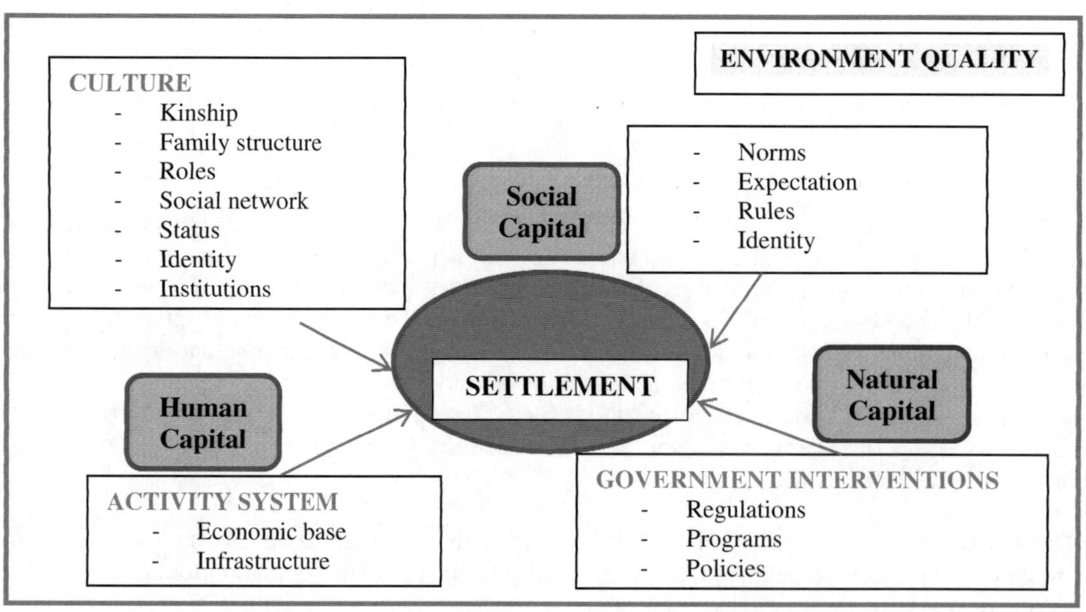

Figure 4. Environment Behaviour Relation in Human Settlement, Modified from Rapopport 2010

4. Conclusions

This research confirms that human environment relationship is needed to be maintained in mutual way to perform settlement sustainability. The result also reveals that adaptation process should be done simultaneously inside the house and in the environment. As a process, it is conducted by learning by doing approach. Exploring local wisdom and cultural value is one of the successful factors for Mentaya riverine settlement adaptation. Furthermore, positive behaviour from the residents and collaborative works between neighbourhoods should be preserved as the main cultural value that keeps the adaptation process harmoniously.

5. Acknowledgement

The authors would like to express gratitude for the Basic Research Grant from Engineering Faculty of Diponegoro University for their financial support in conducting this research.

6. References

[1] Rapoport A and Rapoport A 2010 Theory , Culture and Housing Theory , Culture and Housing 37–41
[2] Budihardjo, Eko and Sujarto D 1999 *Kota berkelanjutan* (Alumni Bandung)
[3] Maclaren V W 2007 Urban Sustainability Reporting *J. Am. Plan. Assoc.* 37–41
[4] Rapoport, Amos; Altman I W F J 1980 Volume 4 Environment and Culture *Environment and Culture*
[5] Brocklesby M a. and Fisher E 2003 Community development in sustainable livelihoods approaches - an introduction *Community Dev. J.* **38** 185–198
[6] Diem A 2012 Wisdom of The Locality (Sebuah Kajian : Kearifan Lokal dalam Arsitektur Tradisional Palembang) **2** 299–305
[7] Widodo J 2012 Urban Environment and Human Behaviour: Learning from History and Local Wisdom *Procedia - Soc. Behav. Sci.* **42** 6–11
[8] Taylor J 2015 A tale of two cities: comparing alternative approaches to reducing the vulnerability of riverbank communities in two Indonesian cities *Environ. Urban.* **27** 621–636
[9] Saragih, S. J L dan A R 2007 Kerangka Penghidupan Berkelanjutan Sustainable Livelihood Framework
[10] Surjono 2011 Lessons Learned from Traditional Settlement Patterns for Housing *J. Basic Appl. Sci. Res.* **1** 759–763
[11] Pacione M 2011 Applied Geography : Principles and Praxis vol 28pp 7–28
[12] Karlsson M and Bryceson I 2016 Continuity and change: understanding livelihood shifts and adaptation in coastal Belize 1830–2012 *Local Environ.* **21** 137–156
[13] Sampit B K 2015 *Sistem Informasi Permukiman Perkotaan Kota Sampit*
[14] Rijanta R, Setiawan B, Marfai M A, Geografi F, Geografi J, Geografi F, Mada U G, Geografi P J, Geografi F, Mada U G, Jurusan P, Teknik F, Gadjah U, Kota S and Pahandut K 2014 Urban Riverside Settlement Model Case : Kahayan Riverside Settlement , Palangkaraya **9** 17–27
[15] Dovey K and King R 2011 Forms of informality: Morphology and visibility of informal settlements *Built Environ.* **37** 11–29
[16] Sasongko I 2005 Pembentukan Struktur Ruang Permukiman Berbasis Budaya (Studi Kasus : Desa Puyung - Lombok Tengah) *Dimens. Tek. Arsit. Vol. 33, No. 1, Juli 2005 1 - 8* **33** 1–8
[17] Stedman R C 2002 Toward a Social Psychology of Place *Environ. Behav.* **34** 561–581
[18] Serrat O 2017 Knowledge Solutions 21–26

IOP Conf. Series: Earth and Environmental Science **123** (2018) 012040 doi:10.1088/1755-1315/123/1/012040

A Socio-spatial Dimension of Local Creative Industry Development in Semarang and Kudus Batik Clusters

P Nugroho[1]

[1]Department of Urban and Regional Planning, Diponegoro University,Indonesia

Email: prihadi.nugroho@pwk.undip.ac.id

Abstract. Creative industries existence is inseparable from the underlying social construct which provides sources for creativity and innovation. The working of social capital in a society facilitates information exchange, knowledge transfer and technology acquisition within the industry through social networks. As a result, a socio-spatial divide exists in directing the growth of the creative industries. This paper aims to examine how such a socio-spatial divide contributes to the local creative industry development in Semarang and Kudus batik clusters. Explanatory sequential mixed methods approach covering a quantitative approach followed by a qualitative approach is chosen to understand better the interplay between tangible and intangible variables in the local batik clusters. Surveys on secondary data taken from the government statistics and reports, previous studies, and media exposures are completed in the former approach to identify clustering pattern of the local batik industry and the local embeddedness factors which have shaped the existing business environment. In-depth interviews, content analysis, and field observations are engaged in the latter approach to explore reciprocal relationships between the elements of social capital and the local batik cluster development. The result demonstrates that particular social ties have determined the forms of spatial proximity manifested in forward and backward business linkages. Trust, shared norms, and inherited traditions are the key social capital attributes that lead to such a socio-spatial divide. Therefore, the intermediating roles of the bridging actors are necessary to encouraging cooperation among the participating stakeholders for a better cluster development.

Keywords: Creative Industry, Socio-Spatial Dimension, Batik Cluster, Semarang Municipality, Kudus Regency

1. Introduction

Creative industries have emerged as the key driver of the modern economy for cities and regions to increase competitive advantages in pursuit of economic growth and social welfare[1–4]. According to United Nations Development Programme (UNDP) and United Nations Educational Scientific and Cultural Organization (UNESCO) [5], world trade in creative goods and services doubled to US$624 billion during 2002-2011 with the annual growth rate 8.8 percent on average. Foord [2] observed that the annual growth rates of employment in creative industries in some world cities growing faster than those in any other sector, i.e. 7 percent (1998-2002) in New York, 5 percent (1995-2000) in London, 5.7 percent (1996-2003) in Amsterdam, 6 percent (1999-2003) in Vienna, 22 percent (1999-2001) in Glasgow, 13.4 percent (1986-2000) in Singapore, and 17 percent (1999-2003) in Shanghai. In Indonesia, it shared about 7 percent to gross domestic product (GDP) engaging more than 10 percent employment during 2010-2013 [6,7].

IOP Conf. Series: Earth and Environmental Science **123** (2018) 012040 doi:10.1088/1755-1315/123/1/012040

Creative industries are often entangled with competitiveness following a series of economic development shifts from ahigh dependency on natural resources endowment and material possessions to human capabilities, specializations and innovations [8]. The importance of creative industries has increasingly gained its popularity since the 1990s in line with competitiveness campaign of Porter's diamond model worldwide which promotes the increase of productivity and innovation through industrial clustering[9,10]. In fact, creative industries discourse originated in 1994 when the Australian Government publicly released new cultural policy "Creative Nation: Commonwealth Cultural Policy" proposing an AUS$250 million fund to support cultural institutions for wealth creation; and then followed by the United Kingdom Department for Culture, Media and Sport (DCMS) which launched "Creative Industries Mapping Document 1998" suggesting the important role of individual talents as a potential source of wealth and job creation through the generation and exploitation of intellectual property [1,11,12]. Later, it spread rapidly to numerous developed countries in Europe and the United States [11,13] and developing countries in Asia [14,15].

Despite the growing attention (mostly) from government agencies in implementing the concept, there remains a broad range of pros and cons in defining and measuring it. On the conceptual ground, the debates on creative industries embark a problematic marriage of culture with economic activity. In many literature the term creative industries is interchangeably used with cultural industries i.e.[13,16], and is often linked with creative class e.g.[17,18], creative cities e.g.[19,20], creative or knowledge economy e.g.[16,21] and industrial clustering e.g.[22,23]. Referring to Department for Culture, Media, and Sport (DCMS) [24], the definition of creative industries is:

> *"Those industries which have their origin in individual creativity, skill and talent and which have a potential for wealth and job creation through the generation and exploitation of intellectual property."*

Those include 13 types of industries, i.e. advertising, art and antique market, architecture, crafts, design, designer fashion, film and video, interactive leisure software, music, performing arts, publishing, software and computer services and television and radio. Creativity as the main ingredient of the concept serves as a bridge between the culture and economy, where both have brought opposite values in nature [13,14]. Talking about culture, it is strongly associated with a set of customary values, beliefs and thoughts representing shared identity and common practices of determined conduct in a society to which any attempts to sustaining, protecting, developing, reproducing and preserving it are likely to further locally based heritage and civilization over generations. Thus, culture accounts for a differentiated product and process of articulating symbolized meanings of particular ways of life manifested in a diversity of race, ethnicity and religion. In this sense, culture is supposedly taken for granted as a foundation of economic activities [25].

On the contrary, economy deals with the process or system of producing, distributing and consuming goods and services through the efficient use of capital accumulation for the sake of wealth creation. It constitutes (mostly) profit-taking commercialization of resources in a society through various channels of market mechanisms by which individuals and interest groups may obtain better-off living conditions. The contradictory conception of creative industries rests on the incorporation of cultural activities and events into a market economy where monetary transactions create disruptive replacement over the sacred admiration of cultural practices. *What you pay is what you get.* As a result, the commodification of cultural institutions is simply transferable by the value of money to meet consumer satisfaction gained from the cultural attractions and events instead of promoting the intrinsic values of the culture itself [26–28].

Such conflicting seedbed has contributed to a vague and ambiguous rather than a rigorous definition of the notion of creative industries. Regarding the DCMS [24] definition of creative industries mentioned before, two key attributes of creativity are surrendered to skilled and talented individuals and intellectual property. This definition is too broad because any activity which involves human creativity could be considered as part of creative industries [13,29]. The boundaries of definition may confine an artistic expression to a scientific-technical process, an imaginative man-made creation to a machinery-based technology adoption; an intuitive tacit knowledge mastery to a codified knowledge internship; and a subjective symbolic meaning purpose to an objective practical use. All these attributes are packed into a single thing called innovation as a product of creative processes [27,30]. Despite the growing confusion in taking up the notion into the practice, some key elements used to identify the creative industries consist of creativity, intellectual property, symbolic meaning, use value, terminological clutter, and production methods [25,29]. These could help the authorities in classifying sectors that fall into creative industries grouping which is necessary to assess its contribution to the economy.

Yet, the story does not end up here since there remains problematic measurement of the effects of the identified creative industries to the economy and the rest of society. The first problem relates to the isolation of certain manufacturing sectors to be categorized whether as part of creative industries grouping – for example, advertising, publishing and broadcasting (television and radio) – or simply 'conventional' manufacturing sectors such as petroleum and chemical production, textile production and garments and food and beverage industry. There are some difficulties to make a clear cutting demarcation between these two as exemplified by the intersection between fashion industry which belongs to creative industries and textile production and garments industry which belong to the mainstream manufacturing industries. Secondly, the economic contribution of creative industries – and so does to the society – actually is measurable at the firm level even though the creativity stems from (skilled and talented) individuals. Direct effects of the creativity-led firms could be measured in the economy through employment size, firm activity, gross output and export variables [11]. This does not fit to individual art performers or sculpture creators, for example, since their earnings could not be considered as a basis for their contribution through taxation or any other levies. In many cases, their earnings collected from cultural events are not the primary source of income; rather, their creative activities are just a matter of hobbies and/or voluntary activities away from profit-making purpose. Similarly, some public and philanthropic organizations may undertake various exhibitions in museums, galleries and convention halls at affordable ticketing prices but, again, these events are not intended to maximize profits. So, how their contribution to the economy could be measured precisely?

The third problem is associated with the intellectual property right of the creative industries operations. Even though it is useful to protecting the creativity process either at the individual or firm levels, the law enforcement to ensure it is quite ineffective. In the case of computer software, for example, the legal purchase of original software products with high-tech security system still cannot cut-off the production and distribution of imitated and fraudulent products. In the case of fashion design, the creators cannot claim for copyright products of their own entirely if the market still provides welcoming access to imitated low-quality products. As long as there are many consumers of such illegal products, the supply-demand law of the market economy still exists. Last but not least, poor data recording and consistency on the governmentside, in particular, has led to inaccurate statistics of the creative industries profile in the overall economy. Registration of business permits is not well-organized, particularly due to the lack of reliable documenting system, qualified human resources, and stringent law enforcement. On the business players' side, the existing business climate does not always provide a convenient environment for business continuation. Some external factors like free trade regime and bilateral and multilateral agreements may impede the opportunities for small and medium enterprises (SMEs) and individuals to producing particular creative goods and services. In addition, the internal factors associated with knowledge and technology acquisition and organizational capacity do contribute in limiting upscaling process and market penetration. The mixture of all these factors has created obstacles in reporting updated business activities.

This paper aims to examine how a socio-spatial divide contributes to the local creative industry development in Semarang and Kudus batik clusters. Dealing with the wicked problems surrounding the measurement of the effects of creative industries operations to the economy and society, I will argue for the importance of localized social construct in determining batik industry development. It is argued that the localized social construct is not static; rather, it is fluid to create a socio-spatial configuration of the prevailing business environment which encourages typical competition and cooperation among the firms and individuals engaged in the creative industries. The different social construct is likely to result in different path dependence of the creative industry development. The key social capital attributes, i.e. trust, shared norms and inherited traditions, therefore determine the formation of socio-spatial divide where the prevailing social construct takes place.

Most previous studies have paid much attention to the outputs and outcomes of the creative industries towards the economy and society. Inevitably, this tendency has attracted many scholars, industrialists and policymakers due to the very promising benefits of the notion implementation for increasing industrial competitiveness as well as economic growth performance at the local to the global levels. All these concerns are condensed in the policy making real politic. This study focuses on the opposite direction and the role of social capital in the creative industry development. In fact, there is a close relationship between social capital and creative industries. Clare [31] suggested that specific territorial localities enable concentrations of people, resources, and infrastructure. Thus, the embeddedness of social capital in local regions encourages the nurturing of social networks necessary to stimulating creativity and innovation. Similarly, Bontje & Musterd [32] highlighted the development of attractive places is as important as the creative industries themselves to provide the creative knowledge workers, or the creative class in Florida's terminology [17,33,34], a liveable (mostly) urban spaces for acquiring, reproducing and transmitting skilful knowledge and technology adoption useful to new innovations. The location specific creative industries create a selective confinement of recursive innovations rather than contagious effects of replicative innovations ubiquitously [35–37]. Therefore, the better understanding of the location specific creative industries will bring forward heterogeneity of the local culture and identity instead of polarizing the products of creative industries into a single direction of performance.

2. Data and Methods

2.1. Data Collection

This study examined a set of data related to the characteristics of batik industry clusters, i.e. the concentration of batik firms and their spatial networks, batik firm size and production scale, and the common practices of knowledge sharing as well as innovation processes, including responsible actors and their roles in enabling such transmissions. This data was used to depict various forms of physical, economic, and social linkages between the batik firms and within the batik clusters. Meantime, the data related to social capital were observed through the enactment of determined rules and guidelines, customary behaviors and attitudes, actors' positions and influential roles within the industry and society, the patterns of competition and cooperation between firms and the forms of collective actions sustained by the firms and society. From this sort of data, we could capture the underlying institutions which encourage the clustered batik firms to exist, interact, and develop individually or collectively at the society level. Following the idea of 'sticky places in the slippery space' from Markusen [38], some locations could be so attractive than the other places for individual firms to co-locate because of the available business environment that has provided structural incentives for the clustered firms to maximise the benefits of external economies. Such a sticky place has performed a dynamic transforming production space resulted from a complex institutional milieu comprising corporate strategies, industrial structures, profit cycles, state priorities, and local and national politics which have involved both endogenous and external forces.

The data collection was completed in three months starting from May 2017 until July 2017. Surveys on secondary data taken from the government statistics and reports, previous studies, and media exposures were completed to lead and enrich the interview phase. Unfortunately, I could not gather firm size and production scale data of each batik firm precisely due to the lack of quality statistical data and government reports, particularly from the Local Agency for Industry and Trade Affairs (*Dinas Perindustrian dan Perdagangan*/DISPERINDAG) and the Local Agency for Cooperatives and Micro, Small, and Medium Enterprises (*Dinas Koperasi dan Usaha Mikro, Kecil, dan Menengah*/DISKOP-UMKM). On the other hand, the unwillingness of the firm owner to share and the poor recording of business operations created difficulties in data acquisition from the first hand. Thus, I combined with previous studies and media exposures for helping us to identify the key actors and their roles in the local batik industry for proceeding the following in-depth interviews and field observations. The in-depth interviews engaged several key informants from the Economic Division of the Local Planning Authority (*Badan Perencanaan Pembangunan Daerah*/BAPPEDA), the Head of Forum for Economic Development and Employment Promotion (FEDEP), and prominent local batik entrepreneurs to provide valuable insights on understanding the dynamics of the local batik industry development in Semarang Municipality and Kudus Regency. Lastly, field observations were undertaken to verify data from the previous phase and to obtain fuller and more comprehensive impressions on what actually happens on the field and how the creativity and innovation processes take place during the industrial business operations.

2.2. Methods

The study applied explanatory sequential mixed-methods approach where quantitative approach comes first to provide a background understanding of the research phenomena useful to stepping further to the following qualitative approach for exploring deeper comprehension of the prior phase [39]. At the beginning, the present data of the batik firms gathered from the survey were mapped out to analyzethe spatial distribution of batik clusters in Semarang Municipality and Kudus Regency. From this point,we could identify clustering pattern of the local batik industry. By comparing with the existing socio-economic background of the firm locations, we could obtain an initial impression to lead us in understanding the local embeddedness factors which have provided a so-called business environment for the batik firms to co-locate. Enriched with content analysis on the government reportssuch as*Rencana Pembangunan Jangka Menengah Daerah* (The Local Mid-Term Development Plan) of Semarang Municipality (2016-2021) [40] and Kudus Regency (2013-2018) [41], *Rencana Strategis BAPPEDA* (The Strategic Plan of the Local Planning Authority) of Semarang Municipality (2010-2015) [42] and Kudus Regency (2013-2018) [43],previous studies, and media exposures, we could analyze how the local batik cluster development being contextualized into the policymaking process.

At the qualitative phase, I explored the reciprocal relationships between the elements of social capital and the local batik cluster development. Arguably, the underlying social construct as a product of social capital mobilization provides various channels of information sharing, knowledge transfer, and technology adaptation required by the clustered firms to generate creativity and innovation [44–46]. A tendency of firms to establish particular clustering pattern is associated with the degree of proximity of firms to take the most advantages from colocation [47,48]. In this sense, social relations and networks play an important role in the formation of physical and economic linkages between firms within the respective cluster. Therefore, the positions and roles of actors in the industry were analyzed to figure out how the prevailing social construct is maintained to flourish the business environment.

3. Results and Discussion

Currently, there are nearly 100 batik firms in Semarang spread into almost 16 *kecamatan* (sub districts) but Gajahmungkur Sub district. The actual size of batik industry in Semarang remains unknown, however, because of poor statistical data availability, unregistered business operation, and the unstable business cycle of firms. There are only a small number of batik production firms, approximately 10-20 firms, while the rest is batik trading firms. All of them are new business ventures established in the period of 2000s, and none has carried out the prior generations' inheritance.

Historically, the embryonic batik industry in Semarang emerged in the nineteenth century as indicated by Budha Manjusri sculpture found in Ngemplak Simongan area. Batik clothing with *liris* or *lereng* motif was crafted into the sculpture. Unfortunately, there is no further information regarding the early development of Semarang batik industry until the historical archives found during the Dutch colonization period (1596-1942). In 1919, there were 25 batik firms with 58 male batik workers, and then it increased to 107 batik firms employing 268 male and 223 female workers in 1925. The industry originated in Kampung Bubakan then spread into several settlements in Bugangan, Rejosari, Kulitan, Kampung Melayu and Kampung Darat. There was a famous high-end product batik firm, Batikkerij Tan Kong Tien, which was named by the Indo-Chinese owner, Tan Kong Tien. He married to a daughter of the former Yogyakarta Court royal family Sultan Hamengku Buwono III, Raden Ayu Dinartiningsih, from which batik mastery had been turned into market scale production. The firm applied specialized division of labor ranging from design to coloring processes, involving many batik workers from Rejosari, Kampung Darat, Kintelan, Karang Doro, Mlaten Trenggulun, and Layur.When the massive settlement burning occurred under the Japanese colonization period (1942-1945), most batik firms ruined and leaving a few batik firms to survive including Batikkerij Tan Kong Tien, which run the business until the 1970s under the second generation successor, Raden Nganten Sri Murdijanti. Later, she was privileged by the Indonesian Batik Cooperatives Union (*Gabungan Koperasi Batik Indonesia*/GKBI) to manage batik trading monopoly in Central Java Province. In the 1950s, a new batik firm Asaco was established in Senjoyo, and then followed by Neni Asmarayani who operated a home-based batik gallery in Seroja in the 1970s, and Sri Retno batik firm established in Jatingaleh by a couple Oentoeng Suwardi and Tamsiyati who run the business during 1973-1982. However, the emergence of these late-coming batik firms did not affect much to the resurgence of the industry due to sporadic and unorganized industry management vis-à-vis the rise of printing batik firms since the 1970s which could produce batik fabrics much faster and much cheaper. Despite the competitiveness issue, their short-lived business operation was also caused by the absence of business successors [49–51].

After a 25-year business decline, the recent initiatives of Semarang batik industry resurgence have started in 2006 by Umi Adi Susilo who built a new batik firm Batik Semarang 16 in Tembalang. She promoted natural coloring batik production and registered no less than 11 batik motif patents representing iconic symbols of Semarang place identity, i.e. *Lawang Sewu Ngawang*, *Merak Mlerok Latar Asem*, *Merak Mangu*, *Cheng Ho Neng Klenteng*, *Asem Arang*, *Gambang Semarangan*, *Tugu Muda Kekiteran Sulur*, *Blekok Srondol*, and *Merak Njeprak*. Another initiative by a couple Eko and Indah started in the same year through the establishment of batik firm Semarang Indah in Kampung Gedong, Rejomulyo. They combined both natural and synthetic coloring processes in batik production. Following their success, the Local Agency for Industry and Trade Affairs in Semarang Municipality Government was keen to support the local batik industry development by replicating various technical programs including training, financial assistance, promotion, and marketing and exhibition in other places. As a result, no less than 32 batik firms supported by 188 batik workers existed in 2010 [50]. Furthermore, the local government has endorsed the former batik kampong in Bubakan as a cultural heritage preservation site through *Peraturan Daerah* (Local Regulation) No. 14 Year 2011 concerning *Rencana Tata Ruang Wilayah Kota Semarang Tahun 2011-2031* (Spatial Plan of Semarang Municipality Year 2011-2031). According to Article 69 of the regulation, the preservation of cultural heritage site may cover preserving the socio-cultural pattern of the local society, regulating any change in the existing building size and shape, and developing on-site tourism activities. This implies on public responsibilities in managing kampong development carefully to abide with pre-emptive socio-cultural values preservation [52].

To date, the growth of batik firms in Semarang has been driven externally instead of encouraging the local potentials from within. Some key points that could be discussed here are related to agent, place, activity, and contribution. The prominent agent in promoting the local batik industry is the external actors. They are new batik entrepreneurs who do not possess any family business background in the industry and have obtained government training to start up new batik business ventures. They work individually to build business linkages and therefore do not maintain connections with previous generations. As a result, a truncated intergenerational business continuation occurs since the business cycle tends to end up until the second generation only. The place of origin to run the business does not depend on the existing industrial agglomeration pattern. Even though batik kampong in Bubakan has been acknowledged as the oldest batik industry center in Semarang and officially supported by the local government policy, actually it is not quite attractive to the new business ventures to operate nearby. Rather, the new batik entrepreneurs could establish new firms elsewhere in the city sporadically. Some have built new firms close to city centers and densely populated settlements while the rest chose other places in suburbs such as Mijen, Banyumanik, Tembalang, and Genuk Subdistricts (see Figure 1). For example, Batik Semarang 16 has started the business in Tembalang and run galleries in some other places. Another example is Zie Batik owned by Marheno and Salma Batik owned by Umi Salamah who run business in Kampung Malon of Mijen Subdistrict and currently encourage the local kampong community to learn batik production and entrepreneurship skills. However, all these batik firms have shared a similarity in terms of home-based industry preference to start up new business ventures.

The potential market demand in manufacturing and trading activities in the city plays the key driver of such batik industry revival. Both sectors have provided significant shares to the local GDP consecutively at 29.75 percent and 30.81 percent (2000), 24.79 percent and 16.19 percent (2010), and 27.45 percent and 14.06 percent (2016). With a market size of more than 1.7 million population and annual growth rate at 1.66 percent (2016), the city provides a promising economic viability for both sectors. It is not surprising if more than 80 percent of the existing batik firms focusing on batik trading activities rather than batik production. Thus, the economic profitability has stimulated the local batik industry resurgence recently, not the socio-cultural tradition preservation intentions. Considering that the industry has never been rooted strongly enough in the entire society or even in the local batik kampong neighborhood, its contribution serves more to economic reasoning while the socio-cultural interests are somewhat left beyond. As exemplified by batik kampong in Bubakan, almost all of the existing batik firms are owned by immigrants or temporary tenants, and ironically, more than 90 percent batik products available supplied from Pekalongan batik producers and traders. Even though it has been endorsed as a cultural heritage site, Kampung Bubakan merely performs as a marketplace beautified with colorful batik ornaments and signage. The so-called place branding in this site has therefore failed to recreate certain batik business environment which authenticates Semarang batik tradition.

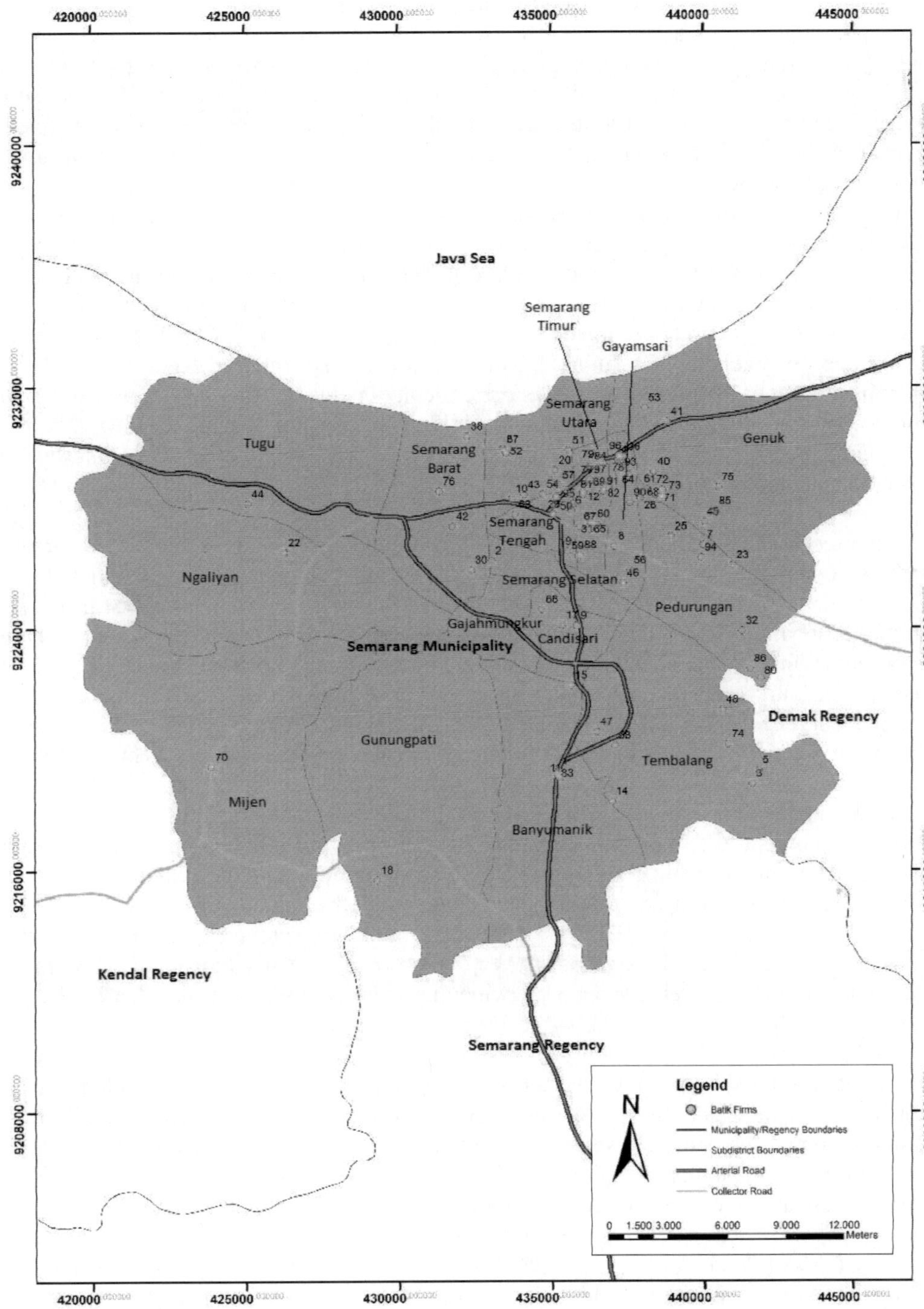

Source: Author's analysis, 2017

Figure 1. Spatial Distribution of Batik Firms in Semarang Municipality, 2017

The present social construct in Semarang batik industry is opened to incoming permanent residents and temporary visitors. By history, it has been shaped by a viable mercantilism tradition for centuries, thanks to the city's strategic location at the intersection of two national roads connecting the country's capital Jakarta and the second largest city Surabaya and the capital cities of Yogyakarta-

Surakarta-Semarang (JOGLOSEMAR) growth triangle. Equipped with an international freight and passenger seaport Tanjung Emas, an international airport Ahmad Yani, and national scale railway stations and bus terminals, the city has transformed to become an important hub city for both international and domestic flows of people, goods and services, and information. It is not surprising that both manufacturing and trade activities have grown rapidly, creating a potential marketplace for a broad range of economic activities. However, the society has carried out a long-standing mixture of Javanese patriarchal tradition and borrowing foreign influence of Dutch, Islamic, and Chinese tradition brought by foreign imperialists, traders, and migrating dwellers. To some extent, such socio-cultural assimilation is good for the formation of opened economy and plural democracy but it also creates some adverse impacts in forms of individual opportunism and hierarchical patronage structuralism. Short-term profit-taking behavior has featured the society where the business cliques are concealed with collusion and nepotism practices inside the government bureaucracy. Simply business and power relations patronage is likely to determine business growth performance. As a result, the current business environment has fostered individualistic behavior inside the batik industry with less cooperation among the entrepreneurs in dealing with high competition market. In turn, this leads to a high-cost economy that has inhibited knowledge sharing and innovation necessary to improve product competitiveness and place branding.

Clearly, the recent development of batik industry in Semarang has demonstrated a structural shift from bonding to linking social capital as a product of truncated intergenerational business continuation. As the older family business firms went out, the newer batik firms have emerged to take the lead of the local batik industry development by introducing modified path dependence. The linking social capital actors involve some migrating residents in Semarang and batik masters in Pekalongan. Neither the former batik ancestors nor their successors have got involved during the early stage of batik industry resurgence. With a forceful role of the local government through various policies, the industry has grown rapidly to spread across the city. However, such circumstance may jeopardize the future developments of the industry due to high dependency on a few charismatic local champions to whom the local batik industry sustainability rests on. Indeed, their roles in generating local economy and empowering people at the grassroots level are advantageous for stimulating collective actions. A community-based learning organization could be built for promoting creativity and innovation further to foster the local batik industry development. On the other hand, if such initiative is not institutionalized carefully, it may fail to produce new batik entrepreneurs and business start-ups. Furthermore, it may create an exclusive business network within locally built batik clusters either at neighborhood or community levels so that a citywide industrial competitiveness is hard to be achieved due to the prevailing short-term profit-taking business opportunism.

In contrast, the birth of batik industry in Kudus Regency is often associated with Islamic taught propagation era in Java Island by a prominent *majelisdakwah* (Islamic teaching institution) Wali Songo. Many references and narratives have pointed two Wali Songo missionaries, i.e. Sunan Kudus and Sunan Muria, who had inspired the local batik makers to apply calligraphic motifs, Islamic nuanced ornaments, and symbolic religious stories into batik fabrics since the 1600s. At the early period, batik production was merely a spare-time activity for women living in a settlement nearby *Masjid Menara Kudus* (Kudus Minaret Mosque) complex functioned as a center for Islamic teachings and studies, i.e. Kampung Langgar Dalem. Later, this settlement was also known as Kudus Kulon precinct. From this place, batik fabrics were used to be subsistent goods to fulfill the clothing needs of family members and relatives. In the 1800s, batik production turned into a home-based industry following the increasing demand for batik products domestically. The industry lasted until the 1880s and faced a business slowdown due to the rise of *kretek* (clove) cigarette industry in Kudus which had caused a massive inter-industry employment shift [53,54].

During the 1920s-1930s the industry survived as a result of intensified batik fabrication by the Indo-Chinese entrepreneurs. While many indigenous entrepreneurs collapsed or switched to clove cigarette industry and other types of business, the Indo-Chinese entrepreneurs began to take over the local batik industry dominance. Since then, the assimilation with Chinese culture in socio-economic life brought back the industry into a new form of business environment. A few batik firms owned by the Indo-Chinese entrepreneurs employed the indigenous batik workers to introduce new batik motifs representing a multicultural blend of Hindu, Arabic, and Chinese culture. They also recruited batik workers from Pekalongan from which the local batik workers' skills and production technology got improved. Some popular batik motifs such as *Kapal Kandas*, *Parijotho*, *Sekar Jagad*, *Beras Kecer*, *Rama Kembang*, and *Pakis Haji* illustrated a naturalistic meticulous design which carried out philosophical values of the local wisdom and folklore narratives. The batik industry center was no longer at city center in Langgar Dalem but spread to Tanjung Karang, Dawe, and Gebog precincts. Some prominent Indo-Chinese entrepreneurs were GS Liem, TS Ing, Pho An Nyo, and Liem Boe In. In the 1950s, some new Indo-Chinese fellowsjoined the industry such as Liem Boen Gan, Kwe Suiauw, Ok Hwa, and Gan Tjioe Gwat [53–56].

Despite such transitional industrial dynamics, the actual contribution of these newly built batik firms to the economy was not well-identified yet. There was insufficient data and information regarding industrial size and growth performance. As a consequence, we could not figure out what actually happened during this slowdown period in contrast to the heightened dominance of the Indo-Chinese batik firms over the indigenous competitors. This situation continued until the dark period of the industry during the 1980s-2000s when the local batik industry in Kudus suddenly disappeared. The missing local batik tradition rested on the hands of a very few old batik workers like a 74-year-old(deceased) female worker Niamah[54,57]. One possible reason is that the batik industry was less prioritized by the local government compared to tobacco industry which shared more than 70 percent to the local GDP and occupied more than 25 percent labor force. Batik industry seemed lack of economic viability for the local economy as well as entire society. As for the batik workers, it was not reckoned as a primary source of household income due to low and unstable labor wage earnings. Generally, batik industry whose mostly belong to female workers was being treated an additional income source during planting and harvesting periods break. This practice is very common in agriculture-based society particularly in Java to collect extra money from undertaking batik labor as a side job. In addition, a more prospective income earning and social status improvement in clove cigarette industry stimulated the batik workers for doing employment shift easily.

Recently, the batik industry in Kudus has re-emerged from the new generation of batik entrepreneurs. Similar to Semarang case, the local batik industry resurgence was initiated by newcomers who have neither attachment to batik production background nor family business circle. At the beginning, these initiators started to learn batik techniques independently with the old batik makers at various places, including batik masters in Surakarta and Pekalongan, before taking part into several government-driven training. It was Ummu Asyiati the owner of Alfa Batik firm and Yuli Astuti the owner of Muria Batik firm, both living in Gebog Subdistrict, who started the business in 2006. They worked relentlessly to discover authentic local batik motifs and succeeded to file patent register for some motifs. Since 2007 the Local Agency for Industry, Trade, Cooperatives and Micro, Small, and Medium Enterprises Affairs (*Dinas Perindustrian, Perdagangan,Koperasi dan Usaha Mikro, Kecil, dan Menengah*/DISPERINDAGKOP-UMKM) has supported these initiatives through various training and financial assistance until the present days. Unfortunately, only a few training participants have continued to establish new business start-ups while the rest preferred to become blue collar workers or find more prospective jobs[53,54]. As a result, the current local batik industry grows slowly with 25 batik firms where less than half producing their own products and the rest runs batik trading businesses. In Figure 2 we may observe the spatial distribution of batik firms in Kudus. The batik production firms locate along the axial road towards Mount Muria in Gebog Subdistrict while the batik trading firms concentrate in the city center area.

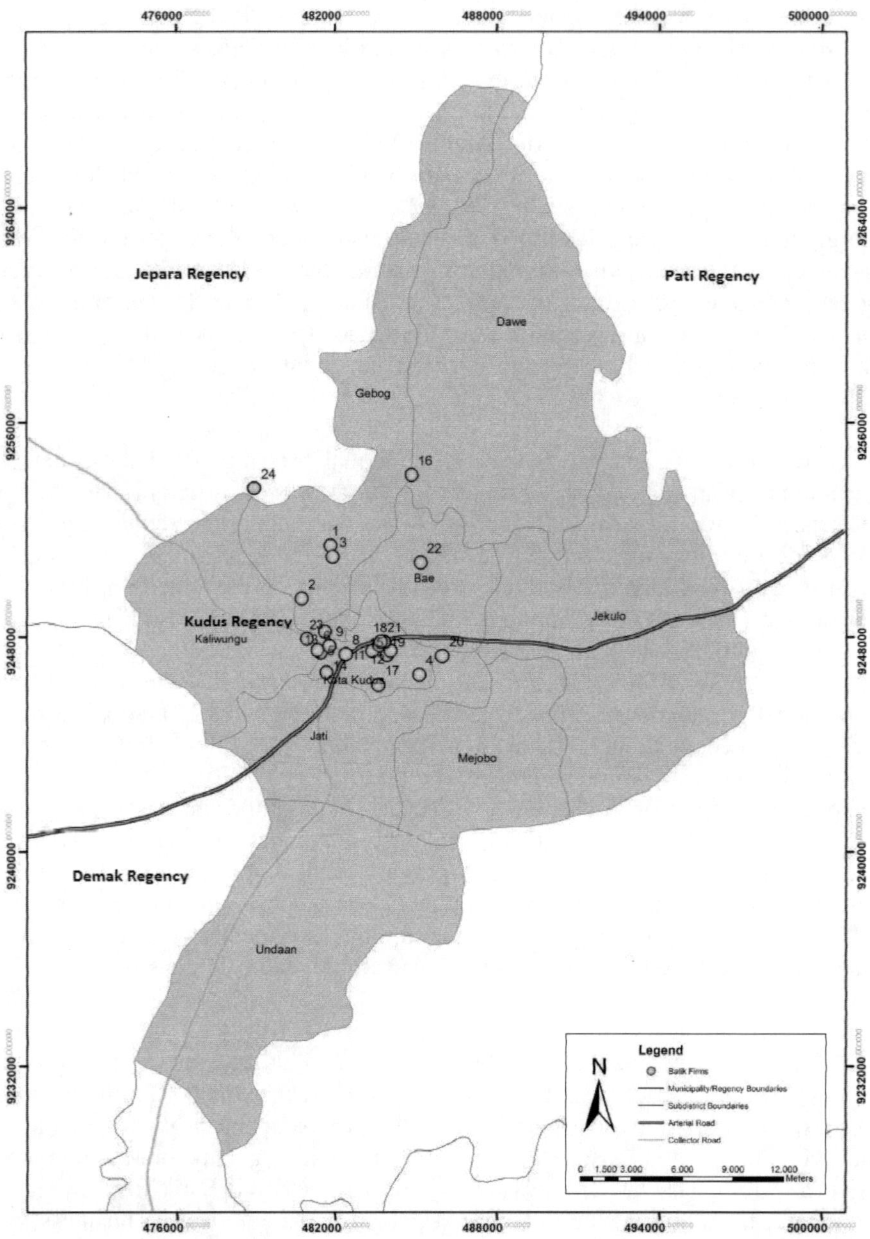

Source: Author's analysis, 2017

Figure 2. Spatial Distribution of Batik Firms in Kudus Regency, 2017

Inevitably, the prominent agent of batik industry revival in Kudus is the local citizens. They started to learn batik production processes and entrepreneurial skills at their own costs to regenerate the local batik tradition. The local government support is very useful to multiply such endogenous initiative for promoting new entrepreneurs and business start-ups. However, this government routine is not quite effective in negotiating the young generation's preference and mentality to change from industrial paid workers to self-employed persons. Another support comes from the large-scale clove cigarette industry, PT. Djarum, which provides a corporate social responsibility (CSR) program through Bakti Budaya Djarum Foundation. The company has facilitated talented individuals, groups, and organizations in promoting cultural tradition and events. Ummu Asyiati is one of program beneficiaries to whom the Foundation has provided funding, technical assistance, exhibition support, and business connections for promoting the local batik tradition richness. The similar way has been provided to Miranti Serad Ginanjar – a homegrown batik scholar – to conduct research and documentation on the local batik tradition as published in her book entitled "Batik Kudus the Heritage". In the meantime, supports from the external actors are also available from various interest groups such as fashion designer community, vertical government agencies, research institutes, and batik artists and collectors, making the industry more attractive for public consumers and tourism visitors [55–57].

The preferred place of the current batik industry to operate is no longer attached to its origin in Kampung Langgar Dalem. A typically specialized localization exists when the batik trading firms concentrate in city center precincts and the batik production firms move outwards to suburbs. As for the batik trading firms, it is not required to locate close to Kampung Langgar Dalem but the national road *Daendelsweg*, which is stretched across the region connecting Jati, Kota Kudus, Bae, and Jekulo Subdistricts. As for the batik production firms, proximity to roads is also important for ensuring input-output linkages during the production processes. Therefore, they tend to concentrate along the axial road to Mount Muria, which is not only accessible from the city center but also connected with popular tourism destination in Colo resort at Mount Muria.

Such industrial growth tendency is associated with the existing place branding of Kudus as *Kota Kretek* (Clove Cigarette City), making the city a potential marketplace for the growth of supporting industries. In addition, growing demand for pilgrimage tourism has also attracted visitors for purchasing local souvenirs including batik products. Ironically, the domestic trade market is overwhelmed by cheaper batik products from other places, particularly Pekalongan, which is mostly consumed by the locals. The local government policy support which encourages the civil servants to wear embroidery or batik clothing on certain working days is less useful to boost the batik industry growth because it does not specify the prioritized use of the local products only. As a result, there is a missing trade economy where the locals buy batik products from the local batik industry competitors, and the local batik producers sell the products to visitors and potential buyers in distant regions. Slightly different to the Semarang case, the emerging trade market opportunities in Kudus have not implied much on the batik industry growth. Instead of creating greater multiplier effects locally, the industry has intensified dependency on input factors as well as marketing destination towards other regions. To make it worse, since the locals are preferred to buy cheaper imported batik products, the industry seems to create deficit exchange to the economy. Therefore, similar to the Semarang case, the short-term economic rationality of the new batik entrepreneurs has emphasized economic profitability against the socio-cultural tradition preservation.

The prevailing social construct that leads to acurrent business environment in Kudus is primarily shaped by industrialist hardworking society, thanks to the harmonious socio-cultural assimilation of Hindu, Arabic, and Chinese culture in the past. The local society has brought forward agriculture-based peasant economy into a modern manufacturing industry. Strong attachment to homeland and natural resources cultivation has fostered typical family business relations so that limited business cliques in high competition market exist. As a result, the recent batik industry development has encouraged bridging and linking social capital actors when limited collective actions by the local actors represented by the home-grown batik initiators, scholars, private companies, and local government are blossoming, so that broader participation by the external agencies come later. Collaboration between the local batik firms and private companies through CSR programs has indeed restrained the local batik industry development down to earth. This is useful to rediscover the local batik tradition and identity with complimentary moderate support from the local government.

Unfortunately, short-term profit-taking behavior by the local batik entrepreneurs is more favored, particularly by those who running batik trading activities. Mostly these batik traders import cheaper batik products from Pekalongan and any other places to be sold in the local market. Instead of promoting local products, such business activity causes intensified competition driven by the high-cost economy in the local market. High dependency on factor inputs by the local batik producers against high dependency on cheaper final products by the local batik traders has worsened the existing business environment. Moreover, the lack of government policy support has contributed to such institutional obstacles where business opportunism exceeds inter-firm business cooperation which is counterproductive towards batik industry resurgence. Therefore, the batik firm's marketing strategy and product differentiation play the key role in stimulating collective actions. Building business alliance could be the most preferred choice for tackling these obstacles. Creativity and innovation of participating firms could be intensified through resources exchange among the alliance members. Eventually, product competitiveness and batik tradition identity may be achieved even though the local batik industry development is underpinned by a few groups of selected batik production firms and supporting stakeholders.

Drawing lessons from both cases in Semarang Municipality and Kudus Regency, we can figure out a distinguished development roadmap of the local batik industry. The Semarang case represents a territorial clustering approach at neighborhood or community levels while the Kudus case promotes a functional clustering approach. Even though they both have shared socio-cultural background similarities, the different local institutional setting has brought forward each industry to opposite directions for increasing local competitive advantage. According to Kitson et al. [3], the regional competitive advantage is indicated by the achievement of regional productivity, employment, and standard of living (see Figure 3). This output is resulted from the mobilization of six types of capital, i.e. human, social-institutional, cultural, infrastructural, knowledge/creative, and productive capitals. Human capital refers to the quality and skills of the labor force; social-institutional capital refers to the extent, depth, and orientation of social networks and institutional forms; cultural capital refers to the range and quality of cultural facilities and assets; knowledge/creative capital refers to the presence of an innovative and creative class; infrastructural capital refers to the scale and quality of public infrastructure; and productive capital refers to a cumulative product of these capitals which are mutually supporting and underpinning an efficient productive base to the regional economy.

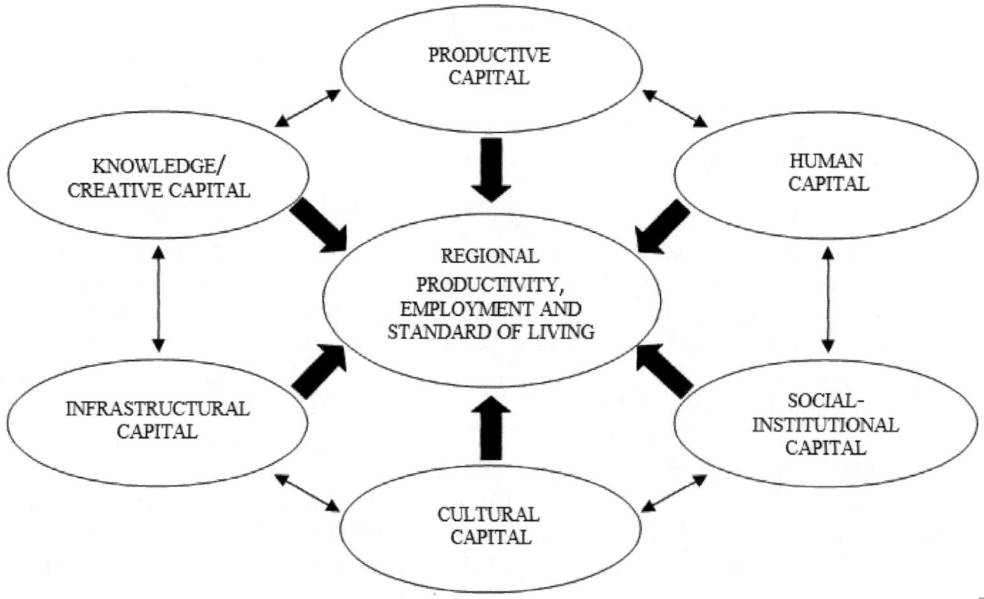

Figure 3. Bases of Regional Competitive Advantage [3]

In this study, I have partially utilized this regional competitive advantage framework to examine social-institutional capital in contributing to the social construct formation locally. Concerned with the local batik industry resurgence in both locations, the growing number of batik firms since the mid-2000s, regardless of their primary business type either as batik production or batik trading firms, has indicated increasing productivity and employment size levels. Based on triangulation process by evaluating interview results, government reports, previous studies, media exposures, and field observations, we may obtain a sense of performance change even though the accurate figures are missing. In addition, most batik firms belong to SMEs group in which poor bookkeeping applies, and the owners are less cooperative to share firm performance honestly. Regarding the improvement on the standard of living, I did not examine that far as the research focus took place at society level for examining the role of the localized social construct in creating a socio-spatial setting of the prevailing business environment. To some extent, the social construct in both locations has shared socio-cultural similarities in terms of the underlying Javanese culture that carries out opened economy and society. Socio-cultural assimilation between the indigenous citizens and foreigners has somehow determined the local batik industry development roadmap which cultivated high competition and opportunistic behavior in the local business environment. However, the local social construct has been adaptive and evolving towards inextricable endogenous factors and external influences along the periods of time. As a result, the recent batik industry revival has encouraged each case to take opposite directions in shaping the local industrial competitiveness.

4. Conclusion

I shall begin to conclude this study by addressing a question why local industrial competitiveness state may be different between localities although the existing social construct has stemmed from the similar socio-cultural root of tradition. More interestingly, when the recent batik industry revival in different locations has opportunities to achieve industrial development convergence, the prevailing business environment seems reluctant to exist even though each has brought forward less modified prolonged path dependence. Clearly, the local social construct is not static; rather, it evolves to deal with external influences while it also accommodates the adaptation of the local society to respond such a changing environment. The adaptive capacity of the participating actors during the local batik industry revival reshapes and determines how the local institutional setting effectuates the batik industry development initiatives. The contesting roles and responsibilities among the actors in managing competition and cooperation in the market provide the key for driving the prevailing business environment. Excessive roles of particular parties may result in socio-spatial separation rather than competitive industrial integration. On the contrary, fewer roles may lead to exclusion in both economic and social standings. In turn, such socio-spatial configuration lubricates a tension between individualistic behavior and collective actions. When individualism is greater than collectivism, a limited territorial closure of the local batik industry tends to emerge. Conversely, a selective cross-border functional business network may occur when collectivism exceeds the individualism.

Finally, I would like to emphasize that regardless of which industrial development preference emerge, they carry out their own virtues and drawbacks. Interestingly, each has a potential to direct the local batik industry to promote local competitiveness through different ways. On the other hand, they may also promote business exclusiveness which is counter-productive towards the local batik industry sustainability. However, this claim needs further research to explore substantive key drivers within the existing social construct that has shaped the adaptive capacity of the participating actors in the local batik industry development.

5. Acknowledgments

This study is available because of the research grant under *Penelitian Berbasis Kompetensi2017* (the 2017 Competence Based Research) scheme entitled *"Local Embeddedness Factors pada Variasi Perkembangan Klaster Industri Batik di Wilayah Cepat Tumbuh Provinsi Jawa Tengah: Studi Kasus*

IOP Conf. Series: Earth and Environmental Science **123** (2018) 012040 doi:10.1088/1755-1315/123/1/012040

Kota Semarang dan Kabupaten Kudus" (Local Embeddedness Factors on the Variations of the Batik Industry Cluster Development in the Fast Growing Regions of Central Java Province: the Case Study of Semarang Municipality and Kudus Regency). I am indebted to the Director of Research and Community Services at the Indonesian Ministry of Research, Technology and Higher Education (DITLITABMAS-KEMENRISTEKDIKTI) for providing the grant and advisory support during the research. I also would like to gratitude the Head of Research and Community Services Institution at Diponegoro University (LPPM-UNDIP) who have relentlessly encouraged the staff for undertaking sustained research and many kinds of scientific activities in favor of either personal career development or institutional advancement.

6. References

[1] Chapain C, Clifton N and Comunian R 2013 Understanding creative regions: Bridging the gap between global discourses and regional and national contexts *Reg. Stud.* **47** 131–134
[2] Foord J 2009 Strategies for creative industries: An international review *Creat. Ind. J.* **1** 91–113
[3] Kitson M, Martin R and Tyler P 2004 Regional competitiveness: An elusive yet key concept? *Reg. Stud.* **38** 991–999
[4] O'Connor J and Xin G 2006 A new modernity? *Int. J. Cult. Stud.* **9** 271–283
[5] United Nations Development Programme and United Nations Educational Scientific and Cultural Organization 2013 *Creative Economy Report 2013 Special Edition: Widening Local Development Pathways* (New York and Paris: United Nations Development Programme and United Nations Educational, Scientific and Cultural Organization)
[6] Fahmi F Z and Koster S 2017 Creative industries and regional productivity growth in the developing economy: Evidence from Indonesia *Growth Change* **0** 25–28
[7] Pirmana V, Maulana N and Panjaitan N A *How important is creative industry to Indonesian economy?: Preliminary findings from extended national input-output table for creative industry* (Bandung)
[8] Dörry S, Rosol M and Thissen F 2016 The significance of creative industry policy narratives for Zurich's transformation toward a post-industrial city *Cities* **58** 137–142
[9] Porter M E 1990 *The competitive advantage of nations*
[10] Porter M E 1998 *The Adam Smith address: Location, clusters, and the""new" microeconomics of competition*
[11] Higgs P and Cunningham S 2008 Creative Industries Mapping: Where have we come from and where are we going? *Creat. Ind. J.* **1** 7–30
[12] Moore I 2014 Cultural and creative industries Concept – A historical perspective *Procedia - Soc. Behav. Sci.* **110** 738–746
[13] Banks M and O'Connor J 2009 After the creative industries *Int. J. Cult. Policy* **15** 365–373
[14] O'Connor J and Kong L 2009 Introduction *Creative Economies, Creative Cities: Asian-European Perspectives* ed L Kong and J O'Connor (Dordrecht: Springer) pp 1–8
[15] Yusuf S and Nabeshima K 2005 Creative industries in East Asia *Cities* **22** 109–122
[16] Markusen A, Wassall G H, DeNatale D and Cohen R 2008 Defining the creative economy: Industry and occupational approaches *Econ. Dev. Q.* **22** 24–45
[17] Florida R 2005 *Cities and the Creative Class* (New York: Routledge)
[18] Peck J 2005 Struggling with the rise of the creative class *Int. J. Urban Reg. Res.* **29** 740–770
[19] Pratt A C 2008 Creative cities: The cultural industries and the creative class *Geogr. Ann. Ser. B Hum. Geogr.* **90** 107–117
[20] Scott A J 2006 Creative cities: Conceptual issues and policy questions *J. Urban Aff.* **28** 1–17
[21] Cooke P, De Laurentis C, Tödtling F and Trippl M 2007 *Regional Knowledge Economies: Markets, Clusters and Innovation* (Cheltenham: Edward Elgar Publishing Limited)
[22] Martin R, Florida R, Pogue M and Mellander C 2015 Creativity, clusters and the competitive advantage of cities *Compet. Rev.* **25** 482–496
[23] Nathan M and Overman H 2013 Agglomeration, clusters, and industrial policy *Oxford Rev.*

Econ. Policy **29** 383–404
[24] Department for Culture Media and Sport 1998 *Creative Industries Mapping Document 1998* (London)
[25] Galloway S and Dunlop S 2007 A critique of definitions of the cultural and creative industries in public policy *Int. J. Cult. Policy* **13** 17–31
[26] Castells M 2010 *The Power of Identity* (Chichester, UK: Wiley-Blackwell)
[27] Pratt A and Jeffcutt P 2009 Creativity, innovation and the cultural economy: Snake oil for the twenty-first century? *Creativity, innovation and the cultural economy* ed A C Pratt and P Jeffcutt (Oxon: Routledge) pp 3–20
[28] Scott A J 2008 *Social Economy of the Metropolis: Cognitive-Cultural Capitalism and the Global Resurgence of Cities* (Oxford: Oxford University Press)
[29] Pitts F H 2015 A hidden history: Defining and specifying the role of the creative industries *Creat. Ind. J.* **8** 73–84
[30] Ketels C 2013 Recent research on competitiveness and clusters: What are the implications for regional policy? *Cambridge J. Reg. Econ. Soc.* **6** 269–284
[31] Clare K 2013 The essential role of place within the creative industries: Boundaries, networks and play *Cities* **34** 52–57
[32] Bontje M and Musterd S 2009 Creative industries, creative class and competitiveness: Expert opinions critically appraised *Geoforum* **40** 843–852
[33] Florida R 2012 *The Rise of the Creative Class Revisited* (New York: Basic Books)
[34] Glaeser E L 2005 Review of Richard Florida's *Reg. Sci. Urban Econ.* 593–596
[35] Eklinder-Frick J, Eriksson L T and Hallén L 2014 Multidimensional social capital as a boost or a bar to innovativeness *Ind. Mark. Manag.* **43** 460–472
[36] Chuluunbaatar E, Ottavia, Luh D-B and Kung S-F 2014 The Role of Cluster and Social Capital in Cultural and Creative Industries Development *Procedia - Soc. Behav. Sci.* **109** 552–557
[37] Trüby J, Rammer C and Müller K 2008 *The Role of Creative Industries in Industrial Innovation* vol 08–109
[38] Markusen A 1996 Sticky places in slippery space: A typology of industrial districts *Econ. Geogr.* **72** 293–313
[39] Creswell J W 2014 *Research Design: Qualitative, Quantitative, and Mixed Methods Approaches* (Thousand Oaks: Sage Publications, Inc.)
[40] Pemerintah Kota Semarang 2016 *Rencana Pembangunan Jangka Menengah Daerah Kota Semarang Tahun 2016-2021* (Semarang, Indonesia)
[41] Pemerintah Kabupaten Kudus 2014 *Rencana Pembangunan Jangka Menengah Daerah Kabupaten Kudus Tahun 2013-2018* (Kudus, Indonesia)
[42] Badan Perencanaan Pembangunan Daerah Kota Semarang 2011 *Rencana Strategis Badan Perencanaan Pembangunan Daerah Kota Semarang Tahun 2010-2015* (Semarang, Indonesia)
[43] Badan Perencanaan dan Pembangunan Daerah Kabupaten Kudus 2014 *Rencana Strategis Badan Perencanaan Pembangunan Daerah Kabupaten Kudus* (Kudus, Indonesia)
[44] Asheim B, Cooke P and Martin R 2006 The rise of the cluster concept in regional analysis and policy: A critical assessment *Clusters and regional development: Critical reflections and explorations* ed B Asheim, P Cooke and R Martin (London: Routledge)
[45] Balland P-A, Belso-Martínez J A and Morrison A 2016 The dynamics of technical and business knowledge networks in industrial clusters: Embeddedness, status or proximity? *Econ. Geogr.* **92** 35–60
[46] Camps S and Marques P 2014 Exploring how social capital facilitates innovation: The role of innovation enablers *Technol. Forecast. Soc. Change* **88** 325–348
[47] Boschma R 2005 Proximity and innovation: A critical assessment *Reg. Stud.* **39** 61–74
[48] Molina-Morales F X, López-Navarro M Á and Guia-Julve J 2002 The role of local institutions as intermediary agents in the industrial district *Eur. Urban Reg. Stud.* **9** 315–329
[49] Batik Semarangan pernah menghilang 2017 Batik Semarangan Pernah Menghilang *Suara Merdeka*

[50] Manggarani D P 2010 *Peran Pemerintah Daerah dalam Pengembangan Industri Batik Semarang* (Universitas Diponegoro)
[51] Nugroho P 2014 *The Role of Local Institutions in Industrial Cluster Development in Indonesia:* (Technische Universität Dortmund)
[52] Pemerintah Kota Semarang 2011 *Rencana Tata Ruang Wilayah Kota Semarang Tahun 2011-2031* (Semarang, Indonesia)
[53] Faizin A 2012 Sejarah Batik Kudus
[54] Pemerintah Kabupaten Kudus 2012 Eksotika Batik Kudus
[55] Priherdityo E 2015 Batik Kudus, Media Penyebaran Islam hingga Promosi Rokok *CNN Indones.*
[56] Indriani R and Nodia F 2015 Buku Ini Ungkap Sejarah Batik Kudus dari Masa ke Masa
[57] PT. Kompas Media Nusantara 2017 Selisik Batik Kompas: Sejarah Panjang Kebangkitan

Preservation "Kauman Kampong Qur'an" as Religious Educational Tourism

R Kurniati[1] and N Nurini[1]

[1]Department of Urban and Regional Planning, Universitas Diponegoro, Indonesia

Email: adalerina_66@yahoo.com

Abstract: Kauman is a historic kampong in Semarang and historically inhabited by the Java Moslem community. Characteristic of Kauman is the number of 'santri'(Moslem apprentice) as the centre of Semarang's santri in the past. Kauman Mosque Semarang as a centre of Islamic religion also has an important role in the development of Semarang until now. Kauman's main characteristic is their Great Mosque, which around the square are the centre of government, and traditional markets. Kauman as a centre of Islamic culture, has the Great Mosque Kauman as a relic of Ki Ageng Pandan Arang, it is the oldest mosque in Semarang. The religional tradition is 'dugderan' activities undertaken ahead of the month of Ramadhan to inform the public that tomorrow is already entered 1 Syawal or Ramadhan. This tradition is accompanied by a parade or procession through the main streets of Semarang City. Kauman progressing increasingly crowded, with many migrants who need land for dwelling and trade spaces. Therefore Kauman requires conservation efforts to restore the identity of the region to invite people to maintain the culture that still exists. The conservation efforts implemented is still maintaining its culture. Meanwhile, to keep the buildings' Arab architectural style need to support the preservation of the religious educational tourism branding "Kauman Kampong Qur'an". The method used in this research is qualitative approach. Analytical techniques used were descriptive and spatial analysis using GIS to observe Kauman's figure-ground. This research used primary and secondary data with data collection methods through field observation, interview and document review. The result of this research is the arrangement and design of Kauman Kampung Qur'an in the public space around the Great Mosque of Kauman Semarang.

Keywords: Kauman, Educational Tourism Planning, GIS, Spatial Analysis

1. Introduction

Kauman is a Kampong name that almost always exists in the cities of Java Island that is close to a big mosque and the square [1,2]. Kampung Kauman Semarang is a traditional Muslim village located in the heart of Semarang City. Kampong Kauman Semarang has survived as a traditional settlement for the Muslim community for approximately three centuries. The sustainability of Kauman Village in Semarang cannot be separated from the important role of religious leaders (Kyai) and Traders (local businessmen) who maintain the balance of socio-cultural and economic aspects [3]. The development of Islam in Kauman also cannot be separated by the existence of some boarding schools in the region. The place of religious education, whether in the form of "pesantren" or Quran Education Park is the most dominant building in Kauman, beside other buildings such as housing, "wakaf" area, worship place or working place [4]. The location of pesantren cottages that spread in various areas within Kauman made the boarding school as part of the Kauman community [1]. The large number of Qur'an learning places in Kampong Kauman raises the idea of an integrated Qur'an learning Kampong entitled Kauman Kampung Quran [5]. The process of branding Kauman as Kauman Kampung Quran is not only appeared unilaterally from the caretaker of Pesantren and the residents, but it is also known by Semarang City Government. In addition to holding learning Al Quran memorization, in Kauman Kampung Quran also want to teach the Quran's application to everyday life [6].

Kampong Kauman area has historical value for the development of Semarang city, that is as embryo of city growth starting from around Semarang River. The growth of these embryos forms the urban villages organically. Similarly with the development of Kampong Kauman, it started when Ki Ageng Pandan Arang headed to the area to spread Islam with his students. Physically, Kampong Kauman looks very different in line with the development of Semarang City, the kampong is experiencing rapid growth seen from the mass of the building that is more solid until it does not have green open space. Because of Kampong Kauman is located in the Cental Business District (CBD) of Semarang City, making it crowded with shops that strengthen the function of this area as a trade and service area, especially with Johar market which is the center of trade and services activities.

Currently what remains of Kauman's historic is the Kauman Great Mosque with the little side of Square that used for parking lots. The Square that was once widespread now diminishes due to the development of Johar market. Changes in land use that occurred in the area of Square make the identity of Kauman area is reduced. Although the Kauman Great Mosque still exists as the identity of the Kauman area, the sense of place is reduced by looking at the existing buildings that has changed to the modern facade, so that the visiting people are less aware of Kauman's historic. In addition, some buildings in Kauman still have the architectural style of Arab and Javanese fusion. They are one of the historical and uniqueness aspects of Kauman. Therefore, to support the branding of Kauman as a religious educational village namely "Kauman Kampong Qur'an", it is necessary to preserve the area to maintain its historical value and increase the sense of place as one of the religious tourism attractions.

2. Data and Methods

2.1. Methods

The method used in this research was a qualitative approach by using descriptive analysis techniques and spatial analysis by using Geography Information System [7,8]. The data used in the form of secondary data and primary data with data collection techniques used were field observation, document review, and interview. Spatial analysis using GIS was the basis for the site analysis in the form of a figure ground and boundary of study area [6,9,10]. There are several analysis of tread used in this research that is accessibility analysis, environmental and vegetation analysis, view analysis, and also road network analysis that will produce zoning area for physical plan of Kauman center for supporting the branding "Kauman Kampong Qur'an" as Religious Educational Tourism.

2.2. Data

2.2.1. Kampong Kauman History

The history of the Kauman began when Ki Ageng Pandan Arang went down the mountain to the coastal area to spread Islam with his students (santri). After obtaining the right area as the center of the spread of Islam, then he built a mosque. This mosque is located on the west side of Semarang River. The "santri" settled around the mosque and this area by Ki Ageng Pandan Arang named as "Kemesjidan" area. After the death of Ki Ageng Pandan Arang, his position was replaced by his son. Gradually the "Kemesjidan" area became crowded. This became the concern of Sultan Demak, and in the end, the son of Ki Ageng Pandan Arang was crowned the first regent by obtaining the title Ki Ageng Pandan Arang II. Having been crowned as a regent, Ki Ageng Pandan Arang II built palace now known as "Kanjengan".

The Semarang City at that time has been inhabited by a number of ethnic groups. In the map, clearly the Mosque (javanese temple) built by Ki Ageng Pandan Arang was in the Pedamaran area. In the southwest of the Mosque, there is a complex of "dalem" which is the beginning of Semarang regency. Around the "dalem" there had to be settlements of the courtiers. From the mosque to the east across the Semarang river there is a settlement of ethnic Chinese (now Pekojan). To the east again is now known as "Bubakan", depicted a garden with a square shape that was designed very neatly. To the east of the mosque and Chinese settlement, on the east side of the Semarang river, depicted a fort with a pentagon shape. This fort was built by the Dutch government in 1679. In this 1695 map, the Blenduk Church area and its surroundings which by Semarang known as the old city, have not been illustrated. This clearly proves that the Blenduk Church area and its surroundings are not Semarang's old town. Based on this map, it is clear that the area around "dalem" (now in the form of Kanjengan shops) is the site of the Semarang old city.

In 1740 there was a revolt of the Chinese against the VOC. In this revolt, Chinese ethnic settlements were destroyed and the Chinese developed settlements in the area now known as Chinatown. As a result of the rebellion, the Mosque built by Ki Ageng Pandan Arang and the occupancy of his santri were also destroyed and burned. In 1741, the mosque was rebuilt by the Regent of Suromenggolo and the mosque was built in an area known as the Grand Mosque of Kauman. The orientation of the Mosque building which was built by the Regent of Suromenggolo to the west. While the students who are descendants of the santri Ki Ageng Pandan Arang settled around the mosque and this settlement is known as Kampung Kauman.

Figure 1. Kauman 1880 [7]

Figure 2. Kauman 1892-1913 [7]

Figure 3. Kauman 1913 [7]

Figure 4. Kauman Recent Year [2]

Source: Suprapti, 1997 and Bappeda Semarang, 2010

2.2.2. *Kampung Kauman Preservation*

Preservation intended in urban design is the protection of existing residential neighborhoods and existing urban places and its characteristic, as well as protection against historic buildings. Preservation involves the maintenance of conservation buildings, maintenance of design elements (parks, billboards, etc.), taking into economic and cultural values.

Currently, what remains as Kauman's historic is the Kauman Great Mosque with a little side of the square that used as a parking lot. Changes in land use that occurred in the square area make the identity of Kauman area reduced. Although the Kauman Great Mosque still exists as the identity of Kauman, the sense of space has been reduced by the existence of buildings that are turning into modern fasades, so the historical side of Kauman is less visible.

Buildings or parts of the city are preserved as they represent special achievements in a particular historical style. This criterion is associated with high aesthetic and architectonic value in the form, structure, layout of the ornament. Kauman has a mixture of Javanese and Arabic architecture, because in the days of Ki Ageng Pandan Arang there were many Javanese and Arab communities residing in this area. The residence is

IOP Conf. Series: Earth and Environmental Science **123** (2018) 012041 doi:10.1088/1755-1315/123/1/012041

occupied by the current society, typical style with three large and widths doors that functioned at once for the window of the house.

Source: Observation, 2015

Figure 5. Traditional House in Kauman

In addition to the house with Java-Arabic architecture, buildings that will always be remembered by the community and as a landmark of Kauman is Kauman Great Mosque. The building also has Javanese and Arabic styles.

Source: Observation, 2015

Figure 6. Kauman Great Mosque

Based on these matters, Kauman has an aesthetic value that must be maintained and preserved so that the identity of the area is not lost in the future. Kauman is a history for Semarang City, the beginning of the village formed from associations of santri and finally developed with the arrival of Arabs until hereditary until now. The existence of building relics in Kauman formed an Arab and Javanese fusion arrangement to make this area worthy to be conserved, maintained, and revived the identity of the region as a historical and religious area. In preservation of Kauman area also looks at three other aspects namely economic, religy, and socio-cultural activities.

2.2.3. Economy Activities

Early Kauman is a settlement dominated by santri and Arabic descent. Then, along with the development of Semarang City, and located also in the CBD of Semarang City changes in the environment as the alun-alun which disappeared and replaced with commercial buildings. Therefore, in Kauman currently the most activities are trading and services activities, such as religious shops, building shops, trophy shops, etc. This development has an impact on the people of Kampung Kauman which most of their livelihoods are traders, they transform their house into shops / stalls. Trades and services in the Kauman Village area is also dominated by Chinese people who are in Kranggan area, this is because the location of this Kauman Village adjacent to Chinatown and Kranggan where most of its trade in the form of gold shop. The existence of shops has not been matched by the availability of adequate parking space so that traffic becomes blocked and crowded.

Source: Observation, 2015

Figure 7. Shops in Kauman

2.2.4. Socio-Cultural Activity

Kauman has strong cultural activity related to Islam. This tradition is held every year and before Ramadhan namely Dugderan. Since 2004, the dugderan tradition has returned to its original place, at the Great Mosque of Semarang / Kauman after several years moving around. The existence of this tradition is as a gathering of people who bring economic and social impacts. The economic impact is reflected in the presence of sellers scattered around the Square, whereas the social impact is the blending of the entire society from various ethnicities, all indigenous (Javanese), Arabic, and Chinese.

Dugderan tradition has been embedded in the hearts of the people of Semarang since a long times, it is necessary to purifying steps so that the historical, philosophical, and symbolic values that are inherent in this tradition can continue to survive. The people especially the young generations are expected to imitate the religious values in this tradition, namely "remember dugderan = remember Ramadan; Remembering Ramadan = remembering the mosque; Continue to remember the mosque is meant remember Allah SWT ". Hence, Dugderan, Ramadan, and Kauman Great Mosque are as implementation ways considering Allah SWT (Media Semarang, Ed 06/06/2011). Dugderan organized by ta'mir of Kauman Great Mosque, cooperation with Semarang City Government.

Source: Observation, 2015

Figure 8. Dugderan in Kampung Kauman

Until now dugderan still known by the people of Semarang and outside Semarang, there is also a night market that enlivens the dugderan. Many games are presented for this night market event, so the people of Semarang or outside Semarang are enthusiastic to arrive in Kauman. Similarly with the sellers who trade various goods or toys made of clay. This tradition makes Kauman still widely known.

2.2.5. Religious Activity

Kauman is famous for the Arabian homes because as an Islam spread place in Semarang, especially supported by the Great Mosque of Kauman. Most of the religious activities took place at the Great Mosque of Kauman, and there were also small teachings in each village in Kauman. Praying activity conducted in

IOP Conf. Series: Earth and Environmental Science **123** (2018) 012041 doi:10.1088/1755-1315/123/1/012041

early months or "mauludan" performed by the community Kauman or outside Kauman. In addition, the women also participated in praying activity that is usually done on the morning, the praying activity between men and women is done separately, as well as when there is praying activity at the Great Mosque Kauman men and women also have their respective places in accordance with Islamic culture.

Source: Observation, 2015

Figure 9. Praying activity in "Gang Butulan" at the beginning of the month

Religious activities held in this area to add social relationships in Kauman because they interact directly with each other. In Kauman can still found many people who wear Islamic clothes. In this Kampong at mahgrib time, we can hear the chanting holy verses of the Qur'an. This is what characterizes Kauman compared to other regions.

3. Result and Discussion

3.1. Kauman Preservation Effort as Religious Educational Tourism Kampong

The effort to restore the identity of Kauman, although not fully returning like the beginning, at least the community, especially the community outside the Kauman has to participate in maintaining the culture that still exists. The preservation concept of Kauman is based on economic, religious and socio-cultural activities. Including the physical condition of the buildings and the environment in Kampung Kauman area must be considered, this is the sustainability concept of Kauman to be preserved, including the characteristic of traditional buildings and the names of alleys that have their own history. With the comfortable Kauman physical circumstances will attract visitors to visit, and with the concept can make visitors to deepen the culture and religion that exist in Kauman.

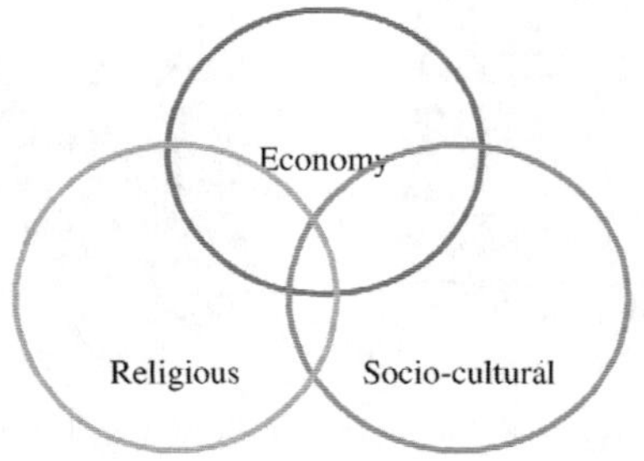

Figure 10. Sustainable Concept "Kauman Kampung Qur'an" based on community towards Tourism Region Educational Religius

The conservation effort process in order to realize Kampong Kauman as a religious educational tourism area by looking at the concept of preservation can be done as follows:

a. Preserving and maintaining the traditional forms and architecture of existing buildings, in collaboration with local government.
b. Conducting religious activities that can attract people outside the region to revive Kauman.

IOP Conf. Series: Earth and Environmental Science **123** (2018) 012041 doi:10.1088/1755-1315/123/1/012041

c. Implement routine religious activities by the community around Kauman, to keep maintaining the nuances of Islam in the Kampong from individuals as well as people from outside the region. Such as paguyuban, religious activities, and other praying activities.
d. Enliven youth assosiation so that the teenagers in the Kauman neighborhood know / explore the culture / tradition in Kauman.
e. Keep maintaining trade and services selling typical arab trinkets, such as perfumes and Islamic religious fixtures.
f. The government provides socialization for people who live permanently in Kauman and community outside Kauman area, so that the distinctiveness of the Kampung is not lost. As with conducting religious gatherings (assemblies), marawis and barzanji events.
g. Structuring or designing the center area of Kauman, one of them is circulation and parking.

3.2. *Physical analysis to support "Kauman Kampung Qur'an" as Religious Educational Tourism Kampong*

To support Kauman Kampung Qur'an concept as a religious educational tourism area, the arrangement of the environment around the mosque as the center of activities to create comfort and beauty in the core location of Kauman Kampung Qur'an is. Here is an analysis to know the zoning area at the design location.

3.2.1. Accessibility Analysis

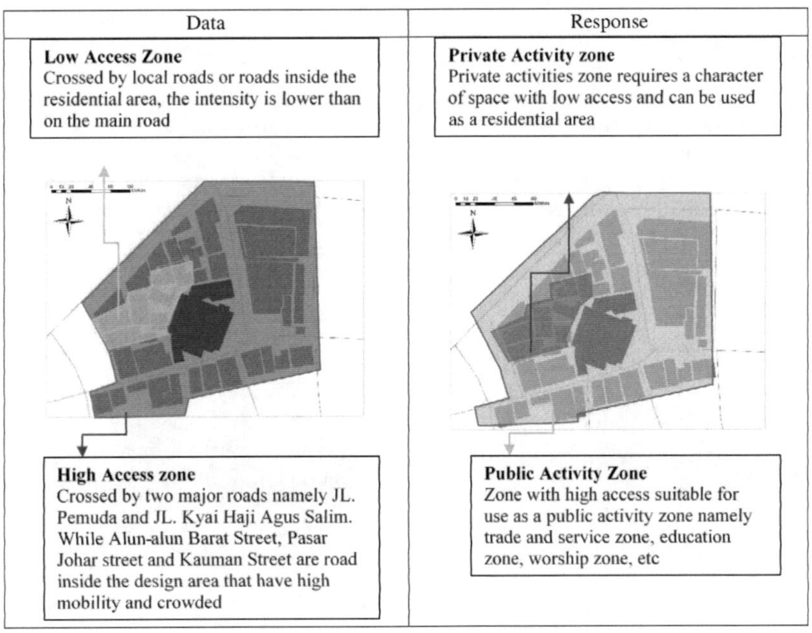

Figure 11. Accessibility Analysis in Kauman

In the existing condition, the design area in Kauman is divided into low access zones namely residential area and high access zones namely trade and service zone passed by two main roads that have high mobility and crowded. So the response to these conditions, for low access zones suitable for private activity zone such as residential area and for high access zones appropriate for public activity zone such as trade and services area, education, worship and others.

3.2.2. Neighborhood Analysis

Neighborhood around the design area is divided into two types namely trade and service zone and residential area so that the response to the condition is that in the design area is divided into four zones namely trade and service zone, worship, residential and supporting zone.

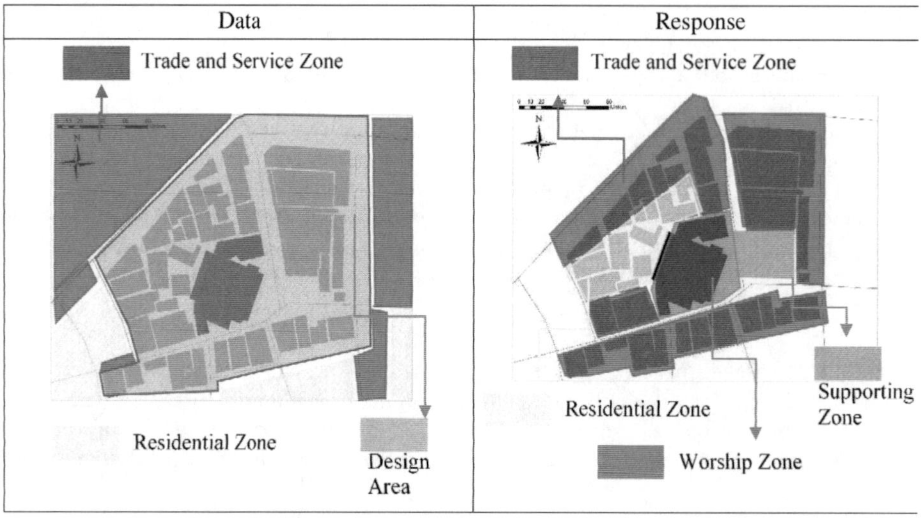

Figure 12. Neighborhood Analysis in Kauman

3.2.3. Vegetation Analysis

Existing conditions in the design area, green open space is very limited, there are only a few trees in front of Ya'ik Market and around the square. Response to the situation will be added a green line along the main road and the expansion of the Kauman Square and the addition of vegetation in it. Planting vegetation in addition to adding the beauty of the area can also play a role to cool the air temperature, considering Kauman located in northern Semarang where during the day especially in dry season was very hot.

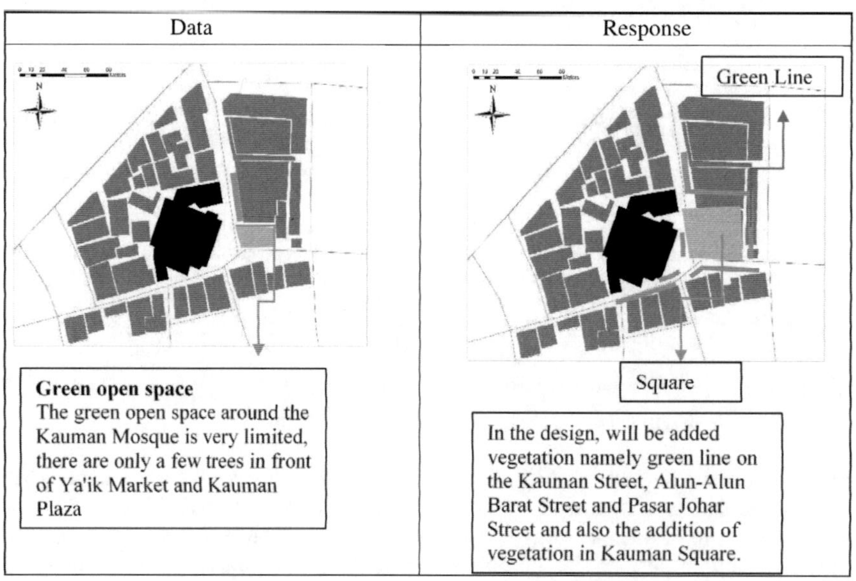

Figure 13. Vegetation Analysis in Kauman

3.2.4. View Analysis

View that can be seen from inside design area is trade and service zone includes shops, markets and roads, while view to site from the surrounding area is Kauman Grand Mosque. Response to the existing condition is to plan the expansion of Kauman square to restore Square function of and add the beauty of the area so that the main view Kauman area is no longer shops and markets but the Kauman Great Mosque along with the square and vegetation in it.

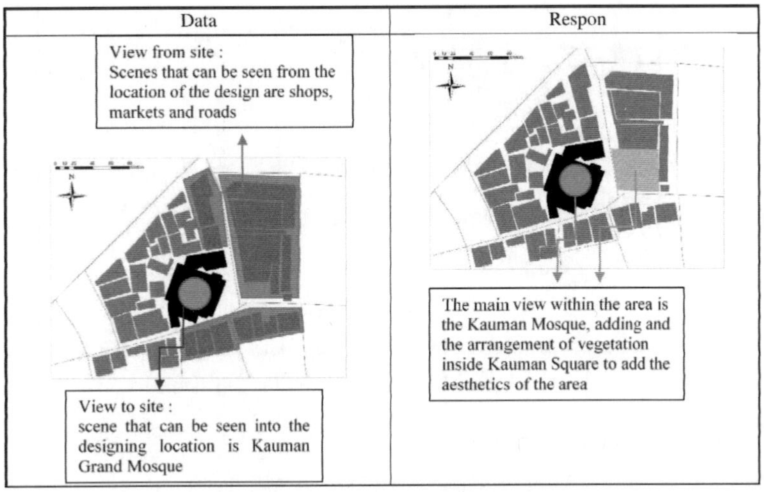

Figure 14. View Analysis in Kauman

3.2.5. *Accessibility Analysis*

The road conditions in Kauman are good but do not have decent pedestrian path so the response is adding pedestrian paths on the main roads namely Alun-Alun Barat Street, Kauman Street and Pasar Johar Street. Not only adding the pedestrian paths, the plan to redevelop the square that cuts the road until the front of Kauman Great Mosque makes a change in the road pattern, especially in Jalan Alun-alun Barat, so the road is turned around the Square.

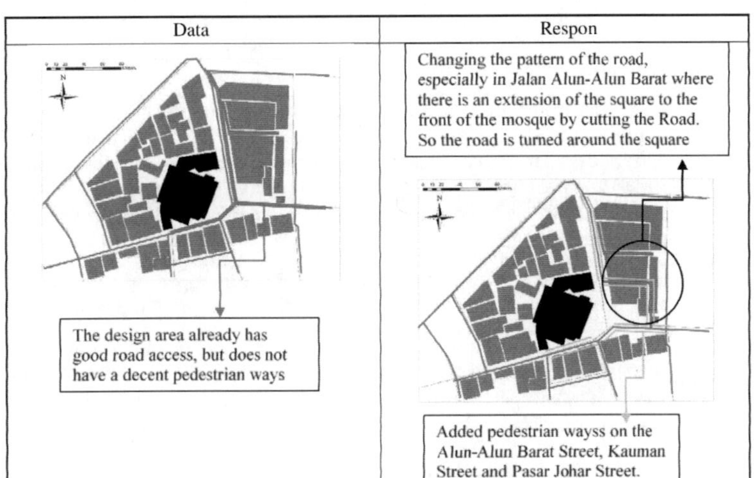

Figure 15. Accessibility Analysis in Kauman

3.2.6. *Zoning Area*

Based on the above five analyzes obtained zoning area in Kauman namely trade and services zone, worship zone, supporting zone, residential zone and green line. Zoning area will be used as a reference in the preparation of site plan design area.

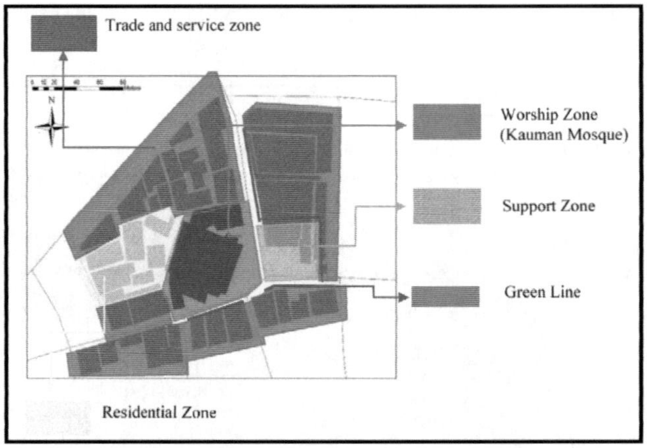

Figure 16. Zoning Area in Kauman

So the concept of the design area obtained as follows:

Table 1. Concept of Design Area in Kauman

Concept	Explanation
Arrangement of Kauman Square	In the existing condition, Kauman square is used as a parking lot without having a sitting area. In planning, the Kauman square will be arranged and expanded to Ya'ik Market so that the function of the square can run properly. On the inside of the square is added trees, landmarks, benches, lights, bins and walkways. During the normal days of the function of the square other than as a place to relax, discuss and gathering between community or students. Then for the celebration of the big day like Eid al-Fitr and Eid al-Adha, the square can be used to accommodate Eid prayer worshipers.
Changes in Road Patterns	There is a change in the road pattern, especially Alun-alun Barat Street. With the expansion of the square cutting off the Alun-Alun Barat street, therefore the road is deflected around the square using the land which now is the Ya'ik Market. The road deflection is also intended to break down the density at the front of Mosque, so activities within the mosque and the square are more conducive.
Pedestrian Path Provision	Pedestrian paths are provided on the right and left side along the main roads namely Alun-alun Barat Street, Kauman Street and Pasar Johar Street in accordance with the road pattern. The pedestrian path is made from ceramic material and there can be no ladder if the pedestrian path is cut off and continues the next few meters. It is intended that the pedestrian path can also accommodate disabled users, making it easier for wheelchair users or walking stick users.
Street Furniture Provision	Street Furniture is a complementary facility in open space and pedestrian path. Street furniture added include bins, benches / sitting area, street lights and parking arrangement
Provision of Green Open Space and Vegetation	Green open spaces are provided in the form of a square surrounded by shade trees. Trees scattered in front of Ya'ik market, Kauman square and in front of Kauman Mosque. Planting shade trees and parks in addition to adding to the beauty of the area, can also play a role to cool the air temperature, especially during the day because the Kauman is in the North Semarang so the air during the dry season was hot.
Drainage	The drainage pattern follows the street pattern in Kauman. Drainage in the form of closed drainage that is under the pedestrian path. The direction of the drainage flow leads to Kali Semarang
Traffic Access Lane	The traffic access point in this area is one-way street for main roads namely Alun-Alun Barat Street, Kauman Street and Pasar Johar Street. This is done considering the density of traffic flow coupled with trading and services activities and also the condition of a narrow road so it is not possible to be passed in two directions. While the road in two directions namely the road inside the settlement area, Pemuda Street and K.H Agus Salim Street.

Source: Concept Analysis, 2016

From the concept, the central area design is as follows:

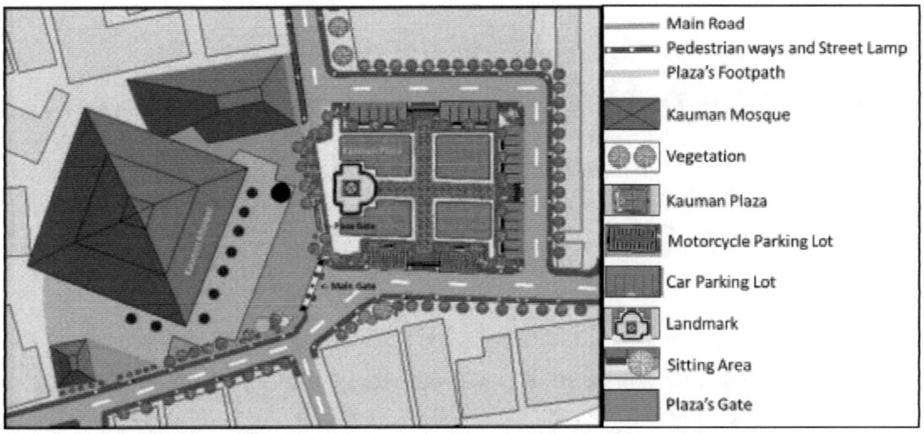

Figure 17. Design for The Center Area of Kauman Kampong Qur'an

4. Conclusion

The preservation concept of Kampong Kauman is based on economic, religious and socio-cultural activities. Including the physical condition of the buildings and the environment in the area of Kampung Kauman must be considered. Conservation efforts in order to realize Kampung Kauman as a religious educational tourism area is by conserving the form of traditional buildings, conducting religious activities to add attraction, revive the youth assosiation, maintain the trade activities of Arabic goods, socialize the importance of maintaining the characteristics of the village, Parties ie the community and government. Then needed the arrangement/planning of Kauman Mosque's central area to restore support Kauman Kampun Qur'an as well as restore the sense of place of the region. The plans include re-expansion of Kauman square, roadblocking, parking arrangement, street furniture pedestrian path provision and reforestation of the area.

5. Acknowledgments

We would like to thank the team of Diponegoro University for Science Techno Tourism Development (UFST2D) who have collaborated and encouraged in the dedication activities to the Kauman communities. We also thanks the board and community of Kauman Semarang Mosque which has helped in realizing the program of area arrangement to "Kauman Kampong Qur'an" as Religious Educational Tourism.

6. References

[1] Darban A 2000 *Sejarah Kauman: menguak identitas kampung Muhammadiyah* (Tarawang)
[2] Suprapti B A, Budihardjo E, Kistanto N H and Tungka A E 2010 Ethnography-Architecture in Kampong Kauman Semarang: A Comprehension of Cultural Toward Space *Am. J. Eng. Appl. Sci.* **3**
[3] Buchori A 2012 *Pengaruh Intensitas Menghafal Al-Qur'an terhadap Moralitas Santri Remaja di Pondok Pesantren Roudlotul Qur'an Kauman Johar Semarang* (IAIN Walisongo)
[4] Bappeda Kota Semarang 2010 *Semarang Spatial Planning Doccument*
[5] Budiarto A S 20016 Concept of Self-Conservation in a Traditional Village: an Empirical Study of Kampong Kauman Semarang Indonesia *Int. Trans. J. Eng. Manag. Appl. Sci. Technol.* **7** 21–34
[6] Clarke G and Stillwell J C H 2004 *Applied GIS and spatial analysis* (John Wiley & Sons)
[7] Zhang Y, Li X (Robert) and Su Q 2017 Does spatial layout matter to theme park tourism carrying capacity? *Tour. Manag.* **61** 82–95
[8] Sugiri A, Buchori I and Soetomo S 2011 Sustainable metropolitan development: Towards an operational model for Semarang Metropolitan Region *Int. J. Environ. Cult. Econ. Soc. Sustain.* **7** 301–323
[9] Goodchild M F, Parks B O and Steyaert L T 1993 Geographic Information Systems and environmental modeling
[10] Steinberg S J and Steinberg S L 2005 *Geographic information systems for the social sciences: investigating space and place* (Sage Publications)

Spatial Analysis in Determining Physical Factors of Pedestrian Space Livability, Case Study: Pedestrian Space on Jalan Kemasan, Yogyakarta

A F Fauzi[1] and A Aditianata[1]

[1]Department of Urban and Regional Planning, Faculty of Engineering, Universitas Esa Unggul, Indonesia

Email: a.faisfauzi@yahoo.co.id

Abstract. The existence of street as a place to perform various human activities becomes an important issue nowadays. In the last few decades, cars and motorcycles dominate streets in various cities in the world. On the other hand, human activity on the street is the determinant of the city livability. Previous research has pointed out that if there is lots of human activity in the street, then the city will be interesting. Otherwise, if the street has no activity, then the city will be boring. Learning from that statement, now various cities in the world are developing the concept of livable streets. Livable streets shown by diversity of human activities conducted in the streets' pedestrian space. In Yogyakarta, one of the streets shown diversity of human activities is Jalan Kemasan. This study attempts to determine the physical factors of pedestrian space affecting the livability in Jalan Kemasan Yogyakarta through spatial analysis. Spatial analysis was performed by overlay technique between liveable point (activity diversity) distribution map and variable distribution map. Those physical pedestrian space research variable included element of shading, street vendors, building setback, seat location, divider between street and pedestrian way, and mixed use building function. More diverse the activity of one variable, then those variable are more affected then others. Overlay result then strengthened by field observation to qualitatively ensure the deduction. In the end, this research will provide valuable input for street and pedestrian space planning that is comfortable for human activities.

Keywords: Pedestrian space livability, spatial analysis, Yogyakarta

1. Introduction

People utilize streets for their activities on daily basis, as streets are the main public spaces in cities [1,2].Therefore, to underpin its function, streets should accommodate many aspects such as economic, aesthetic and social aspects [3,4]. Economic aspects related to commodity circulation, aesthetic aspects related to positive image and visual element; and social aspects related to diversity of human activities conducted in the streets' pedestrian space. Jacobs [5] has pointed out that if there is lots of human activity in the street, then the city will be interesting. However, reality shown that usually streets are dominated by vehicles rather than human activities. It is shown that most of the city has lost its attractiveness.

Based on previous mentioned facts, various cities in the world are now developing the concept of livable streets. They enchanted pedestrian space to increase activities conducted there. According to

Jacobs, Jacobs, and Krier [5–7], livable streets shown by diversity of human activities conducted in the streets' pedestrian space in various range of time. Livable Street is increasingly becoming a vital factor of city attractiveness. Given the importance, this paper calls into question what is the main factor that influence streets' livability. Many proceeding research focusing on general concept of livable street theory and its purpose, therefore this paper focus in how physical aspect of pedestrian space influence streets' livability which is indicated by activity diversity. The research located is in Jalan Kemasan, Kotagede, Yogyakarta, because it shows diversity of activities in various range of time.

2. Methods

In order to analyze factors that influence streets' livability, deductive approach was selected because it is one of the most practical methods that represent a viable alternative to inductive approach. It allowed researcher to observe conceptual variable directly. Conceptual variable mentioned consist of three attributes: safety, comfort and building function. The selection of these three attributes based on the most commonly attributed determinant by experts as can be seen in Table 1 and Conceptual variables then break down into operational variable as display in Table 2. Spatial analysis was performed by overlay technique between livable point distribution map and operating variable distribution map (Figure 1). The results then strengthened by field observation to find the physical factors of pedestrian space livability.

:

Table 1. Livable streets conceptual variables based on experts

Livable Streets Conceptual Variable	Appleyard(1981)	Jacobs, J (1961)	Jacobs, A (1993)	Lusher et. al (2008)	*The American Institute of Architects* (2005)	*Schmitz and Scully* (2006)	Simonds (1994)
Safety	v	v	v			v	v
Comfort	v	v	v				v
Building function		v		v	v	v	
Variation of accessibility					v	v	
Social relation	v		v				
Economic improvement factor		v		v			
Level of identity strength			v		v		
Vehicle speed	v			v			
Pedestrian dimension			v	v			
Integral circulation			v				v
Traffic volume	v				v		
Visual interest		v					
Public health				v			
Landscape conservation					v		

Table 2. Livable streets operating variable

Conceptual Variable	Operating Variable	Measure
Comfort	1. Element of shading	Shading vegetation and roof shade or building canopy
	2. Street Vendors	Street vendors that giving access to pedestrian
	3. Building setback	Short setback building
	4. Availability of seat in pedestrian way	Seat or bench or stool available
Safety	5. Divider between street and pedestrian way	On street parking and tree lined curb
Building function	6. Mixed use building function	Vertical and horizontal mixed use building function

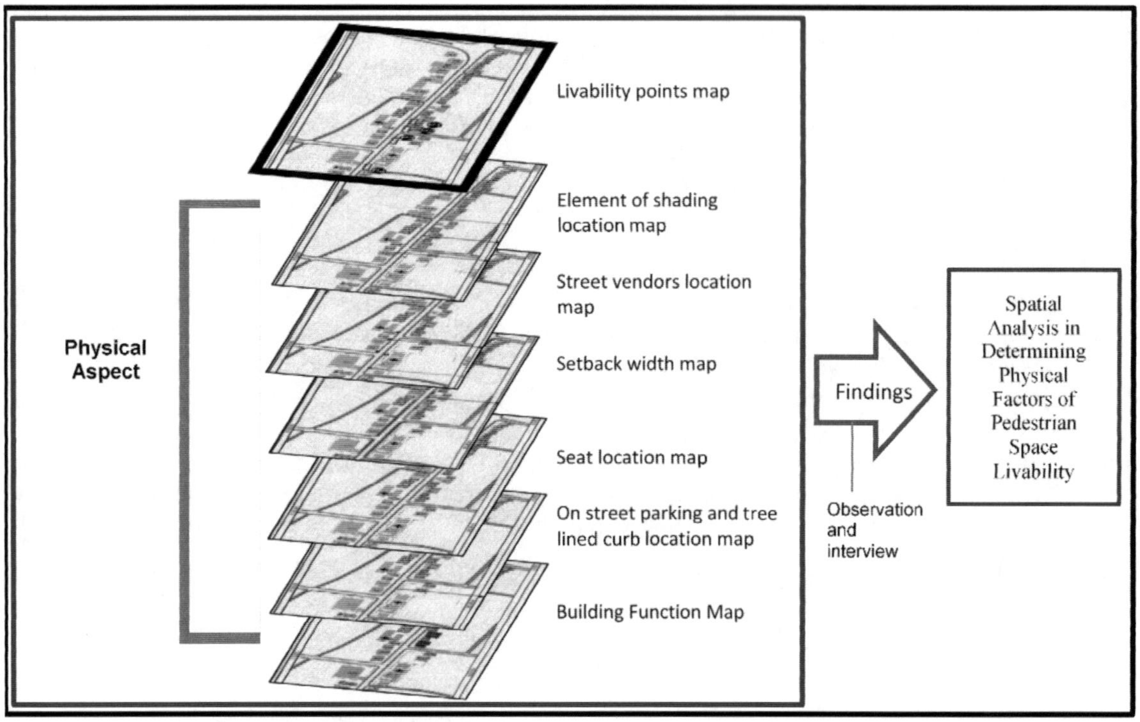

Figure 1. Spatial analysis method in determining physical factors of pedestrian space livability on Jalan Kemasan

Observation points of this study were determined through mapping on the most vociferous point or the most active point in Jalan Kemasan. Moudon [8] noted that those activities grouped into two kind of activities, dynamic and static activity. Whereas, the ideal number of pedestrian ranged from 8 to 18 pedestrians per minute per meter. Observation points which have less than 8 pedestrians per minute tend to be inanimate, while the spot which has 18 pedestrians felt over crowded. Observation time divided into two periods, morning and afternoon. Observation in the morning performed between 07.00-11.00 WIB, while in the afternoon observation performed between 12.00-16.00 WIB. It was chosen because most of the activities happened around that time. Observation conducted on weekday to minimize count up tourists from outside Yogyakarta.

3. Result and Discussion

Observation shows that static and dynamic activity occurs at Jalan Kemasan. Static activities consist of sitting, talking, eating, standing, or waiting for others. While dynamic activity shown by walking. All these activities occur all day long, either morning or afternoon. There were 10 most livable points based on the number of activity occurred. Those points shown in the figure 2 and figure 3. Spatial analysis performed by overlay technique between livable point distribution map and operating variable distribution map. The more variable found in livable points, the more it affects the livability. Overlay results shown in the figure 4.

IOP Conf. Series: Earth and Environmental Science **123** (2018) 012042 doi:10.1088/1755-1315/123/1/012042

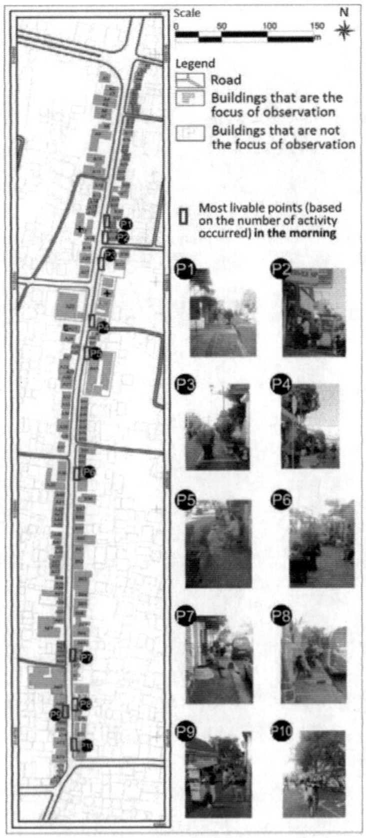

Figure 2. Most livable points map in the morning

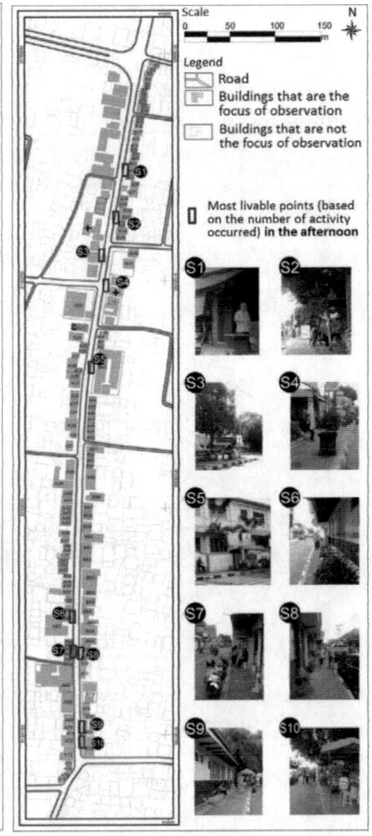

Figure 3. Most livable points map in the afternoon

P1 – P10 dots show most livable points in the morning, whereas S1-S10 dots show most livable points in the afternoon. Quantity and kind of activity occurred in those points are shown in table 3.

Table 3. Number and kind of activity occurred in most livable points (people/meter/minute)

Points	Dynamic Activities	Static Activities								Quantity
	walking	sitting	standing	talking	squatting	sweeping	eating	snacking/shopping	playing	
Most livable points map in the morning										
P1	5	1	2							8
P2	5	1					1	2		9
P3	5	2	1							8
P4	5	3	1	3	1		1			14
P5	6	3		2						11
P6	6	1		2	1			3		13
P7	9	3	1	2	1					16
P8	8	2		2				2		14
P9	7	1	1	2			2	2		15
P10	9	2		2			2	3		18
Most livable points map in the afternoon										
S1	5	1	1			1				8
S2	4	1	1						2	8
S3	3	2		2	1			1		9
S4	5	1	1					1		8
S5	4	2						2		8
S6	4	2		2						8
S7	6	1	1	2						10
S8	5		2			1				8
S9	6	2	2	2						12
S10	7	2	1	2			1	1		14

IOP Conf. Series: Earth and Environmental Science **123** (2018) 012042 doi:10.1088/1755-1315/123/1/012042

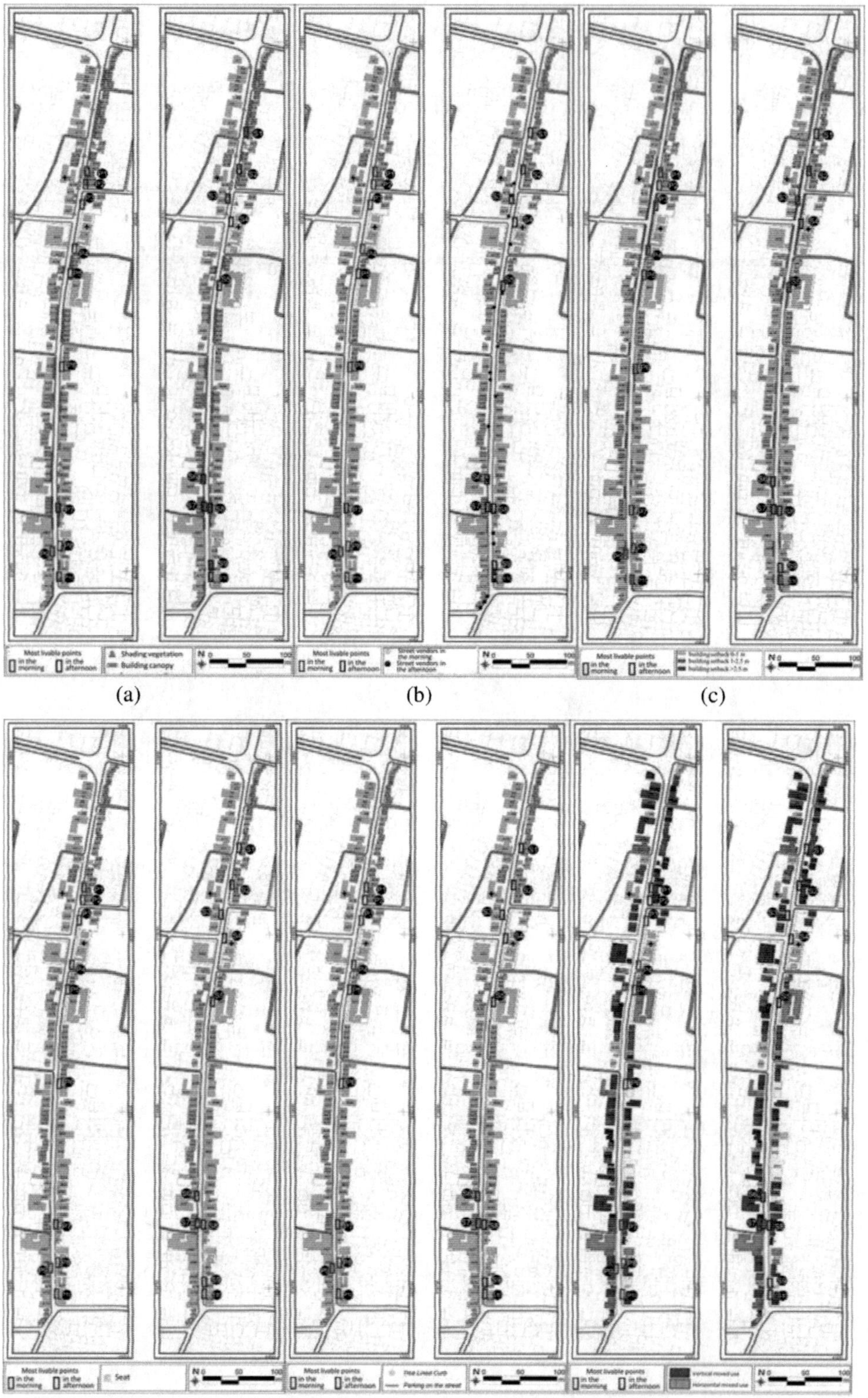

Figure 4. Overlay technique between livable point distribution map and operating variable distribution map: (a) element of shading; (b) street vendors; (c) building setback (d) seat location; (e) divider between street and pedestrian way; (f) mixed use building function

Furthermore, in order to obtain outcome, conductive overlay done between livable points map and all variable distribution map (Figure 5). Overlay results shown in the Table 4.

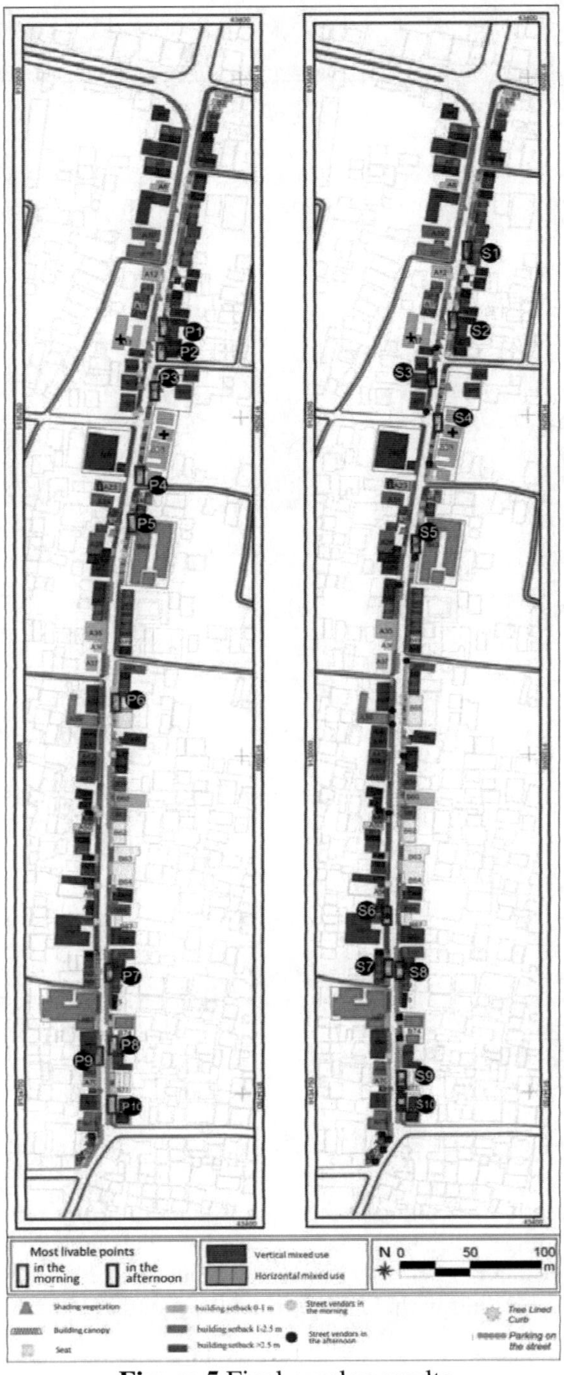

Figure 5. Final overlay results

Table 4. The linkage between livable points (activities) and all variables (physical factors)

Points	Dynamic Activities	Static Activities	Total Activit ies	Element of Shading		Street vendors that giving access to pedestrian	Building setback			Seat	divider between street and pedestrian way		Mixed use building
				Shading Vegetation	Building Canopy		Building setback 0-1 m	Building setback 1-2.5 m	Building setback > 2-.5 m		Parking on street	Tree lined curb	
in the morning													
P1.	5	3	8	v	v		v					v	v
P2	5	4	9		v	v	v			v			v
P3	5	3	8	v					v			v	v
P4	5	9	14	v		v		v		v		v	v
P5	6	5	11	v				v		v		v	v
P6	6	7	13			v		v		v		v	v
P7	9	7	16		v		v			v	v		v
P8	8	6	14			v		v		v	v		v
P9	7	8	15			v		v		v	v		v
P10	9	9	18	v		v		v		v	v		v
in the afternoon													
S1	5	3	8	v	v		v					v	v
S2	4	4	8	v				v					v
S3	3	6	9	v		v			v	v			v
S4	5	3	8			v		v		v		v	v
S5	4	4	8	v		v		v		v		v	v
S6	4	4	8		v	v	v			v			v
S7	6	4	10		v		v			v	v		v
S8	5	3	8		v		v				v		v
S9	6	6	12		v		v			v	v		v
S10	7	7	14	v		v		v		v	v		v

Based on the results of overlay and morning-afternoon observation, it revealed 3 physical factors that affect Jalan Kemasan livability the most. Those are: divider between street and pedestrian way, element of shading, and availability of seat in pedestrian way. The linkage illustrated on the figures 6.

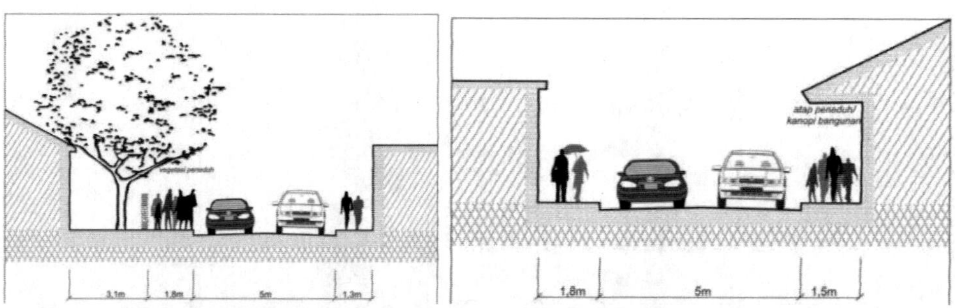

Figure 6. The linkage between element of shading and activity

Figure 6 shown that street with shading, either by a tree or building roof, have much more activity than other part without shading. It relates to sun exposure, which pedestrian feels more comfortable perform their activity under shading. In the afternoon, where sun exposure is higher, almost every activity done under shading. In line with Jacobs [9] and Simonds [10] that said trees and any other shading material is substantial to create pedestrian comfortness.

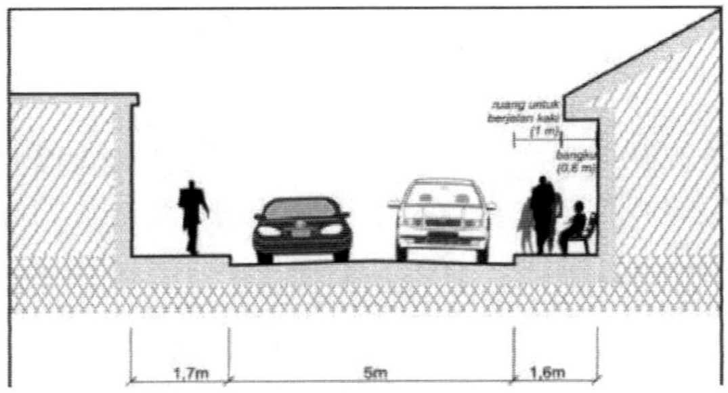

Figure 7. The linkage between availability of seat and activity

Figure 7 shows that the side of the street with seat has more activity than the side that do not. In line with what is said by Jacob [5]and Appleyard [6], availability of seat is important because it relates to pedestrian comfort. It functions as spot to sit, chat and rest.

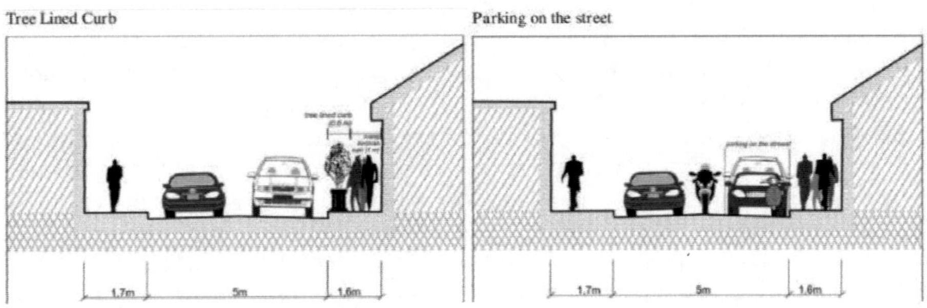

Figure 8. The linkage between divider of pedestrian way and activity

Pedestrian way with divider such as tree lined curb and parking-on-the-street vehicle have much more activity than the one which do not (Figure 8). The divider between street and pedestrian way improves sense of security from traffic accident. This sense of security also arises from the difference height between pedestrian and road. However, Jalan Kemasan pedestrian and road have relatively the same high, spots with tree-lined curb and parking-on-the-street vehicle make act as divider giving the same security sense. In addition, the presence of other aspects such as street vendors, building setbacks, and mixed use building function also affect pedestrian space livability (Figure 9). Although, those aspects effect are not as big as the first three.

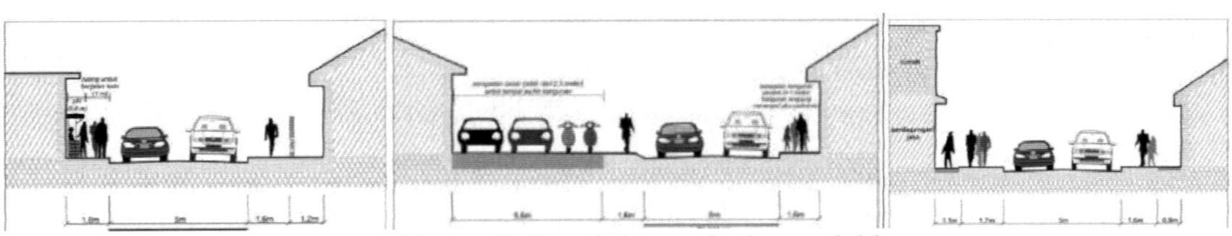

Figure 9. The linkage of other variable
(street vendors, building setbacks, and mixed use building function)

The presence of street vendors relates to the ability to attract people to engage and interact. In accordance to Jacobs [5] that street vendors in pedestrian space able to intensify activity numbers. However, in Jalan Kemasan, it needs to underpin that street vendors who support the sense of security are street vendors who do not use the entire pedestrian way and still give access to pedestrian. Whereas, street vendors who use all pedestrian way raises sense of discontent and harbors accidents.

Furthermore, the existence of a short setback building (0-1 meters) correlates with the security of the pedestrian. It provides pedestrian way sustainability and heat shading. Whereas, mixed use building function relates to make Jalan Kemasan more interesting by doing various activities. This is in line with Jacobs [5] and Schmitz and Scully [11] that said mixed use building more affect diversity of pedestrian then non mixed use building.

4. Conclusion

The research revealed that presence physical factor such as element of divider between street and pedestrian way, element of shading, and availability of seat greatly affects Jalan Kemasan pedestrian way livability. In addition, other factors such as street vendors, building setbacks, and mixed use building function also effect livability only in small portion. In general, these findings suggest more human pedestrian way planning. Pedestrian way need to be more livable and gives space to create activity. This research leads us to better way to plan pedestrian way by considering these things:(A) Continuous pattern of shading to protect pedestrian from sun exposure. In the city center such as Jalan Kemasan, building roof is an effective example of shade. The widths of the building's roof located next to the street should be at least 1.25 meters to fully cover pedestrian activity.(B) Increasing pedestrian security by adding differentiation of altitudes between the pedestrian way and the street, arrange tree lined curb orderly and allowing on-the-street parking. On-the-street parking is only permitted if there is suitable width and brings no potential traffic issues. (C) Providing comfortable seat or bench in a pedestrian way in order to accommodate pedestrian who need to take a rest. Observations showed that many pedestrian sit on pots, hallways, or pedestrian pavement. This shows the importance of seat availability.(D) Buildings with large setback should be avoided, especially those that do not have fence and permanent parking garage. It will courage inaccessibility. Building with short setback is very good for pedestrian. It creates the continuity of the pedestrian way and provides shading to protect pedestrian from sun exposure. (E) Arranging street vendor to assist pedestrian continuity. Space provide by street vendor should courage comfort and safety. (F) Buildings located in the next the street should be made in order to have mixed use function This research shows that mixed use buildings have more varied function, thus it courage more activies.

5. References

[1] Carmona M, Heath T, Oc T and Tiesdell S 2003 Public spaces. Urban spaces *Dimens. Urban Des.* **1** 46–47
[2] Trancik R 1986 *Finding lost space: theories of urban design* (John Wiley & Sons)
[3] Gehl J 2007 *Three types of outdoor activities; outdoor activities and quality of outdoor space* (na)
[4] Krier R and Rowe C 1979 *Urban space* (Academy editions London)
[5] Jacobs A B 1993 Great streets *ACCESS Mag.* **1**
[6] Appleyard D 1980 Livable streets: protected neighborhoods? *Ann. Am. Acad. Pol. Soc. Sci.* **451** 106–117
[7] Lusher L, Seaman M and others 2008 Streets to live by: how liveable street design can bring economic, health and quality-of-life benefits to New York City
[8] Moudon A V 1987 Public streets for public use
[9] Jacobs J 2016 *The death and life of great American cities* (Vintage)
[10] Simonds J O and others 1994 *Garden cities 21: creating a livable urban environment.* (McGraw-Hill, Inc.)
[11] Schmitz A and Scully J 2006 *Creating walkable places: compact mixed-use solutions*

IOP Conf. Series: Earth and Environmental Science **123** (2018) 012043 doi:10.1088/1755-1315/123/1/012043

Sustainable Revitalization in Cultural Heritage Kampong Kauman Surakarta Supported by Spatial Analysis

M Musyawaroh[1,2], T S Pitana[2], M Masykuri[3], and Nandariyah[4]

[1]Doctoral Candidate Study Program of Environmental Science, Graduate School of Universitas Sebelas Maret, Surakarta, Indonesia
[2]Study Program of Architecture, Faculty of Engineering, Universitas Sebelas Maret, Surakarta, Indonesia
[3]Study Program of Environmental Science, Graduate School of Universitas Sebelas Maret, Surakarta, Indonesia
[4]Study Program of Agronomy, Graduate School of Universitas Sebelas Maret, Surakarta, Indonesia

Email: musysofwan@gmail.com

Abstract. Revitalization is a much-needed for a historic kampong as a settlement, place of business, and as tourist destinations. The research was conducted in Kauman as one of the cultural heritage kampong which was formerly as a residence of abdidalemulamaKeraton who also work as batik entrepreneurs. This study aims to formulate a sustainable revitalization step based on the character of the area and the building. Aspects of sustainable revitalization that analyzed are the physical and non-physical condition of the environment. This research is an applied research with qualitative rationalistic approach supported with spatial distribution analysis through satellite imagery and Arch-GIS. The results revealed that sustainable revitalization for Kaumancan be done through: 1) Physical condition of the environment consists of land and building use, green open space, recreational park and sport activities, streets, drainage network, sewer network, the garbage disposal network; 2) Non-physical of the environment consists of economy, heritage socio-cultural, and the engagement of relevant stakeholders. The difference of this study with others is, this study is a continuation of the Kauman revitalization assistance program which involves community participation to produce a more appropriate solution for the problem of kampong.

Keywords: Sustainable Revitalization, Cultural Heritage Kampong, Spatial Analysis

1. Introduction

The conflict of interest between maintaining cultural heritage and conservative development is an important topic for society around the world [1]. This opinion is also stated by Moosavi [2], the planning of a city's development must take into account history, culture and meaning. The attempt to revitalize history, culture and identity in the city will face two conflicting challenges. On the one hand, the old city with its unique is irreplaceable monument, history, culture and identity, and on the other hand, contemporary urban life with its technological requirements.

The effort to overcome this is by sustainable revitalization, which is an integration of technologies and solutions to meet new needs as the impact of socio-economic dynamics by enhancing the vitality of historical potential of urban area in adaptive reuse [3]. Cultural value development activities aim to improve the welfare of the community [4,5]. Revitalization covers the management of the physical aspects as well as the social and economic impacts of the region concerned.

The revitalization of land and buildings as spatial elements can inform the specific architecture of region, architecture can take part in the creation of heterogeneous space in environmental built based on culture [6]. Sustainable revitalization of urban, physical, economic and socio-cultural historic areas is done to counter rapid changes due to the impact of globalization and urbanization. Sustainable revitalization use to prevent environmental degradation, and can only be done with a sustainable strategy involving all stakeholders related with holistic approach [7].

From the description above, it can be interpreted that sustainable revitalization is needed to preserve the historic areas that have decreased environmental quality, which includes physical, economic and socio-cultural aspects. In addition to these aspects, relevant stakeholders also play an important role in sustainable revitalization. Revitalization action is a phase of a non-linear process involving frequent and intense negotiations between stakeholders [8]. Without good coordination, conservation measures will not be able to achieve optimal results.

The role of stakeholders bringing change such as local and central government, state-owned and private companies, and the public, is indispensable for the sustainable development of legacy heritage sites in the future. The top-down regional management system without considering the will of the local community has failed, and has caused difficulties in implementing the revitalization plan of the historic area [9]. Governments should seek to protect cultural heritage assets by considering the wishes of the community, so that conservation and development can proceed in a balanced [1].

Buildings and heritage sites are crucial elements of the city development; they can either act as an image of the city or tourist attractions. One of Surakarta City Government's efforts to revive the traditional territories is by proclaiming the Surakarta City's tourism vision as "The future Solo is the old Solo". The vision proclaimed by Mayor Joko Widodo on 18 February 2009, coincided with the 264th anniversary of Surakarta City.

As an old town, Solo has 81 buildings, 17 monuments, 5 parks and cemeteries along with 6 cultural heritage sites as listed in the Mayor Decree No. 646/116/1/1997 on the stipulation of the buildings and heritage sites and is complemented by the Decree of the Head of the City Spatial Planning No. 646/40/I/2014 on the stipulation of buildings deemed eligible as cultural heritage.

However, in Surakarta many ancient buildings are not maintained, torn down replaced by other commercial buildings, mortgaged to seek loans to the bank, or dismantled for sale retail [10]. Investors and building owners tend to raze the historical objects for more commercial purposes. Thus, revitalizing the heritage sites with all its activities is crucial to do, so that Solo will not lose its identity as a cultural center of Java.

The conservation of heritage sites does not actually hinder the growth of new buildings, but aims to harmonize old and new buildings in the region with cultural and economic activities to become the identity of a leading district in the city. According to Chung [8], harmonization will form the identity of the city that presents a series of events through a series of traces of building culture, architectural details and regional colors on the next generation.

There are four cultural heritage areas in Surakarta namely Baluwarti, Laweyan, Loji Wetan and Kauman. Kauman was chosen as the object of research because of its close relation with Keraton Surakarta (the palace of Surakarta), its potential as a kampong of santri that still exist [11], and alot of environmental problems in it. The location of Kampong Kauman blends with Masjid Agung (the Great

IOP Conf. Series: Earth and Environmental Science **123** (2018) 012043 doi:10.1088/1755-1315/123/1/012043

Mosque), forming an Islamic character in this community. The development of the kampong began in the mid-1800s, when batik entrepreneurs built traditional Javanese and Indische houses [12].

The quality of Kauman's environment began to decrease alongside with the bankruptcy of batik industries in 1930-1970s. The area then turned into slums and the old buildings began to collapse [13]. Sustainable revitalization of Kauman is needed before the area collapses and disappears, particularly because of the location of this kampong which has a potential to be developed into a tourism asset of Surakarta.

Kampong revitalization program has been conducted by the writer with the team from Department of Architecture Faculty of Engineering Universitas Sebelas Maret since 2006. The strategies implemented in the revitalization are: short-term, reviving the batik business; Medium-term, revive the village of santri; And long-term, making Kauman as a tourist kampong of religious and batik [12]. Since 2008 The municipal, provincial and central governments have provided financial support to revitalize the area, but to date the results have not been maximized despite an increase in economic aspects. Sustainable revitalization is needed so Kauman can be more comfortable and healthy to live in and can be used as religious tourism and batik.

This research is a continuation of the kampong revitalization assistance program that has been conducted since 2006, involving community participation and relevant stakeholders. The purpose of this study is to find a sustainable revitalization step that must be done to make the environment more healthy, comfortable to live and can be developed into a religious tourism kampong and batik. Issues raised in this study are:

a) How is the existing condition of environmental in cultural heritage Kampong Kauman nowadays?
b) How is the sustainable revitalization program that should be applied for Kauman?

2. Data and Method

2.1. Data

Kelurahan Kauman is located in the Pasar Kliwon district, at the city center of Surakarta (figure 1). In this district there is *Masjid Agung* (the Great Mosque), that becomes the center of the spatial orientation of kampong. Kauman is a *santri* and batik kampong, marked by old traditional Javanese and Indo-European (called Indische) architecture, as well as Islamic community events.

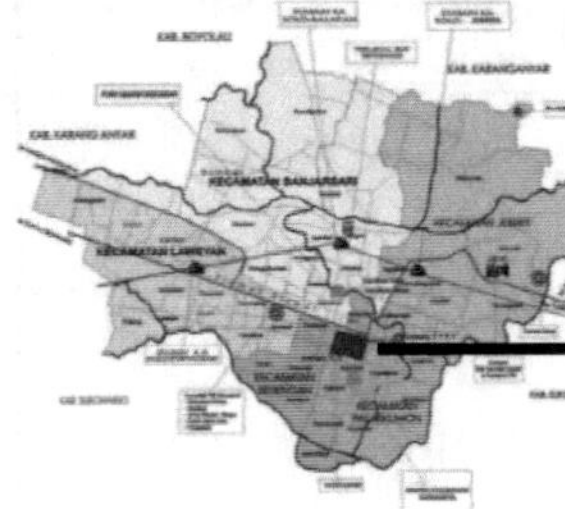

Source : www.surakarta.go.id, 2011

Figure 1. The Location of Kauman in Surakarta

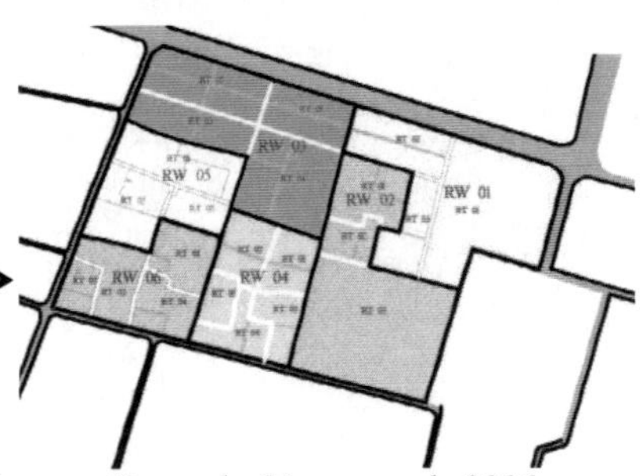

Source : Drawn by Musyawaroh, 2016

Figure 2. The divided area of Kauman

Kelurahan Kauman is divided into 7 *Rukun Warga* (Community Association) and 22 *Rukun Tetangga* (Neighborhood Association). The area of *Kelurahan* Kauman (based on Bakosurtanal topographical maps, 2001) is 20.497 hectares, with most of area covered by buildings and pavement (see figure 2).

IOP Conf. Series: Earth and Environmental Science **123** (2018) 012043 doi:10.1088/1755-1315/123/1/012043

2.2. *Method*

This research was conducted in Kelurahan Kauman in the city of Surakarta, considering the existing problem to revitalize the cultural heritage kampong in Surakarta. This is an applied research, providing solutions to continue the revitalization program that has been implemented, so that environmental issues can be solved better.

This research used a rationalistic qualitative approach, providing interpretation of the phenomena that occur in the field by using the theory developed derived from research and community assistance program previous. The research was carried out by collecting data from literature study, interviews (with residents of Kauman, Pamong Praja and related institutions), satellite imagery observation and field surveys. The data that collected validated with field surveys.

Sampling was done purposively, selected samples are the house of abdi dalem keraton (cleric servant of the palace and another), the Kauman community activities and the environment physical conditions of Kauman. The data obtained were identified, classified and analysed. Data on the physical condition of the environment were analyzed based on National Standard on the Regulation SNI 03-6981-2004 of the simple housing environment is not structured in urban areas and SNI 03-1733-2004 on the planning of urban housing environment. Data on non-physical environmental conditions were analyzed using interactive model analysis continuously until the data were saturated (figure 3). Activities in the analysis include data reduction, display data, conclusion drawing and verification.

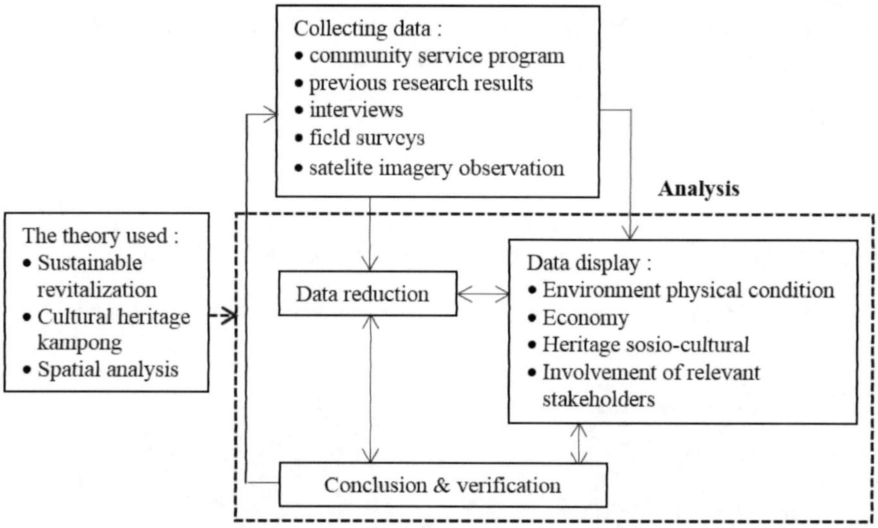

Figure 3. Interactive Model Analysis

3. Result and Discussion

The environment existing condition has gradually changed since 2006, because the revitalization program that has been done by community of Kampong Wisata Batik Kauman supported by revitalization team from UNS Architecture Department. Evaluation of Kauman existing environment condition is needed, covering some aspects as follows.

3.1. The Physical Condition of Environment in Kauman

3.1.1. Land and Buildings Conditions

As described above, the area of Kauman is 20.497 hectares, with most of area covered by buildings and pavement. The total number of buildings is 735 and there are 16 market buildings [14], see figure 4.

Source: Google Earth 2016 &Musyawaroh, 2016 : 22

Figure 4. Figure Ground of Kauman.

Source: Google Earth 2016 &Musyawaroh, 2016 : 25

Figure 5. Green Open Space in Kauman

From the figure ground image above, it can be observed that the area covered by buildings is 128.622 m^2 or 62.75% with an open space of 76.350 m^2 or 37.25%. The open space consists of hardscape (streets and channels), and soft scape (green open space). Based on the result of field surveys and satellite imagery, the open space consists of 60% pavement area or 45.810 m^2 and 40% of green open space or 30.540 m^2.

3.1.2. Green Open Space Conditions

From the 40% green open space in Kauman (figure 5), 438.249 m^2 or 14% of it is private green open space in public buildings property and 857.726 m^2 or 27.4% of it is private green open space at the yard of houses. There is no public green open space in Kauman.

3.1.3. Recreational Park and Sport Activities

In Kauman there is no park for recreation, or facilities to exercise at all. The community uses the Masid Agung yard, the alley in the kampong and the "city walk" on Slamet Riyadi street for sporting activities and children's playground. The Masid Agung yard is used for martial arts sports in the evenings, "city walk" for jogging activities and healthy walk in the morning, and alleys for children's playground in the afternoon.

3.1.4. Streets Conditions

The existing road network in Kauman according to SNI 03-1733-2004 namely local secondary roads type II with the asphalt road 6 m, local road/secondary road III with asphalt/paving block/concrete road 3 m, the environment I road with paving block/concrete road 2 m, and the environment II road with paving block/concrete road 1.2 m. The two-way roads that made of asphalt and are passable by motorized vehicles and non-motorized vehicles are Trisula Street, Kalimosodo Street, Masjid Agung Street, Wijaya Kusuma Street and Cakra Street.

One way road is made of asphalt and is passable by motorized vehicles (motorcycles and cars) and non-motorized vehicles is Hasyim Ashari Street. Others are alleys that made of paving blocks or concrete, that only passable by motorcycles and non-motorized vehicles.

3.1.5. Drainage Network Conditions

Most of Kampong Kauman area does not have drainage channels, and in other parts, drainage channels are mixed with household waste channels that cause environmental pollution. Kauman has only 50% of proper drainage channels, the rest (especially those in the small alleys) let rainwater flow to lower lands, and partly to small and non-permanent channels that are created by the local community.

The drainage channel causing most problems in Kauman is at Wijaya Kusuma Street. The channel is located in the middle of the road, and collects all types of wastewater such as rain water, domestic waste water (including fecal), and batik wastewater. These channels release a bad odor and disturb the neighborhood. Furthermore, the channel concrete cover has many holes and is bumpy, making it uncomfortable to pass.

3.1.6. The Sewer Network Conditions

There are two kinds of sewers in Kauman, household wastewater sewers (domestic) and *batik* wastewater sewers. Most of the community in Kauman does not have a septic tank because of the limited land, and they are still not aware of the usage of PDAM's interceptor (the sewer network of Local Water Company) as the channel to dump the feces. Even though it has been informed, it is still poorly implemented.

Interceptors channels for domestic waste are only available on some roads in the Eastern region of Kauman (east area of Wijaya Kusuma Street), the western region of Kauman has no particular domestic sewer network. The domestic waste channels still merge with the drainage channels.

3.1.7. *The Garbage Disposal Network Conditions*

For garbage disposal network, the citizens provide a simple garbage bin at each house. The garbage is taken every morning by garbage men to be thrown away to garbage dump in the form of garbage truck provided between 6.00 to 8.00 a.m. at Kalimosodo street (located in the westof Bank Central Asia at Gladag) Kauman itself does not have a garbage dump. The nearest garbage dump is located at Gajahan district, in the west of Klewer Market.

3.2. The non-physical condition of environment in Kauman

3.2.1. Santri'sHeritage Socio-cultural

According to the data from the *Kelurahan* Office in February 2016, the population of Kauman currently consists of, 95% Muslims (3,367 people), 47 Christians, 95 Catholics, and 36 Buddhists. The non-Muslim population consists of immigrants (mostly of Chinese descent) who live on the outer edge of Kauman. As a cultural heritage kampong,Kauman has heritage potentials as follow.

a) The settlement kampong of *abdi dalem ulama keraton*, another *abdi dalem* and *kawulo dalem.*
b) Kampong where batik businessmen live and run their business.
c) Kampong where *Sekaten* ceremony and *Grebeg* take place.

There are 6 *langgar* (small mosque) in Kauman, one of the *langgar* (LanggarSememen) was converted into Sememen Mosque. The culture of *santri* is still dominant until today which can also be identified by many pengajian (religious activities) done in this kampong (figure 5).

IOP Conf. Series: Earth and Environmental Science **123** (2018) 012043 doi:10.1088/1755-1315/123/1/012043

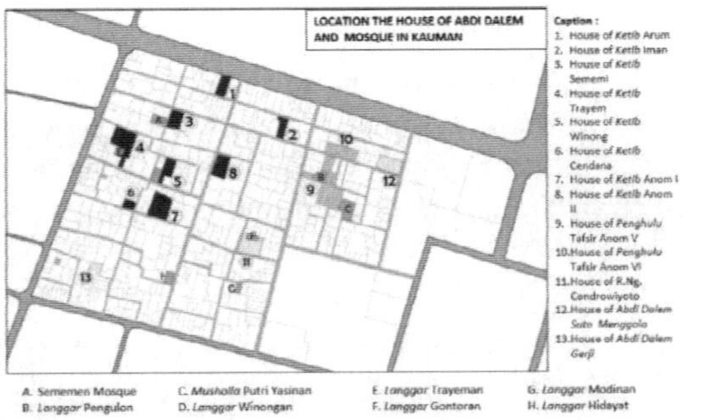

Source: Musyawaroh, 2016 : 40

Figure 6. Location of *abdi dalem* houses &*langgar* in Kauman.

Source : Musyawaroh, 2016 : 22

Figure 7. Location of ancient buildings in Kauman.

All langgar/small mosque can still survive up to this day except Pengulon Mosque and Cendanan Mosque. As for the keraton servant clerics' houses which still survive include house of Ketib Iman's, Tafsir Anom VI's, Ketib Anom II's and R.Ng. Condrowiyoto's. Meanwhile, the house damaged is house of Ketib Anom I's. Ketib Sememi's house has already transformed into Nahdhatul Muslimat School and Ketib Arum's house turned into a new building.

3.2.2. Batik Heritage Socio-cultural

The batik business has been maintained for generations in Kauman, although it went bankrupt in 1939 until the 1970s [13] but began to rise again since 2006. In 2006 batik business which only live 8 [12], now increased to more than 100 entrepreneurs. Business activities have evolved since community empowerment intensified in Kauman with various ways of socialization to the community.

Some of the ancient houses in Kauman are used as commercial buildings such as shops, batik shops, convection stores, bookstores, florists, jewelry stores, restaurants, stationery and office stores. Especially ancient buildings along Wijaya Kusuma street and Cakra street. In some alleys, such as Wijaya Kusuma III Road, there is a row of batik house showroom facing directly to the road, so the atmosphere is like being in the middle of the batik market (figure 6).

Besides the religious activities and batik, everyday cultural activities conducted in Kauman in example *Sekaten* (on the ward *Pradonggo*of *Masjid Agung*), *hadrah* (on *Langgar* Modinan), *karawitan* (in the House of Mr. Bintoro RT 02 RW 08), Grebeg in *Kuncungan* (the canopy of *Masjid Agung*) and Keroncong (at RT 02 RW 01).

3.2.3. Economic Aspects

Kauman is a dense settlement with population density on March 2016 was 3,553 inhabitants that consist of 721 householders with 735 houses (Musyawaroh, 2016). At present, most citizens livelihoods as a traders (719 people), construction workers (111 people), industry labors (155 people), entrepreneurs (348 people), transportation (65), pensioners (74 people), civil servants/the army/police (49 people) and others (1,341 people). Batik business in Kauman has rise again, initially only 8 are active now become more than 100 entrepreneurs [14].

The association of Batik Tourism Kampong Kauman has three shared showrooms, established to achieve a vision to encourage Kauman people to sell and produce batik. This association's office is located at Cakra Street No 14 Kauman, Solo.

3.2.4. Engagement Aspects

The community has been actively involved in revitalizing and reviving their area since 2006 by establishing the Association of Tourism Kampong of Batik Kauman. In addition, the community embraces the competent parties in their fields such as universities (UNS) and private institutions (Association of Experts Management and Association of Transportation Entrepreneurs).

In addition to the authors, several universities had conducted research in the Kauman area. Nevertheless, this research is not followed up with a concrete program to develop this area with the community. Aid programs have been conducted by researchers and teams since 2006, continuing the research findings in 2001.

The local government's agencies involved in physical treatment of Kauman were the Spatial and Urban Planning Agency in 2009 in the form of street furniture planning and program of the Tourism Agency in 2006-2007. Other activities with other institutions were merely training and seminar/FGD invitation. The region revitalization which was done by the Ministry of Public Work in 2007-2008 was DED creation and in 2011 was drainage planning which was then repaired in 2012.

3.3. Sustainable Revitalization for Physical Environment of Kauman

3.3.1. Sustainable Revitalization for Land and Building Use

Ancient buildings are the cultural tourism assets in Kauman (figure 6). To preserve this building, socio-economic functions need to be cultivated so that it can benefit the owner, as regulated in Law no. 11 of 2010 Chapter I Article 1 Paragraph 22. Not all ancient houses in Kauman have been used as commercial buildings. Due to very expensive maintenance costs, the building is then sold/ destroyed and replaced with new commercial buildings that are more profitable to the owner.

The utilization of ancient building is undertaken by employing adaptive-reuse, with receiving a new function without destroying the original building's facade. Thus the ancient building can change from cost center to profit center [3,15]. In addition to functioning as a residence, ancient house can be utilized as batik showroom, homestay, book store, food stall and others. With the improvement of building function, it is hoped that people can preserve their ancient land and buildings independently.

3.3.2. Sustainable Revitalization for Green Open Space

In this case, the total inhabitant of Kauman is 3.545 inhabitants. Therefore, the open green space needed is 3.545 x 0,3 m^2 = 1.063,5 m^2. Open space available in Kauman is 30.540 m^2, thus the width of green open space in Kauman has fulfilled the standard. However the green open space in Kauman is dominated by green open space in the area of the Great Mosque of Surakarta and BCA. There is no public green open space inside the kampong.

In the afternoon, the weather in Kauman is so hot that reaches 42° C (according to measurement result in Musyawaroh [14] since the region is too dense without greening. Alongside the roads it is not possible to greening because the road is too narrow as well as at house yard which is mostly used for carport. The greening of this region can only be done by planting in pot, hanging pot and creeping plants over pergola.

3.3.3. Sustainable Revitalization for Recreational Park and Sport

According to SNI 03-6981-2004 and SNI 03-1733-2004, each village unit inhabited by 30,000 inhabitants needs park and sport field to serve society's activity needs, for instance sport match, ceremony and other activities. Park and sport field for a village unit with 30,000 inhabitants need 9,000 m^2 space in size or by the standard of 0,3 m^2/inhabitant.

Because in Kauman there is absolutely no space available for recreational space, children's play and sports activities, so the activity is done in the area of roads / alley at the region in the morning /evening, and in the courtyard of the Great Mosque in the Night, or along the city walk in the morning.

The space used for the activity is organized as follows:

a) Endeavored in the morning and afternoon (before 08.00 am and after 4 pm pm) motor vehicles do not enter this area (except the occupants or unloading / loading), so as not to interfere with children's play/recreation/sports activities of communities.
b) Children play area/recreation/sports on the alley planted with greening, so that the air becomes fresher, reduce heat and more beautiful.
c) Certain places are provided with garden chairs and pergolas.

3.3.4. Sustainable Revitalization for Street.

In Kauman, the widest road is 6 m whereas the narrowest road is 1.2 m. In order to avoid traffic congestion, the traffic arrangement is made as figure 7. This is done to avoid the chaos of traffic flow and congestion within the kampong.

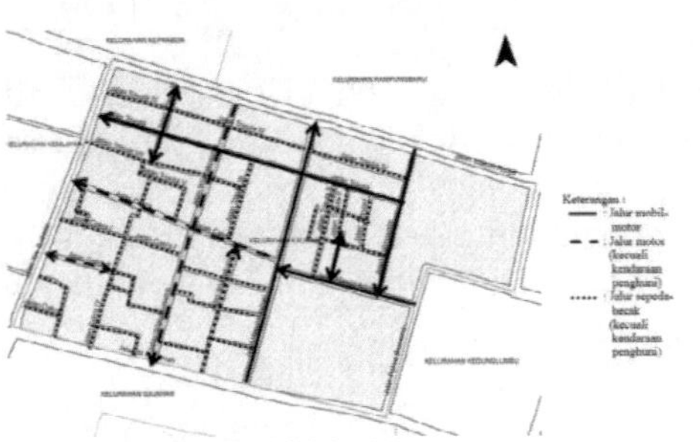

Source: Musyawaroh, 2016 : 26

Figure 8 Traffic management in Kauman

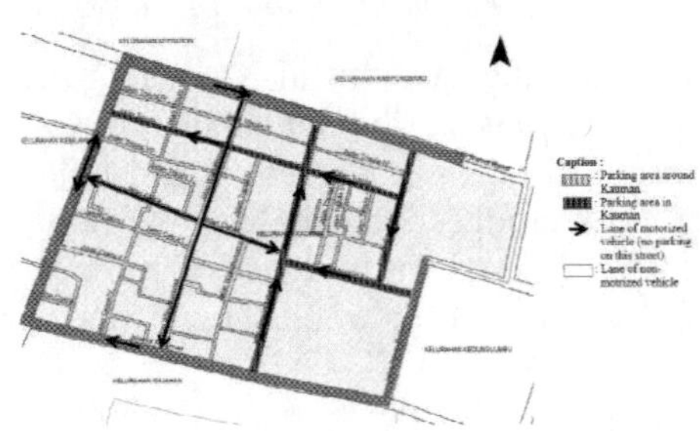

Source: Google Earth 2016 & Musyawaroh, 2016 : 25

Figure 9. Parking management in Kauman

Parking area can not be constructed inside the Kauman area, because the road width is very limited. For that car Kauman visitors parked on the edge of the road around the area, or in the area on the edge of Jl. Hasyim Ashari and Jl. Trident (figure 8).

3.3.5. Sustainable Revitalization for Drainage Network.

Drainage channel on Wijaya Kusuma street according to Muttaqin's calculation (in Musyawaroh, (2016) has filled the capacity, but becomes a source of problems in Kauman. This channel should only be rain water sewer.On the other hand, domestic wastewater is channeled into the interceptor of Regional Water Utility Company (PDAM).Meanwhile, batik waste water is firstly processed and then channelled into these drains.

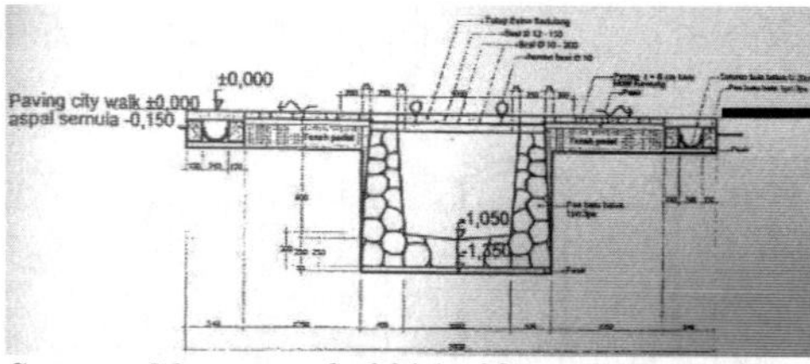

Source : Musyawaroh, 2016 : 28

Figure 10. Existing drainage network

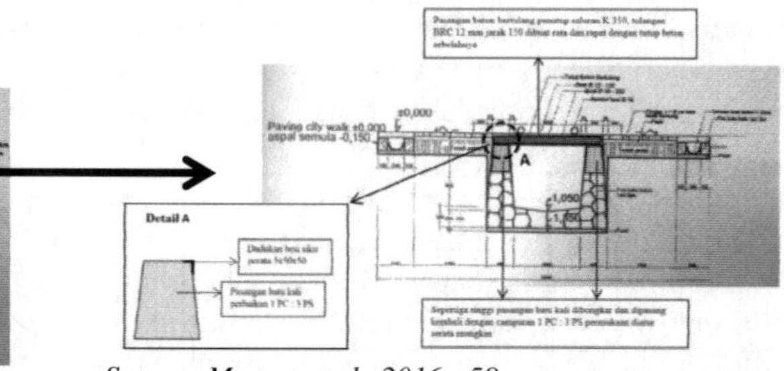

Source: Musyawaroh, 2016 : 58

Figure 11. Design improvements drainage network.

The damaged of concrete cover can be repaired on the high embankments buffer 1/3 of the channel edge. For more thorough perfect, changed the old drain that overlapped with the pre-cast concrete channel whose diameter is 1 m.

3.3.6. Sustainable Revitalization for Sewer Network

Currently, the capacity of Kauman's batik wastewater has reduced greatly. Almost every batik-creating process is carried out outside of Kampong Kauman, such as in Bekonang, Laweyan, Masaran and others. This is done because Kauman does not have an integrated batik wastewater processing yet, and the batik producers do not have their own wastewater treatment unit. Only at certain times, the full process of creating batik is done in Kauman (especially for tourism promotion purposes), but they use natural dyes from Teger wood, Tingi wood, Jambal wood and so on.

In 2012, a portable batik waste treatment unit had been designed for Kauman by Utomo, Musyawaroh & Sawitri [16] that called Upal-RE, based on the result development of the research by Musyawaroh, Utomo & Triratma in 2009 . Upal-RE reduced the level of pollutants COD by 95.92% and color by 75.90%, and has been tested in Kauman on 14th January 2013. But Upal-RE cannot be optimal used, need much electric power to operate it. The development of this tool is still being carried out.

3.3.7. Sustainable Revitalization for The Garbage Network

According to SNI 03-6981-2004 and SNI 03-1733-2004, the municipal waste disposal network is a large waste bin for temporary dump (TPS) of 12 m3. However, in Kauman there is no land available for that, in order to not troublesome the following steps should be taken:

a) Provides a simple, easy-to-use, portable and durable trash can in front of residents' homes.
b) The existing landfill for Trisula Street is maintained and repaired.
c) Garbage dump with garbage truck is still needed.

3.4. Sustainable Revitalization for Non-physical Environment of Kauman

3.4.1. Sustainable Revitalization Planed for Heritage Socio-Culture

Sustainable revitalization for Santri culture that planned for Kauman is as follows:

a) Preserving the homes of abdi dalem ulama and other abdi dalem remaining for commercial buildings.
This effort needs to be optimizing by utilizing higher commercial value in accordance with current needs such as cafes, boutiques, homestays and others. The house used for business place can bring economic benefits for its owner, so that it can be preserved.

b) Providing an area of expansion langgar by installing awnings, to accommodate the need for increased activities such as Tarawih prayer, Islamic studies and others.
   This effort needs to be improved by sharing space management with the alley around to make it more convenient and rewarding. While, Islamic study activities can be done in the local communities' house.

For batik activities, since the revitalization program implemented and community empowerment in 2006 batik business has increased to more than 100 entrepreneurs. Most of the houses along the street / alley in Kauman transformed into batik showrooms. But, this condition is slightly changed when Pasar Klewer burned in 2014, batik entrepreneurs in Kauman decreased again. Sustainable revitalization for batik culture namely needed to improve wastewater management facilities and village infrastructure so that businesses in this area will rapidly increase again.

3.4.2. Sustainable Revitalization Planed for Economy

The analysis on economic aspect was done by economic valuation, namely to assess the economic benefits of heritage resource management. The method used for economic valuation of cultural heritage Kauman was a merger between the theory proposed by Suparmoko [17], Soeroso and Susilo [18]. These values include: 1) direct, indirect and alternative use values; 2) value without use of which includes inheritance and existence.

From the assessment, shows that the economic benefits of Kauman region is currently very high but without maintenance within a period of 50 years, the condition of the area will be damaged by the amount of quality degradation value which is much higher than the value of the economic benefit [14], Therefore, it is necessary to undertake revitalization.

3.4.3. Sustainable Revitalization Planed for Engagement

Revitalization activities carried out involve relevant stakeholders, consist of local community, university (UNS) and the government. The cooperation with capital owners of private sector is not yet accomplished. The program has not been implemented with coordinated well, its still partially, overlapping and ineffective. For a better management, all stakeholders must coordinate so that the program undertaken is more integrated and useful for the community.

4. Conclusion

From the above description it can be seen that more than 60% of the Kauman area is covered by buildings, most of which are ancient buildings. In Kauman there are no green open spaces, parks for recreation and sports. Most of the roads in Kauman have a width of less than 4 meters, with heavy traffic. The drainage system is on the edge / bottom of this road network, mostly mixed with sewerage.

The santri culture in Kauman is still well preserved, but the house of the abdi dalem has been damaged a lot. Meanwhile, batik culture has increased, economic activity has also increased but not followed by adequate infrastructure facilities. There has been regional conservation action by related stakeholders, but not well coordinated, so that the implementation of each other overlap. Sustainable revitalization that should be applied to land and buildings is to further increase the use tailored to the needs of the present, in order to bring more profits for the owner.

Because there is no land left, recreation place, children play and sport is done on the road / gang in the area. The area is planted with green planting in pots, hanging pots and pergola, to reduce heat, refresh the air and to make it more beautiful. Drainage channels, household wastewater and batik waste water are separated. Repair of damaged drainage channels. Household waste channels are channeled into PDAM interconnector networks, as well as batik wastewater channels (once processed first). For garbage disposal network, an improvement on the TPS in Trisula street and keep using trucks for temporary dumps.

5. Acknowledgments

The authors would like to thank (1) Association of Kauman Batik Village Kampung Wisata Batik Kauman, (2) Civil Service of Kauman District, (3) Kauman society who has been willing to provide all the information we need in this study, (4) LPPM of Universitas Sebelas Maret which has provided financial support and the opportunity to conduct research in 2016, (5) Related institutions (DTRK, Bappeda, Department of Tourism).

6. References

[1] Tam V W Y, Fung I W H and Sing M C P 2016 Adaptive reuse in sustainable development: An empirical study of a Lui Seng Chun building in Hong Kong *Renew. Sustain. Energy Rev.* **65** 635–642

[2] Moosavi M S 2011 An Analysis to Challenges and Contradictions in Revitalization of Historic Center of Cities in Iran *Interdiscip. Themes J.* **3**

[3] Boeri A, Gaspari J, Gianfrate V, Longo D and Pussetti C 2016 The adaptive reuse of historic city centres. Bologna and Lisbon: solutions for urban regeneration *TECHNE-Journal Technol. Archit. Environ.* 230–237

[4] Ayiran N 2011 Achitectural continuity towards cultural sustainability in bodrum. *Open House Int.* **36**

[5] Claudia T and Luigi P 2016 A novel paradigm to achieve sustainable regeneration in historical centres with cultural heritage *Procedia-Social Behav. Sci.* **223** 693–697

[6] Hensel M U and Turko J P 2015 *Grounds and Envelopes: Reshaping Architecture and the Built Environment* (Routledge)

[7] Irani M 2014 Sustainable Development of Iranian Historical Cities with Focus on Cultural, Historical and Architectural Identities *Mediterr. J. Soc. Sci.* **5** 647

[8] Chung T 2009 Valuing heritage in Macau: On contexts and processes of urban conservation *J. Curr. Chinese Aff.* **38** 129–160

[9] Cheng S, Yu Y and Li K 2017 Historic conservation in rapid urbanization: a case study of the Hankow historic concession area *J. Urban Des.* 1–22

[10] Budi M 2016 Banyak Bangunan Kuno di Solo Jadi Agunan Bank dan Dijual Eceran *Detik News*

[11] Setyaningsih W, Iswati T Y, Nuryanti W, Prayitno B, Sarwadi A and others 2015 Low-Impact-Development as an Implementation of the Eco-Green-Tourism Concept to Develop Kampung towards Sustainable City *Procedia-Social Behav. Sci.* **179** 109–117

[12] Musyawaroh M 2009 Strategy To Revitalize Slum On Heritage Kampong Kauman Surakarta *Proceedings International. Conference CIB, International Conference on Informal Settlement and Affordable Housing.*

[13] Darban A 2000 *Sejarah Kauman: menguak identitas kampung Muhammadiyah* (Tarawang)

[14] Musyawaroh M 2016 Model Pengelolaan Revitalisasi Kawasan Cagar Budaya Berbasis Arsitektur Berkelanjutan Studi Kasus Kelurahan Kauman Surakarta *Universitas Sebelas Maret: LPPM - Penelitian Disertasi dan Doktor Baru*

[15] Priatmodjo D 2009 Revitalisasi bangunan Cagar Budaya: Konservasi+ Nilai Ekonomi+ Manfaat Bagi Masyarakat Luas *Bull. Tata Ruang* 1571–1978

[16] Utomo B, Sawitri H S R and others 2012 Perancangan Unit Pengolahan Air Limbah dengan Reaktor Eloktrokimia (UPAL-RE) untuk Melayani Home Industry Batik

[17] Suparmoko M 2002 *Ekonomi Publik untuk Keuangan dan Pembangunan Daerah* (Andi)

[18] Soeroso A and Susilo Y S 2009 Nilai Ekonomi Barang Kebudayaan *Modus J. Ekon. dan Bisnis* **21**

IOP Conf. Series: Earth and Environmental Science **123** (2018) 012044 doi:10.1088/1755-1315/123/1/012044

Sustainable Street Vendors Spatial Zoning Models in Surakarta

M J Rahayu[1,2], R A Putri[1,2], and E F Rini[1,2]

[1]Urban and Regional Planning Study Program, Sebelas Maret University
[2]Information and Regional Development Centre, LPPM, Sebelas Maret University

Email: mjanirahayu@gmail.com

Abstract. Various strategies that have been carried out by Surakarta's government to organize street vendors have not achieved the goal of street vendors' arrangement comprehensively. The street vendors arrangement strategy consists of physical (spatial) and non-physical. One of the physical arrangements is to define the street vendor's zoning. Based on the street vendors' characteristics, there are two alternative locations of stabilization (as one kind of street vendors' arrangement) that can be used. The aim of this study is to examine those alternative locations to set the street vendor's zoning models. Quatitative method is used to formulate the spatial zoning model. The street vendor's zoning models are formulated based on two approaches, which are the distance to their residences and previous trading locations. Geographic information system is used to indicate all street vendors' residences and trading locations based on their type of goods. Through proximity point distance tool on ArcGIS, we find the closeness of residential location and previous trading location with the alternative location of street vendors' stabilization. The result shows that the location was chosen by the street vendors to sell their goods mainly consider the proximity to their homes. It also shows street vendor's zoning models which based on the type of street vendor's goods.

Keywords: Sustainable street vendors, Spatial zoning, GIS

1. Introduction

The high poverty rate in a country is directly proportional to the growth of the informal sector, including the street vendors. Street vendors (referred to as hawkers) can be interpreted as a merchant offering goods or services, where the trading locations used are public spaces such as roads, sidewalks or parks [1]. Streets vendors always have to serve their consumer needs, who are mostly from the poor sector [2]. Street vendors sell goods at low prices [3]. Various strategies that have been carried out by Surakarta's government to organize street vendors (PKL) have not achieved the goal of street vendors' arrangement comprehensively. In fact, street vendors have contributed significantly to the populist economy. Therefore, it is important to strive for locations that can ensure the sustainability of their livelihoods.

Structuring in the city of Surakarta has several forms of Relocation and Stabilization [4]. This is in accordance with the local act of action delivered by McGee and Yeung [1]. According to him, relocation is a temporary removal of the street vendors because the location is being built or restored. In this sense means that once the location is completed, the street vendors will return to occupy the old location. In contrast to the concept of Relocation that is conducted by the city government of

Surakarta. Relocation is to move the street vendors from the old location that is considered not the appropriate allocation to other locations in the form of a market so that the street vendors changed the status of a trade market/ formal merchants. Relocating street vendor always accompanied by clashes between governments and the street vendor triggered by the difference of interests [5]. Stabilization (or in Surakarta city is better known by the local term 'shelterisasi') is placing street vendors in the same location or not far from their original trading location by occupying public space such as sidewalks, shoulders, and parks by providing identical trading facilities or shelter and some supporting infrastructure such as electricity, clean water, waste management, parking and security. Structuring street vendors in the form of stabilization will be more applied by Surakarta Government because it is considered more flexible. While the status of street vendors' will remain as informal traders, although they charged for retribution by government (Sub-Field of Street Vendors Management).

Street vendors are generally located in the most profitable areas in the city centre [6]. Street vendors always locate near public areas (parks, bus or railway terminals, offices, schools, and so on) where people can easily buy necessary goods from them [7]. They utilize space based on the strategic location, accessibility, main activity, and space comfort [8]. But the street vendors also usually locate themselves without any guideline [9]. There are 19 alternative locations of street vendors stabilization in Surakarta based on proximity to activities, government land availability, road noise, and road function [10]. From the 19 alternative stabilization sites scattered in the city of Surakarta, some further consideration needs to be used to determine whether these locations can be acted upon as sites for stabilizing street vendors. Those considerations, inter alia, the need to see the proximity of the alternative location with the street vendors' residence and previous trading location. Therefore, this study formulates a zoning model that can be allocated for street vendors based on proximity to the residence and previous trading locations.

2. Methods

Research conducted by Widodo [11] proves that street vendors are located with consideration of proximity to their houses, city hustle, and accessibility. Proximity to residences is also one of the factors considered in choosing a location [4,12–14]. Based on Mc Gee and Yeung [1] street vendors will choose a location that: 1) not far from home; and 2) approaching the consumer or not far from the old location. It becomes the basis of zoning consideration to locate the unregulated street vendors to the alternative stabilization location. In this research, street vendor's zoning models are formulated based on two approaches, which are:

1) Distance to their residences
   The first approach is based on perceptions of the street vendors that they will theoretically be located close to where they live
2) Previous trading locations
   The second approach is, based on the main factor to determine the location of street vendors' stabilization that has been pursued by Surakarta Government, considering the proximity to the previous trading location factors

Geographic information system is used to indicate all street vendors' residences and trading locations based on their type of goods. Through proximity point distance tool on ArcGIS, we found the closeness of residential location and previous trading location with the alternative location of street vendors' stabilization. The number of street vendor (of trading locations or residential) which has the closest distance to a stabilization alternative location showed the tendency of that potential alternative stabilization location to be a stabilization location for street vendors. Technical processing of zoning models using ArcGIS tools are:

1) Spatial Zoning Model 1
   a. Mapping out stabilization alternative location of street vendors with ArcGis

IOP Conf. Series: Earth and Environmental Science **123** (2018) 012044 doi:10.1088/1755-1315/123/1/012044

b. Mapping the trading location and residence of street vendor respondents per type of goods in Surakarta city through ArcGIS software
c. Determining Model 1 by measuring the closest distance of street vendors' residence to street vendors' stabilization alternative location, which is done by using ArcGIS Proximity-Point Distance analysis tool. With these analytical tools, the location of the street vendors' residence which close to a particular stabilization location alternative can be identified.
d. Conducting a classification based on alternative stabilization location which close to street vendors' residence

2) Spatial Zoning Model 2
a. Mapping out stabilization alternative location of street vendors with ArcGis
b. Mapping the trading location and residence of street vendor respondents per type of goods in Surakarta city through ArcGIS software.
c. Determining Model 2 by measuring the closest distance of street vendors' previous trading location to street vendors' stabilization alternative location, which is done by using ArcGIS Proximity-Point Distance analysis tool. With these analytical tools, the location of the street vendors' previous trading location which close to a particular stabilization location alternative can be identified.
d. Conducting a classification based on alternative stabilization location which close to street vendors' previous trading location (Figure 1).

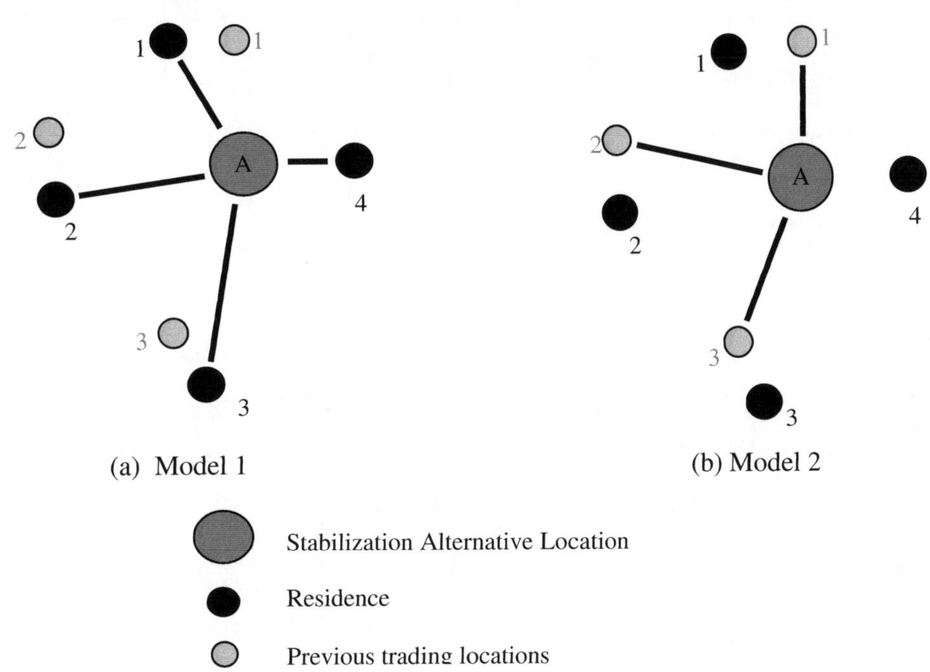

Figure 1. Visualization of Street Vendors' Stabilization Alternative Location Proximity with Residence and Previous Trading Locations

Then mapping technique of ArcGIS was used to map the entire criteria of spatial zonation of "ideal" street vendors in the Surakarta city. Furthermore, efforts were done to clarify the result of research, namely the model of sustainable street vendors, in the form of interviews with the government (market service sub field of PKL/ street vendors structuring).

IOP Conf. Series: Earth and Environmental Science **123** (2018) 012044 doi:10.1088/1755-1315/123/1/012044

3. Results and Discussions

The zoning model is based on two approaches, distance to street vendors' residence and previous trading locations.

3.1. Spatial zoning model 1

The determination of this zoning model is based on proximity to the street vendors' residence [1]. The survey which conducted on street vendors in Surakarta City by the type of goods shows that most of the street vendors' residence do not far from their trading locations. It shows that in determining their trading location, street vendors consider the proximity to their residence. Based on this, spatial zoning model analysis of street vendors' location was based on the proximity of the residence shown in the table 1.

Table 1. Proximity of Street Vendor's Residence with Stabilization Alternative Location

No	Alternative Location	Raw Food	Dine in fast food	Take away fast food	Non-food	Servic e	Average
1	Around Kartopuran Field	11.9	21.54	**25.86**	17.24	15	**18.73**
2	In front of Dinas Pendidikan dan Olahraga (Dispora)	7.14	12.31	8.62	10.34	11.67	**10.25**
3	East of AUB and UTP Mojosongo	7.14	4.62	6.92	10.34	8.33	7.42
4	South of Tax Office	4.76	7.69	10.34	5.17	6.67	7.07
5	Front of Bonoloyo Public Grave	9.52	6.15	8.62	5.17	6.67	7.07
6	East of State Land Office	**19.05**	1.54	5.17	6.9	3.33	6.36
7	West and North of Senior High School 6 Surakarta (SMA 6 Surakarta)	7.14	6.15	5.17	3.45	8.33	6.01
8	St. Yosef High School (SMA St. Yosef)	7.14	6.15	5.17	5.17	1.67	4.95
9	West of Laweyan District Office	4.76	1.54	5.17	5.17	6.67	4.59
10	East of Moewardi Hospital	9.52	1.54	3.45	3.45	6.67	4.59
11	West of Junior High School 14 Surakarta (SMP 14 Surakarta)	0	7.69	3.45	6.9	1.67	4.24
12	South of Lotte Mart Tipes	2.38	3.08	3.45	5.17	5	3.89
13	South of Pedaringan warehouse (Jalan KH. Dewantara)	0	6.15	1.72	3.45	6.67	3.89
14	Front of Samsat	4.76	4.62	3.45	5.17	0	3.53
15	East of Kasih Ibu Hospital	2.38	3.08	1.72	3.45	5	3.18
16	Surakarta State Hospital	0	1.54	1.72	1.72	3.33	1.77
17	East of Sala View Hotel	0	3.08	0	0	1.67	1.06
18	North of Tax Office	2.38	1.54	0	1.72	0	1.06
19	Adi Sucipto Fruit Market	0	0	0	0	1.67	0.35
Total		**100**	**100**	**100**	**100**	**100**	**100**

The table above showed a sequence ranging from large to small, related to the percentage of street vendors who have proximity to the residence with stabilization alternative location. Based on the proximity factor to the residence, 18.73% of the street vendors in Surakarta were closest to the stabilization alternative location of Kartopuran Field. Spatially, there was no street vendors' stabilization location in the southern part of Surakarta. The number of street vendors in the south was much higher than the number of street vendors in the north Surakarta. Therefore, for street vendors who live in the southern city of Surakarta, the stabilization alternative location of the Kartopuran Field was the closest location from their residences.

In addition, 10.25% of street vendors in Surakarta City have the closest distance between their residence and stabilization alternative location in front of Dispora. Spatially, this alternative location was one of strategic alternative location for street vendors to reach from any direction in Surakarta. There were also many major activities in the area, but the area would be quiet after the afternoon. Therefore, there should be an effort to enliven the area at night by optimizing the street vendors' potential. Furthermore, creative events that can revive the area permanently is needed.

The less percentage of street vendors who have close proximity to their homes with stabilization alternative location indicates that there were few / no street vendor whose houses are close to these alternative locations. In addition, when viewed by type of goods, for take-away fast food there are 25.86% of street vendors whose homes were close to stabilization alternative location in Kartopuran Field. This condition indicated that some take away fast food street vendors was trading in a location not far from the location of the Kartopuran field because the locations were close to commercial activities, offices, services, and education. The dominance of street vendors Surakarta who trades in Southern Surakarta was very high if compared in Northern Surakarta. Most other street vendors' type of goods have the same pattern, which was the closest alternative stabilization location with street vendors' residence is the Kartopuran Field, except for raw-food street vendors.

Raw Food street vendors amounted to 19.05% had the closest residence with East BPN Surakarta alternative location. This indicated that the location was close to street vendors' residences which were currently located in the alternative location. This was caused by street vendors of raw food products come from out of town (majority coming from Karanganyar regency) which went to Surakarta via Jalan Ir Sutami. The Raw Food Street vendors' product from outside the city always stay trading in a location until their goods ran out, then they return to their residence to bring the other goodson the next day. From all of type of goods, dine-in fast food street vendors' residences have high proximity to almost all stabilization alternative locations (except Adi Sucipto fruit market). This indicates that this type of street vendors will always choose the trading location (previous trading locations) close to their residenceswhich has economic activity that can be served by their goods [1]. The spatial zoning model 1 shown in figure 2.

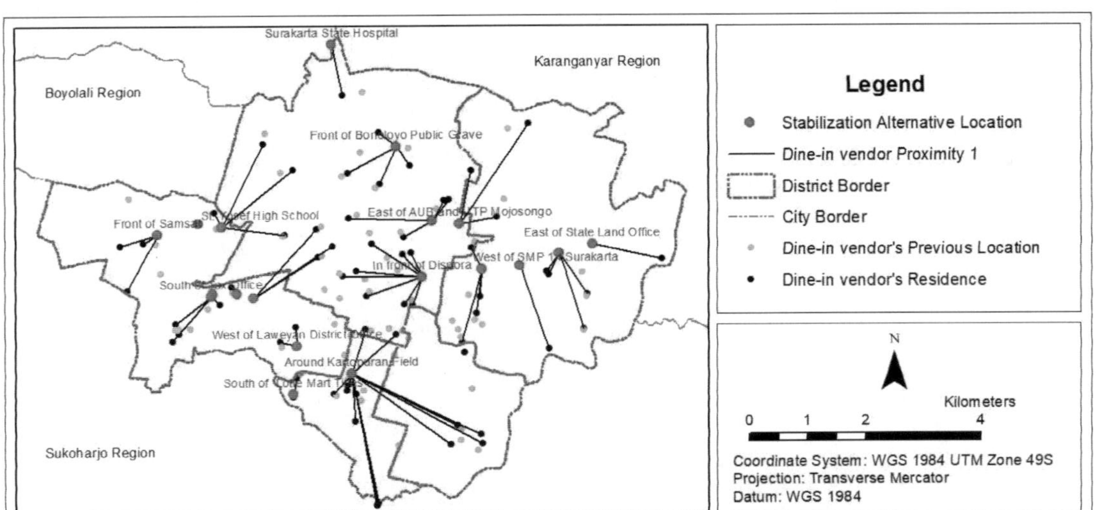

Figure 2. Spatial Zoning Model 1 Based on Proximity to Street Vendors' Residences

3.2. Spatial zoning model 2

Determination of this zonation model is based on proximity to the previous trading location (existing). Several precedents in structuring street vendors in Surakarta city is based on the problems that arise as a result of street vendors. The concept of stabilization arrangement of street vendors in Surakarta City is implemented by arranging street vendors at the previous trading locations. This shows that Surakarta City Government in choosing the location of stabilization is based on the location of the existing street vendors. This method is taken so street vendors do not lose their customers. Based on this, the following table shows analysis of street vendors' previous trading locations proximity to the stabilization alternative locations (Table 2).

Table 2. Proximity of Previous Trading Locations with Alternative Locations

No	Alternative Location	Raw Food	Dine in fast food	Take away fast food	Non-food	Service	Average
1	Around Kartopuran Field	21.43	20	25.86	17.24	23.33	**21.55**
2	In front of DinasPendidikandanOlahraga (Dispora)	9.52	20	12.07	17.24	15	**15.19**
3	West and North of Senior High School 6 Surakarta (SMA 6 Surakarta)	11.9	7.69	5.17	15.52	10	9.89
4	West of Laweyan District Office	2.38	6.15	12.07	5.17	8.33	7.07
5	Front of Bonoloyo Public Grave	11.9	6.15	10.34	5.17	1.67	6.71
6	St. Yosef High School (SMA St. Yosef)	9.52	6.15	6.9	3.45	5	6.01
7	West of Junior High School 14 Surakarta (SMP 14 Surakarta)	4.76	4.62	1.72	10.34	8.33	6.01
8	East of AUB and UTP Mojosongo	4.76	4.62	3.45	5.17	6.67	4.95
9	South of Tax Office	7.14	6.15	1.72	5.17	1.67	4.24
10	South of Pedaringan warehouse (Jalan KH. Dewantara)	0	4.62	5.17	3.45	5	3.89
11	East of KasihIbu Hospital	0	3.08	5.17	1.72	6.67	3.53
12	East of State Land Office	9.52	0	3.45	1.72	0	2.47
13	North of Tax Office	2.38	1.54	1.72	1.72	1.67	1.77
14	East of Moewardi Hospital	2.38	0	3.45	1.72	1.67	1.77
15	Front of Samsat	0	4.62	0	1.72	1.67	1.77
16	East of Sala View Hotel	2.38	1.54	0	1.72	1.67	1.41
17	South of Lotte Mart Tipes	0	1.54	1.72	1.72	0	1.06
18	AdiSucipto Fruit Market	0	0	0	0	1.67	0.35
19	Surakarta State Hospital	0	1.54	0	0	0	0.35
Amount		**100**	**100**	**100**	**100**	**100**	**100**

Based on the above table, the zonation model result did not show significant difference with the first zonation model. There are 21.55% of street vendors have the closest distance between their existing trading locations and alternative stabilization locations in Kartopuran Field. This was caused spatially, only stabilization alternative location of Kartopuran Field which was located in the southern city of Surakarta. Therefore, the presence of street vendors in the southern city of Surakarta that was numerous, closest to the stabilization alternative location of Kartopuran Field. A total of 25.86% of take-away fast food street vendors closely related to the Kartopuran Field. This meant that take-away fast food vendors widely scattered around the Kartopuran Field.

When viewed for each type of goods, in addition to the stabilization alternative location of Kartopuran Field, 11.90% of the Raw Food vendors were each located near the alternative location stabilization Bonoloyo Public Grave and West and North SMA 6. This indicated that the distribution of raw food street vendors in Surakarta City was mostly located around these locations. In thestabilization alternative location of AdiSucipto Street Fruit Market, East Sala View, and Surakarta State Hospital, there was not a single take-away fast food street vendor that trade adjacent to the location. This was because at this time AdiSucipto Street Fruit Market was a form of street vendors arrangement made by the Government of Surakarta.Previously, it used by raw food street vendors (fruit) in the city of Surakarta. This meant that there are no unregistered street vendors near the location.

While for street vendors of dine in fast food majority located around the stabilization alternative location of Kartopuran Field and front of Dispora Surakarta City. On the other hand, for street vendors of take away fast food, 25.68% of the street vendors were located near the stabilization alternative location of Kartopuran Field, 12.07% located near the stabilization alternative location in front of Dispora and 12.07% street vendors located near the stabilization alternative location of Western Laweyan District. This indicated that the location distribution of take-away fast food vendors tend to be in the city centre. However, these conditions were different with non-food street vendors, most of them were located close to the stabilization alternative location in front of Dispora and Kartopuran Field. This showed the different location distribution with take-away fast food street vendors. Non-food street vendors were more likely to spread to the northern Surakarta. These differences resulted in different models of zonation. However, Non-food street vendors have a zoning model that was not significantly different from the service street vendors.

The location of south Pedaringan has specificity. This was because; this location had the exact same value both when viewed from the proximity with the house of street vendors or from previous trading locations. This indicated that the location of the south Pedaringan was adjacent to street vendors' residence and with the previous location. There were many street vendors who trade in the area and live not far from the location of the Pedaringan. The spatial zoning model 2 of dine-in street vendor is shown in Figure 3.

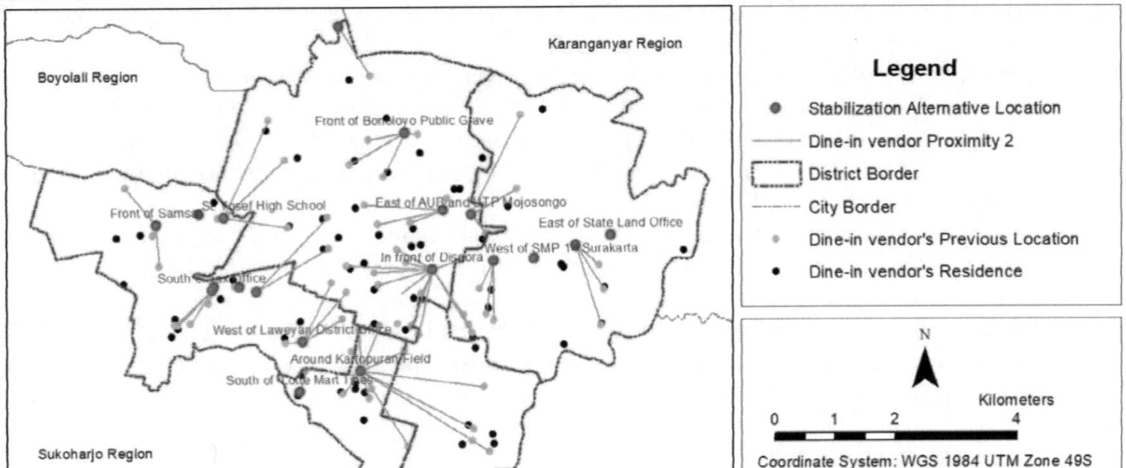

Figure 3. Spatial Zoning Model 2 Based on Proximity to Street Vendors' Previous Location.

4. Conclusion

Spatial zonation model of street vendors' location is obtained based on spatial data of street vendors' existence with 2 models that are based on proximity to the location of street vendors' residence and proximity to the previous location of street vendors. This is to maintain the closeness of street vendors with their customers and with their residences.

However, from 2 zonation models obtained, there is no significant difference in alternative location produced. This further reinforces that the previous (existing) location selected by street vendors to trade has considered closeness to their residential and consumer location. In other words, street vendors always choose a location close to home and close to their customers.

5. References

[1] McGee T G and Yeung Y 1977 *Hawkers in Southeast Asian cities: Planning for the bazaar economy* (IDRC, Ottawa, ON, CA)
[2] Bodhikong K and Moongjongklang B 2003 Problems and Possible Solutions for Stall Sellers and Street Vendors The Case of Stall Sellers and Street Vendors in front of Chulalongkorn Hospital *Case Study Report - Bangkok, Thailand*
[3] Bhowmik S K 2005 Street Vendors in Asia: A Review Economic and Political Weekly
[4] Rahayu M J and Musyawaroh M 2013 Stabilisasi Sebagai Bentuk Penataan PKL Makanan Siap Saji Di Kota Surakarta *TATALOKA* **15** 39–52
[5] Permatasari A, Putro U S and Nuraeni S 2014 Strategic Analysis Relocating Street Vendor through 3D Negotiation Case Study: Street Vendor Surakarta, Indonesia *Procedia-Social Behav. Sci.* **115** 436–443
[6] Manning C and Effendi T N 1985 *Urbanisasi, pengangguran dan sektor informal di kota* (Diterbitkan untuk Yayasan Obor Indonesia dan Pusat Penelitian dan Studi Kependudukan UGM [oleh] Gramedia)
[7] Bromley R 2000 Street vending and public policy: a global review *Int. J. Sociol. Soc. Policy* **20** 1–28
[8] Widjajanti R 2016 The Space Utilization by Street Vendors Based on the Location Characteristics in the Education Area of Tembalang, Semarang *Procedia-Social Behav. Sci.* **227** 186–193
[9] Onyango J, Olima W and Onyango L 2012 Dynamics of street vending phenomenon in the Kisumu Municipality, Kenya *Int. J. Arts Commer.* **1** 107–120
[10] Rahayu M J 2013 Strategi Pemberdayaan PKL secara Partisipatif dalam Mendorong Ekonomi yang Berkelanjutan di Kota Surakarta *UNS Surakarta*
[11] Widodo A 2000 *Faktor-Faktor yang Mempengaruhi Pemilihan Lokasi Usaha Pedagang Kaki Lima (Studi Kalus: Kota Semarang)* (Program Pascasarjana Universitas Diponegoro)
[12] Ray C N, Sharma S B and Bhowmik S 2009 *Urban planning hawkers and hawking space: a study of hawkers of commercial centers in Delhi* (Center for Environmental Planning and Technology University (CEPT))
[13] Rahayu M J, Buchori I and Widjajanti R 2017 Study on Characteristic of the locations of Street Vendors Stabilization in Surakarta *4th International Seminar on Sustainable Urban Development, Trisakti, Jakarta*
[14] Rahayu M J, Werdiningtyas R R and Musyawaroh M 2017 Faktor-Faktor yang Mempengaruhi Keberhasilan Penataan PKL Sebagai Strategi Penataan Ruang Kota Surakarta *Reg. J. Pembang. Wil. dan Perenc. Partisipatif* **7** 109–122

Space Livability of Street Vendors in Simpang Lima Public Space, Semarang

R Widjajanti[1] and H Wahyono[1]

[1]Department of Urban and Regional Planning, Universitas Diponegoro, Indonesia,

Email: retno.widjajanti@pwk.undip.ac.id

Abstract. Street vendors in Semarang have been growing rapidly and uncontrolled. They always use public space such as public roads, sidewalks, parks and fields as trading locations. The street vendors' activities in the public space are considered as the cause of declining on environmental quality and aesthetics of the city. All these years, the government often evicted the street vendors than organized and provides adequate space for them. As one of the actual urban activities, the street vendors' activities should be accommodated by the government and the location for them is managed in the urban spatial plan. Street vendors need spaces which livable and suitable to their activities' requirements, has a relationship with users (street vendors' doers and consumers) and the activities of street vendors themselves. Research on the aspect of space for street vendors is still less in quantity, whereas space for them is an urgent matter for the government in managing their activities. This study aims to identify the livability of space based on the street vendors' behavior in their location. This research used descriptive quantitative method with questionnaires and GIS as the mapping tool for street vendors' location. The result of the research shows that the livability of street vendor space is based on the activity of street vendors (type of merchandise, trading places' size, trade place assessment, space dimension, trading time, duration and period) and space conditions (access, natural elements, safety and parking space).

Keywords: Street vendor, Simpang lima, GIS

1. Introduction

The increasing urban population growth is often not matched by the expansion of employment. As a result the population of working age is greater than the existing employment opportunity [1]. The narrowness of employment in the formal sector encourages an increase in the population to enter the informal sector [2,3]. One form of informal sector in trading sector is the street vendors. They are able to provide primary and secondary goods at affordable prices for low-income communities. They also play a role in reducing unemployment, crime, and as potential tourism activities [4,5]. In addition to potentially positive, the street vendors who always use public space such as roadsides, sidewalks, parks, and fields often cause various problems.

Street vendors who occupy public space are considered to have degraded the quality of the environment and changed its function. Consequently, street vendors are often regarded as an annoying element or even as unplanned element [6–8]. One of the cities with the development of street vendors quite rapidly, namely the City of Semarang. The existence of street vendors in Semarang brought changes to the development of the city, especially in the area of Simpang Lima. The number of street vendors in Simpang Lima is increasing and occupying public space. Their merchandises lined up causing a disturbance for covering the view to the building facade behind him. The mass of means of street vendors is mixed and unorganized so it looks chaotic and impressed slum. One of the activities of street vendors is the problem that is in the front of Baiturrahman Mosque because it caused a traffic jam.

So far, in overcoming the street vendors, the government more often displaced than to organize and provide adequate space for them. Relocation efforts by the government also always fail. The street vendors who are transferred to places that have been built by the government always return to their original place. The location of the relocation is considered not in accordance with the characteristics of its activities. Street vendors always choose places that approached consumers, but blocked the user's path of functional activities [9]. Often street vendors lose customers after relocation. Relocation places that do not consider the nature of street vendors only make the street vendors feel uncomfortable because it is far from the concentration of consumers.

In contrast to the handling of street vendors in other areas, the street vendors in Simpang Lima were overcome by stabilization, except for street vendors located in front of Baiturrahman Mosque and Ciputra Hotel and Mall. Stabilization is the re-arrangement of the informal sector to harmonize and not disturb the function of the city in the surrounding environment [10]. The street vendors in Simpang Lima remain in the public sphere, organized and controlled. The government builds street vendors in public spaces, builds infrastructure, and there are no evictions. As Shirvani [11] explains that activity always needs a place that matches the character of its activity. Street vendors who already feel comfortable with the place because it is in accordance with the characteristics of its activities, will be very difficult if moved to another place. Thus, the provision of space for street vendors should really consider the convenience of space for street vendors.

Research on livable space for street vendors is still rarely done, so far research has focused only on economic, social policy, trade characteristic and physical arrangement. Whereas space for street vendors is an urgent problem for the government in managing the activities of street vendors. Lefebvre [12] states the ideal city planning is that which can accommodate all real urban activity. As one of the real urban activities, the activities of street vendors should be accommodated by the Government by arranging the place and its location in the city spatial plan. Discussion about the livability of the space for street vendors becomes very important for street vendors' activities are no longer considered as a problem and can be viewed as the potential of the urban economy.

2. Data and Methods

This research approach using quantitative methods. Research based on deductive thinking that is verification of the existing theory. Theories used in the study include the concept of street vendors, the theory of activity and space, the theory of production space and thirdspace and space behavior. After that's done data collection through questionnaires, interviews and observation. A sample of this research is determined by a random sampling method that is disproportionate stratified random sampling. Disproportionate stratified random sampling is a sampling technique for populations with stratified but less proportional members [13].

Respondents in the research are street vendors who occupy public spaces. The population of street vendors is a stratified but not proportional population, as the number of street vendors selling fast food is not the same as the number of street vendors selling locally processed food. Questionnaires are shared in the form of closed questions and there are also open questions. In addition to questionnaires, interviews were also conducted to find out the reasons for choosing the respondents' locations. Then the observation is done by looking at how much the level of consumer visits.

The analysis is undertaken by using cross tabulation by crossing the public space type variables on the behavior of space. Then, using an explanatory descriptive approach to get a comprehensive explanation about the convenience of street vendors to the place of activity or space livability based on the behavior of street vendors in the public space of Simpang Lima, Semarang.

3. Results and Discussion

This analysis is conducted to identify the space in accordance with the perception of street vendors to be used as a place of activity, in order to be livable, that is to move safely and comfortably, based on the consideration that is the main activities in Simpang area influencing and encouraging street vendors to utilize space for activities In the public space, thus providing a characteristic of street vendors in choosing a place to trade. Utilization of public space as a place of street vendor activity based on the characteristics of space will be discussed based on the activity of street vendors with place, shape of street vendors in public space, activity and time, suitability of activity with condition of place in public space.

The public spaces in the Simpang Lima area referred to in this discussion are public spaces used by street vendors to trade, such as on sidewalks based on statements of 100% street vendors who move around the Simpang Lima road.

3.1. Analysis of Street Activities with Street Places on Sidewalks Around Road Area in Simpang Lima Area

The Simpang Lima area is a downtown area with major activities such as trading activities and services, offices, trade, entertainment, recreation and sports. These diverse main activities resulted in the emergence of many trade and service activities around the road, arising from internal linkage activity within the area, as described. Meanwhile, the emergence of internal linkage activity is also influenced by the emergence of many people who come to visit in the region. This condition gives rise to the creation of street vendors to provide goods that are in accordance with the characteristics of these activities. The diversity of these activities relates to the types of merchandise sold by street vendors, leading to the dominance of food and beverages (as prepared food) and nonfood items (clothing, toys, grocery, cigarettes, magazines and newspapers). The types of merchandise, food and drinks, as well as nonfood items, affecting and relating to the trading activities of street vendors, namely mendisplai merchandise, serving customers and preparing and processing the merchandise, so that those activities which dominate trading activity street vendor.

The activity of street vendors with this type of goods is closely linked to the type of merchandise used by street vendors [9,14], resulting in space requirements as described by Shirvani [11], that activity requires a place or container and character from This container corresponds to the character of the activity it occupies. Trading activities of street vendors with the dominance of food and beverage products and nonfood items, the majority of street vendors stated that the type of merchandise used is tent carts and display cabinets with shelters, and requires a space of 12 m^2 and 6 m^2. This wide measure by street vendors is considered large enough to accommodate trade activities with the type of food and beverage merchandise as well as services with the type of shopping cart wares and display cabinets with shelter.

The findings of this study indicate that street vendors think that there is enough space available in the sidewalk, so the street vendors are interested in utilizing sidewalks around the Simpang Lima street as a place to trade. Street vendor considers the footprint on the sidewalk to accommodate any repetitive trading activities that start from arranging the merchandise, organizing, mendisplai, presenting and processing the merchandise, and can freely move to serve the buyer, as well as a place that is free for the consumer.

Thus, since this study shows that the activities of street vendors are related to the suitability of places related to their space needs, it can be said that the street vendors choose a place to trade on the sidewalk because it assumes that on the sidewalk there is enough space for trade (Figure 1).

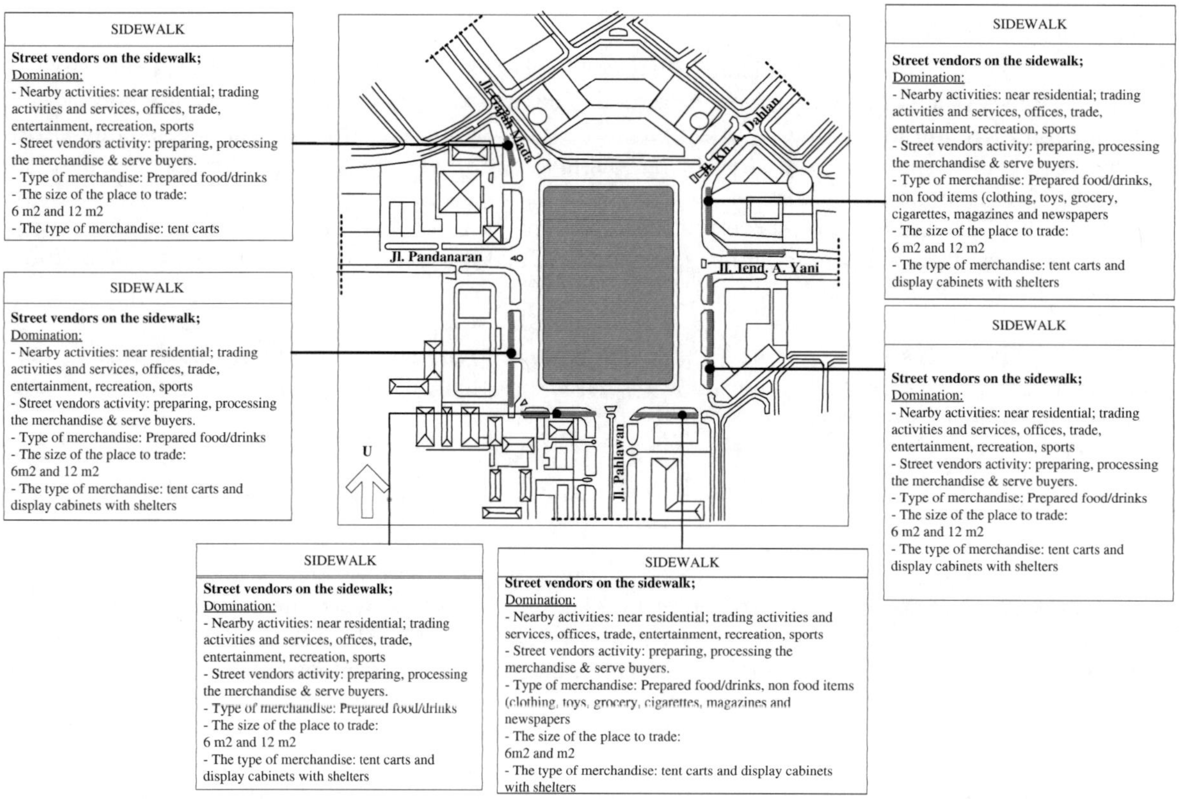

Figure 1. Street activities on sidewalks around road area in Simpang Lima area

3.2. Analysis of Spatial Values in Public Space (with Real and Unreal Boundaries) on Sidewalks Around Simpang Lima Roadces.

In the Simpanglima area, the majority of street vendors stated that their food and beverage type (as prepared food), as well as nonfood items (clothing, toys, grocery, cigarettes, magazines and newspapers) that affect or closely related to the activities of street vendors is mendisplai merchandise - serve buyers and prepare and process merchandise. The street vendors with these types of merchandise and trading activities, assume on the pavement are available a large enough place in accordance with its activities, thus attracting street vendors to take advantage of the public space. On the other hand, the types of merchandise and trading activities of street vendors also affect the form of space needed. The shape of the space is a form of space with a real limit or form of space with unreal limits [15]. While this form of trading space is related to the availability of space in the public space as described by Marcella [15] and Lang [16], that the limitation of a place of activity is to show the extent to which a behavior stops, this boundary is a real boundary, or an unreal limit. The type of delimiter tailored to the needs and character of the activity. This form of space to adjust to the activities of street vendors, in addition to the assessment of the availability of places in the public space, the street vendors can take advantage of places in the public space with a form of space in accordance with its activities.

The majority of street vendors on the sidewalks around the streets in Simpang Lima stated that the form of trading (Figure 2) space is: a) the limit is not real (open, no massive or permanent limit) (52.2%), the space limit is the order of the merchandise, Consumer street vendors can enter easily, easily seen by consumers; And to enable street vendors to exchange and cooperate to sell merchandise with other street vendors; b) real limits (there are limits with other street vendors, transparent limits of fabric and tents) (47.8%) so that consumers are comfortable not disturbed by other activities around the street vendors and there is a clear boundary of selling places with other street vendors. In addition, street vendors also consider the availability of space in the sidewalk, the street vendors can take advantage of the place in the form of space in accordance with its activities, namely the form of space that the limit is not real (open, without a massive or permanent barrier).

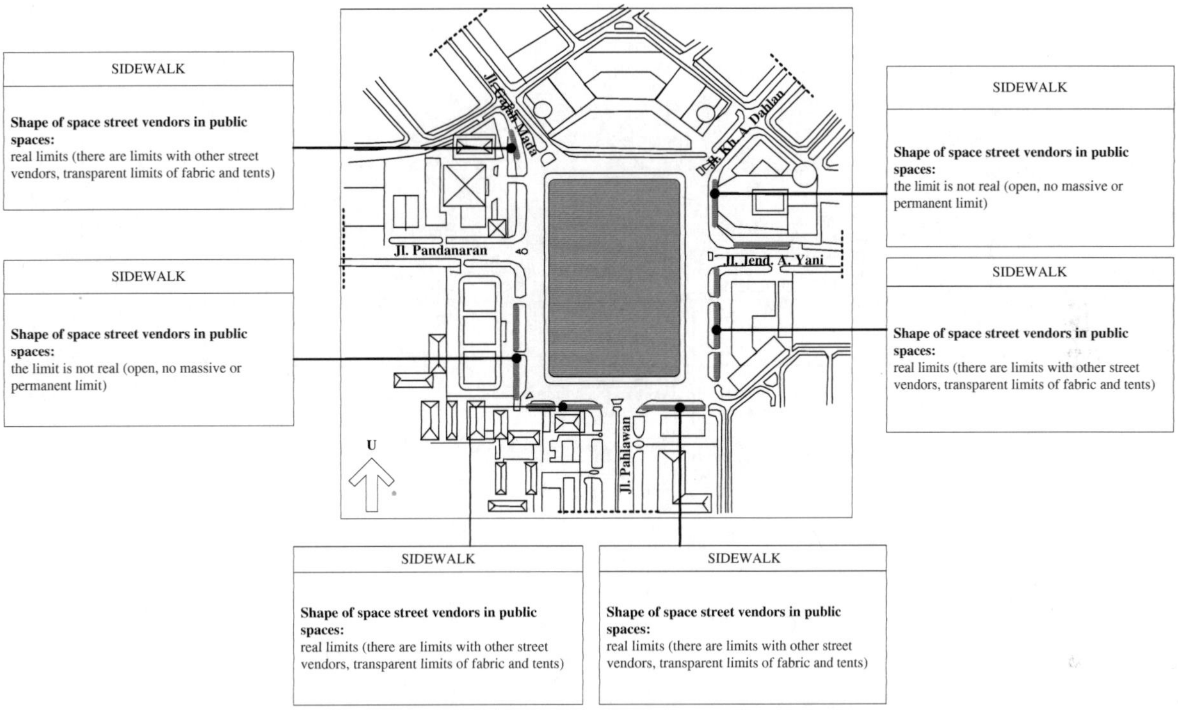

Figure 2. Spatial values in public space (with real and unreal boundaries) on sidewalks around Simpang Lima roadces

3.3. Analysis of The Activities and Time on The Sidewalks Around The Simpang Lima Road Section

The majority of street vendors is on the sidewalk around the Simpang Lima road, because there are trade and service activities, entertainment, recreation, sports. So the opening and closing time of street vendors is diverse at 04:00 to 15:00, at 07:00 to 15:00, at 07:00 to 24:00, at 15:00 to 24:00, and 17:00 to 5:00 o'clock. The diversity of open and close trading time is to follow the activities of residents, schools, shops and offices around the street vendors. The opening and closing time of the street vendors is related to the provision of daily necessities for the surrounding community, explained by the reasons of street vendors, namely:

1) From 7 am to 3 pm; 07.00 -24.00; from 15:00 to 24:00; 5pm to 5pm
   a. Food and drink, the reason for providing breakfast until lunch, lunch until dinner, dinner community visitors' area.
   b. Non food items (clothing, toys, grocery, cigarettes, magazines, newspapers), to meet the needs of consumers whose time is uncertain.

2) In addition, the majority of street vendors, assumes that the trading time is 24 hours, in the afternoon and in the evening, and the period of trading activity is daily. This condition follows the activities of the community and visitors of Simpang Lima and its surroundings, due to Simpang Lima as the center of the city, and is one of the nodes in Semarang city, the life of the region goes on and on throughout the day.

Meanwhile, it is explained by McGee and Yeung [9] that the activity time of street vendor follows the rhythm and time or life characteristics of the residents and surrounding communities. So the time of street vendor activity on the sidewalk around the Simpang Lima road is following the times of community activities and visitors in the area (Figure 3).

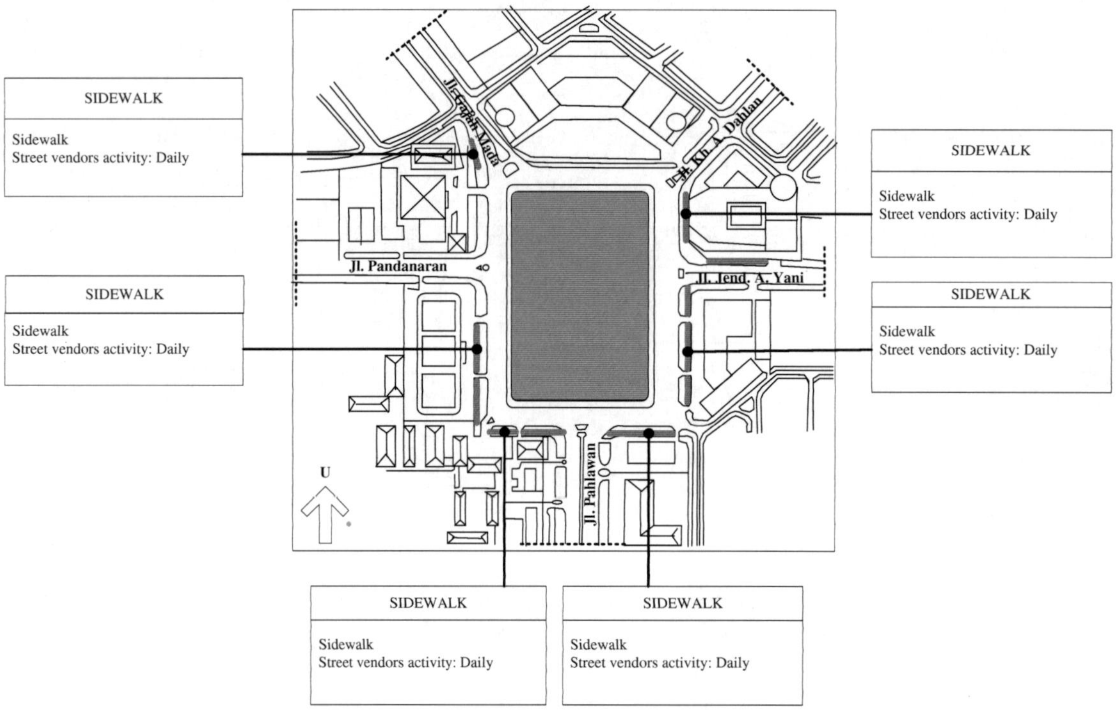

Figure 3. Activities and time on the sidewalks around the Simpang Lima road section

3.4. Analysis of Conformity of Activities With Condition Place in Public Spaces on Sidewalks Around Simpang Lima Road Section

This analysis discusses the appropriateness of strain vendors' activities to the element of space, comfort or livability of space, which consists of access aspect, natural element element, space security (activity place permit), parking space (Figure 4).

According to Shirvani [11], convenience measured by the quality of the place. On the other hand, McKay, Singh, Singh, Good, Osborne [17] and Lang [16] explained that the quality of the comfort of the place is tailored to the characteristics of the activity and user behavior.

The convenience measures of each aspect are based on:

1) Access, explaining to easily reach the object, then the object should be easy to see and easy to achieve [11]

IOP Conf. Series: Earth and Environmental Science **123** (2018) 012045 doi:10.1088/1755-1315/123/1/012045

2) Elements of natural elements, explaining the shade and brightness cause a sense of comfort for its users, and a comfortable place of shade and light attracts people to use it as a gathering / activity place [17].
3) Space security (permission), is the use of places permitted by government regulations, making sense of security and comfort for space use [11]
4) Parking space, Shirvani [11] explains the availability of parking spaces, allowing users to park their vehicles is a convenient element for space users.

The result of the research shows that the street vendors on the sidewalks around the Simpang Lima street segment stated that the food and beverage products, non food items nonfood items (clothing, toys, grocery, cigarettes, magazines and newspapers. The activities of street vendors display merchandise, serve buyers, prepare and process merchandise; display merchandise activity serving buyers [18].

In addition, the street vendors where they trade on the sidewalks state that:

1) Access, obtained that the street vendors declare the place: a) easily visible to visitors and consumers, and not hindered by other buildings or activities; b) It is easily accessible to visitors and consumers from the side of the road, and easily accessible from public and private ride and ride.

   Street vendor is located because it is easily seen and reached from the side of the road and from up and down the vehicle, is giving explanations and illustrations that the street vendors consider this place convenient to the place of activity. This condition is related to the character of street vendors who cut or block consumer channels [9,19]. Street vendors interested and take advantage of places or sidewalks around the Simpang Lima street as a place to trade.

2) Elements of natural elements, street vendors declare that trading on the sidewalk because: a) the place is shaded with the reason, because there are shade trees, quite shady and cool so buyers are not hot and feel comfortable; b) place quite bright, the reason is because the place is quite open so For daylight enough bright lighting and for the night there are lights lighting.

   The statements and reasons stated by the street vendors can illustrate that the majority of street vendors thinks the place they use is convenient for activities and gatherings.

3) Security of the space (permit), obtained a statement that the street vendors are located because the location permitted by the government, the reason for having received permission from the government, pay retribution and order as security guarantees, so there is never raid and control of the government [19]

   With the reasons of the street vendors, it illustrates that the street vendors consider the place and sidewalk around the Simpang Lima road is safe to be used as a place of activity [11].

   At the beginning of the growth and development of street vendors around the area of Simpang Lima carried out the action and control, but on the development and the determination that the area Simpang Lima developed for the culinary tourism center of Semarang city, the Government accommodates the needs of street vendors in the public space around the Simpang Lima road, So that street vendors are not dealt with curbed, but carried out stabilization structuring street vendors [11].

4) Parking space, from the results obtained by the majority of street vendors stated there is a place to stop the vehicle and parking lot, the reason is because there is space on the side of the road for a stop on the road, there is a parking lot in the yard of shops near or around street vendors, Which is enough to park at the ends of the sidewalk. Based on the explanations of street vendors, then illustrate that although parking on the side of the road and in the store yard that is nearby, street vendors consider there is space for parking lot visitors [20].

Related to Shirvani's [11] explanation that available space and parking lot to facilitate the user, as a measure of comfort. Street vendor thinks places on the sidewalk can be used as a place of trade, although there is no parking available. Street vendors can also be said that there is available parking or parking space is not available, street vendors will still move on the sidewalk in the public sphere.

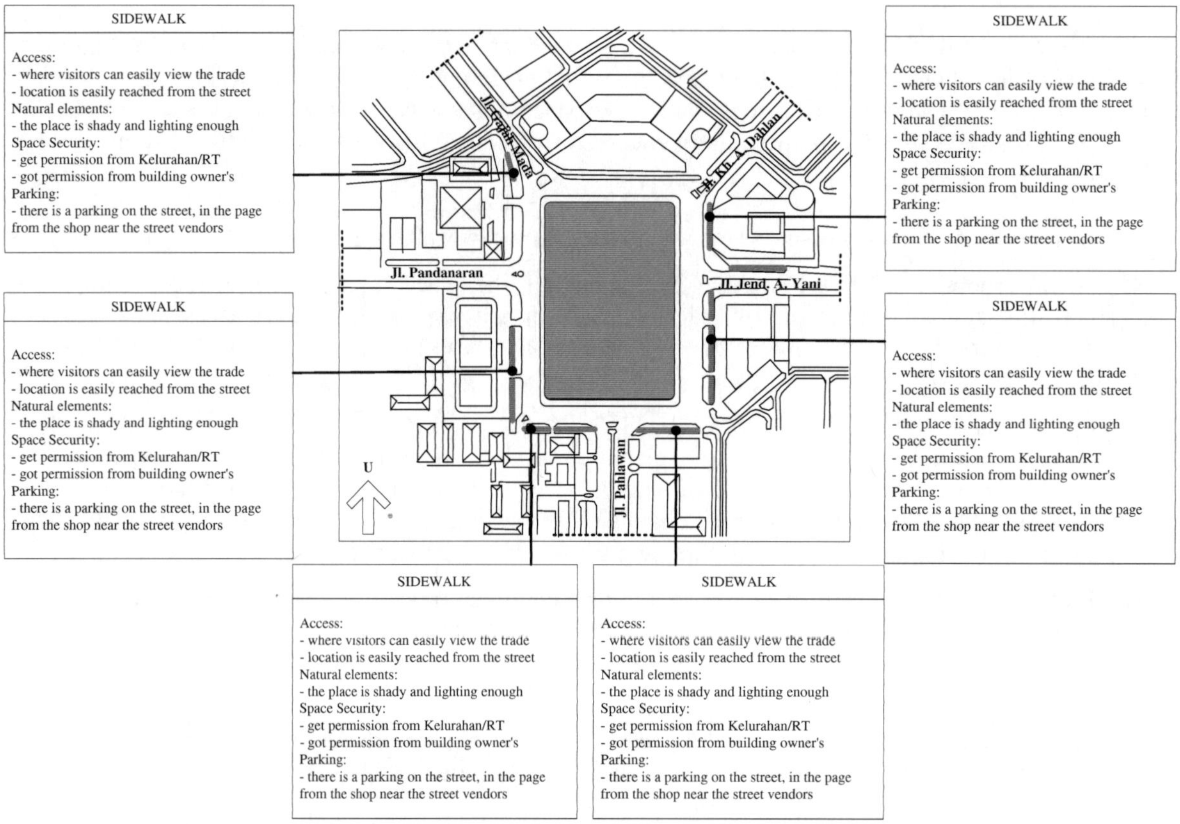

Figure 4. Conformity of activities with condition place in public spaces on sidewalks around Simpang Lima road section

4. Conclusion

Based on the results of research where the activities of street vendors that fit their characteristics in the public space in Simpang Lima is utilizing sidewalks around the Simpang Lima road section. Specifically, it can be concluded that the street vendors in determining the livable place to trade are: (1) Activities and places, according to street vendors, (2) According to street vendors pavements as a place of trade, can be arranged in the form of space without a real barrier (no barrier of the massive or permanent), the barrier between traders can be an arrangement of merchandise. Relationships between traders can support each other and work together, and facilitate visitors and consumers to get the desired type of merchandise, (3) When activities are in place, street vendors can take advantage of the sidewalk to keep up with visitors and consumers in the region throughout the day, for 24 hours, and every day, (4) Security and comfort of space, street vendors utilize sidewalks to trade on the basis that:

a. Accessible places: a) where visitors can easily view, and are not hindered by other buildings or activities; b) Sidewalks are easily accessible to visitors and consumers from the side of the road, and within easy reach of the public and private ride.

b. Comfortable places: a) Shade because there are shade trees, quite shady and cool to make buyers are not hot and feel comfortable; b) The place is quite bright, because the place is quite open during the daytime lighting is quite bright and at night there are lights of illumination.
c. A safe place to trade, because the location is permitted.
d. Government, street vendors feel safe to move and never get raided and control by government.
e. Street vendor considers available parking space, consumers can park on the road.
f. On the sideline near and around the street vendors, at the ends of the sidewalk.

5. References

[1] Todaro M P and Stilkind J 1985 Dilema Urbanisasi. Dalam Buku Urbanisasi, Pengangguran, dan Sektor Informal di Kota (Chris Manning dan Tadjuddin Noer Effendi) *PT. Gramedia*
[2] Bromley R A Y 1979 Introduction {\textendash} The Urban Informal Sector: Why Is It Worth Discussing? *The Urban Informal Sector* (Elsevier) pp 1033–1039
[3] Martinez L, Short J R and Estrada D 2017 The urban informal economy: Street vendors in Cali, Colombia *Cities* **66** 34–43
[4] Forkuor J B, Samuelsen H, Yeboah E H, Rheinländer T and Akuoko K O 2017 The Regulation of Street Foods: Experiences of Front-Line Regulators in Ghana *Urban Forum* **28** 251–269
[5] Rachbini D J and Hamid A 1994 *Ekonomi informal perkotaan: gejala involusi gelombang kedua* (Pustaka LP3ES Indonesia)
[6] Creswell J W 2003 *Research design: Quantitative, qualitative, and mixed methods approaches (2nd ed.)* (Thousand Oaks, CA: Sage)
[7] Yatmo Y A 2008 Street Vendors as `Out of Place' Urban Elements *J. Urban Des.* **13** 387–402
[8] Hough M 1990 *Out of place: Restoring identity to the regional landscape* (Yale University Press)
[9] McGee T G and Yeung Y 1977 *Hawkers in Southeast Asian cities: Planning for the bazaar economy* (IDRC, Ottawa, ON, CA)
[10] Widjajanti R 2000 Penataan Fisik Kegiatan Pedagang Kaki Lima pada Kawasan Komersial di Pusat Kota (Studi Kasus: Simpang Lima Semarang) *Master Thesis, Bandung Institute of Technology*
[11] Shirvani H 1985 *The urban design process* (Van Nostrand Reinhold Company)
[12] Lefebvre H 1991 *The production of space* vol 142(Oxford Blackwell)
[13] Sugiyono 2014 *Metode Penelitian Kuantitatif Kualitatif dan R&D* (Bandung: Alfabeta)
[14] Deguchi A 2005 Re-evaluating street vendors in Asian cities and Asian urbanism *8th International conference of the Asian Planning Schools Association*
[15] Marcella J 2004 *Arsitektur & Perilaku Manusia* (Grasindo)
[16] Lang J T 1987 *Creating architectural theory: The role of the behavioral sciences in environmental design* (JSTOR)
[17] McKay F H, Singh A, Singh S, Good S and Osborne R H 2016 Street vendors in Patna, India: Understanding the socio-economic profile, livelihood and hygiene practices *Food Control* **70** 281–285
[18] Khongtong J, Karim M S, Othman M and Bolong J Bin 2016 The mediation effects of consumers' need recognition and pre-purchase evaluation in consumers' decision-making on purchasing safe street food: The case in Nakhon Si Thammarat, Thailand *J. Foodserv. Bus. Res.* **20** 192–203
[19] Shukla A and Navratra N 2017 Streets as public spaces: A case study of Manek chowk *Int. J. Civil. Eng. Technol.* **8** 1367–1376
[20] von Lampe K and Kurti M 2016 The illegal cigarette trade in New York City *Trends Organ. Crime* **19** 329–350

Policy Analysis of Poverty Alleviation in Semarang City Using Spatial and Sectoral Approach

M Muktiali[1]

[1]Department of Urban and Regional Planning, Diponegoro University, Semarang 50275, Indonesia

Email: mukti.ali@pwk.undip.ac.id.

Abstract. Poverty is a multidimensional problem. Therefore, poverty reduction policy is not only related to the increase of income, but also various other dimensions such as improvement of education, health, quality of life, access to electricity, access to sanitation and water supply. Semarang City Government in 2012 initiated a policy of poverty reduction synergy program called "Gerdu Kempling". Gerdu Kempling is an integrated policy which gives priority to addressing poverty in each village and sub-district in Semarang based on aspects of health, economy, education, infrastructure, and environment. Based on the results of Budget Allocation Analysis and Analysis of Geographic Information Systems (Poverty Mapping), it can be concluded that the program and budget allocation for poverty alleviation in Semarang City are not synchronized either spatially or sectorally.

Keywords: Poverty, Spatial Mapping, Policy Analysis

1. Introduction

Poverty is a multidimensional phenomenon that encompasses the lack of opportunities, lack of empowerment, lack of security and malnutrition and poor health. Poverty not only has the connotation of the shortage of economic factors but also includes missing out on opportunities, lack of access to or exclusion from social services and other social deprivation [1–8].

Poverty is a state of lack of goods and services needed to attain a decent standard of living, since the standard of living is different, so there is no universal concept of poverty Levitan in Effendi [9]. The problems of poverty, especially in urban areas, are generally characterized by the inability of the poor to access basic infrastructure and facilities. In addition, inadequate environmental conditions with quality housing and settlements are far below the standards of eligibility and uncertain livelihoods. Lack of access to these resources will result in poverty being difficult to overcome and the poor will be trapped in a culture of poverty resulting in behavioral attitudes that tend to be fatalistic and helpless. Therefore, poverty reduction policy is not only related to the increase of income, but also various other dimensions such as improvement of education, health, quality of life, access to electricity, access to sanitation and water supply.

Poverty is a spatially heterogeneous phenomenon, i.e. poor people tend to be clustered in specific places. The geographic variation in the incidence and magnitude of poverty is often due to factors with spatial dimensions, such as natural resource endowments, and access to services such as health care and education [10]. As part of the wider debate about the causes of poverty and specifically the interventions needed to alleviate poverty, researchers, development practitioners and policy makers are increasingly using poverty maps to visualize the spatial dimensions of poverty [10–13].

Recent development of the poverty mapping methodology has enabled usto estimate poverty indicators at a level of small areas and analyze the local-level relationship between poverty and other geographic factors. Some of the evidence of increased use of poverty maps can be found in Henninger and Snel [14] Poverty maps are an increasingly popular mode of visualizing the spatial dimension of poverty. They help guide priority-setting and target poverty-alleviation interventions. The utility of poverty maps can be enhanced by spatially disaggregating the underlying causes of poverty.

The research was conducted to assess the budget allocation for each poverty indicator (based on Mayor's Regulation No. 18C of 2009) [15] for each region (sub-district). Indicators of poverty include vulnerable to the economy, prone to clean water and prone to sanitation. Ideally between sectoral and spatial aspects should be synchronized, meaning the higher the level of vulnerability of an indicator poverty in a region (sub - district), then the higher the budget allocated for the region, vice a versa.

2. Data and Methods

Data on the population that categorized as vulnerable to the economy, vulnerable to clean water and sanitation prone as well as data on the budget allocation for each sector for each sub-district source from the Regional Development Planning Board of Semarang City. This research is a quantitative approach and using Budget Analysis and Geographic Information System (Poverty Maps). Both analysis methods are used to analyze the synchronization between the amount of budget allocation and the level of vulnerability (economy, sanitation and clean water) in each sub-district in Semarang City.

The economically vulnerable population is seen from the indicator of job type, amount of income, and ability to meet basic needs. Population prone to clean water are seen from the indicators of household access to clean water. Sanitary prone populations are seen from indicators of basic sanitation facilities owned by households.

3. Poverty Profile of Semarang City

3.1. The Amount of Poor House Hold and Poor Population by Sub District and Village

The highest number of poor people is in North Semarang Subdistrict which is 46.100 inhabitants. Second , West Subdistrict which is 40.043 inhabitants The majority of the poor population are in the coastal area of Semarang City (Tabel 1).

3.2. Povery Allevation Program of Semarang City

Semarang City Government accelerated poverty reduction (poverty reduction strategy) through GERDU KEMPLING (Integrated Movement in Health, Economy, Education, Infrastructure and Environment). The importance of poverty reduction program is because high poverty can also affect economic development in Semarang City. Gerdu Kempling's description is an Integrated Movement involving all stakeholders in this case City Government, Universities, State-Owned Enterprises, Private Companies, Banking which together work together to solve the problem of poverty covering all aspects and summarized in 5 areas: Health, Economy, Education, Infrastructure and Environment. The objectives of Gerdu Kempling as a strategy to accelerate poverty reduction in Semarang City by synergizing the city government program with the existing stakeholders, namely PTN and PTS, NGO, banking, BUMN, public figure. In order to optimize all potential in Semarang City in accelerating poverty alleviation so that poverty reduction program goals and targets can be achieved efficiently and effectively. Another goal is to minimize obstacles and problems in accelerating the achievement of integrated, integrated, synergy and sustainable poverty alleviation programs. Gerdu Kempling program is implemented by providing capital assistance, goods assistance, and skills training. In addition, there are also rescue, empowerment and strengthening programs realized through health, economic, education, infrastructure development and environmental development programs

Table 1. The Amount of Poor House Hold and Poor Population by Sub District

No	Sub district	Poor House Hold (HH)	%	Poor Population (Person)	%
1	Semarang Tengah	6,088	5.2%	18,390	4.8%
2	Semarang Utara	13,775	11.7%	46,100	12.1%
3	Semarang Timur	6,425	5.4%	20,610	5.4%
4	Gayamsari	6,715	5.7%	22,109	5.8%
5	Genuk	8,167	6.9%	26,264	6.9%
6	Pedurungan	7,745	6.6%	25,832	6.8%
7	Semarang Selatan	6,943	5.9%	21,371	5.6%
8	Candisari	7,662	6.5%	24,635	6.5%
9	Gajahmungkur	5,054	4.3%	15,901	4.2%
10	Tembalang	11,194	9.5%	36,823	9.7%
11	Banyumanik	4,642	3.9%	14,643	3.9%
12	Gunungpati	6,171	5.2%	19,188	5.1%
13	Semarang Barat	12,287	10.4%	40,043	10.5%
14	Mijen	4,868	4.1%	15,067	4.0%
15	Ngaliyan	6,925	5.9%	21,951	5.8%
16	Tugu	3,272	2.8%	10,709	2.8%
	Total	**117,933**	**100.0%**	**379,636**	**100.0%**

4. Results and Discussion

4.1. Analysis of Vulnerable Populations

4.1.1. Economic Vulnerability

The distribution of poor people categorized as vulnerable to the economy is in North Semarang Subdistrict, 13,371 inhabitants. The high economic vulnerability in North Semarang Sub-district due to the lack of available sources of income in the sub-district. The absence of facilities as a place to earn income due to environmental problems North Sub District Semarang often experience rob and flood, thus causing the absence of centers of activity. In addition, the source of income obtained by local residents is also not sufficient in every day / month. For details can be seen in Figure 1-2.

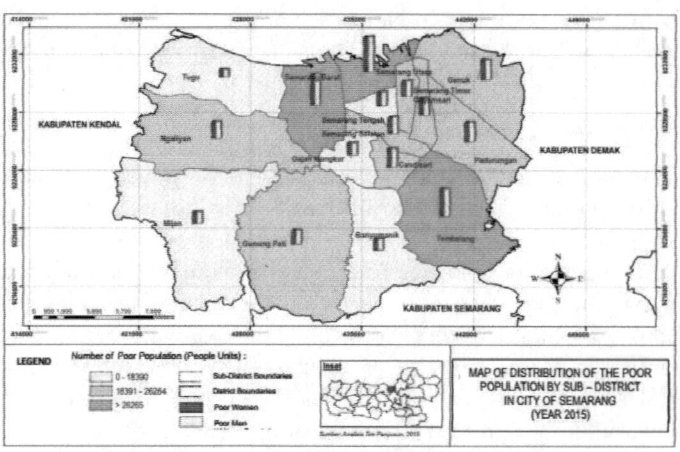

Figure 1. Distribution of Poor Population

4.1.2. Clean Water Vulnerability

Districts with poor people who are prone to access to clean water are North Semarang Subdistrict (largest number), East Semarang, Semarang Barat, South Semarang, and Tembalang. The number of poor people who are prone to access to clean water in North Semarang Sub-district is 4,816 people, not much different from West Semarang Sub-District, which is 4,602 people.

4.1.3. Sanitation Vulnerability

Kecamatan with the number of poor people prone to sanitation is in North Semarang District. The low quality of the environment in North Semarang Sub-district is one of the consequences of poor sanitation in residential areas. Not only in North Semarang Subdistrict, similar conditions are found in other coastal areas and some slums in Semarang City.

4.2. Analysis of Budgeting

4.2.1. Alocation of Budgetting for Economy Sector

Sub-districts that receive the highest budget allocation in economic sector are Kecamatan Tugu, Mijen Sub-district and Genuk Sub-district. The lowest allocation for the sector is Candisari and West Semarang. For details can be seen in Table 2 and Figure 3.

Table 2. The Aloction of Budgetting for Economy Sector

No	Sub District	Amount (IDR)	Percentage
1	Semarang Selatan	261,128,000	2.9%
2	Semarang Utara	298,983,000	3.4%
3	Semarang Timur	385,062,000	4.3%
4	Semarang Barat	256,238,000	2.9%
5	Semarang Tengah	737,572,000	8.3%
6	Genuk	793,212,000	8.9%
7	Pedurungan	609,857,000	6.9%
8	Gunungpati	711,514,000	8.0%
9	Candisari	167,930,000	1.9%
10	Gajahmungkur	229,401,000	2.6%
11	Banyumanik	335,276,000	3.8%
12	Gayamsari	365,127,000	4.1%
13	Mijen	911,780,000	10.2%
14	Ngaliyan	576,119,000	6.5%
15	Tugu	1,568,285,000	17.6%
16	Tembalang	691,905,000	7.8%
	Total	**8,899,389,000**	**100.0%**

4.2.2. Alocation of Budgetting for Clean Water

Sub-districts that receive the highest budget allocation in economic sector are Kecamatan Tembalang Sub-district, Ngaliyan Sub-district and Candisari Sub-district. The lowest allocation for the sector is Semarang Selatan and Semarang Utara. For details can be seen in Table 3 and Figure 4.

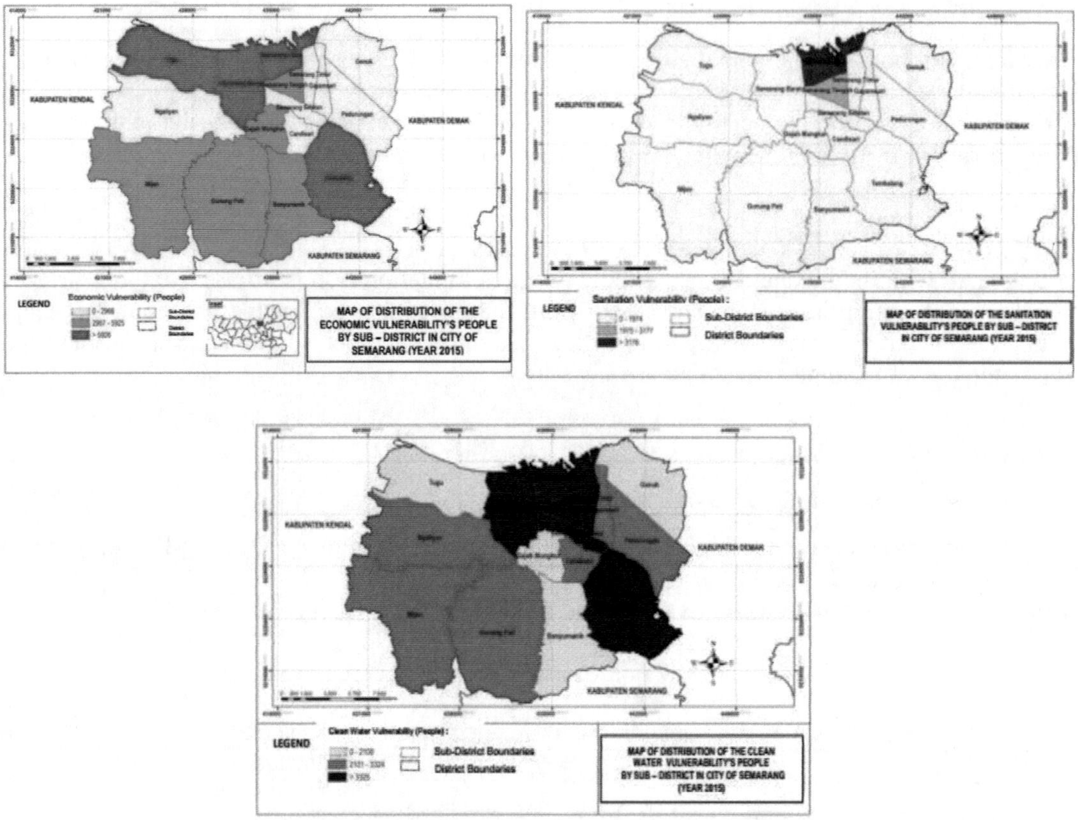

Figure 2. Map of Vulnerable Population (Economy Sector, Clean Water and Sanitation)

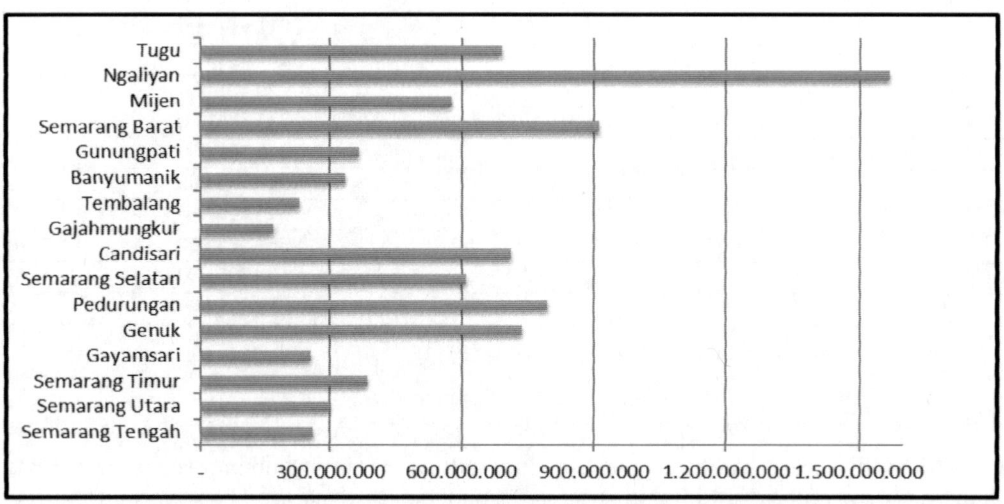

Figure 3. The Alocation of Budgetting for Economy Sector

Table 3. The Aloction of Budgetting for Clean Water

No	Sub District	Amount (IDR)	Percentage
1	Semarang Selatan	0	0.0%
2	Semarang Utara	0	0.0%
3	Semarang Timur	509,104,000	3.4%
4	Semarang Barat	129,298,000	0.9%
5	Semarang Tengah	35,000,000	0.2%
6	Genuk	556,804,000	3.8%
7	Pedurungan	466,436,000	3.2%
8	Gunungpati	1,545,323,000	10.4%
9	Candisari	2,058,747,000	13.9%
10	Gajahmungkur	67,700,000	0.5%
11	Banyumanik	1,072,739,000	7.3%
12	Gayamsari	638,437,000	4.3%
13	Mijen	1,235,540,000	8.4%
14	Ngaliyan	2,669,719,000	18.0%
15	Tugu	1,035,769,000	7.0%
16	Tembalang	2,772,067,000	18.7%
	Total	14,792,683,000	100.0%

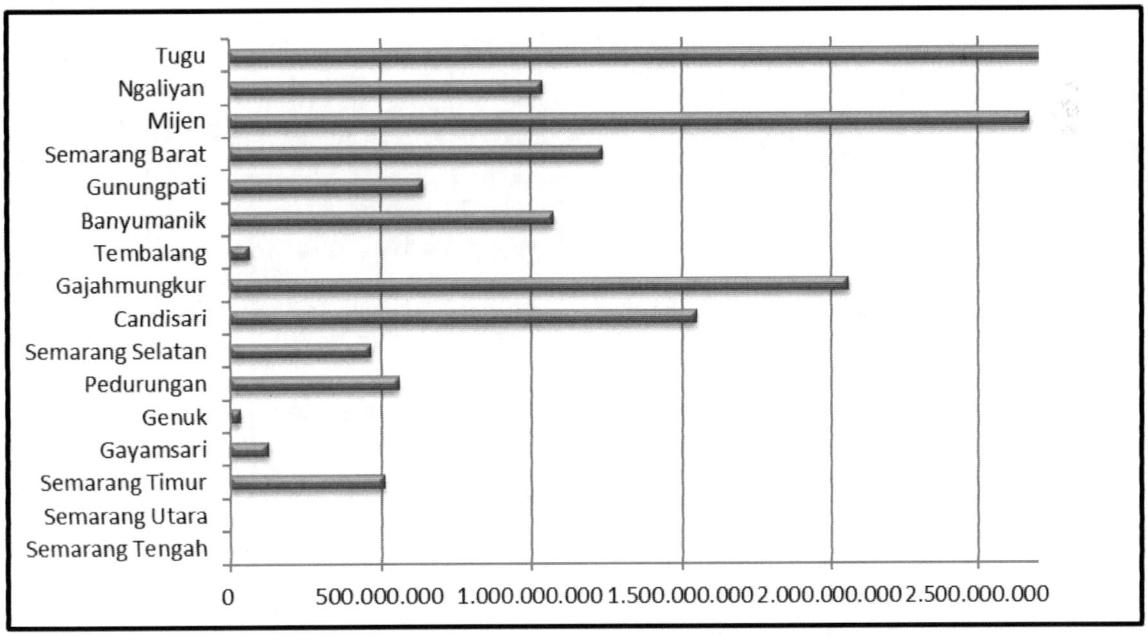

Figure 4. The Aloction of Budgetting for Clean Water

IOP Conf. Series: Earth and Environmental Science **123** (2018) 012046 doi:10.1088/1755-1315/123/1/012046

4.2.3. Alocation of Budgetting for Sanitation

Sub-districts that receive the highest budget allocation in economic sector are Kecamatan Tembalang Sub-district, Ngaliyan Sub-district and Candisari Sub-district. The lowest allocation for the sector is Semarang Selatan and Semarang Utara. For details can be seen in Table 4 and Figure 5.

Table 4. The Aloction of Budgetting for Sanitation

No	Sub District	Amount (IDR)	Percentage
1	Semarang Selatan	139,306,000	8.7%
2	Semarang Utara	2,224,000	0.1%
3	Semarang Timur	51,806,000	3.3%
4	Semarang Barat	30,277,000	1.9%
5	Semarang Tengah	7,306,000	0.5%
6	Genuk	421,620,000	26.5%
7	Pedurungan	421,620,000	26.5%
8	Gunungpati	422,944,000	26.5%
9	Candisari	2,306,000	0.1%
10	Gajahmungkur	2,306,000	0.1%
11	Banyumanik	10,630,000	0.7%
12	Gayamsari	2,306,000	0.1%
13	Mijen	44,880,000	2.8%
14	Ngaliyan	2,306,000	0.1%
15	Tugu	2,306,000	0.1%
16	Tembalang	28,903,000	1.8%
	Total	**1,593,046,000**	**100.0%**

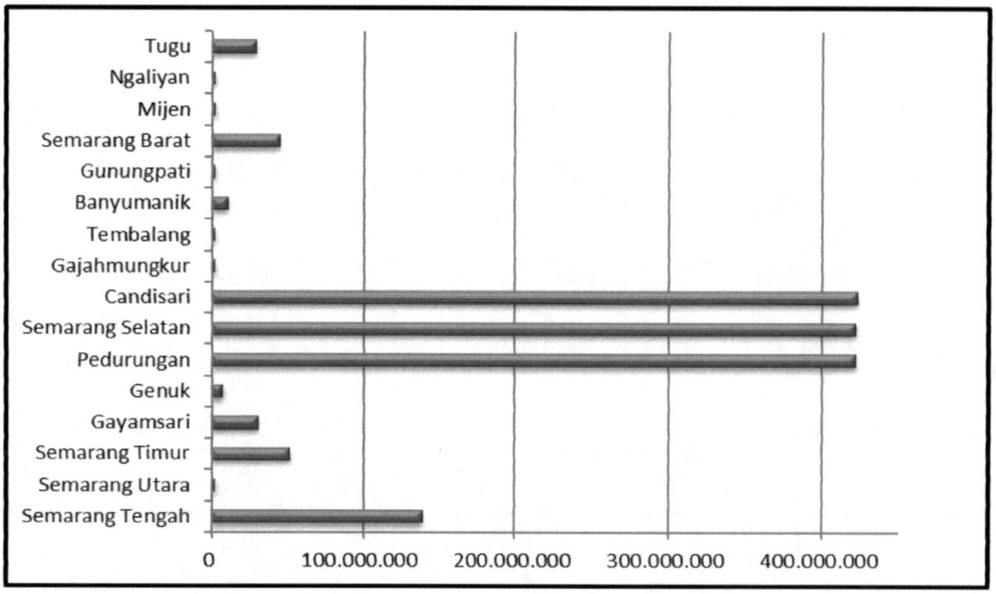

Figure 5. The Aloction of Budgetting for Sanitation

IOP Conf. Series: Earth and Environmental Science **123** (2018) 012046 doi:10.1088/1755-1315/123/1/012046

4.3. Overlay Between The Level of Vulnerability & Budget Allocation

4.3.1. Economy Sector

The right target budget allocation is in Tugu Sub-district, because the area is a sub-district with high economic vulnerability and get a high economic aid budget as well. The right target budget is also in Kecamatan Mijen and Gunungpati. Both sub-districts are sub-districts with moderate vulnerable categories and obtain a moderate budget. Sub-district is not the right target is in North Semarang District, because the district is a district with high economic vulnerability, but get low funds. For details can be seen in Figure 6.

4.3.2. Clean Water

The target budget is in Tembalang, Gajahmungkur and Genuk sub-districts. Tembalang Sub-district is a sub-district with high water prone category and get high budget also. Gajahmungkur and Genuk sub-districts are low-risk sub-districts and have low budgets. The budget that is not on target is in North Semarang Subdistrict, West Semarang, East Semarang, Central Semarang and South Semarang. These five sub-districts are high-risk sub-districts but have low budgets. For details can be seen in Figure 7.

4.3.3. Sanitation

The budget is not on target, such as in North Semarang Subdistrict, Genuk, Gunungpati, and Pedurungan. It can be seen that North Semarang Subdistrict is a sub-district that has high sanitation prone category, but it gets low budget. Sub-district of Genuk, Gunungpati and Pedurungan are sub-district with low sanitation prone category, but get high budget. For details can be seen in Figure 8.

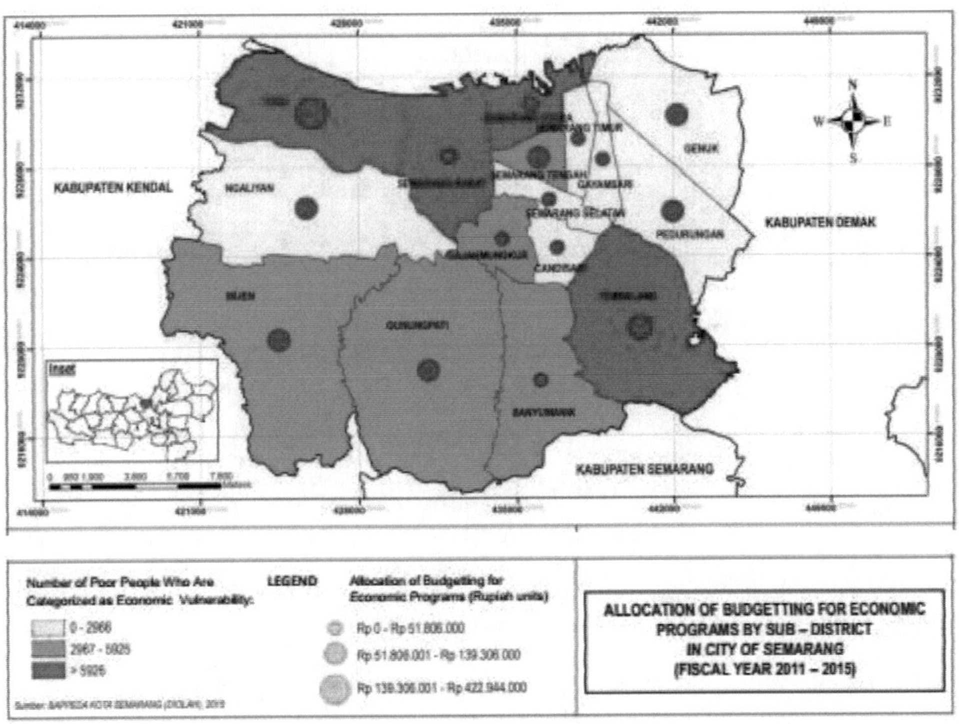

Figure 6. Overlay between the Level of Economy Vulnerability and Budget Allocation

IOP Conf. Series: Earth and Environmental Science **123** (2018) 012046 doi:10.1088/1755-1315/123/1/012046

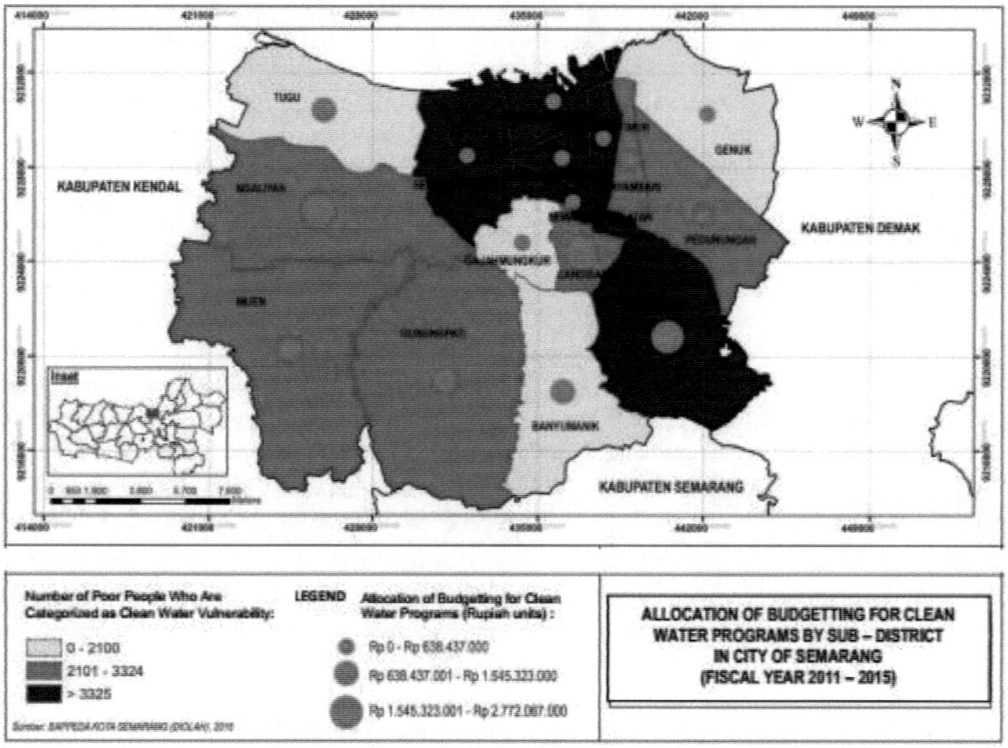

Figure 7. Overlay between the Level of Clean Water Vulnerability and Budget Allocation

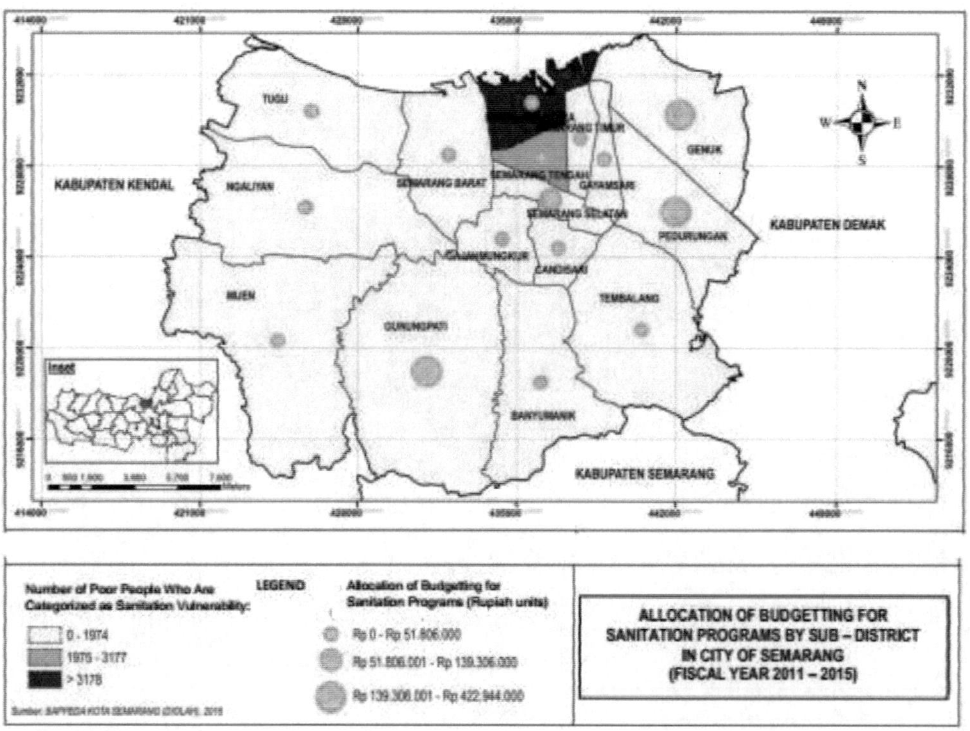

Figure 8. Overlay between the Level of Sanitation Vulnerability and Budget Allocation

4.4. Synchronization Analysis Results

Table 5. Synchronization Analysis Results

No	Sub - Disrict	Economic Vulnerability	Budget Allocation for Economic	Clean Water Vulnerability	Budget Allocation for Clean Water	Sanitation Vulnerability	Budget Allocation for Sanitation
1	Banyumanik	Low	Medium	Low	Medium	Low	Medium
2	Candisari	Medium	Medium	Medium	Medium	Medium	High
3	Gajahmungkur	Low	Low	Low	High	Low	Low
4	Gayamsari	Medium	Low	Low	Low	Low	Medium
5	Genuk	High	Medium	Medium	Low	Medium	Medium
6	Gunungpati	Low	Medium	Medium	Medium	Medium	Low
7	Mijen	Low	Medium	Medium	High	Medium	Low
8	Ngaliyan	Medium	High	High	Medium	Medium	Low
9	Pedurungan	Medium	High	Medium	Medium	Low	High
10	Semarang Barat	High	High	Medium	Medium	Medium	Medium
11	Semarang Selatan	Medium	Medium	High	Medium	Medium	High
12	Semarang Tengah	Medium	Low	High	Low	High	Medium
13	Semarang Timur	Medium	Medium	Medium	Medium	High	Medium
14	Semarang Utara	High	Medium	Low	Low	High	Low
15	Tembalang	High	Low	Medium	Low	Medium	Low
16	Tugu	Low	Medium	Low	High	Low	Medium

Based on the result of poverty map analysis and budget which have been done, hence seen existence of mismatch between level of vulnerability of poor people in a sub district with amount of budget received. Such as in North Semarang Subdistrict, Genuk, Gunungpati, and Pedurungan. It can be seen that North Semarang Subdistrict is a sub-district that has high sanitation prone category, but it gets low budget. Overall, the budget allocation in accordance with the level of vulnerability to the economic sector only reached 31%; For clean water by 44% and the smallest for sanitation which only reached 19%. For detail can be seen in Table 5 and 6.

Table 6. Recapitulation of Synchronization Analysis Results

No	Sub - Disrict	Economic Sector	Clean Water	Sanitation
1	Banyumanik	Not appropriate	Not appropriate	Not appropriate
2	Candisari	Appropriate	Appropriate	Not appropriate
3	Gajahmungkur	Appropriate	Not appropriate	Appropriate
4	Gayamsari	Not appropriate	Appropriate	Not appropriate
5	Genuk	Not appropriate	Not appropriate	Appropriate
6	Gunungpati	Not appropriate	Appropriate	Not appropriate
7	Mijen	Not appropriate	Not appropriate	Not appropriate
8	Ngaliyan	Not appropriate	Not appropriate	Not appropriate
9	Pedurungan	Not appropriate	Appropriate	Not appropriate
10	Semarang Barat	Appropriate	Appropriate	Appropriate
11	Semarang Selatan	Appropriate	Not appropriate	Not appropriate
12	Semarang Tengah	Not appropriate	Not appropriate	Not appropriate
13	Semarang Timur	Appropriate	Appropriate	Not appropriate
14	Semarang Utara	Not appropriate	Appropriate	Not appropriate
15	Tembalang	Not appropriate	Not appropriate	Not appropriate
16	Tugu	Not appropriate	Not appropriate	Not appropriate
	Apropriate	**31%**	**44%**	**19%**
	Not apropriate	**69%**	**56%**	**81%**

5. Conclusion

Based on the results of Budget Allocation Analysis and Analysis of Geographic Information Systems (Poverty Mapping), it can be concluded that the program and budget allocation for poverty alleviation in Semarang City are not synchronized either spatially or sectorally. In conclusion, for next implementation, local government should take into account an integrated approach between budget and the region as beneficiaries. Sectors of program should accommodate the potency and problems of regions.

Sub-districts with high economic vulnerability should have a higher proportion of the budget for the field than in the previous period, such as in North Semarang District. Districts with high water prone areas should have higher budget proportions for the field compared to the previous period, such as in Semarang Utara, Semarang Barat, Semarang Timur, Semarang Tengah and South Semarang. Districts with high sanitation prone should have higher budget proportions for the field compared to previous periods, such as in Semarang Utara and Central Semarang Sub-districts.

6. References

[1] Wagle U 2002 Rethinking poverty: definition and measurement *Int. Soc. Sci. J.* **54** 155–165
[2] Park J, Turnbull A P and Turnbull III H R 2002 Impacts of poverty on quality of life in families of children with disabilities *Except. Child.* **68** 151–170
[3] Alkire S and Foster J 2011 Counting and multidimensional poverty measurement *J. Public Econ.* **95** 476–487
[4] Alkire S and Sumner A 2013 Multidimensional poverty and the post-2015 MDGs *Development* **56** 46–51
[5] Bossert W, Chakravarty S R and D'Ambrosio C 2013 Multidimensional poverty and material deprivation with discrete data *Rev. Income Wealth* **59** 29–43
[6] Alkire S and Santos M E 2014 Measuring acute poverty in the developing world: Robustness and scope of the multidimensional poverty index *World Dev.* **59** 251–274
[7] Alkire S and Seth S 2015 Multidimensional poverty reduction in India between 1999 and 2006: Where and how? *World Dev.* **72** 93–108
[8] Liu Y, Liu J and Zhou Y 2017 Spatio-temporal patterns of rural poverty in China and targeted poverty alleviation strategies *J. Rural Stud.* **52** 66–75
[9] Manning C and Effendi T N 1985 *Urbanisasi, pengangguran dan sektor informal di kota* (Diterbitkan untuk Yayasan Obor Indonesia dan Pusat Penelitian dan Studi Kependudukan UGM [oleh] Gramedia)
[10] Erenstein O, Hellin J and Chandna P 2010 Poverty mapping based on livelihood assets: A meso-level application in the Indo-Gangetic Plains, India *Appl. Geogr.* **30** 112–125
[11] Amarasinghe U, Samad M and Anputhas M 2005 *Locating the poor: spatially disaggregated poverty maps for Sri Lanka* vol 96(IWMI)
[12] Bellon M R, Hodson D, Bergvinson D, Beck D, Martinez-Romero E and Montoya Y 2005 Targeting agricultural research to benefit poor farmers: Relating poverty mapping to maize environments in Mexico *Food Policy* **30** 476–492
[13] Bigman D and Fofack H 2000 Geographical targeting for poverty alleviation: An introduction to the special issue *World Bank Econ. Rev.* **14** 129–145
[14] Henninger N and Snel M 2002 Where are the poor? Experiences with the development and use of poverty maps.
[15] Pemerintah Kota Semarang 2009 *Peraturan Walikota Semarang No 18 C Tahun 2009 tentang Indikator, Kriteria dan Klasifikasi Warga Miskin Kota Semarang.*

Urbanization and the Resulting Peripheralization in Solo Raya, Indonesia

W Pradoto[1], F H Mardiansjah[1], O R Manullang[1], and A A Putra[1]

[1]Department of Urban and Regional Planning, Universitas Diponegoro, Indonesia

Email: wisnu.pradoto@gmail.com

Abstract. Dynamic urbanization in *Solo Raya,* a local term for Surakarta Metropolitan, amongst rapid regional based-urbanization in Indonesia, shows the unbalance pattern of growth. A number of Surakarta City's peripherals become the newly growing area which is characterized by a well-facilitated region, while the former urbanized areas next to the city center present the declining process. Different socioeconomic development triggers a unique mosaic of socio-spatial pattern, on which the phenomena of peripheralization could be investigated. Urban investment that boosted by the political will of both the national and local government has led to a shift in demographic condition. A relatively massive in-migration has been attracted to the peripheral and creates the new landscape of urban-rural society. Complex dynamic of metropolitan growth and the resulting peripheralization reminds that socio-spatial pattern calls the challenges for managing the rapid change of land use and space use. The pattern of urbanization that differs upon the surrounding areas of Surakarta City would be interesting to be explored. This paper will discuss the conceptual framework of peripheral urbanization and the methodological approach. It is actually the part of ongoing research on peripheralisation in Solo Raya.

Keywords: Urbanization, Metropolitan Growth, Peripheralization, Unbalance Development

1. Introduction

Economic growth and investment have fueled the rapid development of small towns and surrounding rural areas. Changes in social structure and spatial patterns in the suburbs signal the emergence of new service centers [1]. The growth of urban areas has penetrated the agricultural area and changed the social and economic characteristics of rural areas [2–4]. In some parts looks very fast development, while the other part grows slowly. Meanwhile, the old city area that has evolved earlier shows the dynamics of the decline. The decline is characterized by a decrease in the population. On the other hand, marginal areas show symptoms of population growth. Improving the quality of infrastructure and completeness of facilities is another fact that marks developments in the periphery. In certain areas even show symptoms of increasing service scale very rapidly. The region grew into a new growth center and created a multi-core pattern on a regional scale. Urbanization on a regional scale is becoming a contemporary phenomenon that flourished in Asia. Fisher [5] and Ainsaar [6] studied the pattern of suburban development and built typology to illustrate the differences in social and spatial dynamics.

Differences in the dynamics of development that lead to differences in space growth and the level of investment lead to the emergence of differences in the level of progress of a region. The gap between center and periphery begins to change shape. New areas of development on the periphery tend to be better, while some parts of the central region are reversed. The term periperalization is used to explain the phenomenon of regional inequality or inequality. The focus of the study is more on social and spatial aspects [7]. Research on peripheralization, as Lang [8] has done in some countries in Eastern Europe, is often associated with advanced region dichotomies that are considered to win competition and backward loses. The political change from socialism to capitalism has triggered a major change not only in the social aspect but also in the space aspect. Physically, progress is more often associated with the rate of development of physical construction and socioeconomic facilities. While socially associated with changes in the economic structure of the communist to liberalism. In subsequent developments, referring to Tahir and Naumann [9], the dependence of the periphery to the center gradually declines, and even the population tends to move from the central to the periphery. In this context, the central dominance of the periphery has shifted the European countries

In the context of urbanization and metropolitan development in Indonesia, periperalization occurs due to the rapid growth of rural areas. The availability of land becomes one of the considerations of the development of newly built areas on the outskirts of major cities. In some areas there is still the dominance of big cities on their new growth areas, but some dynamics show that the periphery is growing faster. Despite the decline in agricultural sector contribution, as explained by Hudalah [10], the rapid development driven largely by private investment has led to an increase in the role of small towns in the periphery. The urban service center began to shift from the big city to the periphery. Among the new growth areas, there are even more developed beyond the center of the city, as happened in Sukoharjo, one of the neighboring districts of Surakarta. The dynamics of urbanization and the emergence of symptoms of periperalization becomes an interesting issue to study. The symptoms illustrate how the role of small towns grow and vice versa the dominance of big cities is declining. The marginalization initially experienced by the periphery gradually shifted to the central region. The displacement of the central population has had an impact on the declining quality of service in the central region. Parts of the central region began to lag.

Observations on socio-economic changes and spatial areas can be linked to assess the adaptation and resilience of local communities in responding to urban development. What are the factors that influence people to stay in areas that have changed due to the influence of urban growth and merge into the urban socio-economic culture. And what makes the local people to shift into the interior and maintain their socio-economic culture.

2. The Impact of Urbanization on The Development of The Periphery

The development of investment spur the growth of urban space. The best-developed periphery areas are growing first. Economic and demographic changes encourage social shifts and spatial patterns. The entry of middle economic groups in rural areas has triggered social polarization. The dominant peasant communities are marginalized and occupy deeper areas. Some of them remain among the new settlers and create social hybrids that have not fully blended. Fragmentation of space according to economy class becomes a new character in the peri-urban and sub-urban areas. Changes in social, economic and spatial dimensions vary depending on the socioeconomic and cultural characteristics of the community and the politics of local government development. Agglomeration of metropolitan areas presents variations in structural change. The visible differences reflect the socio-economic and spatial attractiveness of the region (see Norgard [11] and Ravetz et al. [12]).

The phenomenon of peri-urbanization and sub-urbanization has been the focus of several studies. The shift in development to the periphery marks the contemporary urbanization that is spreading in third world countries. The suburbs are the main target of middle-class migration in the city. This phenomenon by Ainsaar [6] is called deconcentration, namely the process of migration of population to areas of lower density. Social and economic factors behind the migration of the population to the suburbs [13]. In the context of investment, the availability of sufficient quantity of land at an economical price is a consideration for investors to penetrate the area. Some researchers focus on the classification of urbanization types (see Ainsaar [6] and Fisher [5]), and others focus on the migration process. Champion [14] observed the migration process and the concentration of the population that occurred in the periphery. He named the process as concentration diffusion. In his research in some peri-urban areas in several states in Australia, Fisher [5] divides urbanization typology into four types, namely Sub-Urbanization; Counter Urbanization; Population Retention; and Centripetal Migration.

2.1. Sub-Urbanisation Dynamics

The term sub-urbanisation is used to name a new region whose proximity and dependence on the central area is still high. This region generally has a high commuting level. Sub-urban areas are generally new settlements for the city's established groups. The group moved to the outskirts to get a better, more natural and better residential neighborhood and still have a high degree of affordability from the city. This area is generally developed by utilizing existing road access. In the Indonesian context, the tollgate entrance is a potential area for sub-urban development due to the ease of achievement. In a short time this area is packed with high-rise settlements. Groups living in the region are generally young professional groups and young families looking to find land that is still relatively cheaper than in the city, but has a high level of ease of achievement. The tendency of suburbanization at the same time trigger land issues is speculation. Areas along the main corridor are subjected to speculators to control large amounts of land.

2.2. Counter Urbanization Dynamics

This dynamic is associated with a phenomenon in which the migration to the periphery occurs in an area relatively far from the city. The linkage or dependence on the city is also relatively low. Target areas are generally rural areas that are still dominated by green land with a moderate level of ease of achievement. Access is no longer a major factor, but rather a complete facility and rural atmosphere that offers a natural environment. The migrant groups that make this choice are generally established families or retired retirees. The economic establishment encourages them to live in relatively beautiful places and opens the opportunity to have large enough land. The countryside atmosphere is the main attraction because they are looking for a relatively quiet location and quite far from the noise of the city. Some of the migrants in this area have deliberately started a new life by opening a business in an area far from the city. Generally are those who have been at relatively high career levels in some corporations. The existence of this upper middle class makes the region grow and open up new job opportunities for local communities.

2.3. Population Retention Dynamics

This term is used to characterize areas that are still physically dominated by rural culture. This area is clearly located at a relatively far distance from the city. In the aspect of population, this area is characterized by the presence of local people with a relatively equal proportion or greater than the group of immigrants. Local people choose to survive in this region due to adequate economic development. These economic developments were triggered by the presence of established urban migrants who opened businesses and offered opportunities for local people to work in the administrative, factory, maid, security, hotel servants, plantation workers as well as other informal sectors such as informal transporter, tailor clothes, construction workers, and others. In this region, the completeness of facilities is not the main requirement. It is precisely the nature of the rural atmosphere to be the main attraction. In this region, it usually develops tourism and agribusiness sectors. The existence of businesses that generally rely on rural potentials opens employment opportunities for local communities to increase incomes. In addition, vocational education is also evolving, especially relevant to emerging sectors. This indirectly encourages the awareness of rural communities to improve skills and education. This region is generally within a certain range of counter-urbanisation areas, as some areas prove that the presence of counter-urbanisation areas pushes the surrounding area into a population retention.

2.4. Centripetal Migration Dynamics

This term is used to describe the dynamics of the region where people around the area are interested in migrating to the region due to certain attractions. Community interest dominated by the village community is generally driven by the condition of less fertile villages, limited economic potential, and inadequate social facilities. While they are interested in migrating is the development of medium or large scale industries, the construction of trade facilities and services, the development of tourism objects and others that offer employment opportunities. This phenomenon is generally also a further process that is preceded by population retention. The development of a rural area due to the migration of established economic groups and investors in turn will open up new economic opportunities. The area is the target of migration for people living around it. Therefore, population retention in subsequent developments push the region into a new growth center that will evolve into economic independence. This possibility is particularly the case in areas experiencing rapid growth due to large-scale investments, such as the development of industrial zones or special economic zones fueled by relatively cheap land prices [10]. In time, the region will develop into a new, self-contained city that will become the forerunner to subsequent sub-urbanization and countermeasure of urbanization. Thus it can be observed that the process of urbanization is a continuous cycle that evolves from one place to another continuously, where agricultural activity is increasingly shifted and replaced by the urban economic sector, ie industry and services [15].

3. Peripheralization as a form of imbalance development

Peripheralization is not always associated with peri-urban. This terminology is used to describe the condition in which an area is inequality compared to other areas in the vicinity. Peripheralization is not determined by distance or accessibility, but to indicate the difference in the degree of development of a region and the degree of backwardness of one region to another [16]. Other experts review from a socio-political point of view. Peripheralization is associated with discrimination of certain groups or communities in a region [7,17]. The factors behind the peripheralization are quite varied, each expert has a different point of view.

Weck and Beisswenger [1] examine how migration triggers socio-economic changes that result in peripheralization. The economic dynamics of migration is generally the beginning of how a region develops. Economic opportunities will attract migration. In subsequent developments, migration to a region will affect the development of the region. Migration by the upper middle class will be followed by significant developments due to investment. Construction of new facilities and infrastructure will occur in the region. While the abandoned areas, although located in the downtown area actually decreased quality. In this case, migration has a central role that determines the development of a region as well as the decline of a region. The conditions under which the dynamics of development and decline occur in an area called peripheralization. The peripheral region in this case is characterized by declining population, especially productive groups and the stagnation of the construction of facilities or infrastructure. This phenomenon is one of the dimensions that characterize ubanization and urban growth. Michelini and Pintos [18] examine the dynamics of peripheralization occurring in some regions of Latin America. Their research proves that peripheralization occurs due to seizure of land access and facilities. Disadvantaged communities will suffer from lack of access and completeness of facilities.

Based on his research on several countries in Eastern Europe, Lang [8] concluded that peripheralization is caused by several things, namely: concentration of facilities in certain areas that result in interest in populations from other regions that ultimately lead to setbacks in the abandoned areas; a decline in the rate of economic growth resulting in increased dependence of a declining region to a more developed region; development policies that prioritize metropolitan areas; development of infrastructure that is only concentrated in certain areas that trigger the gap. Broadly speaking, the results of Lang's study were confirmed by Kuhn [16] which explains that peripheralization is a form of regional disparity due to economic polarization, different strata of social development and weak political power and the bargaining position of government or community that result in the decline of a region.

4. Research Methods

The study methodology study will cover three things: research approach; data retrieval methods; and data analysis methods. In general, the approach applied in this research is phenomenology approach. This approach is chosen because it is considered the most appropriate approach to understanding the phenomenon of urbanization and observes the process of peripheralization. With phenomenology approach, it is expected to get a detailed and deep understanding of the patterns of urbanization that occur and the dynamics that make up the periperalization. The objective of the research is to observe the urbanization pattern in Metropolitan Surakarta and the peripheralization due to the development of urban area. Survey method will be conducted to dig secondary and primary data. Some speakers will be selected by snow balling to complete the primary data. The collected secondary and primary data were analyzed both quantitatively and qualitatively. Population data will be observed for the period 1980 to 2016. This is intended to see the dynamics that occur, from the beginning of development to the present.

The research area covers Solo Raya area, namely urban agglomeration area of Surakarta city. Geographically, the city of Surakarta as a metropolitan center is surrounded by six districts of Sukoharjo, Boyolali, Karanganyar, Wonogiri, Sragen, and Klaten. Among the six districts, agglomeration of urban areas due to the development of Metropolitan Surakarta occurs only in Sukoharjo, Karanganyar and Boyolali districts. Thus, observation of urbanization pattern will be done in all three districts. Fisher typology described above will be a reference to see the patterns of urbanization that occur. From these three districts, will be selected kecamatan that shows the impact of urbanization development. Each sub-district will be examined from several aspects, namely social, economic and spatial changes. The socio-economic transformation will be tested for the impact on changes in its space structure. More specifically, the changes to be observed are the level of urban intensity that occurs. Rural socioeconomic changes leading to urban areas will be observed to the extent that they alter their spatial order. The socioeconomic characteristics of the community, both local communities and migrants in each new urban area are analyzed to assess whether there is a relationship between socio-economic characteristics and the characteristics of space. To understand the contrast, regions with high degrees of change from village to city will be compared with low degree of change.

Based on research conducted Nuriasari [19], the influence of Surakarta urban can be observed up to a radius of 8 km from the city of Surakarta. In this study the distance will be re-tested to ascertain whether the influence of the urban is still the same or widened in the area further away from the city of Surakarta. Observed urban influences, as described earlier, include aspects of population, land use change, and socio-economic structure change. In the three selected districts will be seen which represents the highest and lowest intensity of the urban. This is done to test whether the effect of urbanization occurs evenly or not. If uneven, it is possible that a kabpaten will show a high intensity of change while the other districts do not show the same thing. This dynamic will certainly provide a more detailed explanation of how the distribution of the effects of urbanization and what factors lie behind the different levels of urbanization intensity. Sub-districts to be observed in this study include (see Figure 1) Baki, Gatak, Kartasura, Grogol, Mojolaban (Sukoharjo) District; District Colomadu, Gondangrejo, Jaten (Karanganyar District); and District Ngemplak (Boyolali District).

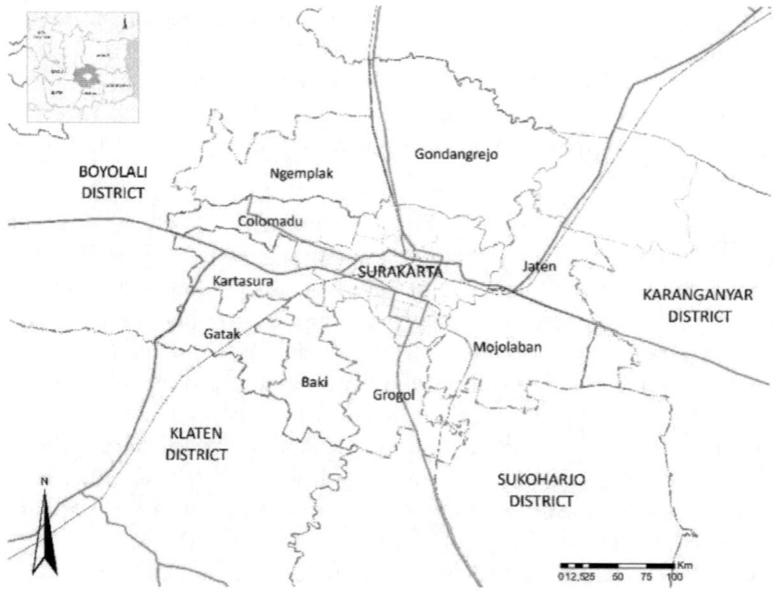

Figure 1. Map of Scope of Research Area

Data analysis in this research will use descriptive analysis method. The assessment results of each aspect derived from the literature review will be assessed and measured. Data to be analyzed include (i) land use data, (ii) population migration, (iii) demographic data, (iv) economic data, and (v) availability of basic facilities data. Spatial analysis to see the dynamics of land use will use data from satellite imagery. The data is then analyzed by overlay method with the help of Geographic Information System. Spatial analysis will use the series data in the form of land use map from 1980 to 2016. It is expected to observe the spatial development stage.

Meanwhile, the analysis of peripherals will also use data relating to economic, social and infrastructure aspects and public facilities. Scoring method will be used for data processing. The size of the score shows the high low inequality that occurs. In the scoring analysis, the value to be given for each indicator is as follow.

Table 1 : Score of Peripheralisation's Indicators

Scores of Indicators	Scores of Migration Rate and Number of Agricultural Household
5 : Very Good, percentage>81% of the average	5 : Very Good, percentage<20%of the average
4 : Good, percentage 61%-80% of the average	4 : Good, percentage21%-40% of the average
3 : Moderate, percentage 41%-60% of the average	3 : Moderate,percentage 41%-60% of the average
2 : Bad, percentage 21%-40% of the average	2 : Bad, percentage 61%-80% of the average
1 : Very Bad, percentage<20% of the average	1 : Very Bad, percentage>81% of the average

After obtained the value of each indicator, then done index score. The index value of each indicator has a range of values 0-1. With the calculation formula is as follows.

$$\frac{\sum counted\ score}{\sum highest\ score}$$

For each indicator, the result of calculating the index value will be divided into three criteria, i.e.:

0 – 0,33 : Bad; 0,34 – 0,66 : Moderate ; 0,67 – 1 : Good

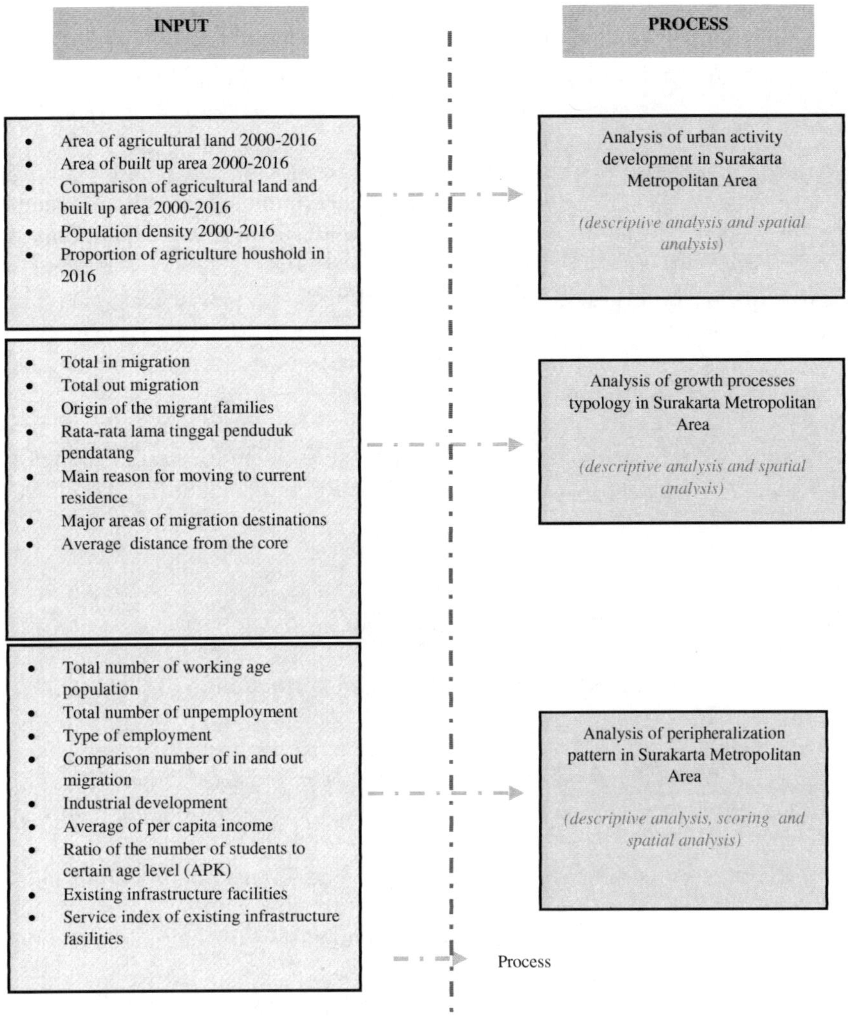

Figure 2. Stages of Research Analysis

5. Conclusion

According to the former researches, peripheral development becomes the contemporary form of urbanization. The different pattern of development in the peripheral regions of the city expresses the particular characteristic in which the rural-urban transformation could be investigated. The distinguish pattern of spatial change shows how the economic forces has driven and how the interplaying role of development stakeholders have been bargaining. The trade-off between public and private interest implies the different impact on physical dimension. A number of studies realize that social and cultural aspects still play as the determinant driver. In additional, the characteristic of urban society that moving into the peripheral regions will create the different kind of social integration. In this matter, we can see that in some extent the change bring the betterment of living condition, but in other situation the change brings the unbalance growth. The expansion of urban infrastructure triggers the peripheralization, in which the former neighborhoods are sometimes decline while the new neighborhood tends to be predominated.

The dynamics of rural-urban integration in the surrounding area of a primate city shows the four different pattern of change that represents the different level of urbanization. The phenomenon of suburbanization performs how the regions nearby the city have been strongly transformed into the new urban areas. Meanwhile, the counter urbanization explains how the emerging of new rural-urban region represents a kind of less dependent area to the city centre. These areas become the new destination of migration in which the inhabitant, especially the former urban society has decided to engage with the new livelihood. In the respective period, these regions are continuously developing its urban characteristic.

The gap between urban and rural that so called peripheralization which is strongly expressed in suburban becomes less intensive in these regions. The other is called population retention, that characterized by the resilient rural area that experience the new opportunity of economy due to the influx of urban middle class into this sites. It can still be seen the existence of farmer household cultivating their land while the family member has engaged with the new urban jobs. This kind of additional income has increased their living standard. Last but not least is the centripetal migration. This type of urbanization happens when the new growing urban areas attracts the surrounding rural people to come, mostly in search of employment. In this region, we can also see how the gap between affluent people who live in a better environment comparing to those who are just came in that mostly living in the relatively substandard area. It is common that those area are located in the periphery that less facilitated. In this matter, the phenomenon of peripheralization is also could be examined.

6. References

[1] Weck S and Beißwenger S 2013 Coping with Peripheralization: Governance Response in Two German Small Cities *Eur. Plan. Stud.* **22** 2156–2171
[2] Winarso H, Hudalah D and Firman T 2015 Peri-urban transformation in the Jakarta metropolitan area *Habitat Int.* **49** 221–229
[3] Winarso H and Hudalah D 2006 Metropolitan di Indonesia: Kenyataan dan Tantangan dalam Penataan Ruang *Direktorat Jenderal Penataan Ruang, Dep. Pekerj. Umum*
[4] Pradoto W 2012 Development patterns and socioeconomic transformation in peri-urban area *Univerlagtuberlin*
[5] Fisher T 2003 Differentiation of Growth Processes in the Peri-urban Region: An Australian Case Study *Urban Stud.* **40** 551–565
[6] Ainsaar M 2002 Reasons for move: A study on trends and reasons of internal migration with particular interest in Estonia, 1989--2000.
[7] Bernt M and Colini L 2013 *Exclusion, marginalization and peripheralization*
[8] Lang T 2012 Shrinkage, Metropolization and Peripheralization in East Germany *Eur. Plan. Stud.* **20** 1747–1754
[9] Fischer-Tahir A and Naumann M 2013 Introduction:Peripheralization as the Social Production of Spatial Dependencies and Injustice *Peripheralization* (Springer Fachmedien Wiesbaden) pp 9–26
[10] Hudalah D, Winarso H and Woltjer J 2007 Peri-urbanisation in East Asia:A new challenge for planning? *Int. Dev. Plan. Rev.* **29** 503–519
[11] Nørgaard H 2011 Futures of rural and peripheral areas-challenges, strategies and policies
[12] Ravetz J, Fertner C and Nielsen T S 2012 The Dynamics of Peri-Urbanization *Peri-urban futures: Scenarios and models for land use change in Europe* (Springer Berlin Heidelberg) pp 13–44
[13] Hosszú S 2009 *Counterurbanization: A literature study* (Danish Institute of Rural Research and Development (IFUL))
[14] Champion T, Hugo G and others 2004 *New forms of urbanization: beyond the urban-rural dichotomy.* (Ashgate Publishing Ltd.)
[15] Webster D, Cai J and Muller L 2014 The New Face of Peri-Urbanization in East Asia: Modern Production Zones, Middle-Class Lifesytles, and Rising Expectations *J. Urban Aff.* **36** 315–333
[16] Kühn M 2014 Peripheralization: Theoretical Concepts Explaining Socio-Spatial Inequalities *Eur. Plan. Stud.* **23** 367–378
[17] Bernt M 2013 Exclusion, marginalization and peripheralization: Conceptual concerns in the

study of urban inequalities
[18] Michelini J J and Pintos P 2016 Metropolitan expansion and new socio-spatial segregation scenarios in contemporary Argentina. The case of Nordelta-Las Tunas (Buenos Aires) *Habitat Int.* **54** 40–49
[19] Nuriasari A 2012 Dinamika Perkembangan Wilayah Peri Urban Surakarta: Kajian Dalam Aspek Kesejahteraan Masyarakat dan Ketimpangan *Undergraduate Thesis, Diponegoro University*

Regional Phenomena of Vertical Deformation in Southern Part of Indonesia

D A Sarsito[1], Susilo[2], H Andreas[1], D Pradipta[1], and I Gumilar[1]

[1] Geodesy Research Division Faculty of Earth Science and Technology, Institute of Technology Bandung, LABTEX IXC Jl. Ganesha 10 Bandung 40132 – Indonesia
[2] Geospatial Information Agency, Jl. Raya Jakarta-Bogor KM.46, Cibinong 16911-Indonesia

Email: dsarsito@gd.itb.ac.id

Abstract. Distribution of present-day horizontal and vertical deformation across the Southern Part of Indonesia at Java, Bali and Nusa Tenggara now days can be determined from continuous and campaign types of GNSS GPS data monitoring. For vertical deformation in this case we use the continuous types since they are give better quality of data consistency compare to campaign type. Continuous Global Positioning System (CGPS) are maintaining by Geospatial Information Agency for more than a decade. The vertical displacements or velocity rates are estimated from time series analysis after multi-baseline GPS processing using GAMIT-GLOBK software with respect to the latest International Terrestrial Reference Frame. The result shows some interesting phenomena where the northern part of research area majority have negative value that may indicate land subsidence with or without tectonic subsidence combination. In the middle part, the uplift phenomena are clearly shown and in the southern part show combine pattern between uplift and subsidence. The impacts of those phenomena would be discuss also in this paper since many population and infrastructure are located in the areas that will need more protection planning to reduce the negative impact such as earthquake and flooding.

Keywords: Vertical Deformation, GNSS-GPS, GAMIT-GLOBK, land subsidence

1. Introduction

The vertical deformation phenomenon at the moment is very interesting because it has a significant impact. The scope of vertical deformation is similar to horizontal deformation, ie spatial variation is a global, regional or local event. While in temporal variation can take place from the long term duration of annual, seasonal or instantaneous. For long term duration with global impact, Chaussard [1] and Andreas [2] already mentioned that the global climate change situation has been realized for the few decades with some indicators are worldwide increasing temperature, decreasing volume of ice in Antarctica and the sea level rise. Relating to the decreased of ice volume and the sea level rise, this situation has been predicted to endanger the living at the coastal area in the future. Another phenomenon with long term duration but with only regional coverage is caused by uplift or subside geological events of structural activity and / or gravity attraction. The local vertical deformation phenomenon with a duration of several years can be caused also by earthquake or tsunami events as happened in Aceh due to the 2014 mega earthquake as a result of a combination of co-seismic and post-seismic events. Local events that are seasonal or continuous can also occur eg rob, land subsidence due to groundwater retrieval or ground collapse and volcanic activity. Figure 1 shows examples of some of the effects of vertical deformation phenomena in coastal areas that cause flooding. The territory of Indonesia is a complex area in terms of vertical deformation. As an archipelago, Indonesia is affected by global sea level rise. While tectonically, the Indonesian region is a collision zones of 3 macro plates that cause the activity of horizontal and vertical deformation derived from various geological phenomenon mentioned above. When viewed from local activities, the possibility of vertical deformation also occurs due to groundwater retrieval or liquid material retrieval that causes compaction. Therefore, it is important to study the vertical deformation in this case especially to know the varying causes parameters for each zone for its prevention and reduction of negative impacts.

Distribution of present-day horizontal and vertical deformation across the Southern Part of Indonesia at Java, Bali and Nusa Tenggara now days can be determined from continuous and campaign types of GNSS GPS data monitoring. For vertical deformation in this case we use the continuous types since they are give better quality of data consistency compare to campaign type. Continuous Global Positioning System (CGPS) are maintaining by Geospatial Information Agency for more than a decade.

Figure 1. Example of several subsidence phenomena at northern part of Java

2. Data and Methods

The vertical deformation phenomenon in this study was obtained from GPS monitoring from two types observations, namely continuous type and episodic / campaign type. Continuous Global Positioning System (CGPS) in this research are maintaining by Geospatial Information Agency (BIG) for more than a decade. While the campaign type is obtained from a combination of various GPS measurements ever performed by geodesy research group FEST - ITB. Figure 2 shows example of GNSS GPS station monitoring for continuous and campaign type. The vertical displacements or velocity rates are estimated from time series analysis after multi-baseline GPS processing using GAMIT-GLOBK software with respect to the latest International Terrestrial Reference Frame.

Figure 2. Example of GNSS GPS station monitoring for continuous and campaign type

GAMIT/GLOBK 10.6 is used to analyzed the GNSS-GPS data of all observation campaigns and continuous types [3]. Daily solutions were calculated in 12/24-hours session with 30 second sampling rate using GAMIT for each of measurement, and in each session the theoretical values for phase and pseudorange observable are modeled. Station coordinates, phase biases and zenith delay parameter are adjusted by a least square method. The antenna phase center variations we're model using the IGS table and IGS orbits were kept fixed for each individual session. To transform the solution to ITRF, we include rinexdata from the IGS station located surround the research area as shown in Figure 3.

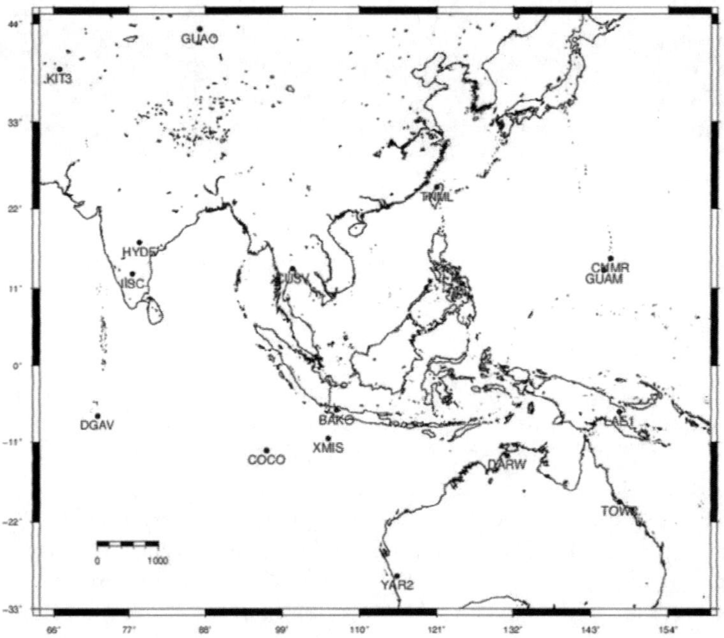

Figure 3. Global IGS station for ITRF transformation

The observable were examined in the ionosphere-free combination (LQ forming double differences to eliminate clock errors and tropospheric parameter were estimated using Saastamoinen model every one hour for each station. The phase ambiguities are estimated as real values in the first step and it was attempted to resolve the phase ambiguity using a routine developed by Dong and Bock [4]. The quality assessment of the individual solution is base on the repeatability of the independent daily estimation for each baseline component. And finally, the multi session free network solution with loose constrain on position and fixed IGS orbits were established, and GLOBK Kalman filter method were applied to the analysis of solution vectors and associated covariance matrices generated during daily solutions. The coordinates were mapped in the ITRF solution of 2008 by constraining the IGS solutions included in the analysis to their positions in ITRF-2008. The displacement rates were computed by subtracting the ITRF-2008 mapped the campaign solutions.

3. Results and Discussion

As is known, the accuracy of positioning results using GNSS GPS has a precision of 3 to 5 times lower for vertical components compare to horizontal components. Thus, the accuracy of the vertical position and its change is highly dependent on the position of the center of the earth's mass and the estimated influence of the atmosphere (ionosphere and troposhere). That is why the determination of vertical deformation becomes more sensitive than horizontal deformation. Figure 4 shows examples for detecting possible biases and possible seismic effects (inter seismic, co seismic and post seismic). After cleaning and filtering data processing, data obtained are free from biased and we can obtain linear trend of vertical displacement or velocity. This analysis method can only be done for continuous type data, while for episodic / campaign type data, analysis is done by using linear assumption with shifting by interpolation of earthquake impact to nearby station.

The average accuracy value obtained for the north, east and up components is 2 mm/yr, 3 mm/yr and 6 mm/yr for continuous type. Figure 5 shows the results of the time series analysis and the resulting linear velocity results . We use CMAT at Lombok Island and CSRJ at Bali Island that already clean from earthquake effect and bias. The CMAT deviation, which is relatively larger than the station caused by the stationary position of this station, is very close to the tall buildings, so the multipath effect predominates.

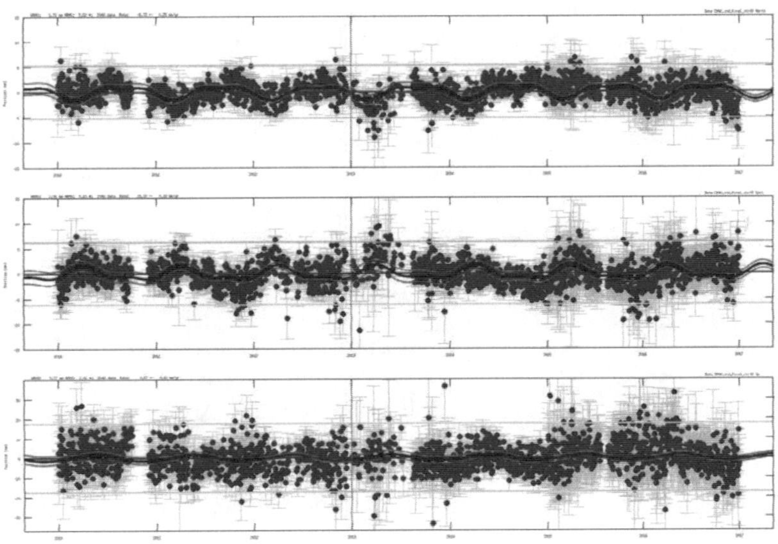

Figure 4. Time Series Analysis for Bias Analysis and to Detect Earthquake

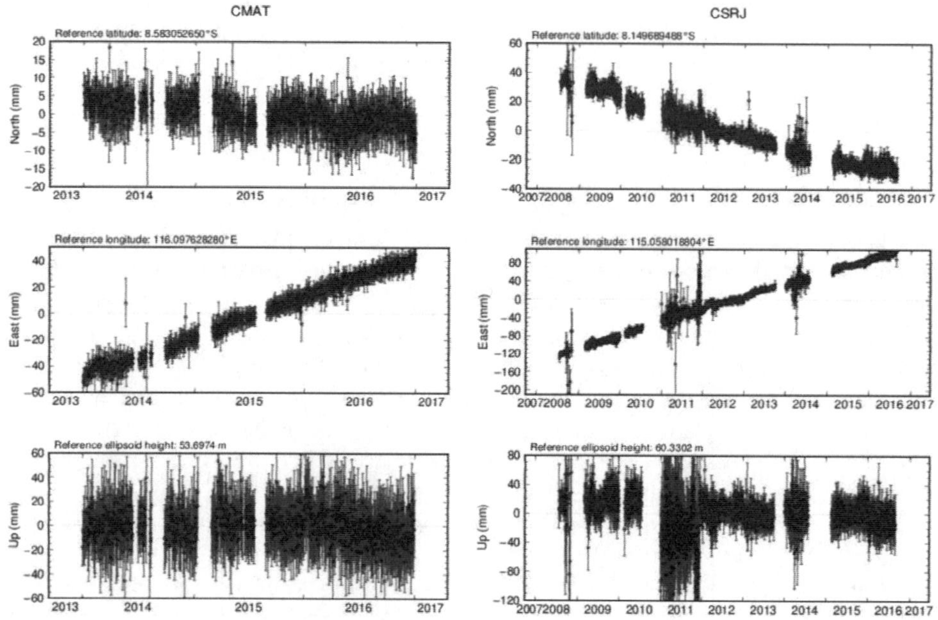

Figure 5. Time series of continuous GPS

After a time series analysis, each station is mapped to ITRF2008 via the Helmert transform procedure, and the results are shown in Fig. 6 and Fig. 7. In this study, we do not analysis with local transformation to an arbitrary plate due to the reference datum we used for vertical deformation in here is the center of the mass of the earth. This is done in order to detect vertical deformation on a regional scale as well as the global impact of sea level rise. The continuous station at Figure 6 shows better results than episodic stations that leads to more reliable analysis for continuous stations. The average negative vertical deformation is about 2.5 mm/yr, and smaller for about 2 mm/yr for positive vertical deformation. Figure 7 shows a more heterogeneous result, due to the small amount of data used in the 3-epoch range of the bonding that causes the results to be analyzed in detail. In order to be able to do further analysis for this type of episodic considering the phenomenon of vertical deformation slowly lasting in duration, then it takes at least 2 epoch observations again.

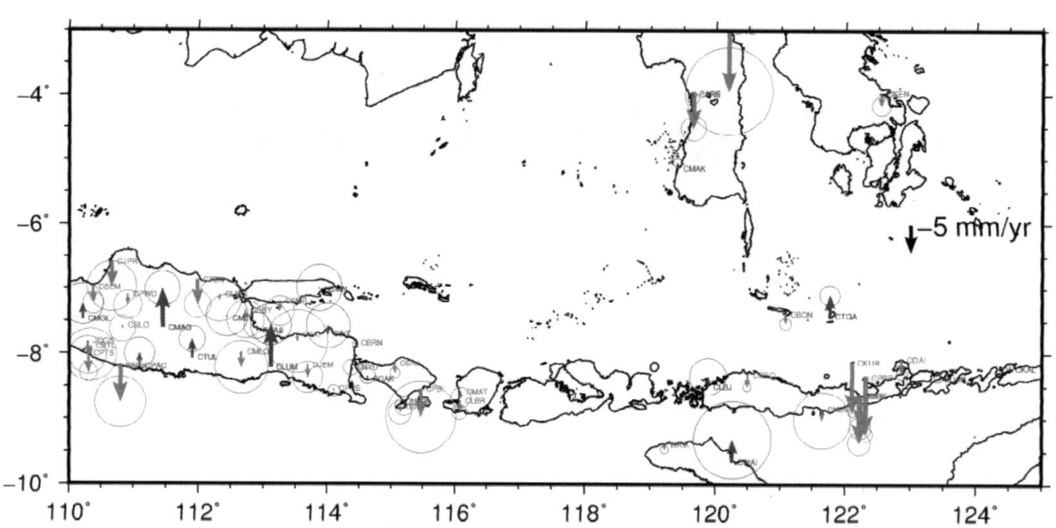

Figure 6. Vertical deformation from continuous GPS

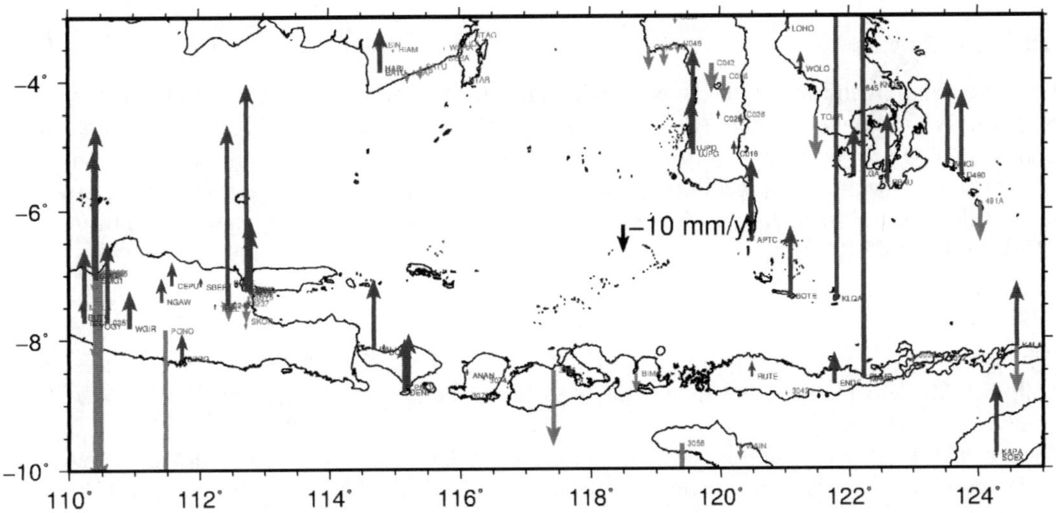

Figure 7. Vertical deformation from episodic/campaign GPS

The result in figure 6 shows some interesting majority phenomena where the northern part of research area majority have negative value that may indicate land subsidence with or without tectonic subsidence combination. This results have positive correlation with tidal inundation at northern part of Java from Andreas [2]. In the middle part, the uplift phenomena are clearly shown and in the southern part show combine pattern between uplift and subsidence. Both last part are dominated by geological force phenomena. In the southern part of the study area there is a subduction zone that dominates the pressure of the Australian plate, so most likely the curvature of the plate due to gravity is the dominant parameter. At central part of East Java, we can see rather large positive vertical deformation that maybe connected with the geological phenomena of Tengger mountain uplift. In Bali Island, still have same pattern, that the northern part and the southern part show negative deformation pattern and in the central part show positive trend. At Lombok Island and Nusa Tenggara Timur, almost all station shows negative vertical rate that maybe connected with geological phenomena of collision between Australian Plate from south with back arc system at northern part of the area from Bali to Nusa Tenggara Timur. Exception are occurred at west part of Flores Island and central part of Sumba Island, and this phenomena need more detail investigation.

The impact of vertical deformation for the northern coastal region at this time should be highly regarded, since the majority is a residential area. Many population and infrastructure are located in the areas that will need more protection planning to reduce the negative impact such as earthquake and flooding.

4. Conclusion

From the results of gps monitoring, it can be seen that vertical deformation phenomenon occurs in all research areas with an average value of 3 mm yr in both positive and negative directions. This deformation is calculated relative to the center of the earth's mass so that the regional vertical deformation pattern will be real displacement. The vertical deformation parameters in the study area were dominated by subsidence in the northern part and geological activity in central and southern regions. To find out in more detail the phenomenon that occurs primarily local, then required further observation and use of physical modeling. For residential areas, especially in the northern region, further studies are needed in an integrated socio-cultural economy in order to prevent or at least minimize the negative impact of vertical deficits

5. Acknowledgments

Deep appreciation and many thanks to Geospatial Information Agency (BIG) and ITB for research collaboration. And deepest gratitute for Prof. Hasanuddin Z. Abidin and Dr. Kosasih Prijatna that has provided knowledge to the author in the understanding of "state of the art" vertical deformation study of physical phenomena using geodetic-geometrically technology and methods. This study was partially funded by the Ministry of Research, Technology, and Higher Education of Indonesia No. 009/SP2H/LT/DPRM/IV/2017.

6. References

[1] Chaussard E, Amelung F, Abidin H and Hong S-H 2013 Sinking cities in Indonesia: ALOS PALSAR detects rapid subsidence due to groundwater and gas extraction *Remote Sens. Environ.* **128** 150–161

[2] Andreas H, Pradipta D, Abidin H Z and Sarsito D A 2017 Early pictures of global climate change impact to the coastal area (North West of Demak Central Java Indonesia) *AIP Conference Proceedings* vol 1857p 90002

[3] Herring T A, King R W and McClusky S C 2010 Introduction to Gamit/Globk *Massachusetts Inst. Technol. Cambridge, Massachusetts*

[4] Dong D and Bock Y 1989 GPS network analysis with phase ambiguity resolution applied crustal deformation studies in California'to be appear in JGR